High-Performance Computing

High-Performance Computing

Edited by

R. J. Allan and
M. F. Guest

HPCI Centre
CLRC Daresbury Laboratory
Daresbury, England

A. D. Simpson and
D. S. Henty

EPCC
University of Edinburgh
Edinburgh, Scotland

and

D. A. Nicole

HPCI Centre
University of Southampton
Southampton, England

KLUWER ACADEMIC / PLENUM PUBLISHERS
NEW YORK, BOSTON, DORDRECHT, LONDON, MOSCOW

Library of Congress Cataloging-in-Publication Data

High-performance computing / edited by R.J. Allan ... [et al.].
        p.   cm.
      "Proceedings of the High-performance computing initiative (HPCI)
   Conference, held January 12-14, 1998, in Manchester, England"--verso
   CIP t.p.
      Includes bibliographical references (p.   ) and index.
      ISBN 0-306-46034-3
      1. High performance computing--Congresses.   I. Allan, R. J.
   II. High-Performance Computing Initiative Conference (1998 :
   Manchester, England)
   QA76.88.H49 1999
   004'.3--dc21   98-46098
                                                                 CIP

Proceedings of the High-Performance Computing Initiative (HPCI) Conference,
held January 12 – 14, 1998, in Manchester, England

ISBN 0-306-46034-3

© 1999 Kluwer Academic / Plenum Publishers, New York
233 Spring Street, New York, N.Y. 10013

10 9 8 7 6 5 4 3 2 1

A C.I.P. record for this book is available from the Library of Congress.

Printed in the United States of America

PREFACE

Over the past decade high performance computing has demonstrated the ability to model and predict accurately a wide range of physical properties and phenomena. Many of these have had an important impact in contributing to wealth creation and improving the quality of life through the development of new products and processes with greater efficacy, efficiency or reduced harmful side effects, and in contributing to our ability to understand and describe the world around us.

Following a survey of the U.K.'s urgent need for a supercomputing facility for academic research (see next chapter), a 256-processor T3D system from Cray Research Inc. went into operation at the University of Edinburgh in the summer of 1994. The High Performance Computing Initiative, HPCI, was established in November 1994 to support and ensure the efficient and effective exploitation of the T3D (and future generations of HPC systems) by a number of consortia working in the "frontier" areas of computational research. The Cray T3D, now comprising 512 processors and total of 32 GB memory, represented a very significant increase in computing power, allowing simulations to move forward on a number of fronts.

The three-fold aims of the HPCI may be summarised as follows; (1) to seek and maintain a world class position in computational science and engineering, (2) to support and promote exploitation of HPC in industry, commerce and business, and (3) to support education and training in HPC and its application.

A number of research grant awards have been made under the auspices of the HPCI, to both (i) establish a number of Consortia of research groups, with provision of post-doctoral research associates (to the so-called "directly supported" consortia), plus travel and computer time (to both "directly" and "indirectly supported" consortia), and (ii) establish three HPCI centres to support new application development and to interact with education, training and technology transfer programmes.

The HPCI Centre at CLRC Daresbury Laboratory, with assistance from the Centres at the Universities of Edinburgh and Southampton, organised the first National HPCI Conference designed to overview the success of this service and to provide a forum for identifying some of the key scientific requirements likely to be encountered during the coming three years.

The event took place at the Manchester Conference Centre Renold's complex on the UMIST campus from 12-14th January 1998. Organisation of the conference had taken a considerable effort and it was therefore pleasing to have nearly 140 registered attendees including representatives of most of the companies active in HPC who, together with NERC and EPSRC, provided generous sponsorship enabling the event to take place.

There were 8 plenary talks, including those from:

Dr. Richard Blake (CLRC), "Background to the U.K. High-Performance Computing Initiative";

Dr. David Dixon (PNNL, Richland, USA), "Computational Environmental Molecular Science: Relevance to DOE Cleanup Problems";

Dr. Alfred Geiger (HLRS, Stuttgart), "Supercomputing and Applications in German Research and Industry";

Prof. Alan O'Neill and Dr. Lois Steenman-Clark (University of Reading), "Modelling Climate Variability on High-Performance Computers and Data Handling Strategies for Environmental Models";

Richard M. Russell (Tera Computer Company, USA), "Early Experience with Tera MTA";

Stuart Ward (EPSRC), "Progress with the HPC'97 Procurement";

Prof. David Crighton FRS (University of Cambridge), "The University of Cambridge High Performance Computing Facility";

Prof. David Walker (University of Cardiff), "Software Standards for High Performance Computing".

In addition a further invited speaker, Ken Turner, entertained us after the Conference Dinner, with a description of the Manchester "Baby" computer rebuild project which is nearing completion. This was especially fitting for not only is 1998 the 50th anniversary of this, the world's first stored-program computer (known as the Small-Scale Experimental Machine, SSEM), but the replica machine is shortly to be exhibited in the Manchester Museum of Science and Industry where the dinner took place.

The ambitious programme also had space for 40 speakers presenting submitted talks in parallel sessions and a further poster session with some 15 contributed papers. Thus we were able to cover the work of all the HPCI-supported consortia plus a number of other related U.K. projects.

This book contains most of the papers presented at the conference. It is intended to act as a record of the scientific achievements of three years usage of the HPCI-supported Cray T3D at EPCC and to indicate a subset of the anticipated scientific requirements over the coming three years.

In addition to the scientific material presented as talks or posters, a number of computer vendors exhibited products at the Conference. These included:

Silicon Graphics/Cray Research, *http://www.cray.com*;

Hitachi Europe Ltd., *http://www.hitachi-eu.com*;

Fujitsu Systems Europe, *http://www.fujitsu.com*;

NEC, *http://www.nec.com*;

IBM UK, *http://www.uk.ibm.com*;

Quadrics Supercomputers World Ltd., *http://www.quadrics.com*;

TERA Computer Company, *http://www.tera.com*;

Digital Equipment Company, *http://www.europe.digital.com*;

Progress Computers, *http://www.progress.co.uk*, and

The Manchester MarkI Rebuild Project, *http://www.computer50.org*;

Finally, we thank the many sponsors of the HPCI Conference, without whom the event could not have taken place. These include, EPSRC; NERC; CLRC; Silicon Graphics/Cray Research; Hitachi Europe Ltd.; Fujitsu Systems Europe; NEC; IBM UK; Digital Equipment Company; NAG Ltd. and Quadrics Supercomputers World.

R. J. Allan, M. F. Guest,
A. D. Simpson, D. S. Henty, and D. A. Nicole
March 1998

Left to right: David Emerson (CLRC), Ameet Davé (SGI/Cray Limited), Robert Allan (CLRC), and Stephen Booth (EPCC).

HPCI 1998 Delegates at the Museum of Science and Industry, Manchester.

CONTENTS

THE U.K. HIGH-PERFORMANCE COMPUTING INITIATIVE

Introduction 3
    R. J. Allan, M. F. Guest, A. D. Simpson, D. S. Henty, and D. A. Nicole

Science Support from the EPCC HPCI Centre 11
    Alan D. Simpson and David S. Henty

The CLRC HPCI Centre at Daresbury Laboratory 21
    R. J. Allan, I. J. Bush, K. Kleese, A. G. Sunderland, and M. F. Guest

Southampton High Performance Computing Centre 33
    Denis Nicole, Kenji Takeda, Ivan Wolton, and Simon Cox

OPTIMISATION, ALGORITHMS AND SOFTWARE

Performance Optimisation on the Cray T3E 45
    Stephen Booth

From FLOPS to UDAPS: Algorithms, Benchmarking, and Tuning 53
    Nick MacLaren

Is Predictive Tracing too late for HPC Users? 57
    Darren J. Kerbyson, Efstathios Papaefstathiou, John S. Harper, Steward C. Perry,
        and G. R. Nudd

Solving Dense Symmetric Eigenproblems on the Cray T3D 69
    K. Murphy, M. Clint, and R. H. Perrott

PARASOL: An Integrated Programming Environment for Parallel Sparse Matrix
    Solvers 79
    Patrick Amestoy, Iain S. Duff, Jean Yves L'Excellent, and Petr Plecháč

Computational Modelling of Multi-Physics Processes on High Performance Parallel
    Computer Systems 91
    M. Cross, K. McManus, S. P. Johnson, C. S. Ierotheou, C. Walshaw, C. Bailey,
        and K. A. Pericleous

Porting Industrial Codes to MPP Systems using HPF 103
    L. M. Delves

Decomposition Independence in Parallel Programs 113
    S. Booth

Software Portability and Maintenance 119
    Kenji Takeda, Ivan Wolton, and Denis Nicole

A Design Environment for Structured Mapping of Signal Processing Applications
    on Parallel Processors 127
    Moe Razaz

MATERIALS CHEMISTRY AND SIMULATION

New Vistas for First-Principles Simulation 137
    G. Ackland, D. Bird, P. Bristowe, M. Finnis, M. J. Gillan, N. M. Harrison,
    V. Heine, P. A. Madden, M. C. Payne, and A. P. Sutton

On the Quasi-Particle Spectra of $YBa_2Cu_3O_7$ 147
    W. M. Temmerman, M. L. Gyorffy, Z. Szotek, O. K. Andersen, and O. Jepsen

Ab Initio Studies of Hydrogen Molecules in Silicon 155
    B. Hourahine, R. Jones, S. Öberg, R. C. Newman, P. R. Briddon, and E. Roduner

Quantum Monte Carlo Simulations of Real Solids 165
    W. M. C. Foulkes, M. Nekovee, R. L. Gaudoin, M. L. Stedman, R. J. Needs,
    R. Q. Hood, G. Rajagopal, M. D. Towler, P. R. C. Kent, Y. Lee, W.-K. Leung,
    A. R. Porter, and S. J. Breuer

Ab Initio Investigations of the Dynamical Properties of Ice 175
    I. Morrison, S. Jenkins, J. C. Li, and D. K. Ross

Phase Separation of Two Immiscible Liquids 185
    S. I. Jury, P. Bladon, S. Krishna, and M. E. Cates

Computer Simulation of Liquid Crystals on the T3D/T3E 193
    Mark R. Wilson, Michael P. Allen, Maureen P. Neal, Christopher M. Care, and
    Douglas J. Cleaver

A First Principles Study of Substitutional Gold in Geremanium 203
    J. Coomer, A. Resende, P. R. Briddon, S. Öberg, and R. Jones

Applications of Self-Interaction Correction to Localized States in Solids 207
    Z. Szotek, W. M. Temmerman, A. Svane, H. Winter, S. V. Beiden, G. A. Gehring,
    S. L. Dudarev, and A. P. Sutton

COMPUTATIONAL CHEMISTRY

Computational Chemistry in the Environmental Molecular Sciences Laboratory 215
    David A. Dixon, Thom Dunning, Jr., Michel Dupuis, David Feller,
    Deborah Gracio, Robert J. Harrison, Donald R. Jones, Ricky A. Kendall,
    Jefferey A. Nichols, Karen Schuchardt, and Tjerek Straatsma

Macromolecular Modelling on the Cray T3D 229
    Matthew D. Cooper, Julia M. Goodfellow, Ian H. Hillier,
      Christopher A. Reynolds, W. Graham Richards, Michael A. Robb,
      Paul Sherwood, and Ian H. Williams

Accurate Configuration Interaction Computations of Potential Energy Surfaces
    using Massively Parallel Computers 237
    Abigail J. Dobbyn and Peter Knowles

Molecular Properties from First Principles 249
    C. J. Adam, S. J. Clark, G. J. Ackland, and J. Crain

Massive Parallelism: The Hardware for Computational Chemistry? 259
    M. F. Guest, P. Sherwood, and J. A. Nichols

ATOMIC PHYSICS

The Multiphoton and Electron Collisions Consortium and the Helium Code 275
    K. T. Taylor, J. S. Parker, and E. S. Smyth

Application of 6DIME: (γ,2e) on He 285
    J. Rasch, Colm T. Whelan, S. P. Lucey, and H. R. J. Walters

Parallelisation of Atomic R-Matrix Scattering Programs 293
    A. Sunderland, P. G. Burke, V. M. Burke, and C. J. Noble

Partial Wave Integrals 301
    J. Rasch and Colm T. Whelan

Molecular Rotation-Vibration Calculations using Massively Parallel Computers 307
    Hamse Y. Mussa, Jonathan Tennyson, C. J. Noble, and R. J. Allan

ENVIRONMENTAL MODELLING

Modelling Climate Variability on HPC Platforms 317
    Lois Steenman-Clark and Alan O'Neill

The U. K. Ocean Circulation and Advanced Modelling Project (OCCAM) 325
    Beverly A. de Cuevas, David J. Webb, Andrew C. Coward,
      Catherine S. Richmond, and Elizabeth Rourke

The Southampton-East Anglia (SEA) Model: A General Purpose Parallel Ocean
    Circulation Model 337
    Matthew Beare

High Resolution Modelling of Airflow over the Isle of Arran 347
    Alan Gadian, Ian Stromberg, and Robert Wood

Development of Portable Shelf Sea Models for Massively Parallel Machines 359
    Roger Proctor, Peter Lockey, and Ian D. James

Satellite Altimeter Data Assimilation in the OCCAM Global Ocean Model 365
    Alan D. Fox, Keith Haines, Beverly A. de Cuevas, and Andrew C. Coward

Parallelisation and Performance of a Stratospheric Chemical Transport Model 371
  Cate Bridgeman

Supercomputing and Applications in German Research and Industry 379
  Alfred Geiger and Roland Rühle

Investigation of Sequencing Effects on the Simulation of Fluid-Structure Interaction . . 385
  J. K. Badcock, G. S. L. Goura, and B. E. Richards

Direct Numerical Simulation of Turbulent Flames 395
  Karl W. Jenkins, W. Kendal Bushe, Laurent L. Leboucher, and I. R. Stewart Cant

Understanding Turbulence in Fluids using Direct Simulation Data 407
  M. Alam, E. Avital, T. J. Craft, S. P. Fiddes, H. P. Horton, R. J. A. Howard,
  D. P. Jones, K. H. Luo, N. D. Sandham, A. M. Savill, T. G. Thomas,
  P. R. Voke, and J. J. R. Williams

Parallel Processing and Direct Simulation of Transient Premixed Laminar Flames
  with Detailed Chemical Kinetics 417
  R. P. Lindstedt and V. Sakthitharan

Time Domain Electromagnetic Scattering Simulations on Unstructured Grids 429
  P. J. Brookes, O. Hassan, K. Morgan, R. Said, and N. P. Weatherill

Large-Eddy Simulation of the Vortex Shedding Process in the Near-Field Wake
  behind a Square Cylinder 437
  F. di Mare and W. P. Jones

Self-Adaptive, Parallel Solution Methods for Complex FEM Problems in CFD and
  Radiation Modelling 449
  Xiao Xu, Christopher C. Pain, Cassiano R. E. de Oliveira, Adrian P. Umpleby,
  and Antony J. H. Goddard

HUMAN SYSTEMS AND INFORMATION

HPC and Human Geographic Social Science Research 457
  Stan Openshaw

Application of Pattern Recognition to Concept Discovery in Geography 467
  Ian Turton

High Performance Computing in Banking 479
  J. A. Keane

Legacy Systems - The Future of HPC 487
  J. A. Keane, M. F. P. O'Boyle, and R. Sakellariou

ASTROPHYSICS AND COSMOLOGY

Simulations of Lattice Quantum Chromodynamics on the Cray T3D and T3E 497
  David Richards

Towards and Understanding of Galaxy Formation 507
    F. R. Pearce, C. S. Frenk, A. R. Jenkins, J. M. Colberg, P. A. Thomas,
      H. M. P. Couchman, S. D. M. White, G. P. Efstathiou, and J. A. Peacock

Massively Parallel Simulations of Cosmic Strings 517
    Mark Hindmarsh and Graham Vincent

The U. K. MHD Consortium: Goals and Recent Achievements 529
    Alan Hood, Tony Arber, Klaus Galsgaard, Axel Brandenberg, Steve Brooks,
      Chris Jones, Graeme Sarson, and Gavin Pringle

N-Body Simulations of Galaxy Formation on a Cray T3E 537
    P. R. Williams and A. H. Nelson

NOVEL METHODS AND APPLICATIONS

Early Experience with Tera MTA System 545
    Richard M. Russell

HPC on DEC Alphas and Windows NT 551
    Denis Nicole, Kenji Takeda, and Ivan Wolton

A Programming Environment for High-Performance Computing in Java 559
    Vladimir Getov, Susan Flynn-Hummel, and Sava Mintchev

High Performance Distributed FDTD Electromagnetic Field Computation for
    Electronic Circuit Design 569
    C. J. Gillan and V. F. Fusco

List of Participants 575

Index 579

THE U.K. HIGH-PERFORMANCE COMPUTING INITIATIVE

INTRODUCTION

R.J. Allan, M.F. Guest,[1] A.D. Simpson, D.S. Henty, [2] and D.A. Nicole [3]

[1]HPCI Centre, CLRC Daresbury Laboratory,
Daresbury, Warrington, WA4 4AD
[2]EPCC, University of Edinburgh,
JCMB - King's Buildings, Mayfield Road, Edinburgh EH9 3JZ
[3]Department of Electronics and Computer Science,
University of Southampton, Southampton SO9 5NH

BACKGROUND

Over the past decade high performance computing has demonstrated the ability to model and predict accurately a wide range of physical properties and phenomena. Many of these have had an important impact in contributing to wealth creation and improving the quality of life through enabling the development of new products and processes with greater efficacy, efficiency or reduced harmful side effects, and in contributing to our ability to understand and describe the world around us. Some highlights have already been presented in various reports [1,2] and the U.K. community is kept up to date by publications such as HPCProfile [3] and HPCNews [4].

A growing trend in computational science and engineering is to study more complex and larger systems and to increase the realism and detail of the modelling. In their 1992 report [6] a Scientific Working Group, chaired by Prof. C.R.A. Catlow of the Royal Institution of Great Britain, presented a scientific case for future supercomputing requirements across many of the most exciting and urgent areas of relevance to the missions of the U.K. Research Councils. The report highlighted a number of fields at the forefront of science in which computational work would play a central role. It was noted that to make progress in these most computationally demanding and challenging problems, increases of two to three orders of magnitude in computing resources such as processing power, memory and storage rates and volume would be required.

The report led ultimately to the procurement of a 256-processor Cray T3D system from Cray Research Inc. which went into operation at the University of Edinburgh in the summer of 1994. The High Performance Computing Initiative (HPCI) was set up in November 1994 to support the efficient and effective exploitation of the T3D, and future generations of systems, by consortia working in the "frontier" areas of computational research.

The Cray T3D, now comprising 512 processors and total of 32 GB memory, represented a very significant increase in computing power allowing simulations to move forward on a number of fronts including:

- improved resolution and accuracy – this is particularly important in grid-based applications where numerical artifacts can distort physical effects and, for example, make CFD simulations more diffusive than the physics dictates;
- moving from two spatial dimensions which can typically be accommodated on a workstation to three spatial dimensions;
- moving from static, steady state or time averaged simulations to dynamic simulations – for example the study of non-equilibrium properties and the calculation of extremal instantaneous behaviour;
- the study of larger systems – for example in materials simulations moving away from periodic simple 3-dimensional systems to more chemically and physically interesting surface and defect systems, modelling systems of sufficient scale to eliminate finite-size computational effects;
- long timescale studies enabling the determination of the low frequency dynamic response of systems;
- increasingly complex or strongly interacting systems – for example non-linear interactions or materials properties, moving beyond mean-field or averaged interactions, to systems under extreme conditions, multi-phase and multi-physics simulations;
- optimisation studies within large parameter spaces, model parameterisations, trends and sensitivity analyses.

We now describe how the development of large-scale computer codes in science and engineering is managed in the U.K. and how efficient exploitation of the national supercomputer facilities is assured through the actions of the Collaborative Computational Projects and the High-Performance Computing Initiative.

THE CCPs

Collaboration in large-scale computer code development for academic research has been the *modus operandum* in the U.K. since the late 1970s. The accepted mechanism is via the Collaborative Computational Projects (CCPs). The CCPs bring together all the major U.K. groups in a given area of computational science to pool ideas and resources on large-scale software developments of mutual interest and general importance. Typically, CCPs implement "flagship" code development projects, maintain and distribute code libraries, organise training in the use of codes, hold meetings and workshops, invite overseas researchers for lecture tours and collaborative visits, and issue regular newsletters.

Each CCP has a chairman and a working group which sets the scientific agenda, decides the work programme, and monitors progress. The current list of CCPs is given in Table 1. The entire CCP programme involves about 240 academic groups in most U.K. universities, and around 150 collaborating groups in the rest of Europe, USA, Japan etc. The list indicates the tremendous breadth of science and engineering covered by the CCP programme as a whole. This is not the place to describe the current interests of CCPs in detail, but much more information is available via the WorldWide Web pages maintained by CLRC
http://www.dci.clrc.ac.uk/ListActivities/CLASS=5; CLASSTYPE=21;

---

*Funding sources
EPSRC – Engineering and Physical Sciences Research Council (previously the SERC)
BBSRC – Biotechnology and Biological Sciences Research Council

Table 1. Current CCPs and their Funding Sources

CCP	Title	Funding Source
CCP1	The Electronic Structure of Molecules	EPSRC
CCP2	Continuum States of Atoms and Molecules	EPSRC
CCP3	Computational Studies of Surfaces	EPSRC
CCP4	Protein Crystallography	BBSRC
CCP5	Computer Simulation of Condensed Phases	EPSRC
CCP6	Heavy Particle Dynamics	EPSRC
CCP7	Analysis of Astronomical Spectra	PPARC
CCP9	Electronic Structure of Solids	EPSRC
CCP11	Biosequence and Structure Analysis	BBSRC
CCP12	High Performance Computational Engineering	EPSRC
CCP13	Fibre Diffraction	BBSRC/EPSRC
CCP14	Powder and Single Crystal Diffraction	EPSRC

Typical resources available to a CCP include the running costs of organising meetings, newsletters, visits, etc. In addition, some CCPs have a post-doctoral research associate to carry out flagship code development (there are five PDRAs currently in place), with the standard funding mechanism being a 3-year research grant. Finally, the CCP Programme is supported by a number of permanent staff at the CLRC's Daresbury Laboratory.

Historically, the entire CCP programme was overseen by a steering panel comprising CCP chairmen and independent overseas members (most recently Profs. Giovanni Ciccotti and Roberto Car). Its main activities were to provide a forum for discussion of general computational issues, to encourage inter-working of CCPs (where appropriate), to review the work of associated permanent CLRC staff, and to filter and provide "quality assurance" of CCP grant proposals.

How Successful are the CCPs?

To operate at a word-class level in computational science and engineering involves a major and long-term investment in software development and maintenance, dissemination and training. The key function of the CCPs is to provide a framework within which all the active groups in a given field can work together to make this happen. Achievements include:

- The support of around 300 codes of varying sizes (up to several hundred thousand lines of code) on many different types of hardware;
- High quality publication record; our last survey showed that the CCPs produce an average of about 64 publications per year each, of which about 11% are in the Phys. Rev. Lett. or Nature "hot topics" category;
- Strong presence in "grand challenge" consortia within the HPCI (see below);
- Extensive participation in CEC programmes (HCM, TMR, ESF Networks);
- Considerable interest from industrial research groups.

It is, moreover, worth noting the longevity of the CCPs. The first projects were initiated in the late 1970s and early 1980s, and have maintained their funding within rigorous peer-review processes which have themselves changed fairly frequently. The fascinating account of the birth of the first project, CCP1 [7] has many resonances for other areas of computational science and engineering.

PPARC – Particle Physics and Astronomy Research Council
NERC – Natural Environment Research Council
ESRC – Economic and Social Research Council

The CCPs thus form an active, high quality, co-ordinated and cost-effective programme of computational science and engineering code development – a key element in the national support infrastructure for HPC applications.

THE HIGH-PERFORMANCE COMPUTING INITIATIVE

In 1993, SERC reviewed government programmes supporting the exploitation of HPC systems based on MPP technology and recommended an additional programme of support and the development of a coordinated national strategy. Information on that can be found on the WWW URL, *http://www.epsrc.ac.uk/hpc* .

The review led to the development of the High Performance Computing Initiative (HPCI) which identified a number of medium-term priorities including:

- the establishment of a limited number of consortia in "grand-challenge" areas combining support from the HPC programme and Research Councils;
- the establishment of three centres to support new applications development and to interact with other HPC research, education, training and technology transfer activities.

Consortia were invited to propose computational projects which should be competitive at the global level to exploit the Cray T3D with the aim of advancing scientific or engineering understanding. The consortia are typically organised around flagship codes although a number are developing sets of algorithmically-related codes targeted at different or new application areas. The projects contain a number of short- medium- and long-term activities, ranging from exploiting the system in task-farm mode, through to the development of fine-grained algorithms aimed at exploiting the next generation of systems. Many of the consortia were founded upon existing CCPs.

Three support centres were established, at the Universities of Edinburgh and Southampton and at the CLRC's Daresbury Laboratory, with the aim of supporting the consortia, providing an interface between the projects and providing a focus for the consortia to interact with other HPC programmes. Of the three centres, the one at Daresbury Laboratory is the largest supporting approximately half the consortia.

The following chapters describe the work of these three centres and further information is available via the WWW as follows:

CLRC: *http://www.dci.clrc.ac.uk/Activity.asp?HPCI*
Edinburgh: *http://www.epcc.ed.ac.uk*
Southampton: *http://www.hpcc.ecs.soton.ac.uk*

The following sections list current projects and their funding councils. Highlights from the first two years of these projects can be found in an EPSRC report [2]. Current highlights and future directions of these projects is the subject of the rest of this book.

The HPCI Consortia

In addition to access to the national computer resources, the consortia listed in Table 2 received funding from the national HPC programme for travel and for a postdoctoral research associate to work on their project for an initial 2-year period. The role of the PDRAs was to assist in code porting, optimisation, tactical extension and the development of appropriate algorithms, tools and environments.

In addition to computer time, the consortia listed in Table 3 received funding for travel only and were given additional support from one of the three HPCI centres. Of these, the Turbulence Modelling Project has been discontinued, while the Multiphoton and Electron Collision Projects chose to merge.

Table 2. Directly-funded HPCI Consortia

Title	Funding Source
Global Ocean Circulation Modelling	NERC
UGAMP: Global Atmospheric Modelling	NERC
Macromolecular Modelling	BBSRC/EPSRC
VIRGO: Cosmological Simulations	PPARC
UKQCD: Quantum Chromodynamics	PPARC
Computational Combustion	EPSRC
Large Eddy Simulation	EPSRC
Materials Chemistry	EPSRC
UKCP: Car-Parrinello Materials Modelling	EPSRC
Colloid Hydrodynamics	EPSRC
Chemical Reactions and Energy Exchange	EPSRC
Human Systems Modelling	ESRC

The call for proposals to establish consortia was heavily oversubscribed. A number of projects listed in Table 4 were deemed to be interesting, but still in the development phase. It was hoped that full consortia might be formed to exploit the results of their initial work.

At the present time, after 3 years of intense activity, the Consortia are publishing results of their work and developing new codes which will be brought to bear on grand-challenge applications on the next computer generation.

FUTURE COMPUTING PROVISION

At the time this book goes to press, the U.K. will be nearing the end of the major Joint Research Councils' HPC Procurement, known as HPC'97. HPC'97 is an important project, the aim of which is to provide access to a high performance computing service to support the participating Research Councils' scientific programmes (BBSRC, EPSRC, NERC, ESRC and PPARC for their astronomy community). It is planned that the new service will be in place in 1998.

The EPSRC is the managing agent for the provision of High Performance Computing across the Research Councils. A steering committee, chaired by Prof. P.G. Burke (Queen's University, Belfast) and appointed across the relevant Research Councils, is

Table 3. Indirectly-funded HPCI Consortia

Title	Funding Source
Vegetation-climate Interactions	NERC
Shelf-sea Hydrodynamics	NERC
Turbulence Transition Modelling	EPSRC
Complex Aerodynamics Flows	EPSRC
Direct Numerical Simulation of Fluid Flows	EPSRC
Simulations and Statistical Mechanics of Complex Fluids	EPSRC
UK Micromagnetics	EPSRC
Quantum Monte Carlo Calculations of Materials	EPSRC
Ab initio Simulation of Covalent Materials	EPSRC
Particle Cosmology	PPARC
Atomic Multiphoton Processes	EPSRC
Fundamental Electron Collision Processes	EPSRC

Table 4. Other HPCI Projects

Title	Funding Source
Ab initio modelling of solvation using Monte Carlo methods	EPSRC
Numerical modelling and experimental study of chemistry during shock induced combustion	EPSRC
Self-adaptive, parallel solution methods for complex FEM problems in nuclear safety	EPSRC
Ab initio simulations of catalytic processes at extended metal surfaces	EPSRC
MHD theory of the sun	EPSRC
Seismic modelling of naturally fractured rocks	NERC
HPC in mineral physics	NERC
Flow over complex terrain using adaptive numerical methods	NERC
High resolution modelling of ice sheets and sub-glacial process	NERC
A modelling study of the nutrient dynamics spatial and inter-annual variability of the marine eco-system in U.K. continental shelf seas between 1985 and 1988	NERC
High-resolution 3-dimensional tropospheric chemistry modelling	NERC
Development and use of an intermediate GCM	NERC
Ab initio study of high-pressure and temperature behaviour and stability of liquid and crystalline Fe	NERC
Molecular modelling of pollutant sorption on mineral surfaces using quantum mechanical and classical methods	EPSRC
Computing simulation of macro-molecular interactions	BBSRC
High-level parallelism with predictable performance for unstructured mesh PDE solvers	EPSRC
DNS of turbulent flames using wavelet methods	EPSRC
Refined modelling of an open channel with flood plains	EPSRC
Refined modelling of free surface flows with curvilinear coordinates	EPSRC
The inference of data mapping and scheduling strategies from Fortran 90 programs on the T3D	EPSRC
Automatic generation of parallel visualisation modules	EPSRC

overseeing the project. The Research Councils' project team has appointed Smith System Engineering to provide support to this project.

The initial call for proposals was made in June 1997 with the publication of a notice in the Official Journal of the European Communities. The call invited proposals to provide a high performance computing service to the Research Councils, as described in the Request for Information (RFI), with the procurement expected to be undertaken through the Private Finance Initiative (PFI).

Seventeen potential contractors responded, of whom fifteen met the qualification criteria and were issued with a Statement of Requirements (SOR) in September 1997. The ten responses to the SOR were then reduced to a shortlist of seven organisations who have been invited to respond to an Invitation to Tender for all elements of the required service. The closing date for Invitation to Negotiate (ITN) responses was 16th March 1998.

The organisations shortlisted were:

Technology Providers:
- IBM/Fujitsu
- Silicon Graphics(Cray)
- NEC

Service Providers:
- CLRC

- Edinburgh University
- Manchester University
- CSC Computer Sciences Ltd.

ITN submissions have been requested from consortia containing at least one technology provider and one service provider from the shortlist. The responses will be evaluated and assessed to determine whether the benefits of a service combining the service provision and computing technology under the PFI approach outweigh the normal public procurement route. It is hoped that a decision on the consortia chosen to operate the service will be made in the summer of 1998.

REFERENCES

1.   *A Review of Supercomputing, 1991-1994* EPSRC (1995)
2.   *Computational Research on the Cray T3D - the First Two Years*, EPSRC, ISBN 1-899371-86-9 (October 1996)
3.   HPCProfile is published six times a year by CLRC on behalf of the EPSRC, to disseminate information of interest to HPC and distributed computer users including the CCPs and HPCI consortia. It is available online at *http://www.dci.clrc.ac.uk/Publications* or by free subscription to *HPCProfile@dl.ac.uk*
4.   HPCNews is published by the University of Edinburgh Parallel Computing Centre to provide information on the national T3D and T3E service, related activities and information about the HPCI Consortia. Contact *epcc-support@ed.ac.uk*
5.   M.F. Guest, J. van Lenthe and P. Sherwood, Concurrent Supercomputing at SERC Daresbury Laboratory *Supercomputer* 36 ASFRA (March 1990)
6   *Research Requirements for High Performance Computing*, SERC (September 1992)
7.   S. Smith and B.T. Sutcliffe, The Development of Computational Chemistry in the United Kingdom

SCIENCE SUPPORT FROM THE EPCC HPCI CENTRE

Alan D. Simpson and David S. Henty

EPCC
The University of Edinburgh
JCMB
The King's Buildings
Edinburgh EH9 3JZ
http://www.epcc.ed.ac.uk

INTRODUCTION

EPCC was one of three science support centres set up under HPCI, along with Daresbury Laboratory and the University of Southampton. The major role for the centres was to provide HPC and computational science expertise for the various national consortia.

The first section of this paper puts the work of EPCC's HPCI Centre into the context of the rest of EPCC's activities. We then discuss the scientific support we provide for EPCC's initial consortia, which span a very broad range of science, and the work done in setting up two new consortia. This is followed by a discussion of EPCC's strategic effort under HPCI including technical reports of general interest, additional training courses and a series of multidisciplinary seminars. The final sections include the discussion of the extension of this work to the new Maxwell Institute for HPC Applications at the University of Edinburgh, and a summary of the scientific benefits of EPCC's HPCI activities.

EDINBURGH PARALLEL COMPUTING CENTRE

EPCC was established in 1990 to act as the focus for the University of Edinburgh's long-standing interests in the application of HPC to research problems. The centre currently has around 50 professional staff and provides expertise and services on a range of HPC facilities unrivaled within any European university. Since 1994, EPCC has provided facilities management and applications support for the national services on the Cray T3D and T3E facilities. The importance of its position has been recognised by the CEC, and EPCC is a Large Scale Facility under the Training and Mobility of Researchers programme. EPCC also leads a round table of major European HPC and data centres.

High Performance Computing
Edited by R. J. Allan *et al.*, Kluwer Academic / Plenum Publishers, New York, 1999

EPCC has a thriving technology transfer programme which is funded by the CEC, Scottish Enterprise and direct contracts with industry. EPCC has a wide range of projects with companies ranging from blue chip multinationals to local small and medium enterprises. These strong links with both academia and industry ensure the relevance of the work undertaken within the Centre.

There are three committees designed to oversee and direct the work of the Centre: an Advisory Board which provides external advice on the Centre's activities; an Executive which sets EPCC's strategy and direction; a Management committee which reviews and directs day-to-day operations. EPCC thus has an established management structure which ensures effective and timely delivery of results.

EPCC is structured into a number of groups, with inter-group liaison provided by regular technical meetings and joint project work. The group which includes the HPCI centre is the Applications Group (AG).

Applications Group

AG has more than 30 staff providing consultancy on HPC, computational science and novel computing techniques to both academic and industrial clients. The major activities include supporting the national HPC facilities, HPCI, training, European projects and industrial projects. Almost half the funding comes from industrial work. AG operates a matrix-management approach where tasks are distributed amongst the individuals within the group according to demand and the abilities/interests of the group members. The tasks are scheduled on a 6-weekly timescale and individuals produce a progress report each week. This information is collated into monthly project reports.

HPCI

Initially, EPCC was funded as an HPCI Centre at the level of one FTE to provide support for six HPCI consortia. These consortia represented a very broad range of science and came from four different Research Councils — EPSRC, NERC, PPARC and ESRC. It would have been impossible to find a single individual to provide high quality support for such a broad science base. The matrix-management approach allowed the identification of a named individual for each consortium; this individual was selected from AG to have a close match of interests and skills. This approach is very flexible in that it allows the total effort to vary on a monthly basis. For example, during the initial period when consortia were learning about the T3D (and later T3E) and porting their initial codes, it was possible to put in substantially more than one FTE of effort. In addition, the management of the effort and the workplan can be simpler since there are no significant inter dependencies between the work we carry out with each individual consortium.

During 1996, the core funding level of the EPCC HPCI Centre was increased to two FTEs. This made it possible to help set up and support new consortia while still having effort to put into strategic activities such as HPCI coordination, interdisciplinary seminars and technical reports.

ORIGINAL CONSORTIA

Table 1 lists the original six consortia supported by EPCC and indicates the broad range of science represented. Their requirements and experience varied considerably

Table 1. Original EPCC-Supported Consortia.

Consortium	Scientific Area	Research Council
UKQCD	Quantum Chromodynamics	PPARC
QMC	Many-body Simulations	EPSRC
Colloid	Colloidal Hydrodynamics	EPSRC
Human Systems	Computational Geography	ESRC
Ecology	Ecological Modelling	NERC
VIRGO	Cosmology	PPARC

and it was critical to understand their needs and to tailor the support accordingly. Support for the consortia contained two main strands: detailed porting advice given to the new users of the machine and those with new software, and consultancy on optimisation and data-management issues for the more established consortia with their mature and stable parallel codes. An activity which has become significant more recently is data management and visualisation. EPCC's visualisation facilities are used to produce high-quality images, animations and videos for various consortia. This is an extremely important activity as, in addition to providing valuable qualitative information on large data sets, visualisations serve to publicise and promote the scientific work done on the National Facilities.

The scientific results of EPCC's consortia are primarily detailed in their respective contributions to this conference. The following sections highlight the contributions of EPCC to their success and lists papers to which EPCC made significant contributions; EPCC co-authors are highlighted in bold. The consortia are listed in approximate order of the system size they are investigating to indicate the breadth of the science, from elementary particles, through materials, computational geography and the natural world, to the large-scale structure of the universe.

UKQCD

The UKQCD consortium's internationally-recognised research programme is based on a suite of codes all of which have been developed in close collaboration with members of EPCC staff. These staff have developed a detailed knowledge of T3D/T3E hardware, Cray software and local configuration issues and are therefore able to assist with complex data-handling issues and low-level optimisations such as assembly-language programming. EPCC has provided substantial input from initial code design and implementation right through to optimisation of working codes.

For example, the current GHMC code is one of the most highly optimised T3E codes running anywhere in the world. However, GHMC is still fully portable and has been designed to run on any HPC platform that supports industry standard software. This was proved when GHMC was used as the benchmark for the recent PPARC machine procurement, where it ran efficiently on a wide range of hardware. EPCC's input is aimed at future-proofing UKQCD's research, enabling them to run all their codes for many years to come on whatever machines become available.

- *Lattice QCD simulation programs on the Cray T3D*, **S.P. Booth**, presented at *Optimising of codes for Cray MPP systems* at Pittsburgh Supercomputing Centre, January 1996

- *Lattice QCD on the Cray T3D*, **S.P. Booth**, plenary talk at *The Second European Cray MPP workshop*, EPCC, July 1996.

- *UKQCD Dynamical Fermions Project*, Z. Sroczynski, M. Pickles, **S.P. Booth**, C1 Consortium Project Report.

- *Parallel Implementation of the Hybrid Monte Carlo Algorithm*, **S.P. Booth**, presented at *Stochastic Methods on Parallel Machines*, Cambridge, August 1997.

- *Implementing The Generalised Hybrid Monte Carlo Algorithm*, Z. Sroczynski, M. Pickles, **S.P. Booth**, presented at *Lattice '97, the XV International Symposium on Lattice Field Theory*, July 1997.

QMC

This consortium utilises two main techniques to determine the electronic structure of solids, or more generally atomic systems. These are Quantum Monte Carlo (QMC) simulations and space-time GW quasiparticle calculations, and novel advances have been made to both approaches. QMC calculations are computationally demanding, and the initial vector programs were ported and optimised for the Cray T3D by EPCC; these codes alone resulted in eight publications. EPCC then helped to couple QMC simulation with AIMPRO, a gaussian orbital code, and the entire suite of codes was ported to the T3E when it became available in late 1997. EPCC is also working on the GW approach where perturbative corrections are added to the results of density functional theory. We are in close collaboration with the consortium, aiming to produce a fully working and optimised T3E parallel code. To aid in publicising the consortium's work, EPCC has used its visualisation expertise to produce images and animations of the consortium's large output datasets. A production-quality video of exchange correlations has been produced for a future network television programme.

- *Optimised wavefunctions for quantum Monte Carlo studies of atoms and solids*, A.J. Williamson, S.D. Kenny, G. Rajagopal, A.J. James, R.J. Needs, L.M. Fraser, W.M.C. Foulkes, and **P. Maccallum**, Phys. Rev. B 53, 9640 (1996).

- *Quantum Monte Carlo studies of electronic systems*, R.J. Needs, G. Rajagopal, A.J. Williamson, L.M. Fraser, S.D. Kenny, W.M.C. Foulkes, A.J. James, and **P. Maccallum**, Journal of Korean Physical Society, 1996 Vol.29, No. SS, pp. S116-S120.

- *Load Balancing Diffusion-Based Simulations on Massively Parallel Computers*, **P. Maccallum**, EPCC Technical Report, 1996.

Colloid

EPCC has provided significant training and support for this group and has collaborated closely with them to produce efficient, general purpose parallel codes for the simulation of colloidal systems. These codes form the basis for investigation of many different scientific areas and use a wide range of different techniques. There is considerable industrial interest in these simulations, and Unilever is an active partner.

EPCC's role has included: parallelising new serial codes; developing decomposition strategies for colloid simulations; porting existing codes to the T3D and T3E; adding additional functionality to codes; producing high quality visualisations of the simulations; and optimising the codes for the T3D and T3E. Close contact between EPCC staff and consortium researchers has been critical in ensuring that the researchers have the HPC expertise necessary to exploit the facilities efficiently. The work with this consortium has fed directly into one of the early topics for the new Maxwell Institute, which is discussed later.

- *Scalable Molecular Dynamics of Large Colloidal Suspensions.*, S. Krishna and **A.D. Simpson**, E7 Consortium Technical Report, 1996.

- *ME3D: A Three-Dimensional Code for the Simulation of the Mesoscale Dynamics of Amphiphilic Fluids*, Bruce M. Boghosian, Peter V. Coveney, Sujata Krishna and **Mario A. Antonioletti**, E7 Consortium Technical Report, November 1997.

Human Systems Modelling

In collaboration with EPCC, this consortium has produced a wide range of optimised parallel codes, allowing them to remain at the international forefront of their field: applying HPC techniques to problems in social geography. Areas of interest include: the pattern of daily journeys to work across the UK; detection of spatial clusters of diseases; automatic identification and classification of centres of population; investigation of how results of studies of social data are affected by the way the data is spatially aggregated prior to analysis. All these involve the processing and modelling of huge quantities of information; hence the need for HPC. For example, the 1991 national census contains data for almost every household in the UK. EPCC has worked in all these application areas, for example implementing a genetic optimisation algorithm in parallel on the T3D and developing an interface between High Performance Fortran and the optimised message-passing libraries supplied by Cray. In order to promote the use of HPC in their field, the consortium has developed a Social Science Benchmark code to aid in machine evaluation. EPCC has developed versions of this code using both MPI collective communications and the BSP model, and results have been submitted to Euro-Par'98.

- *Parallel Zone Design Algorithms for Optimising Spatial Representation*, S. Openshaw, I. Turton, S. Alvanides, J. Schmidt, **D.S. Henty**, plenary talk presented at *The 2nd European Cray MPP Workshop*, EPCC, July 1996.

- *Human Systems Modelling: Decomposition Techniques*, **D.S. Henty**, H1 Consortium Technical Report, 1996.

- *Optimisation of a Genetic Programming Code for Non-linear Systems*, **D.S. Henty**, H1 Consortium Technical Report, 1996.

- *Implementation of Social Science Benchmark Code with MPI Collective Communications*, **D.S. Henty**, H1 Consortium Technical Report.

- *A comparison of MPI, HPF and BSP implementations of the Social Science Benchmark*, S. Openshaw, J. Schmidt, **D.S. Henty**, submitted to Euro-Par'98.

Ecological Modelling

After many years of research the Ecology Modelling group has developed Hybrid, a computer model of the complex interactions between vegetation and the atmosphere. Generalised Plant Types (GPTs) approximate the rich diversity of plant life on the surface of the Earth, and competition for resources such as light, water and nitrogen is modelled to predict the distribution of vegetation. Large-scale simulations are very computationally intensive, so EPCC parallelised Hybrid for the T3D to enable the model to be run on a global scale. Subsequent work involved the investigation of novel methods for performing automatic optimisation of the memory layout of the code to improve performance. The results of the parallel Hybrid model can then be used to close the feed-back loop between climate and vegetation. EPCC has integrated the vegetation model with an existing parallel climate-simulation code, CCM3. This research is of direct relevance to current concerns about global warming and its possible effects on the environment.

- *A Genetic Algorithm for Optimising Cache Usage on the Cray T3D*, **J. Malard**, SIAM Annual Meeting 1996, July 1996, Kansas City.

- *Evaluation of a dynamic terrestrial ecosystem model under pre-industrial CO_2 and nitrogen conditions. I. Global distribution of Generalised Plant Types*, A.D. Friend, **J. Malard**, Global Change Biology, August 1996

Virgo

EPCC provides consultancy for the Virgo consortium on Hydra, the code that enables them to perform some of the most ambitious and advanced N-body cosmological simulations in the world. Hydra was initially parallelised on the T3D using a shared memory approach, a strategy that was designed in close collaboration with EPCC. Subsequent work concentrated on optimisation of this code, the design of new parallel algorithms and the development of parallel data-processing tools. The development of a message-passing version of Hydra is currently under way in collaboration with EPCC, and is critical to the future research programme. Support was also provided to allow Virgo to use EPCC's visualisation facilities to make a widely disseminated video that was broadcast on network television.

EPCC staff have significant experience in different parallelisation strategies and have made major contributions to many aspects of code design and implementation. Hydra is highly complex and great care is required to preserve load balance during a simulation as galaxies form out of the initial smooth distribution of matter. A technique called refinement placing combats this problem, but it is extremely difficult to implement efficiently in parallel. An entirely new approach was proposed by EPCC, designed to be efficient on both serial and parallel machines. This work perfectly illustrated the advantages of providing HPC support centrally at EPCC; the algorithm drew on techniques that were known to EPCC staff through their work in Computational Geography, a field that would at first sight appear completely unrelated.

- *Hydra - Resolving a parallel nightmare*, F.R. Pearce, H.M.P. Couchman, A.R. Jenkins, P.A. Thomas presented at *Dynamic Load Balancing on MPP systems*, Daresbury, November 1995

- *Refinement Placing*, **S.P. Booth**, Virgo Consortium Technical Report, October 1997.

NEW CONSORTIA

As well as providing high-quality computational science expertise to major national consortia, a key aim of the HPCI Centres was to help nucleate new consortia which would be likely to be major users of future facilities. EPCC has worked closely with a variety of new user groups and two of these have now been recognised as successful consortia. We helped them produce initial proposals and workplans and ensured that their codes were ported quickly to allow prompt use of the facilities. The first consortium, MHD, is funded by EPSRC and PPARC, while the second is funded under the NERC Special Topics initiative.

MHD

MHD differs from some other consortia in that, when first formed, they were relatively new to HPC. However, advice from EPCC allowed them to make effective use of the T3D as soon as they gained access. A porting strategy was rapidly identified using High Performance Fortran. At their request, members of EPCC staff ran the HPF training course at a remote location convenient for MHD users. Subsequent work at EPCC has concentrated on optimising these HPF codes, and has helped to develop new codes and parallelisation strategies. EPCC are helping MHD develop their future HPC strategy by benchmarking selected codes on various machines.

The fact that MHD now regularly use their full allocation of computer time is an example of how EPCC's extensive training expertise, coupled with the wide scientific and computing experience of its staff, enables researchers new to HPC to make efficient use of world-class HPC facilities.

- *Solar, Astrophysical, Geophysical MHD Computational Consortium*, the MHD Consortium proposal to TRAP, August 1996.

- *Optimising HPF codes for Magnetohydrodynamics*, **A. Ewing**, MHD Consortium Technical Report, May 1997.

- *Performance Comparison of MHD HPF codes on the T3D and T3E*, **G. Pringle**, presented at MHD Consortium meeting, Newcastle, November 1997.

- *The UK MHD Consortium: Goals and Recent Achievements*, A. Hood, T. Arber, K. Galsgaard, A. Brandenberg, S. Brooks, C. Jones, G. Sarson, **G. Pringle**, presented at *HPCI Conference 1998*, Manchester, January 1998.

RHO

Recently, EPCC has helped in the formation of a new HPCI consortium, the RHO consortium, who study the long-range correlations seen in oilfield well-pressure data and in natural seismic activity. There is no current explanation for these observations, which are termed Spearman ρ-correlations (hence the consortium's name). Their research involves sophisticated computer modeling that requires the use of HPC facilities. The results are of direct relevance to industry, and British Petroleum are co-funding the project with NERC via the Connect scheme. EPCC have provided training for relevant RAs, guidance in producing a workplan and have ported the initial codes.

STRATEGIC EFFORT

EPCC has put significant effort into taking a higher level view of national HPC issues and improving the dissemination of HPC awareness and expertise. As well as providing additional training for HPCI researchers, this effort has resulted in a very successful, ongoing series of multidisciplinary seminars and a wide variety of technical reports on important topics for users of the national facilities.

Seminars

As part of HPCI, EPCC is committed to organising regular multidisciplinary seminars across the UK, on computational techniques of interest to Research Council users. This series of seminars is targeted both at specialists and at people who wish to find out more about the benefits of HPC in their own research area. Table 2 lists the seminars coordinated by the EPCC HPCI centre.

Table 2. Seminars coordinated by EPCC HPCI centre.

Title	Date	Location
2nd European Cray MPP Workshop	07/96	Edinburgh
Molecular Dynamics Simulations on MPP Platforms	11/96	London
European MPI Information Meeting	02/97	Edinburgh
Stochastic Methods on Parallel Machines	08/97	Cambridge
HPCI98[a]	01/98	Manchester
Simulation of Fluid Flow on Parallel Machines	06/98	London

[a]Jointly organised with the other HPCI centres.

Rather than focus only on the scientific aspects of the work, these seminars have also covered the techniques and algorithms used to exploit leading-edge supercomputers to maximise their cross-disciplinary relevance. The five seminars which have already taken place have had a combined audience of over 500 people. Reports of all these seminars have been included in *HPC News*, EPCC's newsletter for the users of the national facilities.

Technical Reports

EPCC's HPCI funding has supported a number of strategic technical reports and has helped in the dissemination of the activities of the team providing applications support on the national facilities. These technical reports represent a valuable resource for all users of the facilities.

In the following list, EPCC authors are highlighted in bold.

- *HPF on the Cray T3D: A Comparison with Craft*, **A. Ewing, D.S. Henty**, EPCC Technical Report EPCC-TR95-05.

- *Fast Fourier Transforms on the Cray T3D*, **D.S. Henty**, EPCC Technical Report EPCC-TR96-02.

- *IO on the Cray T3D*, **S.P. Booth**, EPCC Technical Report, EPCC-TR96-05.

- *Optimising HPF: A User's Experience*, **A. Ewing**, presented at *The First HPF User Group Meeting*, Santa Fe, US, February 1997.

- *Performance and Code Writing Bottlenecks on the Cray T3D*, **S.P. Booth**, EPCC Technical Report EPCC-TR97-01.

- *Portability of Performance Optimisations between T3D and T3E Systems*, **D.S. Henty, S.P. Booth**, C. Keable, D. Tanqueray, EPCC Technical Report EPCC-TR97-02.

- *Decomposition Independence in Parallel Programs*, **S.P. Booth**, EPCC Technical Report EPCC-TR97-03.

- *Portability of Performance Optimisations between T3D and T3E Systems*, **D.S. Henty**, presented at *The 3rd European Cray/SGI MPP Workshop*, Paris, September 1997

- *Performance of Data Parallel Programming Models on Cray MPP Systems*, **D.S. Henty, G.J. Pringle**, invited plenary talk at *1st HPaC Seminar*, Delft, Holland, January 1998.

- *Performance Optimisation on the Cray T3E*, **S.P. Booth**, presented at *HPCI Conference 1998*, Manchester, January 1998.

Training

A small amount from the EPCC HPCI Centre's funding is used to increase the number of training courses on the National Facilities, so as to ensure that IIPCI RAs and other appropriate members of the consortia have priority access to courses when required.

Training has always been a major priority for EPCC and this has been recognised by JISC funding as a Centre of Excellence for training in distributed computing and computational science. We provide a whole range of courses covering: introduction to parallel computing; efficient use of the local facilities; major software standards for parallel programming; performance optimisation.

Training has been critical in providing support for the more than 700 users of EPCC's national facilities. Since the start of service on the T3D, we have presented 70 runs of more than 10 different courses written specifically for users of the T3D and T3E, as listed below. The courses range in length from one day to three days and include both introductory and advanced material covering all aspects of programming the facilities. So far, this training represents over 800 student-days. The current portfolio of T3D and T3E courses includes:

- HPC on the Cray T3D and T3E: Introduction

- HPC on the Cray T3D and T3E: Parallel Decomposition

- HPC on the Cray T3D and T3E: Computer Simulation

- MPI Programming on the Cray T3D and T3E

- HPF Programming on the Cray T3D and T3E

- PVM Programming on the Cray T3D

- CRAFT Programming on the Cray T3D

- Vector Processing on the J90

- Performance optimisation on the Cray T3D

- Efficient use of the Cray T3E

- Performance optimisation on the Cray T3E

EPCC also provides more general courses on HPC techniques and applications-oriented material, which are free to all UK academics. These are supplemented by a series of Technology Watch reports covering key topics in HPC.

MAXWELL INSTITUTE

The University of Edinburgh has recently established the Maxwell Institute for HPC Applications, a new international multi-disciplinary centre devoted to the uses of computer simulation in science; EPCC will be closely involved with this, providing HPC consultancy and support which builds on the HPCI work. The objective of the Institute is the advancement of knowledge through the exploitation of high performance computing technology. The advent of HPC has enabled computer simulation to emerge as a new scientific method, alongside theory and experiment. It is the only way to study systems which are too large, too small, too dangerous, too complex, or too expensive for the traditional approaches of theory or experiment.

The Institute will be able to draw upon one of the strongest concentrations of top-quality scientific research in the UK and is well placed to become an international centre of excellence in this new and expanding area. In order to respond to changing scientific priorities and to cover as wide a range of topics as possible, the Maxwell Institute will operate along similar lines to the highly-successful Newton Institute for the Mathematical Sciences at Cambridge; it will host a sequence of six-month focused workshops, each organised by an internationally-renowned Programme Director. Funding is being sought to allow participation from throughout the UK and beyond.

The Maxwell Institute builds on expertise, research projects and funded programmes already available in EPCC, other departments in the University of Edinburgh, and other HEIs. In the long term, new opportunities created by advancing technology will guide the choice of programme topics. However, the following will be important themes in the work of the Institute:

- The new science enabled by computer simulation.

- The new methodology and its rigorous foundation.

- Multi-disciplinary models.

SUMMARY

The research carried out by the EPCC HPCI centre has largely focused on development of computational techniques and algorithms for HPC systems. As such, results obtained during work for a particular consortium are widely applicable to the research of all the consortia, and also to the HPC research community as a whole. Reports have been widely disseminated via conference presentations and the WWW. Since EPCC undertakes many other HPC-related activities, there is significant leverage with, for example, the industrial programme. EPCC is in an ideal position to effect technology transfer amongst all the UK research groups with interest in HPC, and has a proven history of also applying novel computational techniques to the benefit of industry.

The HPCI effort itself is aimed directly at facilitating the scientific programmes of the supported consortia rather than performing independent research, although the contribution of the EPCC HPCI centre is acknowledged wherever appropriate in all consortium publications. With the advent of the Maxwell Institute, however, there is now an opportunity for EPCC to further develop the successes of the HPCI work in new research areas.

THE CLRC HPCI CENTRE AT DARESBURY LABORATORY

R.J. Allan, I.J. Bush, K. Kleese, A.G. Sunderland and M.F. Guest

HPCI Centre, CLRC Daresbury Laboratory,
Daresbury, Warrington, WA4 4AD

INTRODUCTION

This chapter provides an overview of the work of the CLRC Daresbury HPCI Centre, both in support of the HPCI Consortia and in addressing the increasing computational demands from U.K. industry.

We firstly provide a summary of the Consortia associated with the Centre and outline the role of the IBM SP2 development platform in supporting these consortia. The next section addresses the wide range of code development activities undertaken over the past two years that has increased the application base on the Cray T3D/T3E (e.g., MOLSCAT, ANGUS, FELISA, CONQUEST, the R-matrix Propagator Codes, MOLPRO96, DVR3D, and CETEP). Much of this work is built on the associated development and assessment of parallel algorithms and tools, and the following section outlines on-going work in the area of numerical software, including the Global-Array tools, dense-matrix diagonalisation, multi-dimensional block-cyclic FFT algorithms, sparse matrices and related work, and parallel partitioning, scheduling and load balancing.

Finally, we consider the steady uptake of supercomputer technology in industry over the past few years, driven in part by new methods developed in academic research. Application of these methods promises to lead to results for much more complex systems than could previously be treated. Systems which are too difficult, complex or expensive to study in the laboratory can yield important information for design or new models for incorporation into industrial process optimisation. The undoubted potential of computation provides an attractive incentive for academic and industrial scientists and engineers to interact, an interaction that is facilitated by programming support staff and national facilities provided by the U.K. Research Councils. We illustrate this potential by outlining a typical large-scale scientific collaborative project that is exploiting a range of expertise and software to study the complex process of titanium dioxide powder production.

SUPPORT FOR HPCI CONSORTIA

Three support centres were founded at the Universities of Edinburgh and Southampton and at the Central Laboratory of the Research Councils (CLRC), with the aim of

High Performance Computing
Edited by R. J. Allan *et al.*, Kluwer Academic / Plenum Publishers, New York, 1999

Table 1. HPC Consortia supported by the CLRC HPCI Centre

Macromolecular modelling (BBSRC, EPSRC)	Integrating electronic structure and molecular mechanics to study e.g. active sites in enzymes
Computational Combustion (EPSRC)	Simulations of laminar flamelet chemistry and its interaction with turbulent fluid dynamics
Large Eddy Simulations (EPSRC)	of complex engineering industrial flows to develop techniques for use in design of complex materials
Car-Parrinello (EPSRC)	First-principles electronic structure calculations of a wide range of solid-state physics of complex materials
Chemical Reactions and Energy Exchange (EPSRC)	Reactive and non-reactive atomic and molecular scattering
Shelf-sea hydrodynamics (NERC)	Modelling the near-coastal regions with improved accuracy
Transition modelling (EPSRC)	Modelling of fluid flow in canonical geometries to study the change from laminar to turbulent motion
Complex aerodynamics flows (EPSRC)	Developing techniques to model physically and geometrically complex air flows
Micromagnetics (EPSRC)	Developing techniques to study the structure and dynamics of mesoscopic magnetic systems
Ab initio simulation of covalent materials (EPSRC)	Large-scale cluster calculations of the structure and properties of covalent systems
Atomic Multiphoton processes (EPSRC)	Exploring the physics of multi-electron systems subjected to very intense laser fields
Electron Collision Processes (EPSRC)	Developing techniques to model the scattering of electrons from complex atoms and molecules
Parallel Conjugate Gradient methods (EPSRC)	Developing new numerical algorithms for finite-element applications
UK MHD Consortium (EPSRC)	Studies of solar corona, Earth's geodynamo and dynamo action in turbulent plasmas
Cosmological simulations (PPARC)	Simulations of galaxy and star formation to predict observed luminosities and abundancies
Direct numerical simulation of fluid flows (EPSRC)	to guide the development of approximate turbulence transport models
Materials Chemistry (EPSRC)	developing techniques to study the structure, phase diagrams and properties of complex materials

supporting the consortia, providing an interface between the projects and a focus for the consortia to interact with other HPC programmes. Of the three centres, that at the CLRC Daresbury Laboratory is the largest supporting approximately half the consortia, as shown in Table 1.

Of these the transition modelling consortium has been closed down (because the active scientist left the country), while the multiphoton and electron collision consortia chose to merge their activities. The last three projects listed in the table are formally supported by other centres, but some assistance is also provided by CLRC.

IBM SP2 Development Platform

An IBM SP2 provides the Centre's main development platform for distributed-memory parallel programs. Other facilities available at CLRC include the Fujitsu VPP/300, the Cray J932 and DEC 8400. All machines serve a diverse user community which is fully supported by CLRC staff.

The initial SP2 system was procured in July 1995, with part funding from the HPCI. It comprised 16 nodes, two Wide Nodes each with 128 MByte memory, and 14 TN2 (Thin Nodes), each with 64 MByte memory. An interim upgrade of this system was conducted in July 1997, following EPSRC's review of its service level agreement with CLRC's Computational Science and Engineering Programme (CSEP). The machine now comprises two SP2 Wide Nodes (66.7 MHz) with 128 MB memory, 24 P2SC Thin Nodes (120 MHz) with 256 MB memory, 3 GByte disk capacity local to each compute node, and 40 GByte disk capacity on an attached RAID3 system.

This system is used to both develop new parallel codes and algorithms, using methods and lessons learned from the previous three years, and to process a class of intermediate sized production jobs which would not be possible on most local university facilities, but which would nevertheless be considered too small for the national "grand-challenge" system. A few Universities have now purchased similar equipment with funding from the EPSRC/HEFCE Joint Research Equipment Initiative (e.g. the Cambridge Hitachi SR2201, the Belfast and Trinity College Dublin IBM SP2, and Manchester SGI Origin 2000) and they have been offered parallel programming tutorials and other assistance.

We now describe a number of major code-development projects which have been carried out in collaboration with the HPCI Consortia.

NEW CODE DEVELOPMENT

A major activity of the CLRC HPCI Centre features the collaborative development of new computer codes for the most challenging of scientific problems. This builds, in many cases, on the work of the CCPs and is highly successful because of the co-existence of HPCI Centre staff and other scientists supporting the wider EPSRC research programmes. Many other codes supported by these programmes, e.g. DL_POLY [1], GAMESS-UK [2] and CRYSTAL [3] have been parallelised, ported and optimised on the national computing facilities by Centre staff, and are now available for all consortia to use.

MOLSCAT. MOLSCAT is a code to compute probabilities and spectra resulting from energy transfer between colliding molecules in the earth's atmosphere and in space [4]. It was one of the first codes to be ported to the Cray T3D at EPCC and is used by CCP6 and the Consortium for Chemical Reactions and Energy Exchange Processes. A hierarchical decomposition strategy was used to give good parallel performance.

ANGUS. ANGUS is used for direct numerical simulation in the study of flames in a turbulent reacting mixture. The parallelisation of the code employs a 3-dimensional domain decomposition of a regular grid of points [6], and follows work done on a previous benchmark code, FLOW [5]. Over the lifetime of the Centre ANGUS has been highly optimised, allowing simulations to be carried out routinely using over two million grid points; see the chapter by Jenkins et al. in this book.

FELISA. This code is a benchmark for production codes using an irregular tetrahedral mesh of points to compute aerodynamic and electromagnetic properties of objects with complex geometry, such as aircraft components. To develop methodology in this class of applications, the FELISA code was parallelised during the early days of the HPCI as a joint project with the CLRC Engineering Group [7, 8]. This work involved fundamental research into partitioning, scheduling and load-balancing algorithms (see below).

CONQUEST. A significant amount of time has been spent developing the O(N) code CONQUEST with the UKCP Consortium, details of the parallelisation strategy adopted having already been published [9]. Development tasks included, (i) major improvements to the message passing performance; (ii) enhanced memory utilisation, with dynamical allocation of the larger arrays – the code will now automatically make a compromise between memory and performance when the arrays describing the message passing become too large, and (iii) further improvement in the floating point performance, achieved by assembly-coding critical routines.

CONQUEST was tested on a silicon system containing over 6,000 atoms and subsequently ported to the Cray T3E so that "real" calculations on MgO could be undertaken. See the chapter by Gillan et al. in this book.

The R-matrix Propagator Codes. A long-term development project has been to produce an efficient parallel implementation of the FARM [10] intermediate R-matrix propagator code used in accurate atomic physics calculations. The delivery of a complex code involving hierarchical parallelisation techniques in October 1997 highlighted the effectiveness of close inter-working of dedicated HPCI staff and Consortium members. In this case the team comprising A.G. Sunderland, C.J. Noble and V.M. Burke (CLRC) with J. Heggarty, S. Scott, M. Watts and P.G. Burke (Queen's University of Belfast) completed various parts of the project leading to the final chosen code structure. Results and methodology have been published [11]; see also the chapter by Noble et al. in this book.

MOLPRO96. Dr. A. Dobbyn, the PDRA with the Chemical Reaction consortium, spent two years working within the HPCI Centre to to parallelise and port the MOLPRO96 code to the Cray T3D and T3E systems. The major focus was the production of a fully parallel version of the Multi-Reference Configuration Interaction (MRCI) module [12]; it is this section of the code which used the majority of computer time. The integral evaluation section has also been parallelised and scales almost linearly.

The distributed-data parallel implementation was carried out using the Global Array tools from PNNL (see below). The MOLPRO96 code is now being used in production work; a separate paper by Dobbyn and Knowles in this volume contains further information.

DVR3D. A separate paper in this volume by H. Mussa et al. outlines the development and application of the parallel DVR3D code. This was initially produced as a serial code via CCP6 [13] and is used by the Chemical Reaction Consortium to compute complex spectra of highly excited (hot) molecules, such as water in the Jovian atmosphere. Hamse Mussa spent a year working at the HPCI Centre and details of the parallelisation strategy have been published [14].

CETEP. CETEP is the key code used by the UKCP Consortium, one of the largest users of the T3D and T3E systems with applications in many areas of molecular surface and solid-state physics. Support for this consortium, in addition to the work on CONQUEST described above, is provided jointly by the HPCI Centre and the CLRC Materials Group via CCP3. Recent work has focused on performance optimisation using memory-intensive algorithms and a new 3-dimensional block-cyclic FFT package(see below). The organisation of UKCP and diverse applications of CETEP are described by Gillan et al. in this book.

DEVELOPMENT AND ASSESSMENT OF PARALLEL ALGORITHMS AND TOOLS

Numerical Software Evaluation

A variety of Numerical Software has been investigated over the past few years, e.g. eigensolvers [15] (BFG [16], PeIGS [17, 18, 19], and PARPACK [20, 21]), multi-dimensional FFT routines [22], and the proprietary libraries, PESSL [23] and LibSci [24]. Much of this software is being migrated into the application codes and is used in all new development work. Use of ScaLAPACK [25] has also grown, especially with both Cray and IBM providing tuned versions of some of the routines in their proprietary libraries.

The Centre has published a number of reports of its evaluation work indicating the performance characteristics of the codes investigated [26, 15, 22, 27, 28]; research carried out by students visiting the Centre has also been published [29, 30]. Other general information on numerical software and tools for parallel and distributed computers can be found in HPCProfile [31].

Global-Array Tools. The GA Tools implement a *portable* "shared memory" programming model for distributed memory computers [32]. The major advantage of this over a message-passing model is the ease with which it can be applied to even highly complex codes. The key feature of the GA approach is that it enables nodes to asynchronously access data held on other nodes, using a global index space, and without "interrupting" the node on which the data is stored

Dense-Matrix Diagonalisation. To compliment direct-eigensolver software evaluated from other sources, Ian Bush wrote a parallel one-sided block-Jacobi eigensolver following the method outlined by Littlefield and Maschhoff [33]. This is an iterative solver particularly useful in applications where one problem must be solved and information is already available from a similar one.

Multi-Dimensional Block-cyclic FFT Algorithm. A new project started during 1997 to provide an interface to parallel 2D and 3D FFT routines enabling the block-cyclic distribution of data in a way compatible with ScaLAPACK [25]. The uptake of this promises to be of significant benefit, particularly in the molecular-dynamics and Car-Parrinello communities, but also for use in spectral engineering and climate codes and in time-dependent chemical reaction codes.

Sparse Matrices, the Harwell Subroutine Library and Related Work. The Harwell Subroutine Library [34] sparse-matrix solver MA48 has been tried in several codes, particularly the R-Matrix Floquet multiphoton code, where it proved many times faster than the CG algorithm previously employed. Support from the CLRC Numerical Analysis Group (led by Prof. Iain Duff) in our use of this library and in developing new parallel software and providing training courses in the area has proved particularly beneficial,

Parallel Partitioning, Scheduling and Load Balancing. Support for the HPCI Engineering Consortia is provided by both the HPCI Centre and by CCP12 *High Performance Computing in Engineering*. A long-term research project involves the development and application of new parallel algorithms in areas of importance to unstructured mesh codes [35, 36, 37].

TECHNOLOGY TRANSFER TO U.K. INDUSTRIAL RESEARCH AND DESIGN

Industrial production processes, in common with many other "real world" applications e.g. climate forecasting, involve many complex interactions between component subsystems. Our knowledge of the underlying physics and chemistry is probed in many unexpected ways, due to the interlinking of these interactions on the macroscopic scale. Detailed academic studies of individual interactions are useful in learning how to treat the full process, but may need incorporating into a parametric model if calculations are not to be too time consuming. This is particularly true of chemical reaction rates and turbulence which need to be taken into account in complex multi-phase flow calculations. Likewise treatments of complex materials or chemical systems involve the treatment of long and short-range forces or electrostatic interactions in different ways (leading to the so-called Order(N) methods).

Current applications are stepping beyond the capabilities of even two years ago, through combining techniques applicable to the study of phenomena on diverse length and time scales.

Example – Titanium Dioxide Powder Production

This example illustrates a partnership between Daresbury Laboratory, Tioxide Ltd., an EC ESPRIT Project, CCP3, CCP5 and the U.K. Car-Parrinello HPCI Consortium.

Powders comprised of microparticles of inorganic materials are common industrial products. One such material is titanium dioxide TiO_2 which, in the form of rutile, is an important, non-toxic pigment used in the manufacture of many everyday substances, most importantly in paints, where its whiteness and durability make it an essential ingredient. It can also be used as a catalyst and gas sensor. To the worldwide chemical industry this represents some $6 bn annually! Titanium dioxide is mostly manufactured by the so called "chloride" process, in which titanium tetrachloride is burned in oxygen and the dioxide product is deposited as a white powder direct from the gas phase:

$$TiCl_4 + O_2 \rightarrow TiO_2 + 2Cl_2 + heat$$

The ICI/Tioxide plant at Billingham is typical of a gas phase oxidation reactor. The reactor is pressurised and is basically a cylinder with the raw materials fed into one end. The exothermic reaction is started by a plasma arc of several MW power and itself produces several MW of power. The gases pass through the reactor at near sonic speeds and have a residence time of only milli-seconds. This is shown schematically in Figure 1.

Computer modelling of reactor performance is a key component of an efficient optimisation strategy as it reduces the need for empirical tests on expensive reactor components. The choice of modelling technique depends on the features of the reaction to be studied. We treat large scale features of the gas flow in the reactor using fluid dynamics, while the detailed chemical reactions are treated using quantum mechanics. Features at intermediate length and time scales are studied using molecular modelling.

The aim of the HP-PIPES project (High Performance Parallel Computing for Process Engineering Simulation) was to develop computer simulation models to the point where they can be used as realistic simulation tools for the design and operation of reaction and combustion systems which involve both complex internal flow patterns and complex physical and chemical phenomena. Funding was awarded by the CEC

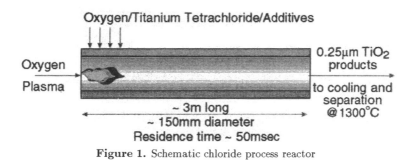

Figure 1. Schematic chloride process reactor

ESPRIT programme for HP-PIPES, a consortium of industrialists from ICI plc, EDP (Portugal), IST (Portugal) and Marex (Netherlands) working with the CLRC computational engineering group. More information about this and related projects are available on a CD-ROM [38].

The core code is an advanced computational fluid dynamics code designed to run on parallel computers. It models the incompressible, turbulent Navier-Stokes equations together with an enthalpy equation describing heat transfer in three-dimensional duct and pipe geometries. This is a basic requirement common to chemical reactors, enclosed combustion and a host of other process engineering applications.

Using this code a set of modules was written which models the complex physics and chemistry involved in the ICI/Tioxide oxidation reactor and the power station combustion and convection chamber at EDP in Setubal. These modules describe such phenomena as particle agglomeration, radiation heat transfer, the chemical reactions involved in the reactors, etc. From these modules the actual models of the reactor or boiler combustion chamber are built. The predictions of these models were tested against the behaviour of the actual plants by using data specially obtained from the plants and from historical records of the behaviour of the plant recorded in the course of process improvement. A validation exercise is essential to the proving of any simulation package.

TiO_2 deposits out of the gas phase as a fine powder that is immediately usable in its many industrial applications. The process by which the particles form consists of at least three distinct stages: (a) nucleation; (b) surface growth; and (c) coagulation and fusion. These sub-processes find expression in the empirical description underpinning the chemical engineering model. Despite the importance of the reaction however, relatively little is known about these processes occurring at the atomistic level, particularly those determining the structure of the microparticles and their growth under the extreme conditions of the reaction. Added to these, the details of the chemical reaction itself, particularly at the surface of the microparticles are also of considerable interest.

The growth of small crystals of TiO_2 in the reactor can be modelled using molecular dynamics [39], with the simulations performed using the DL_POLY code [1]. The molecular simulation groups at CLRC and Tioxide Ltd. are collaborating under the EPSRC's Realising Our Potential Award (ROPA) scheme which rewards academic groups having joint projects with industrial researchers.

A series of molecular dynamics simulations of TiO_2 microclusters, comprising 1245 atoms in vacuo at temperatures ranging from 1000 K to 3000 K, was carried out using the best available force fields. The simulations showed that the microclusters remained in the rutile phase up to the melting point, which was estimated to be approximately 2150 K. From the original spherical microcluster, a clearly faceted form was obtained in the solid phase, exhibiting the (100) and (101) faces of rutile. The integrity of the rutile lattice in the microcluster core was established. In the molten state the microclusters lost both the faceted appearance and planar order in the core, see Figure 2.

Figure 2. Typical TiO$_2$ microcluster

The dynamical aspects studied included the mean-square displacements (MSD) of the constituent ions and the van Hove self correlation functions. The former revealed distinct differences in the Ti and O ion diffusion rates between the surface and core ions of the microcluster in the solid state. A more detailed investigation showed that the diffusion occurred via a hopping mechanism. This study thus revealed several important properties of TiO$_2$ microparticles.

The high surface diffusion of ions was thought particularly important with respect to fusion of microclusters, an important process in the coagulation phase of the growth of TiO$_2$ powder in the Chloride Process. A second series of molecular dynamics studies was designed to reveal the important features of the process. Pairs of 1245 atom TiO$_2$ microclusters were brought together at different temperatures, with different relative velocities and from different starting separations and the dynamics of the collision and the mechanisms of particle fusion studied.

The simulations showed that fusion of microclusters spans several timescales. Stationary microclusters are drawn together by long-range van der Waals forces, a study of the mutual acceleration of the microclusters indicating an approximately $1/r^4$ force law. On contact, there is a rapid (10ps) reduction in overall configuration energy. On a longer timescale (lasting several hundred picoseconds) the configuration energy continues to decrease. The most important process occurring on this time scale appears to be the surface diffusion of the ions leading to effective "cementation" of the fusing particles.

Temperature was found not to be a particularly significant factor in the fusion process, the microclusters fuse irrespective of temperature and the process is not adversely affected by surface roughness or collision velocity. This is in marked contrast to experimental studies of the chloride process, where the "stickiness" of the particles shows a strong temperature dependence. A possible explanation for this disparity suggests that the particle surface could be contaminated by the adsorption of other species, for example chlorine. The increasing stickiness can then be seen to be a consequence of the evaporation of the contaminant to expose the TiO$_2$ surfaces.

Reliable predictions of TiO$_2$ yield and the levels of impurities and contaminants in the final product require a detailed understanding of the chemistry. The detailed chemical reactions which govern the production of TiO$_2$ must be modelled using quantum mechanics. To study this a 3-year collaboration between CCP3 and CCP5 is supported by funding from Cray Research Inc. and is based in the CLRC materials science group.

The properties of oxide surfaces are invariably dependent on the stoichiometry of the material. The stable surface structure may alter radically when reduction takes place, accompanied by changes in catalytic reactivity. In TiO_2 the (100) surface is known to exhibit the 1×3 microfacet reconstruction. However the role played by stoichiometry in this reconstruction had not been demonstrated. The quantum mechanical simulations of this surface showed that for the stoichiometric material there is no driving force towards reconstruction. The most recent simulations suggest that the rutile structure is stable for an oxygen deficient surface. The reduction of the surface also strongly affects the surface electronic structure e.g. spin polarised states localised to particular Ti sites change the colour of the material – an un-desirable property if used as a pigment!

The interaction of water molecules with the TiO_2 (110) surface has been examined in detail using spectroscopic, diffraction and desorption experiments. It is therefore an ideal system with which to probe the chemistry of TiO_2 surfaces using novel quantum mechanical simulations. The constant cross comparison with experiment helps us to explore the reliability of this technique. In the simulations using the CETEP code [40] we position the water molecule above the surface and follow what happens if we "let it go". If the molecule is attracted to the surface, it will be adsorbed there. The molecule may dissociate on adsorption, or remain intact but stuck to the surface (physisorption). The beauty of quantum mechanical simulations is that we do not have to guess the right configuration of the molecule out of the myriad of possibilities – it is found through the thermal motion.

At low coverages the water molecules dissociate and stick to the surface. For higher coverage complex adsorption geometries occur in which molecular water can co-adsorb with dissociation products. This is an unexpected result which resolves many of the unanswered problems in interpreting previous experimental studies.

As revealed in the studies of surface structure, titanium dioxide surfaces are readily reduced and this can have significant consequences for their properties. The surfaces may also be reduced by adsorbing strong reducing agents such as alkali metals. The adsorption of potassium at the TiO_2 (100) surface has been studied by quantum mechanical simulations using the CRYSTAL95 software developed in a collaboration between CLRC and the University of Torino, Italy [3]. Potassium is found to bond strongly to the surface through electron transfer. As in the reduced surface, electrons localise at particular Ti sites near to the surface and the electronic properties of the surface are strongly modified.

CONCLUSIONS

In describing the range of activities undertaken within the CLRC HPCI Centre at Daresbury Laboratory over the past three years, we have pointed to the increasing impact of computational science and engineering in both academia and industry, and shown the crucial role of numerical algorithm and new code development in accelerating this process.

The main significance of the titanium dioxide powder project example, from an industrial viewpoint, is that it has demonstrated that atomistic simulation methods have a part to play in direct modelling of the conditions of a flame reactor. A number of important observations have been made which help clarify or deepen understanding of the fundamental processes. From the Centre viewpoint, the project has seen a fruitful interdisciplinary interaction between industrial and academic researchers. It has brought together computational science and engineering, molecular modelling and quantum mechanical simulation in complementary ways on distinct length and time scales to describe a major industrial process in detail.

Whilst a great deal remains to be done, this innovative project, and others like it, have indicated that real progress is possible through access to modern parallel supercomputers, and to the range of expertise available to CLRC staff and the codes developed jointly by them and their collaborators.

ACKNOWLEDGEMENTS

The computer codes used in the TiO_2 project were all written by staff working with the EPSRC-funded CCPs and are fully supported and maintained at Daresbury Laboratory. These include:

- GAMESS UK — Generalised Atomic and Molecular Electronic Structure System, incorporating a range of accurate properties for first-principles chemistry (CCP1) [2]
- DL_POLY – Parallel package for molecular dynamics, incorporating a range of state-of-the-art techniques (CCP5) [1]
- CRYSTAL95 – accurate self-consistent field calculations for periodic systems (CCP1) [3]
- CETEP – Cambridge and Edinburgh Total Energy Package, using the Car-Parrinello technique and local potentials to establish the equilibrium structure and dynamics of complex materials (CCP5) [40]
- HP-PIPES – High-Performance Parallel Computing for Process Engineering Simulation, a deliverable of EC ESPRIT project 8114.

These and other codes have now been used in a diverse range of applications of academic and industrial relevance and are available, subject to appropriate licensing restrictions. The exception in the above list is HP-PIPES which was developed under an EC ESPRIT grant rather than via the CCP route.

We wish to point out that, as well as supercomputing work, CLRC supports a large range of experimental facilities. Of these the electron synchrotron at Daresbury Laboratory has also been used to study properties of TiO_2.

REFERENCES

1. W. Smith, T.R. Forester, *DL_POLY*, distributed by Daresbury Laboratory on behalf of CCP5 (1996). More information can be found via the WWW at URL *http://www.dl.ac.uk/TCSC/Software/DL_POLY*.
2. M.F. Guest, J.H. van Lenthe, K. Kendrick, K. Schöffel, P. Sherwood, R.D. Amos, R.J. Buenker, M. Dupuis, N.C. Handy, I.H. Hillier, P.J. Knowles, V. Bonacic-Koutecky, W. von Niessen, A.P. Rendell, V.R. Saunders and A.J. Stone, *GAMESS-UK*, distributed by Computing for Science Ltd. on behalf of CCP1 (1996). More information can be found via the WWW at URL *http://www.dl.ac.uk/TCSC/QuantumChem/Codes/GAMESS-UK*.
3. C. Pisani, R. Dovesi and C. Roetti, *Hartree-Fock ab initio treatment of crystalline systems*, Lecture notes in chemistry, Springer-Verlag, Heidelberg (1988).
   CRYSTAL95 is distributed by the University of Torino and Daresbury Laboratory. More information can be found via the WWW at URL
   http://www.dl.ac.uk/TCSC/Software/CRYSTAL.
4. J.M. Hutson and S. Green, *MOLSCAT version 12*, distributed in the U.K. by J.M. Hutson via CCP6 (1993).
   I.J. Bush *User Manual for Parallel MOLSCAT*, Daresbury Laboratory, (1994)
5. R.J. Blake, D.R. Emerson and R.J. Allan, *FLOW: a Parallel Benchmark Code for High speed Air flow: version 1*, report prepared under contract to NPL under the Esprit III PEPS project, Daresbury Laboratory (1992).
6. Y.-F. Hu, D.R. Emerson and R.J. Blake, The communication performance of the Cray T3D and its effect on iterative solvers, *Parallel Computing* 22:928-944 (1996).

7. J. Peiro, J. Peraire and K. Morgan, *FELISA System Version 1.0, User Manual*, University of Wales College Swansea

8. Y.-F. Hu and R.J. Blake, Partitioning and scheduling algorithms and their implementation in FELISA – an unstructured grid Euler solver, in: *Parallel Computational Fluid Dynamics 93: New Trends and Advances*, A. Ecer et al. eds., Elsevier (1993)

9. C.M. Goringe, E. Hernandez, M.J. Gillan and I.J. Bush, Linear-scaling DFT Pseudopotential Calculations on Parallel Computers, *Comp. Phys. Comms.* (1997).

10. V.M. Burke and C.J. Noble FARM – a flexible asymptotic R-matrix package, *Comp. Phys. Comm.* 85:471 (1995)

11. A.G. Sunderland, J.W. Heggarty, C.J. Noble and N.S. Scott, 1D Propagation of Large R-Matrices on Distributed Memory Parallel Computers, *Comp. Phys. Comms.* (1997) in press.

12. A.J. Dobbyn, P.J. Knowles and R.J. Harrison, Parallel Contracted Multi-reference Configuration Interaction, *J. Comp. Chemistry* (1997).

13. J. Tennyson, J.R. Henderson and N.G. Fulton, *Comp. Phys. Comms.*, 86:175 (1995)

14. H.Y. Mussa, J. Tennyson, C.J. Noble and R.J. Allan, Rotation-vibration calculations using massively parallel computers, *Comp. Phys. Comms.* (1997).

15. R.J. Allan and I.J. Bush, *Parallel Application Software on High Performance Computers: Parallel Diagonalisation Routines*, Daresbury Laboratory (1996).

16. I.J. Bush, One-sided Block-factored Jacobi eigensolver available from Daresbury Laboratory. E-mail *i.j.bush@dl.ac.uk*.

17. D. Elwood, G. Fann and R. Littlefield, *Parallel Eigensolver System*, User Manual available from *anonymous@ftp://pnl.gov* PNNL (1994).

18. G. Fann, D. Elwood and R. Littlefield, *Performance of a fully parallel dense real symmetric eigensolver in quantum chemistry applications*, PNNL preprint (1994).

19. G. Fann and R.J. Littlefield, Parallel inverse iteration with reorthogonalisation, in: *Proc. 6th SIAM conference on Parallel Processing for Scientific Computing*, pp409-13 SIAM (1993).

20. K. Maschhoff and D. Sorensen, A portable implementation of ARPACK for distributed memory parallel architectures in: *Proceedings, Copper Mountain Conference on Iterative Methods* (1996).

21. R. Lehoucq, K. Maschhoff, D. Sorensen and C. Yang, *ARPACK*. Parallel ARPACK is available from *anonymous@ftp://ftp.caam.rice.edu/pub/people/kristyn*.

22. R.J. Allan, I.J. Bush, D. Henty and T. Bush, *Parallel Application Software on High Performance Computers: Serial and Parallel FFT Routines*, Daresbury Laboratory (1996).

23. IBM, *Parallel Engineering and Scientific Subroutine Library*, Reference Guide GC23-3836-00.

24. Cray, *Scientific Subroutine Library Documentation*.

25. L.S. Blackford, J. Choi, A. Cleary, E. D'Azevedo, J. Demmel, I. Dhillon, J. Dongarra, S. Hammarling, G. Henry, A. Petitet, K. Stanley, D. Walker and R.C. Whaley, *ScaLAPACK Users' Guide*, SIAM ISBN 0-89871-397-8 (1997). Scalapack Users' Guide is available in html form from WWW URL *http://www.netlib.org/scalapack/slug/scalapack_slug.html*.

26. R.J. Allan and M.F. Guest, *Parallel Application Software on High Performance Computers: The IBM SP2 and Cray T3D*, Daresbury Laboratory (1996).

27. A.G. Sunderland, C.J. Noble, R.J. Allan, I.J. Bush and K. Maguire, *Parallel Application Software on High Performance Computers: A guide to Fortran 90 programming and the Cray T3D*, Daresbury Laboratory (1996).

28. R.J. Allan and P. Lockey, *Parallel Application Software on High-performance Computers: Survey of parallel software ,packages of potential interest in scientific applications. 2nd edition*, Daresbury Laboratory (1996).

29. J.C. Corallo, *Evaluation of HPF Compilation Systems*, M.Sc. Thesis University of Liverpool (1996).

30. R. Ward, J. Heggarty, K. Kleese and R.J. Allan, *Investigation of Parallelisation, Debugging and Performance Optimisation Tools*, Daresbury Laboratory (1997) in preparation.

31. HPCProfile is published six times a year by CLRC to disseminate information of interest to HPC and distributed computer users including the CCPs and HPCI consortia. It is available online at *http://www.dci.clrc.ac.uk/Publications* or by free subscription to *HPCProfile@dl.ac.uk*.

32. J. Nieplocha, R.J. Harrison, and R.J. Littlefield, Global Arrays; A Portable Shared Memory Programming Model for Distributed Memory Computers, in: *Supercomputing '94*, IEEE Computer Society Press, Washington, D.C., (1994).

33. R.J. Littlefield and K.J. Maschhoff, Investigating the performance of parallel eigensolvers for large processor counts, *Theor. Chim. Acta* 84:457-73 (1993).

34. I.S. Duff et al., *Catalogue of Subroutines*, Harwell (June 1993),
    I.S. Duff et al., *Release 11 specification*, Harwell Vol 1. (June 1993) Vol 2. (July 1993).

35.   Y.-F. Hu and R.J. Blake, Numerical experiences with partitioning of unstructured meshes, *Parallel Computing*, 20:815-29 (1994).

36.   Y.-F. Hu and R.J. Blake, Algorithms for scheduling with applications to parallel computing, *Advances in Engineering Software*, 28:563-72 (1997):

37.   Y.-F. Hu and R.J. Blake, An optimal dynamic load balancing algorithm, *Concurrency: Practice and Experience* (1998) in press.

38.   CLRC, IST and Paras Ltd. *High Performance Computing for Process Engineering* APEX ESPRIT Project 20259 CD-ROM 1, Paras Ltd. (1997).

39.   P.J.D. Lindan and N.M. Harrison, And the rest is chemistry! *HPCProfile* 1 (March 1995).

40.   More information can be found via the WWW at URL *http://www.dl.ac.uk/TCSC/projects/UKCP/ukcp.html* .

SOUTHAMPTON HIGH PERFORMANCE COMPUTING CENTRE

Denis Nicole, Kenji Takeda, Ivan Wolton and Simon Cox

Southampton HPCI Centre
Department of Electronics and Computer Science
University of Southampton
Southampton SO17 1BJ

INTRODUCTION

The Southampton HPCI Centre supports several EPSRC, NERC and PPARC consortia to fully utilise national HPC resources. The Centre provides high-level parallel applications support with most of the work being involved in optimising consortia codes for available machines. The Centre has been responsible for managing the benchmarking of prospective machines shortlisted for the HPC97 procurement. This involved selection, porting and validation of a representative suite of benchmark codes and subsequent audit of results.

The HPCI Centre is based within the High Performance Computing Centre in the Department of Electronics and Computer at Southampton University and the HPCI Centre's activities have provided a major focus for the development of Computational Modelling within the University. Along with the Engineering Faculty's Computational Engineering Design Centre, we are developing a wide range of computational techniques at Southampton. The group's research covers three main areas: commercial distributed computing, tools and techniques, and collaborative computational science and engineering. Our strength lies in this mix of sophisticated computer science alongside both commercial and real scientific applications. Our objectives are to design, develop and deploy:

- Cost-effective high performance computing systems.
- Advanced programming environments and architectures for high performance and parallel computing.
- Computational algorithms and application programs within major research programmes in Engineering and the Physical, Environmental and Social Sciences.

HPCI CONSORTIA SUPPORT

The Southampton HPCI Centre supports eight consortia, as shown in Table 1. The work performed ranges from optimisation and porting of programs, to full parallelisation of serial codes and algorithmic development. A brief overview of the work we have performed for these consortia is given in this section.

High Performance Computing
Edited by R. J. Allan *et al.*, Kluwer Academic / Plenum Publishers, New York, 1999

Table 1. Supported Consortia.

Support	Consortia
Direct	Global Ocean Circulation (OCCAM)
	Universities Global Atmospheric Modeling Programme (UGAMP),
	Materials Chemistry
	Mineral Physics
Indirect	UK Particle Cosmology
	Simulation and Statistical Mechanics of Complex Fluids
	Direct Numerical Simulation of Fluid Flow
	Geoscience

The Centre organised and hosted a one-day Workshop on Long-Range Forces attracting 25 delegates from across the sciences. Feedback from attendees was very positive, highlighting the desire of scientists to meet others facing similar computational problems in different disciplines.

Global Ocean Circulation (OCCAM)

The OCCAM consortium has developed two high-resolution models of the World Ocean. These are based on the Bryan-Cox-Semtner ocean with improved advection and free surface capabilities. The consortium makes extensive use of the facilities at Edinburgh. Their consortium RA (Ivan Wolton) was based in the HPCI Centre and resulted in day to day interaction with the team. In mid-1996 Ivan joined the HPCI Centre team permanently.

Universities Global Atmospheric Modeling Programme (UGAMP)

The UGAMP consortium has a close working relationship with the European Centre for Medium Range Weather Forecasting (ECMWF) and they have deployed the IFS simulation code. This is already well parallelised and highly portable; some of the original development was performed by part-time students at Southampton. We have been working with the team at Reading in utilising the IFS/UMAP codes in the HPC'97 procurement benchmarking suite. This application is both compute-intensive and particularly stresses the i/o capabilities of the computer system. A brief description of the particular features of the benchmark code are given in Table 2.

Geosciences

In collaboration with the British Geological Survey and the University of Durham, we have developed programs to model the interaction between rock fractures and fluids. We expect to continue working in this field on advanced seismic techniques for imaging underground fractures. Together, these two technologies should substantially enhance the ability to exploit oil reservoirs. We have developed a post-processing system based on animated GIFs that allows users to visualise the fluid flow without having to scan through individual data files. This is of course platform independent as all that is required for viewing is a WWW-browser.

E2EP: MPP Supercomputing in Materials Chemistry

We are working with a team led by Professor Catlow at the Royal Institution to develop programs for the Cray T3D to provide detailed simulations of crystal growth at dislocations. The materials chemistry consortium had its own directly supported PDRA until mid-1996. Since then the Southampton HPCI Centre has been involved in developing the Metadise code (in conjunction with the Mineral Physics consortium, F2NE). Metadise consisted of a mixture of FORTRAN IV, 66 and 77 code. This was restructured and consolidated into a consistent FORTRAN77 version to improve maintainability considerably. A parallel version of Metadise was also developed using PVM and later MPI running on the SP2 at Southampton.

Current work in progress includes the full parallelisation of GULP1 and GULP2 codes for simulation of 3D periodic solids and bulk defects. GULP1 has a large user base and is designed to handle both molecular and ionic solids through the use of a shell model. GULP2 is a development code which uses a more advanced quantum-mechanical approach.

We are also collaborating with Daresbury Laboratory to develop a fully distributed data version of the DLPOLY molecular dynamics program.

E1EP: Simulation and Statistical Mechanics in Complex Fluids

We are also developing molecular dynamics simulations to provide detailed models of liquid crystal grain boundaries with Dr Mike Allen at the University of Bristol [1]. We have used the CVS code repository system [2] to manage the programs for this consortium. A central repository based in Southampton, with full remote access, is used to store all source code files. Separate modules are used for the GBMESO and GBMEGA codes, and a sample test module is also supplied for users to become familiar with the use of the CVS system. The Complex Fluids consortium is also the main EPSRC user on the Southampton IBM SP2 following porting of their codes onto this platform using MPI.

Initial efforts were directed towards the GBMESO family of codes which consisted of four distinct programs. Consolidation of the GBMESO codes (GBMESONPT, GBMESONPH, GBMESON and GBMESONPVE) into a single program (GBmesoG) was performed. Optimisation of computation and communication routines for GBMESO family of codes was also carried out. This was greatly simplified following the merging of the codes. The resulting increase in performance was significant; together with other more general optimisations, a speedup of between 1.3 (4 PEs) and 2.1 (64 PEs) was achieved.

The GBMEGA code has been updated to use PVM and MPI via a common communication harness. A Cray-SHMEM version has also been implemented. By using general purpose message-passing libraries the program was ported to the IBM SP2 and is now in production on this machine.

D8EP: Direct Numerical Simulation of Fluid Flow

Work performed for the DNS consortium was concerned with optimising data transposition routines and FFTs on the Cray T3D. The code for this was ported to SHMEM along with restructuring the communication and computation pattern.

The original tridiagonal solver in the DNS program contained a large number of division operations; three divisions for every ten floating point operations. In order to improve performance a Cray, vector divide routine was implemented. However, this did not take advantage of the Alpha hardware divide unit so the tridiagonal solver was rewritten to utilise this. The two improvements together yielded a large speedup. The optimised tridiagonal solver increased performance from 4 Mflop/s per processor to 12 Mflop/s on the Cray T3D on a problem size of 144x384x128 (on 32-128 T3D processors).

Current work for the DNS consortium includes re-structuring parallel FFTs to replace the currently used global transpose method and help in parallelising LES code used for simulating turbomachinery flows.

F2PP: UK Particle Cosmology Consortium

Cosmic strings offer an exciting, if speculative link between the distribution of clusters of galaxies and the spatial fluctuations in the cosmic microwave background radiation. The computer simulation of these fast-moving interacting strings on parallel computers is, however, a challenging problem which we are solving with teams at Cambridge and Sussex Universities.

Work for this consortium has mainly consisted of optimising and developing the STRING and ASTRING codes. This has involved development of a farmed version of STRING for collection of statistical data and a distributed memory version for MPP systems. This comprised of full code parallelisation, optimisation and tuning in terms of memory usage and i/o performance. Additions were also made to aid in the data analysis of output. We also produced a benchmark version of STRING which was used to assist in the procurement which resulted in the COSMOS machine at Cambridge. Our current work includes optimisation of the ASTRING code

HPC'97 PROCUREMENT SUPPORT

The Centre has put substantial resources into managing the development, distribution and auditing of benchmark codes for the HPC97 procurement. This was a major support effort.

A significant part of the effort was devoted to investigating which codes were available and desirable as benchmarks, acquiring appropriate codes with suitable test cases and evaluating their characteristics.

The majority of the codes have been developed by existing HPCI consortia for the current T3D system at EPCC, but third party software of major interest academic research users was also considered. Early aspects of this process involved close cooperation with both individual consortia and the Daresbury and Edinburgh Centres.

Candidate benchmarks were evaluated for suitability using a range of criteria. Apart from their importance to existing and future academic research communities, these criteria included:

- Their algorithmic characteristics and how these stressed key areas of machine architecture,
- The amount and characteristics of i/o
- Assessing potential problems with portability within the code, the build process and the actual run.
- The existence of suitable testcases to adequately stress the target systems.
- Ability to verify correct results from testcase runs on different architectures.

The input from this assessment process was used to whittle the candidate benchmarks down from the large range of current production codes to a manageable handful which could be prepared and analysed in more detail. Fifteen potential benchmark codes were evaluated in detail based on suitability using the above criteria. These were built on the Cray T3D and IBM SP2 and run where possible.

Work then concentrated on increasing the robustness and portability of the codes, identifying and correcting, where possible, architecture-specific features. Effort was also put into obtaining larger, more demanding, test data sets. The knowledge gained in this process has enabled us to feedback information to the code originators on potential problems and improvements to their codes. The experience gained in examining a large cross-section of codes for benchmark purposes will benefit future consortia support activities both in general techniques and in specific areas.

Table 2. HPC'97 procurement benchmark codes.

Field	Code	Features
Direct Numerical Simulation (DNS) of turbulent flow	ANGUS	Multi-grid and conjugate-gradient solvers
		High-level BLAS calls
Condensed matter simulation using Car-Parinello approach	CETEP	3D FFTs
		Global communication of complex datatype arrays
		High-level BLAS calls
Molecular Dynamics	DLPOLY	Global EWALD summation
		Replicated data
Finite Element solution of electromagnetic scattering problem	FLITE3D	Indirect addressing
		PVM message-passing
Atmospheric climate modelling	IFS/UMAP	FFTs and data transposes between physical, Legendre and Fourier space
		Semi-Lagrangian advective solver
		Large file i/o requirement
		Tunable vector performance
		Parallel compilation
		Integration of MPP and serial/vector processing

The five production codes selected to form the benchmark suite are listed in Table 2. The main features of these codes, including how these stress key parts of the machine architecture, are also given.

A common build strategy was implemented and the benchmark suite (including all input datafiles) was distributed to the shortlisted vendors on CDROM. We were able to make effective use of our Alpha Windows NT cluster in the partitioning of large FLITE3D grids.

Support for vendors was provided and modifications vetted with official patches being administered by the Centre where necessary. In support of the audit process source code and results files were thoroughly checked and reported on. This included an independent assessment of optimisation changes made by the vendors to ensure that they complied with the benchmark guidelines.

RELATED ACTIVITIES AT SOUTHAMPTON

The High Performance Computing Centre (HPCC) in the Department of Electronics and Computer Science pursues research in distributed computing, tools and techniques and scientific applications. Applying such a three-pronged approach has led to increased collaboration between departments within, and outside the University. Interaction with the computational modelling community has helped to proliferate know-how and leverage postgraduate programmes in simulation and modelling across disciplines. A key to this success is our very popular, weekly, Computational Modeling Community seminar series which focuses on applications and techniques.

The 23-node Southampton IBM SP2 system hosts a variety of research by users at the University, with our main EPSRC users being the Complex Fluids consortium. We also have a share of the Faculty of Engineering's twelve node SGI/Cray Origin. We are developing our own commodity based supercomputing platform based around an eight node Alpha cluster within the group [3, 4]. We are collaborating with Daresbury Laboratory who have a similar Pentium II-based system [5].

The recently established Computational Engineering Design Centre under Professor Andy Keane has consolidated research in the Engineering Faculty in modelling of engineering systems. The primary areas of research are in evolutionary optimisation, fluids, structures and computational biomechanics. They have their own SGI/Cray Power Challenge machine and a major share in the Faculty SGI/Cray Origin 2000 system. Simulations are also run on the SP2 and clusters of workstations where appropriate.

Commodity Supercomputing.

The entry price of supercomputing has traditionally been very high. As processing elements, operating systems, and switch technology become cheap commodity parts, building a powerful supercomputer at a fraction of the price of a proprietary system becomes realistic.

Most recently, in support of both our internal and national collaborations, we have purchased a dedicated computational cluster of DEC Alpha workstations. Our judgement is that they are fully competitive node for node with systems from major vendors such as IBM or SGI/Cray for a wide range of engineering and science applications, and at a cost lower by about a factor of three. Indeed, relative to these vendors' high-end offerings, the only area of under-performance is in inter-node communications bandwidth and latency [3].

The price-performance of a system also depends upon the operating system and compiler costs. Whilst there are free operating systems for Pentiums and Alphas, such as Linux, the best compilers currently run only on Windows or Digital UNIX. Microsoft have recently transferred their FORTRAN power station business to Digital and Digital Visual FORTRAN for the Pentium and Alpha is one of the best, fastest and most useable Windows NT compilers, outperforming the free EGCS g77 compiler by about 60% [4]. We are pursuing the long term goal of delivering an effective remote and local parallel computing service directly under Windows NT, with all the commensurate benefits of commercially supported software and integration into a mainstream commercial IT environment.

HPC Applications

Our research work in applications of HPC is almost all collaborative; we bring our expertise in efficient computational techniques and data handling to bear on a wide range of problems in several disciplines.

Maximum Entropy analysis of the UK National Lottery. For some lotteries the organisers reveal statistical information about players' choices of numbers and combinations of numbers. Such information is of considerable interest in modelling the behaviour of lottery players, and can enable players to increase their expected winnings by choosing unpopular combinations of numbers. For the UK National Lottery, however, only the number of prize winners in each prize category and the total number of players are released for each draw. Working with our Physics Department, we have used the Maximum Entropy method to estimate the probability of each of the 14 million tickets being chosen by players in the UK National Lottery. By choosing unpopular combinations of numbers, one's expected winnings can be doubled for the least popular tickets [6]. For a syndicate buying a large number of tickets each week, it is possible to increase one's winnings by fifty percent compared to that obtained from random choices of ticket.

Cooperation in Groups. Working with our Archaeology Department, we are developing a new class of models for the study of cooperation in groups, based on the repeated Prisoner's Dilemma Game. This study focuses on the manner in which group size, and the sophistication of individual strategies, affects the ability of a group to sustain global cooperative behaviour. Simulations so far have yielded results exhibiting a number of features reminiscent of group behaviour. In particular, we find that as group size is increased, cooperation can be sustained and individuals with longer memories are favoured. This model seems to give theoretical support to field observations of a positive correlation between group size and neocortex size in primates [7].

Liquid Crystals. Collaborating with our Mathematics Department, and the National Academy of Sciences in the Ukraine, we are performing calculations of the light scattering properties of Polymer Dispersed Liquid Crystal films in the long wavelength régime, where the Rayleigh-Gans approximation applies [8, 9, 10, 11, 12].

A PDLC film consists of a random ensemble of micrometre-sized liquid crystal droplets dispersed in an isotropic polymer matrix. In the absence of any external electric field, the film has a milky white translucent appearance. However, when a field is applied, the film becomes clear. In contrast to polarization devices, there is little loss of light in the transparent state, making them highly desirable for a range of applications. The contrast between the off and on states relies on the optical mismatch between the liquid crystal droplets and the polymer matrix, which disappears when the field is applied.

PDLC films combine some useful properties of ceramic and liquid crystalline materials. There are a number of current and potential applications for these devices, ranging from large flexible windows and signs, to high resolution active matrix addressing devices, and direct view displays.

Simulation of 3D High-T_c Superconductors. High temperature superconductors were only discovered in 1987, and at present no satisfactory theory exists to explain their behaviour. By performing computational simulations it may be possible to develop insight into the theoretical basis for high temperature superconducting behaviour. Both high temperature superconductors and conventional superconductors contain vortices; a term used to describe the circulating supercurrents in the material. Pinning of these vortices is believed to allow a superconductor to carry high currents. Our simulations are more concerned with increasing the current carrying capability of superconductors, than their critical temperature. We are developing an $O(N)$ fast multipole hybrid Monte Carlo algorithm to simulate anisotropic three dimensional layered superconductors.

Electrical Impedance Tomography. Medical electrical impedance tomography (EIT) is a non-invasive technique for imaging conductivity changes within a patient from measurements taken from skin mounted electrodes on the patient. One potential use is in the real time imaging of pulmonary ventilation, due to the impedance of the thoracic cavity changing with the volume of air breathed. Due to the relative low-cost of an EIT rig, and its non-invasive nature, this technique has obvious application in long-term monitoring.

In collaboration with our Physics department we are parallelising existing sequential EIT algorithms so that real-time three-dimensional imaging can be performed. This involves a parallel finite element code, which exploits the sparse nature of the system, coupled to a non-linear equation solver.

Ice Sheet Modelling. Very large areas of the present-day ice sheets are drained through fast-flowing ice streams. An understanding of their long-term evolution is therefore crucial in determining the response of ice sheets to climatic change and their contribution to global sea level. In collaboration with the department of Geography, we are developing models of these processes; the resulting non-linear partial differential equations are solved numerically.

Remote Sensing. The landscapes of the high and mid latitudes are littered with glacial features, such as drumlins (small oval hills) and moraines (elongated lateral mounds), left behind by the actions of ancient ice sheets. Interpreting these features provides the key to understanding the dynamics of former ice sheets and therefore changes in climate over the last 10,000 to 2 million years. For this we can gain valuable insight into how the climate might change in the future.

Remote sensing from space offers extensive coverage of large areas with great accuracy. Using a new technique, Interferometric Synthetic Aperture Radar (InSAR), remarkably detailed digital elevation models can be generated which are allowing the study of landscape features for the first time. However, processing InSAR data is computationally intensive due to the large amount of data which must be analysed to provide high resolution coverage of glacial features.

Tools and Techniques

We maintain a strong interest in the development of advanced software systems to support parallel programming.

Performance Estimation. Being able to estimate the execution time of programs has been identified as a key technology that is required to support the next generation of automatic parallelising compilers. Performance estimating tools quantify the performance differences between functionally equivalent code sections in order to determine the optimum data layout strategy. Research in this area has developed a new fast simulation technique that is applicable to a substantial subset of Fortran programs and has made great improvements in accuracy over previous methods by considering the effects of the memory hierarchy. We have been able to apply a related cache-sensitive technique to the IFS code.

Advanced Parallel Usage Checking. This EPSRC-funded project is developing ways of using symbolic model checking, already widely used to validate hardware designs, to validate computer software. We are devising methods of generating large but useful finite state models from the text of parallel computer programs which are then analysed by tools such as SMV and FDR. This EPSRC-funded work is in collaboration with IBM, Formal Systems (Europe) Ltd and SGS-Thomson Microelectronics.

The SPOC occam compiler. Occam is a parallel programming language especially well suited to the development of small embedded systems. Our compiler, implemented using advanced vector dependency analysis technology, is now used in, for example, a ship-borne radar tank level gauge developed by Autonica of Norway.

Lightweight message-passing using Java. The Java programming language is now very fashionable for world-wide-web applications. We have been using Java and its security features to allow users to insert their own Java code into the kernel of a multitasking UNIX operating system. Our kernel Java technology permits ordinary users to gain substantial efficiency advantages by avoiding the overheads of crossing protection domains between input-output devices and their user code. This EPSRC-funded project was in collaboration with Parsys Ltd.

REFERENCES

1. A.V. Lyulin, M.P. Allen, M.R. Wilson and N.K. Allsopp, *Computer Simulation of Liquid Crystalline Polymers*, Presented at British Liquid Crystal Society Meeting (24-26 March 1997).
2. K. Takeda, I.C. Wolton and D.A. Nicole, Software Portability and Maintenance, *Proc. of HPCI '98 Conference, Manchester* Plenum Publishing Company (1998) this volume.

3. D.A. Nicole, K. Takeda and I. Wolton, Running HPC Codes on Alpha-Windows NT Systems, *Proc. of HPCI '98 Conference, Manchester* Plenum Publishing Company (1998) this volume.

4. S.J. Cox, D.A. Nicole and K. Takeda, Commodity High Performance Computing at Commodity Prices, WOTUG-21, *Proc. 21st World Occam and Transputer User Group Technical Meeting* (1998).

5. D. Emerson, D.A. Nicole and K. Takeda, *An Evaluation of Cost Effective Parallel Computers for CFD*, to be presented at Parallel CFD '98, Taiwan (1998).

6. S.J. Cox, G.J. Daniell and D.A. Nicole, *Maximum Entropy, Parallel Computation and Lotteries*, to be presented at 1998 International Conference on Parallel and Distributed Processing Techniques and Applications (1998).

7. S.J. Cox, T.J. Sluckin and T.J. Steele, *Computational Archaeology: Modeling Primate Interactions*, to be presented at Europar '98, Southampton (1998).

8. V.Yu, Reshetnyak, T.J. Sluckin and S.J. Cox, Effective medium theory of polymer dispersed liquid crystal droplet systems I: spherical droplets, *J. Phys. D: Appl. Phys.* 29(9):2459-65 (1996).

9. M.A. Osipov, T.J. Sluckin and S.J. Cox, Influence of permanent molecular dipoles on surface anchoring of nematic liquid crystals, *Phys. Rev. E* 55, No.1 Pt A:464-76 (1997).

10. V.Yu. Reshetnyak, T.J. Sluckin and S.J. Cox, Effective medium theory of polymer dispersed liquid crystal droplet systems II: partially oriented bipolar droplets. *J. Phys. D: Appl. Phys.* 30:3253-66 (1997).

11. S.J. Cox, V.Yu Reshetnyak, and T.J. Sluckin, Theory of Dielectric and Optical Properties of PDLC Films, *Molecular Crystals and Liquid Crystals.* (1998) in press.

12. S.J. Cox., V.Yu. Reshetnyak and T.J. Sluckin, Effective Medium Theory of Light Scattering in PDLC Films, *J. Phys. D: Appl. Phys.* (1998) in press.

OPTIMISATION, ALGORITHMS AND SOFTWARE

PERFORMANCE OPTIMISATION ON THE CRAY T3E

Stephen Booth

EPCC
The University of Edinburgh
Edinburgh
EH9 3JZ

INTRODUCTION

This paper compares the code optimisation techniques used on the Cray T3D and T3E-900 systems. The Edinburgh Parallel Computing Centre (EPCC) runs a national super-computing service on behalf of the academic and research community in the UK including HPCI consortia. This service runs on a 512-PE Cray T3D and has been recently upgraded by the addition of a T3E-900 system with 256 application processors. Significant expertise has been developed using the T3D system since its installation in 1994. Since the installation of the T3E several HPCI consortia have migrated codes from the T3D to the T3E.

The T3E is a follow on to the T3D and the two systems are very similar. Nevertheless there are a number of differences that become very important when optimising application code.

COMPARING THE T3D AND T3E-900

The programming environments on the two systems are very similar. Both systems support the Single Program Multiple Data (SPMD) programming style. Parallel programs can be developed using the PVM or MPI message passing libraries or using the Cray SHMEM remote memory access routines. Application programs can be developed in either Fortran-90 or C++ though the majority of codes tend to use Fortran. The T3D only supports parallel applications that use a power of two processors. The T3E may run jobs of arbitrary size though the local policy is to request users to continue to use power of two job sizes.

T3D Hardware Specification

T3D processing elements are built from DEC AXP 21064 "alpha" processors. These processors are clocked at 150 MHz. The 21064 can issue a maximum of a single floating point instruction per clock cycle and therefore the peak speed of a T3D PE is

150 Mflops. Typically memory accesses are a much more significant performance bottleneck than floating point performance so most applications run significantly slower. Each PE has 64 MB of main memory. The only memory caches on a T3D PE are the two on-chip processor caches. The 8 Kilobyte data cache is a direct mapped level-1 cache organised as 256 cache lines each 32 bytes long. This is a read-allocate write-through cache. Write operations modify both the cache contents and the contents of main memory. If the data being written is not contained in the cache only the main memory is modified. Cache misses only occur on read operations. A separate 8 Kilobyte cache exists to cache instructions.

There is no second level cache between the processor and the main memory. Instead there is the read-ahead buffer. This is a small (32 byte) buffer of fast memory. Whenever a cache miss occurs the read-ahead buffer performs a speculative fetch of the next contiguous cache line in memory. The read-ahead buffer reduces the memory latency for loops that read memory in a single contiguous stream.

T3E-900 Hardware Specification

T3E Processing elements are built from DEC AXP 21164 "alpha" processors. On the T3E-900 these processors are clocked at 450 MHz. The 21064 can issue both a floating point multiply and addition in a single clock cycle and therefore has a peak speed of 900 Mflops.

A variety of memory sizes are supported from 64 Mbytes to 2 Gbytes.

The Level-1 on chip caches are the same as on the T3D however there is an additional Level-2 cache (also on chip). The Level-2 cache is a 96 Kilobyte unified (data and instruction) cache. It is a 3 way set associative cache organised as 512 sets of 3 cache lines. Each cache line is 64 bytes long and corresponds to 2 Level-1 cache lines. The Level-2 cache is read/write allocate. This means that a write operation can cause a L2 cache miss that in turn causes data to be read from main memory to fill the remainder of the L2 cache line.

Instead of a single read-ahead buffer T3E PEs have 6 stream buffers. These work in an analogous fashion to the read-ahead buffer except that up to 6 memory streams may be interleaved. It is important to note that as write operations can cause L2 cache misses data write streams may consume one of the 6 stream buffers.

COMMUNICATION HARDWARE

Both the T3D and the T3E PEs communicate by performing remote memory accesses to the memory spaces of other processors. However the implementations are very different.

T3D Communication Hardware

On the T3D PEs read and write directly to the remote memory using the normal load and store instructions. There is no cache coherency hardware so cache coherency must be maintained by the application programmer.

Reads from remote memory are significantly slower than writes because each read instruction has to wait the complete round trip time to the remote PE before it can complete. On the other hand remote stores can complete as soon as the data has been accepted by the communication hardware.

The address space of the 21064 is too small to simultaneously address all of the memory of a large T3D so the mapping between memory locations and PEs must be changed frequently using expensive DTB-annex operations.

T3E-900 Communication Hardware

On T3E systems communications are implemented using memory mapped hardware called E-registers. PEs never perform load or store instructions to remote memory locations. Instead they read and write data and instructions to local E-registers.

The contents of an E register can be transferred to and from any memory location on any processor. Cache coherency is maintained automatically. Microprocessor load and store local E-register contents directly, bypassing both the caches and the stream buffers.

E-register operations are non blocking and can proceed independently of the microprocessor until the point where the processor tries to access the results of the operation. This removes the performance hit for remote read (get) operations. Get operations can be pipelined by using multiple E registers.

By using a series of paired E register get and put operations global data copies can be performed without the data ever passing through the processor caches. Therefore a processor can perform a communication phase without evicting useful data from its own cache and the communications can proceed unhindered by cache-misses.

PERFORMANCE

Memory System

Both the T3D and the T3E contain special circuitry to optimise contiguous memory accesses. Following a cache miss the next consecutive cache line is read into a fast buffer. If this location also misses before the buffer contents are over-written the data can be read faster than from the underlying memory system. The T3D has a single "read-ahead" buffer. Only loops with a single memory read stream will benefit from this. Loops of the form:

```
DO i=1,n
   a(i) = b(i) + c(i)
END DO
```

will not benefit from the read ahead buffer as cache misses will alternate between the two input arrays. The T3E has six separate streams buffers which considerably increase the number of possible loops that can benefit. However output streams will also require one of the stream buffers if it misses in level-2 cache.

Balance

The compute/memory/communication balance of the T3E is different to that of the T3D. The peak computational speed of a processing element has increased by a factor of 6. However peak floating point performance is rarely the limiting factor on sustained performance and other more significant machine parameters have not improved as much. For example a simple unoptimised data copy written in Fortran only goes twice as fast on the T3E than on the T3D, suggesting that the sustainable memory bandwidth has increased by a smaller factor than the peak floating point performance.

Communication performance has improved:

- `shmem_get` by a factor of 9.

- `shmem_put` by a factor of 3.

- `MPI_SEND` by a factor of 5.

- `MPI_SSEND` by a factor of 2.

These figures reflect the fact that shmem put and get operations are of comparable performance on the T3E but shmem_put was much faster than shmem_get on the T3D.

On the T3D the subroutine call overhead was very significant. This is about 5-6 times faster on the T3E. Therefore subroutine in-lining is a less important optimisation on the T3E.

Memory bandwidth is still a significant performance bottleneck. This can be demonstrated by comparing the performance of the following two loops.

A saxpy call:

```
DO j=1,n
  y(j) = y(j) + a * x(j)
END DO
```

And a sax3py3 call:

```
DO j=1,n
  t1 = y(j)
  t2 = x(j)
  y(j) = (t1*t1*t1) + a * (t2*t2*t2)
END DO
```

The memory access requirements for the two loops are the same but the sax3py3 loop has more floating point multiplies.

The T3E performs significantly better than the T3D on both loops. On the T3D (figure 1) there is a significant drop in performance once the data no longer fits in the L1 cache. On the T3E (figure 2) performance does not drop off until the L2 cache size is exceeded. The sax3py3 call always has the higher performance because of the greater flops to memory access ratio. Both routines have 2 input streams. On the T3E the stream buffers can be utilised easily. On the T3D the read-ahead buffer does not help the performance of these loops though the `libsci` library does contain a hand coded `saxpy` that uses strip-mining to take advantage of read-ahead. The lower subroutine call overhead on the T3E reduces the performance drop-off for small vector lengths.

We can also consider the impact of loop-unrolling on these loops.

Loop un-rolling is an important optimisation technique for both the T3D (figure 3) and the T3E (figure 4). On the T3E there seems to be little advantage to un-rolling loops by hand. It is often better to keep the Fortran loops simple and let the compiler perform the un-rolling. It is particularly important not to introduce more than 6 memory streams as a side effect of the un-rolling

48

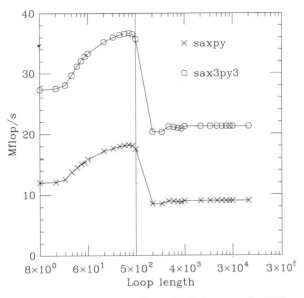

Figure 1 Performance of two simple loops on the T3D

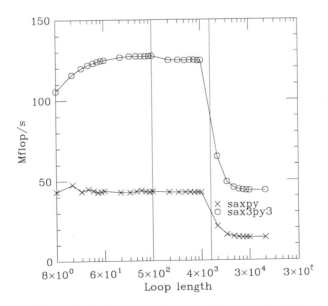

Figure 2. Performance of two simple loops on the T3E

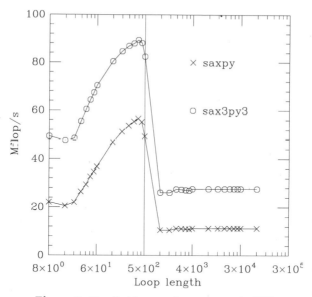

Figure 3. Unrolled loop performance on the T3D

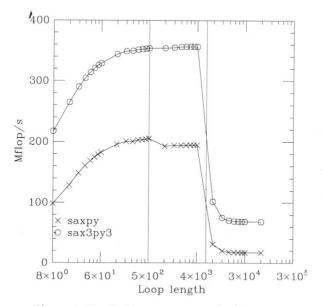

Figure 4. Unrolled loop performance on the T3E

E Registers

The E register hardware on the T3E can also be used to optimise the performance of single processor code. The compiler supports a CACHE_BYPASS directive that can be used to tell the compiler to access a designated array via the E registers. As all E register operations bypass the cache this directive allows local data to be accessed without going through the cache. Using this directive give access times somewhere between the cache miss and cache hit access times and is therefore useful where most or all memory accesses will result in cache misses. For example:

1. Initialising arrays to a constant value. This gives a performance improvement because it avoids the spurious reads for Level-2 cache write misses. Loops of this type can run up to one and a half times faster if CACHE_BYPASS is used.

2. Large stride or random access data patterns where only one element from each cache line is used.

This directive should be used sparingly as most loops will go slower if CACHE_BYPASS is used. Use of E registers to optimise memory access is also particularly useful for hand coded assembly language routines.

The Compilers

The T3E compiler is much more sophisticated than the T3D compiler and it supports several additional features. The T3E f90 compiler has better error checking options allowing the number and type of subroutine arguments to be checked at runtime. The compiler can perform automatic loop splitting replacing a single complex loop with several simpler ones to keep the number of active memory streams below 6. Automatic loop merging is also under development. Automatic subroutine in-lining is supported though the improved subroutine call time on the T3E makes this less of a problem than on the T3D.

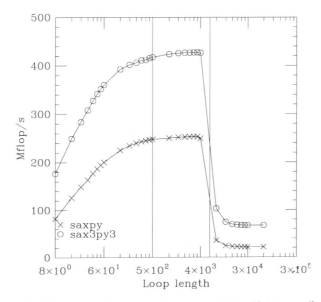

Figure 5. T3E loop performance with **-Ounroll2,pipeline2** compiler flags

The T3E compilers also provide several options to make better use of the floating point pipelines. The floating point pipelines provide a small amount of on-chip parallelism, a number of different floating point operations may be executing simultaneously in different pipeline stages. To utilise the pipelines effectively the compiler has to identify independent expressions that can be executed in parallel. The `pipeline` optimisation flag instructs the compiler to interleave instructions from the end of one loop iteration with those at the start of the next. As can be seen from figure 5 this can significantly increase performance provided though memory access times may be more significant where the data does not fit in cache.

The T3E compiler also attempts "vector" optimisations. These involve re-writing loops that contain intrinsic functions to call alternative vector intrinsic functions. The vector intrinsic functions are typically faster than the normal scalar intrinsics because they can interleave the calculations for different elements of the vector and make better use of the floating point pipelines.

CONCLUSION

On average the T3E seems to be about 4 times faster than the T3D though this varies a lot from code to code. If anything the variation in performance between codes is less on the T3E for the following reasons:

- The impact of cache thrashing is reduced by the level 2 cache.

- A large number of computational loops have less than 6 data streams and can therefore utilise the stream buffers on the T3E. The single read ahead buffer on the T3D was far less useful.

- The smaller subroutine call overhead on the T3E allows poorly structured code to perform better on the T3E than the T3D.

Most of the important optimisation techniques like loop unrolling can now be performed successfully by the compiler. This reduces the need to perform low level code tuning to get good performance out of the T3E. The E-register hardware does provide additional opportunities to optimise single processor code via the CACHE_BYPASS directive.

FROM FLOPS TO UDAPS: ALGORITHMS, BENCHMARKING AND TUNING

Nick Maclaren

University of Cambridge

CHANGES WITH TIME

Back in 1973, floating-point additions and loads from cache took perhaps 0.3 microseconds, floating-point multiplications perhaps 1 and loads from memory perhaps 2. A machine might have 4 KB of cache for 2 MB of memory, and a cache hit ratio of 90% allowed the machine to run at near-peak performance.

In 1998, floating-point additions take perhaps 1 nanosecond, multiplications or loads from primary cache perhaps 2, but loads from secondary cache perhaps 10 and ones from memory perhaps 200. A machine might have 16 KB of primary cache and 1 MB of secondary for 2 GB of memory, and a cache hit ratio of 99% is unsatisfactory.

The key change is that memory has changed from being 'fast' to being very 'slow', especially as far as latency is concerned, and is beginning to bear the same relationship to modern primary caches as disks did to memory in 1973. The simple uniform access time model that has traditionally been used for most theoretical performance analysis is no longer viable. This means that almost all systems from personal computers to supercomputers should be regarded as NUMA (non-uniform memory architecture.)

PERFORMANCE CRITERIA

In 1973, the number of floating-point operations per second was regarded as the key performance indicator. It was not entirely reliable even then, but it was realistic and easy to analyse. This was coupled with minimising the use of expensive operations like division, but a key aspect was that data accesses were regarded as 'free' and were more-or-less ignored. Algorithm and compiler developers concentrated on reducing the number of floating-point instructions executed.

In 1998, data access is the issue, because a reference to main memory is 100 times the cost of a floating-point operation, and the latter can be regarded as 'free'. The worst form of data accesses are unpredicted ones to main memory, because they can cause the instruction stream to 'stall'; most algorithmic, coding and compiler optimisation concentrates on reducing the number of them. To a great extent, hardware performance should now be measured in unpredicted data accesses per second (UDAPS).

High Performance Computing
Edited by R. J. Allan *et al.*, Kluwer Academic / Plenum Publishers, New York, 1999

There are several partial solutions to this. One of which is to regard the primary cache as the working storage and to use the secondary for filling it, but this is feasible only for a few problems. Another common one is to design hardware that can 'stream' memory very fast sequentially into the CPU, rather in the ways that vector machines did. Operand preloading can also be useful and is being used more and more – but the problem is that preloading needs to lead the data use by about 100 instructions.

As far as benchmarking is concerned, John McCalpin's *STREAM* benchmark is extremely relevant, as are the various latency testers (Larry McVoy's *lmbench* and others.) These tend to give better prediction of delivered performance on real problems than the CPU-dominated benchmarks, such as *SpecFP* or even *Linpack*.

PROBLEMS WITH CACHING ETC.

So why is simply increasing the cache size not enough? The reason is that big problems tend to have big data, and the working set exceeds any reasonable secondary cache. Even worse, we need a 99%+ hit rate from primary cache (and even more from secondary), which is almost impossible to achieve. Merely doubling the cache size typically doubles the size of problem at which trouble starts, but has relatively little effect on the hit rate once it does.

And, just to cause confusion, the simple caching models do not match reality. There are problems like TLB misses, memory banking, cache line conflicts etc., just as there were in 1973. But those will not be described here, except to comment that they can easily cause a factor of 5–10 variation in performance, which is extremely system- and code-dependent.

One of the best solutions to this problem is to use 'blocking' methods, which typically break a matrix up into blocks that will fit into the secondary cache. These often give 5–10× improvement, but rarely get anywhere near the peak performance. One reason is that getting peak performance needs most accesses to be to the primary cache, and chopping up a matrix finely enough to do that increases the overheads excessively. The best that is feasible is to stay within secondary cache.

Other methods include loop merging, scanning matrices in different orders and so on. Most of these are somewhat tricky, and very system-dependent, but will often give comparable improvements to blocking methods. Some compilers will help by doing these semi-automatically, but not all.

The most unexpected approach is to replace algorithms designed for working on data in main memory with those originally designed for working on disk! The disk to memory access time ratio in 1973 was perhaps 3,000, whereas the memory to primary cache ratio today is about 100 and steadily increasing. While few current programs use disk algorithms, expect to see more do so as time goes on.

As an indication of typical problems, the following example shows some times for nine transposition codes on a 4096 × 4096 array on various machines – it is not an indicator of relative performance, as the machines were of very different powers and ages, and the numbers have been scaled in strange ways. However, it shows the advantages of blocked methods, and the extent to which the best codes (marked with daggers) will vary between systems.

Table 1. Some Transposition Timings

SPARC	PA-RISC	HARP-1E	ALPHA	MIPS	RS6000
		Unblocked methods			
†248	†112	125	51	†66	†58
452	191	210	60	135	76
340	163	†73	†49	114	78
		Blocked methods			
†81	48	29	†26	†21	†44
121	†39	†27	29	29	52
89	50	31	†27	24	†45
126	†37	†28	31	30	53
93	54	†28	†27	†21	96
100	57	30	†27	23	94

The best methods are marked with daggers

VECTOR CODE IS REBORN

Vector machines have become too expensive to build, and NEC is the last vendor developing any. The SMP and MPP models have won, for now at least. But the same does not apply to vector algorithms and vector code, which are being reborn as solutions for extreme RISC systems.

After some 20 years of development, the optimisation of vector code is well understood and few, if any, directives are needed. The hardware often streams the data automatically, and it is very easy for the compiler to insert preloading instructions a suitable distance ahead. Very high performances can be obtained for moderate effort – 80%+ of peak is often possible for suitable applications if enough bandwidth is available.

It can be done for SMP systems, too, by splitting vectors into sections, though it usually needs some directives and works best on very long vectors. The first vendors to do this in production compilers were IBM (with VS Fortran) and Alliant, but it is now standard technology, and is the model used by High-Performance Fortran (HPF).

Writing vector code can be recommended as the best simple programming style for single CPUs and SMP systems, because it is the one that is most likely to be optimisable on almost all systems. It may not be the fastest, but will usually be fairly good. However, it does not work well on clusters and MPP systems, so message passing methods (MPI etc.) are still best for those.

PROGRAMMING LANGUAGES

When it comes to programming languages, Fortran 66 dominated in 1973, and Fortran 90 leads in 1998. Early Fortran 90 compilers had poor optimisation, but current ones are adequate, though they usually optimise Fortran 77 code best. Future compilers may optimise derived types and other Fortran 90 constructions better than they do today. High-Performance Fortran (HPF) is a Fortran 90 extension for SMP systems, but has got only slight acceptance so far, and there is a competitive proposal (OpenMP).

As far as efficiency goes, C is an evil language and C++ is worse. The main reason is that they allow almost arbitrary aliasing, and so all data accesses must go via memory – very little optimisation is compatible with the C standard. There are some moves to improve this in the future C++ standard, and the next C standard (C9X) should improve C to about the level of Fortran 77.

C++ and Java have another problem in that their implementation of object-orientation involves loading the method address from the object, which is usually seen by the compilers and hardware as an unpredictable branch. This can completely destroy most forms of global optimisation, though it no longer is as incompatible with pipelining as it was on earlier systems.

All this makes it extremely difficult to write efficient C, C++ or Java programs, but there is a practical solution. That is to use those languages for the main structure of the application, which usually accounts for almost all the code but very little of the running time. And then to call optimised libraries (whether BLAS, NAG, the vendor's library, etc.) for the computationally intensive kernels. Alternatively, you can write the computational kernels in Fortran. In the future, C9X may make doing this in C a worthwhile option.

5–10 YEARS IN THE FUTURE

Anyone who predicts 5–10 Years in the future is asking to be proved wrong, but I shall do so.

We can expect memory access to be 5–10× faster, but CPUs to be 10–100× faster, thus making the current problems worse. Also, the CPUs will be heavily multithreaded (10–way or more), which will introduce most of the problems of parallel computing to apparently serial code. We can also expect very complex SMP models, probably with several levels of memory coherence, leading to even more highly non-uniform memory architectures.

This will make developing efficient code even harder, and compiler technology is unlikely to keep pace with the extra complexity. As a consequence, algorithm development will become more and more a specialist activity, and the importance of good libraries will increase. Of course, this neither means that appropriate libraries are certain to be available, nor that users will pay for them, but the need will be there.

MPP has been pronounced dead twice before, but each time has returned from the grave with an increased market penetration – my guess is that it will be back (probably for good) in the period 2005–2010. Unlike SMP etc., MPP scales well and the hardware problems are well-understood – the key to its success is usability, i.e. suitable algorithms, languages and compilers.

More interestingly, I believe that both Fortran and C9X will reach their limits, and that it will become increasingly hard to write efficient code using those languages, except possibly for vector operations. There will be an increasing need for a much more disciplined language, where the programmer can pass much more information to the compiler, and hence to the hardware. But I am not optimistic that we shall see such a language.

And, as far as other developments go, your guess is as good as mine!

IS PREDICTIVE TRACING TOO LATE FOR HPC USERS?

Darren J. Kerbyson, Efstathios Papaefstathiou, John S. Harper,
Stewart C. Perry, and Graham R. Nudd

High Performance Systems Group,
University of Warwick,
Coventry, CV4 7AL, UK
Email {djke,stathis}@dcs.warwick.ac.uk

INTRODUCTION

An underlying goal in the use of high performance systems is to apply the complex resources to achieve rapid application execution times. It is often the case that performance issues are considered late in the application development when major design choices and system choices have already been finalised. Performance tuning tools, including parallel monitoring environments, are useful in these late stages providing a means in which to investigate and visualise the performance effects. However, during the development of an application, certain issues are typically decided upon without reference to their impact on performance (e.g. in the choice of a numerical implementation, or in the choice of a possible mapping to the system). There is a clear need for the study of performance at each and every stage of the development of high performance applications.

The aim of a novel prediction toolset PACE (Performance Analysis and Characterisation Environment), presented here, is to extend the traditional use of performance prediction to cover the full software development cycle. It incorporates facilities for both pre- and post-implementation analysis thus allowing alternatives to be explored prior to the commitment of an application (and its mapping) to a system, and also assists in the performance tuning of existing implementations. The approach is carefully structured consisting of modular performance models that reflect individual parts of the whole system (e.g. software components, parallelisation components, and system components).

In this work, it is shown how PACE can be used to produce predictive traces representing the expected execution behaviour of an application given appropriate workload information. Predictive traces are analogous to traces collected at run-time except that the two key issues of timing information and event ordering information are both determined by the prediction toolset. The trace data can be output from PACE in a commonly accepted format (such as PICL and SDDF) for use in existing performance monitoring environments (including Paragraph[5] and Pablo[12]). Thus, predicted application execution can be viewed and examined within monitoring tools, already familiar to users, in order to identify performance hot-spots before system use.

A discussion of the various forms of performance analysis typically undertaken is given in the next section. The formation of prediction traces from the modular performance prediction toolset, PACE, is then described which illustrates the workload information required in a performance study, along with details on the performance model evaluation procedure to produce predictive traces. Example use of the resultant predictive traces is illustrated utilising two existing parallel performance monitoring environments on an example Financial Option code. The use of the predictive traces can significantly increase efficiency in final implementations when used during the development of application code.

PERFORMANCE PREDICTIONS

There are many performance issues that should be addressed during the development of application code which are not just confined to post-implementation 'tuning' analysis. Performance models can be constructed and used for predictive type analysis throughout. The models can range from 'back of the envelope' type calculations (or complexity analysis), in early software analysis stages, to detailed models in design and implementation stages. The level of abstraction incorporated into these models generally increases towards implementation, and formalistic approaches such as petri-nets, or queuing networks, are often employed to represent the structure of the system[1,2].

It is generally acknowledged that there have been few attempts at tools that provide the use and development of performance models throughout the software life-cycle mimicking the software development itself[3,4,13]. Instead, individual tools are often utilised depending upon the stage of application development. These tools do however, have one thing in common - that is to provide the user with an estimation of execution time for a particular software formulation on a given target platform. Issues that are important in such performance studies include:

> *Execution time*: a prediction of the time to execute the application given a set of application (e.g. problem size) and system parameters (e.g. number of nodes).
> *Scalability*: how an application's performance changes by the increasing of either the application and/or system parameters.
> *Sizing*: determining the size of application that can be processed given constraints on either execution time and/or system resources

Such performance information is important in determining the time of application execution on a system but does not provide sufficient insight into the achieved performance to enable refinement (tuning) of the application in a predictive sense. Further information on the predicted application execution is required in order to promote this process. The understanding of the system further depends upon how information is presented to the user. The visual representation of complex phenomenon aids in the performance understanding and behaviour of the target system. A key performance visualisation from a user's perspective is in the provision of space-time information (see Figure 9 for an example). This, a two dimensional chart, can be used to visualise processor activity (space) as a function of time.

The use of a space-time diagram follows directly from the generation of a trace file at the run-time of an application (after application porting and implementation). The trace file collected in this way is simply a list of events, that occurred during the execution of the application, on and between resources in the system.

A predictive trace, on the other hand, is a list of the same type of events but generated from the use of a suitable performance prediction system. They are analogous to execution

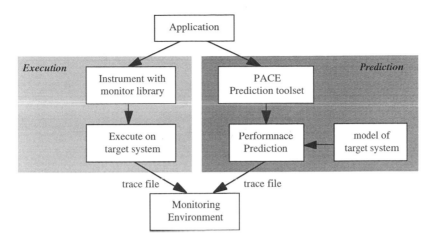

Figure 1. Analogous behaviour between execution and predictive traces.

traces and can be used and manipulated within the same monitoring performance analysis environments, Figure 1. Thus, performance refinement or tuning, can take place in advance of the porting (or even implementation) of an application to a target system.

There are three main features that need to be considered in the generation of predictive traces from a performance prediction system, namely:

Events: The information recorded during the predicted execution (e.g. inter-processor communication, processor idle periods, I/O etc.). If the predictive trace is to accurately portray its run-time counterpart then there should be a one-one mapping between the types of events possible.

Time stamping: Each event has an associated time-stamp to indicate when it occurred. This is simply the processor clock in a run-time system but in a prediction system each time-stamp needs to be individually calculated.

Ordering of events: This follows directly from accurate time-stamping to produce a chronologically ordered list of events.

Most performance prediction tools that have been developed to date have limited analysis capabilities, concentrating on producing overall estimates[6]. In the PACE toolset, described below, it is shown how suitable workload information, concerning both the computational aspects of an application, and their mapping to system resources, can be used in order to provide both realistic performance estimates, and sufficient activity information of the system to provide predictive tracing outputs.

THE PACE TOOLSET

The PACE toolset contains a number of individual components that are used to provide realistic performance predictions of expected application execution which takes into account the operation of the system as a whole. It considers detailed information from the application in terms of its computational cores, its mapping onto a high performance system (including necessary communication costs), and also time costs in terms of the underlying system performance characteristics. This three level structure is a modular approach whereby experimentation on factors such as the choice of the target platform, and also the mapping (parallelisation) that should be used, can be made using a criteria such as that to obtain best application performance.

The PACE toolset is comprehensive in its approach resulting from a period of extensive development, and its use in many different application areas[8,9,10]. In this work, we illustrate the formation of a predictive trace file from PACE when using information from an application code (using a code segment taken from a Financial Option code), combined with predicted time costs on the target platform.

Workload Descriptions

Underlying the PACE toolset is a performance language, CHIP^{3}S (Characterisation Instrumentation for Performance Prediction of Parallel Systems)[11], which provides the necessary syntax and semantics to support workload descriptions for parallel software. This includes: computational control flow information, resource usage information, mapping and communication information amongst others. The core components of PACE are shown in Figure 2.

The compilation of the CHIP^{3}S performance scripts results in output binaries which can be linked to an evaluation engine along with hardware models for the target system. The evaluation engine combines the workload descriptions with the appropriate platform performance characteristics (encapsulated in the hardware models) to produce performance predictions (see below).

The hardware models for a specific system are a combination of: measurements (e.g. micro-benchmark results), models (e.g. statistical, and analytical), and hardware specifications. The hardware components of a system can be modelled at different levels of abstraction depending on the accuracy of predictions required. Models that represent a low-level of abstraction typically provide highly accurate predictions but require detailed workload information and result in long evaluation times.

The workload definitions in CHIP^{3}S are organised into two layers: an application layer, and a parallel template layer. The former includes the workload descriptions for the computational parts of the application while the latter includes descriptions of how the resources of the systems are to be utilised and will interact. CHIP^{3}S is dedicated to provide the necessary syntax to support these descriptions. In order to illustrate these descriptions, a code segment taken from the Alternating Direction Implicit (ADI) solution to a Partial Differential Equation, as shown in Figure 3 will be used. This is written in an SPMD (Single Program multiple data) style parallelism using the PVM message passing interface. Note that the code in Figure 3 is performed on all nodes in the system, and contains communication dependencies between nodes.

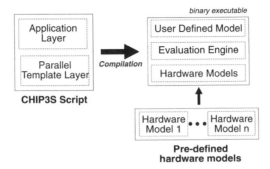

Figure 2. The core components of the PACE toolset.

```
pvm_recv (tids[MY_ID-1], 30);
pvm_upkfloat (&d[i][MY_MIN_J-1], 1, 1);

for (j = 0; j <= MY_SIZE; j++) {
    if ( (j >= 2) && (j != SIZE_J) ) {
        fact = a/b[j-1];
        d[i][j] = d[i][j] - fact*d[i][j-1];
    }
}

pvm_initsend (PvmDataRaw);
pvm_pkfloat (&d[i][MY_MAX_J], 1, 1);
pvm_send (tids[MY_ID+1], 30);
```

Figure 3. Example C code taken from a Partial Differential Equation solution for a Financial Option.

The CHIP³S description for the computational part of the ADI code is shown in Figure 4. The computational control flow, and associated operations, are described in the application layer with the use of *cflow* procedures. These are similar to the procedures in the source code but encapsulate the control flow of the application. Each statement in this script is associated with a *Processor Resource Usage Vector* (PRUV) indicated by the brackets < ... > that represent the operations in the original source code. In this example the operation count is done in terms of input C language operations (clc) with each operation indicated by a four character code. Control flow statements include: compute, loop, and case (conditional execution statements). Note that the case statement has a probability of executing the branch.

```
proc cflow TxEliminate {
    loop ( < is clc, LFOR, CMLL, INLL>, MY_SIZE) {
        compute < is clc, 2*CMLL, ANDL>;
        case ( < is clc, IFBR>) {
            0.9: compute <is clc, ARF1, 3*ARF2, DFSL, 2*TFSL, MFSL,AFSL>;
        }
    }
}
```

Figure 4. An example computational description in CHIP³S contained in the application layer.

The parallel template layer makes reference to the resources used in the system to execute the application. The term *device* is used to refer to the use of any hardware, or software, resources such as hardware devices (CPU, interconnection network, I/O) and message passing libraries (e.g. packing, initialisation). A CHIP³S parallel template is shown in Figure 5 which represents the structure of the original ADI source code in Figure 3. In each step statement, a device is specified, e.g. cpu, pvmrecv (asynchronous receive using PVM) etc. The parameters to each step are specified using the confdev statement. For example, in a communication receive and send, the parameters indicate the source and destination processors respectively. It should be noted that the parameter for the cpu step, is the value predicted from the TxElimate function as defined in the application layer (Figure 4).

```
step pvmrecv              { confdev my_id - 1, my_id; }
step pvmunpack on my_id   { confdev 1, PVM_FLOAT; }
step cpu on my_id         { confdev TxEliminate; }
step pvminitsend on my_id { confdev PVM_DataRaw; }
step pvmpack on my_id     { confdev 1, PVM_FLOAT; }
step pvmsend              { confdev my_id, my_id + 1; }
```

Figure 5. An example parallel template description in CHIP³S.

Model Evaluation

The process of evaluating a PACE model, using the CHIP³S performance scripts, and producing performance predictions is undertaken by the evaluation engine detailed in Figure 6. Initially the application layer is evaluated to produce predicted computational workload information. These are then used in the parallel template step cpu devices

The individual calls to specific devices, and their associated parameters, are passed into the evaluation engine. A dispatcher distributes input, from the workload descriptions, to an event handler and then to the necessary individual hardware models. The event handler is responsible for the construction of an event list for each processor in the system. Although the events can be resolved by the device involved in the step statement, the time spent using the device is still unknown. However, each individual hardware model can produce a time prediction of an event based on its parameters. The resultant prediction is recorded in the event list. When all device requests have been handled, the evaluation engine processes the event list to produce an overall performance estimate for the execution time of the application.

The processing of the event list is a two stage operation. The first is in the construction of the events, and the second is to resolve any ordering dependencies taking into account any contention factors. For example, in predicting the time for a communication, the traffic on the inter-connection network should be known to calculate channel contention. In addition, messages obviously cannot be received until after they are sent. The exception to this type of evaluation is a computational event that involves a single CPU device - this can be predicted in the first stage of evaluation since it does not require interaction with any other events.

Predictive Traces

The ability of PACE to produce predictive traces derives directly from the event list formed during the model evaluation. An example event list produced by the evaluation engine is illustrated in Figure 7 using a space-time representation. The communication dependencies between three processors, using the ADI code segment, are shown. The events, illustrated graphically, correspond to the underlying records in the trace file.

The predictive events produced by the evaluation engine are PACE specific in comparison to general traced events. However, a mapping of these to events to a standard trace file format, recognised by an external monitoring environment, is a straightforward process. Example trace events supported by PACE are shown in Table 1. This is an

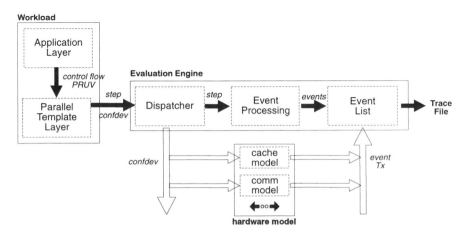

Figure 6. The evaluation process to produce a predictive trace within PACE

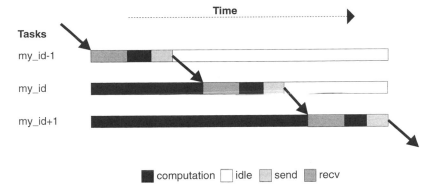

Figure 7. Example event list produced by the evaluation engine

extendible list in which events can be used to represent any information contained within the PACE model. The traces are produced by scanning each processor event list sequentially. During this process the local events are mapped to output trace events and formatted in a standard trace format. The output trace file is sorted according to the event time-stamps.

VISUALISING PREDICTIVE TRACES

PACE can produce overall predictions of execution time, and scaling behaviour, of application code. An example of which is shown in Figure 8 for a monte-carlo simulation code. However, the significant feature of PACE in the generation of a predictive trace allows insight into the time estimation. The predictive traces produced by PACE can be output in one of two formats suitable for use with either the Paragraph[5], or the Pablo environment[12]. They are also quite different in their approach in forming a performance analysis environment. For example, one of Paragraph's main use is to effectively 'replay' a trace file using a number of built-in performance displays which extract appropriate events directly. This is in comparison to Pablo which requires an analysis session to be constructed in a graphical environment first, followed by the use of a number of built-in displays. In this graphical environment, the user effectively relates events from the input trace file to the different displays available.

The two monitoring environments quite different trace file formats. Paragraph uses the PICL[14] (Portable Instrumented Communication Library) format in which each line in the trace file represents an individual event using a pre-specific data coding, thus allowing only certain events to be recorded. Pablo uses the SDDF (Self Defining Data Format) which enables the format of each event to be specified in the trace file along with the actual list of events. This is a more flexible approach, allowing system/application specific events to be defined and recorded, but requires trace-file specific construction of a performance analysis session.

Table 1. Example predictive events used within the PACE Evaluation Engine.

Event (start)	Event (end)	Description
SendBegin	SendEnd	Surrounds an asynchronous message send.
RecvBegin	RecvEnd	Surrounds an message being received.
RecvBlockBegin	RecvBlockEnd	A blocked receive (i.e. processor waiting).
CompBegin	CompEnd	A computation event
Overheadbegin	OverheadEnd	Computation events associated with parallel overheads
TaskBegin	TaskEnd	Surrounds a computational subtask on each processor.

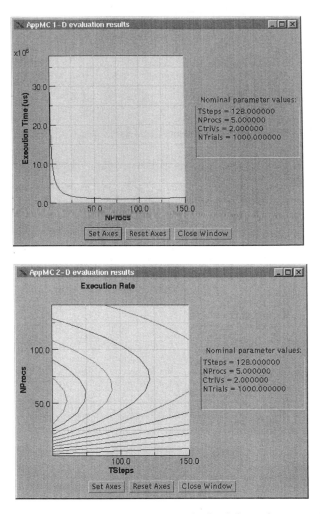

Figure 8. Predicted scaling behaviour of a monte-carlo simulation code as output by PACE

In order to illustrate the use of the predictive trace output from PACE, the core of a financial option code is used, a segment of which was shown in Figure 4. This code requires the solution of a partial differential equation, using the Alternating Direction Implicit (ADI) solution on a dense data grid. The resulting parallel code is a data decomposition of the data grid, but results in much communication between processing nodes.

This code was modelled within PACE, resulting in a binary which, when executed produced performance predictions for the application given a set of input parameters (data grid sizes, and also processor nodes in the system). The example here considered a hardware platform of a cluster of 6 SUN ULTRA workstations. A predictive trace (in either the PICL or SDDF format) can be output from this executable on the setting of an input flag. The overall performance predictions were validated with application run-times, and were found to be within an error bound of 15%. The performance predictions were obtained in a matter of seconds from the binary executable. The speed of evaluation is a significant feature of PACE, not described in detail here, but has been used for on-the-fly performance prediction[8].

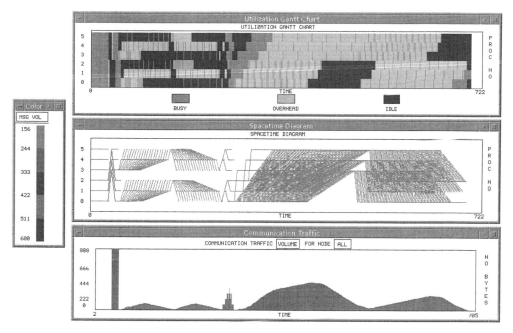

Figure 9. Example of a prediction trace analysed using space-time diagrams (from Paragraph).

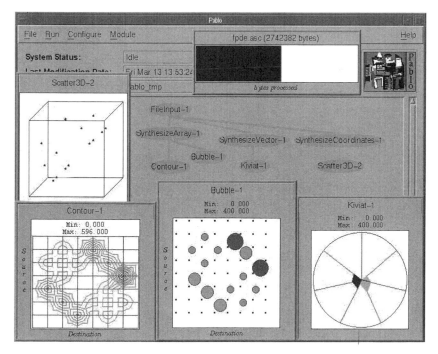

Figure 10. Example of a prediction trace analysed using the Pablo performance analysis environment.

65

Figure 9 shows three example views, derived from Paragraph, illustrating system activity as a function of time, for part of the predicted behaviour of the Financial code. The top chart indicates processor status, either busy, idle, or involved in parallel overheads. The second chart shows the computation/communication activities, colour coded according to message volumes. The final chart indicates the traffic in the system - an effective summation across processors from the middle chart. It can be seen, that there are many regions of both idle and overhead time which may be significant in under-achieving performance. These factors, once identified, could potentially be refined prior to application implementation.

Figure 10 shows an example predictive trace analysis session within Pablo. In the background, an analysis tree refers to an input trace file (at its root node) and, using data manipulation nodes, results in a number of separate displays (leaves in the tree). Four types of displays are shown producing summary information on various aspects of the expected communication behaviour of the Financial code. For example, the display in the lower left indicates the communication between source and destination nodes in the system (using contours to represent traffic), and the middle display shows the same information displayed using 'bubbles' the size and colour of which indicates traffic.

The use of a monitoring environment is not just limited to the analysis of a single trace. Indeed, current work is underway on the identification, and quantification, of changes in performance between one trace and another[8]. Such perturbations in performance may be due to variability in system use, or possibly due to changes in underlying system software (e.g. message passing libraries). The use of PACE can foresee changes in achieved performance, when such system parameters change, but in a predictive sense.

SUMMARY

The development of high performance application code is driven by the need of achieving rapid execution times. At a certain point in the development of the application, the performance issues become the predominant concern. However, typically these concerns are left until after the porting and implementation of the code, with performance analysis relying solely on analysis of executions. Performance analysis is not restricted to this post-implementation analysis, but can (and should) be used during the development of the application using prediction studies. The PACE toolset, as presented here, provides facilities for performance prediction studies to take place. Such studies can aid the implementation and porting of application code by choosing underlying numerical techniques, or guiding the choice of parallelisation, based on expected performance that will be achieved when finally executing the application on the target platform.

A significant feature of PACE is in the provision of a predictive trace output. This, a trace file that contains a list of events that represents the expected behaviour of the system at run-time. The predictive trace mimics the generation of an execution trace (one which is collected at run-time), and can be manipulated in exactly the same way. Thus, available monitoring environments, such as Paragraph and Pablo (familiar to many HPC users) can be utilised to analyse and explore the expected performance of the application code. The use of PACE in outputting predictive traces has been illustrated on a financial option code in this work. The use of the predictive traces can significantly increase efficiency in final implementations when used during the development of application code.

ACKNOWLEDGEMENTS

This work is funded by EPSRC grant GR/L13025, and by DARPA contract N66001-97-C-8530, awarded under the Performance Technology Initiative administered by NOSC.

REFERENCES

1. T. Fahringer, Estimating and Optimizing Performance for Parallel Programs, *IEEE Computer*. 28(11):47 (1995).
2. A. Fercha, A Petri Net Approach for Performance Oriented Parallel Program Design, *Jnl. of Parallel and Distributed Computing*. 15(3) (1992)
3. D.G. Green, C.J. Scott, A. Colbrook, and M. Surridge, HPCN tools: a European perspective, *IEEE Concurrency*. 5(3):38 (1997).
4. I. Gorton and I.E. Jelly, Software engineering for parallel and distributed systems, challenges and opportunities, *IEEE Concurrency*. 5(3):12 (1997).
5. M.T. Heath, A.D. Malony and D.T. Rover, The visual display of parallel performance data, *IEEE Computer*. 28(11):21 (1995).
6. T. Hey, A. Dunlop, and E. Hernandez, Realistic parallel performance estimation, *Parallel Computing*. 23:5 (1997).
7. K.L. Karavanic and B.P. Miller, Experiment management support for performance tuning, in: *Proc. SuperComputing*. (1997).
8. D.J. Kerbyson, E. Papaefstathiou and G.R. Nudd, Application execution steering using on-the-fly performance prediction, in: *High Performance Computing and Networking*, Springer-Verlag, (1998).
9. E. Papaefstathiou, D.J. Kerbyson, G.R. Nudd and T.J. Atherton, An overview of the CHIP^{3}S performance prediction toolset for parallel systems, in *Proc. of 8th ISCA Int. Conf. on Parallel and Distributed Computing Systems*. (1995).
10. E. Papaefstathiou, D. J. Kerbyson, G.R. Nudd, A layered approach to parallel software performance prediction: a case study, in: *Massively Parallel Processing Applications & Development*, L. Dekker, W. Smit, and J.C. Zuidervaart, eds., North-Holland, (1994).
11. E. Papaefstathiou, D.J. Kerbyson, G.R. Nudd and T.J. Atherton, An introduction to the CHIP^{3}S language for characterising parallel systems in performance studies, Research Report RR280, Dep. of Computer Science, University of Warwick (1995).
12. C.U. Smith, *Performance Engineering of Software Systems*, Addison Wesley (1990).
13. D. A. Reed et. al., Scalable performance analysis: the Pablo performance analysis environment, in *Proc. Scalable parallel libraries conf.*, IEEE Press (1993).
14. P.H. Worley, A new PICL trace file format, ORNL/TM-12125, Oak Ridge National Laboratory (1992)

SOLVING DENSE SYMMETRIC EIGENPROBLEMS ON THE CRAY T3D

K. Murphy, M. Clint, and R.H. Perrott

Department of Computer Science
The Queen's University of Belfast
Belfast BT7 1NN, UK

ABSTRACT

The work reported upon here has been carried out within a project one aim of which is to construct, for execution on the Cray T3D, [1,2] efficient implementations of certain key routines selected from the Harwell Subroutine Library [3]. A second aim of the project is to establish guidelines pertaining to the re-engineering of sequential software for parallel execution on the Cray T3D which can be drawn upon by others engaged in similar activity.

In this paper, a method for the eigensolution of real symmetric matrices is addressed. The algorithm and its parallel implementation are described in some detail. Results indicating the quality of performance of the parallel implementation are presented and some observations are made about the efficiency with which certain kernel operations of the implementation may be executed. Some guidelines are provided for applications programmers wishing to exploit distributed memory machines.

INTRODUCTION

The term high performance computing covers the full range of advanced computing systems including parallel systems and supercomputers. It embraces all aspects of a technology which is at the forefront of current developments in computing and which enables many new and complex problems to be solved. Applications software engineers rely on the availability of comprehensive libraries of efficient numerical mathematical routines which furnish them with the basic components from which their systems may be built. It is necessary, therefore, that they be provided with implementations of frequently used numerical routines which exploit efficiently the architectures of the machines on which they are to be executed. Thus, for example, the ScaLAPACK library, a collection of linear algebra routines is available for use on a variety of distributed memory machines and the NAG Parallel Library is currently under construction. The Harwell Subroutine Library (HSL) comprises over 600 sequential Fortran subroutines and functions, designed for incorporation into applications programs. The algorithms included in the library cover a wide range of numerical mathematical areas including differential equations, mathematical functions, linear algebra and numerical integration. In its present form, HSL routines are used by more than 1500 organisations worldwide. However, with the advent of a new generation of computers, massively parallel processors (MPPs) are being introduced into these organisations. Since the routines in the HSL are suitable only for execution on serial machines or on a single processor of a parallel machine, it is required to produce new versions of the routines which are capable of efficiently exploiting the capabilities of the new parallel platforms. In the project so far eight subroutines from the Harwell Library have been parallelised, including subroutines to solve sets of equations, orthogonalise vect ors, and solve linear least squares problems [4]. In this paper the parallel eigensolution of dense symmetric matrices is addressed. The routine from HSL discussed is Subroutine EA06.

THE ALGORITHM

The requirement to compute all of the eigenvalues and their associated eigenvectors of a real, dense symmetric matrix, $A \in \Re^{n \times n}$ arises frequently in scientific and engineering applications. One approach to the solution of the problem gives rise to algorithms which comprise four independent phases. These are:

- reduction of A to tridiagonal form

- computation of the eigenvalues of the reduced form

- computation of the eigenvectors of the reduced form

- recovery of the eigenvectors of A

One widely used algorithm in this class is based on Householder reflections [5]. The phases which comprise the Householder method are now described in more detail.

Reduction to tridiagonal form

The method utilises a sequence of Householder transformations (reflections) to reduce a real symmetric matrix $A \in \Re^{n \times n}$ to tridiagonal form. Householder reflections are orthogonal, similarity transformations which may be used to introduce zeros into a vector *en masse*, for example, by annihilating all but its first component. The method proceeds by systematically applying a sequence of such transformations to reduce, in turn, the columns of A so that on completion A is reduced to tridiagonal form. The eigenvalues of the tridiagonal matrix coincide with those of A. A Householder transformation of order n is a matrix of the form $I - 2u_j u_j^T / u_j^T u_j$ where

$$u_j^T = (0, \ldots, 0, a_{j+1,j} + \text{sign}(a_{j+1,j})s_j, a_{j+2,j}, \ldots, a_{n,j}) \tag{1}$$

with

$$s_j = \left(\sum_{i=j+1}^{n} a_{i,j}^2 \right)^{1/2} \tag{2}$$

where the $a_{i,j}$ are the elements of A. A total of $(n-2)$ such Householder transformations must be applied on the left and right of A to reduce it to tridiagonal form. In practice this is realised by

$$A_j = P_j A_{j-1} P_j \qquad \text{for } j = 1, \ldots, n-2 \tag{3}$$

where $A_0 = A$, P_j is the j-th Householder transformation and A_j is tridiagonal in its first $j+1$ rows and columns. The tridiagonalisation proceeds as follows. Suppose that $P_1, \ldots, P_{j-1}$ have been determined such that A_{j-1} is tridiagonal in rows 1 to j so that it may be expressed as

$$
A_{j-1} = \begin{pmatrix} & & & \vdots & 0 \\ & B & & \vdots & \ldots \\ & & & \vdots & b^T \\ \ldots & \ldots & \ldots & \ldots & \ldots \\ 0 & \vdots & b & \vdots & D \end{pmatrix}
\tag{4}
$$

where $B \in \Re^{(j \times j)}$ is tridiagonal, $D \in \Re^{(n-j) \times (n-j)}$ and $b \in \Re^{(n-j)}$. Applying the next Householder transformation, P_j, gives,

$$
A_j = P_j A_{j-1} P_j = \begin{pmatrix} & & & \vdots & 0 \\ & B & & \vdots & \ldots \\ & & & \vdots & b^T \bar{P}_j \\ \ldots & \ldots & \ldots & \ldots & \ldots \\ 0 & \vdots & \bar{P}_j b & \vdots & \bar{P}_j D \bar{P}_j \end{pmatrix}
\tag{5}
$$

where $\bar{P}_j$ is the Householder matrix defined for the non-zero elements of u_j. The sequence of steps used in the calculation of $\bar{P}_j D \bar{P}_j$ is as follows:

- compute $t_j = \beta D u_j$ where $\beta = 2/u_j^T u_j$

- compute $w_j = t_j - \frac{\beta}{2}(t_j^T u_j) u_j$

- compute $D - u_j w_j^T - w_j u_j^T = \bar{P}_j D \bar{P}_j$

Parallelisation of the reduction phase

Analysis of the reduction phase reveals that it is composed of two basic operations:- matrix-vector products and matrix updates. Both of these operations can be easily parallelised. The matrix A is distributed over the p available processors via a cyclic distribution of columns. That is, each processor is assigned a group of columns in a round-robin fashion. Thus, for p processors, the first processor receives columns 1, $p+1$, $2p+1$, etc. and the j-th processor receives columns j, $p \mid j$, $2p \mid j$, etc. If $n \bmod p = r$ ($\neq 0$), then the first r processors receive an extra column. For matrix-vector products, each processor locally computes the matrix-vector products using the columns of the matrix that have been assigned to it. On a distributed memory machine, like the Cray T3D, it is necessary for these local partial products to be transmitted to the other processors for reduction to the required vector. This step necessitates interprocessor communication. In matrix updates, each processor updates that part of the matrix which has been assigned to it; no communication is required. The main computational load arises in the calculation of the vector t_j and in updating the matrix D. During execution of the Householder algorithm, the elements of the pivot column (that is, the column which is used to construct the current Householder transformation) lying below the main subdiagonal must be transmitted to all of the other processors by the processor which holds the pivot column. This interprocessor communication, together with a global reduction, constitutes the main communication load in the parallel version of the reduction phase.

COMPUTATION OF THE EIGENVALUES

Let $T \in \Re^{n \times n}$ be a real symmetric tridiagonal matrix with the diagonal and subdiagonal elements represented by α_i and β_i respectively.

$$\begin{pmatrix} \alpha_1 & \beta_2 & & & \\ \beta_2 & \alpha_2 & \beta_3 & & \\ & \beta_3 & \ddots & \ddots & \\ & & \ddots & \ddots & \beta_n \\ & & & \beta_n & \alpha_n \end{pmatrix} \tag{6}$$

Assume that no subdiagonal element is zero. If this is not the case, the problem can be partitioned into smaller independent problems. Hence, without loss of generality it may be assumed that none of the β_i is zero. In addition, under this assumption, T has n distinct real eigenvalues [6].

One method for determining the eigenvalues of a real symmetric tridiagonal matrix is based on the use of the Sturm sequence

$$\begin{aligned} p_0(\lambda) &= 1 \\ p_1(\lambda) &= \alpha_1 - \lambda \\ p_i(\lambda) &= (\alpha_i - \lambda)p_{i-1}(\lambda) - \beta_i^2 p_{i-2}(\lambda), \quad i = 2, 3, \ldots, n \end{aligned} \tag{7}$$

In general, the number of disagreements in sign, at the point λ, between consecutive members of the sequence $p_0, p_1, \ldots, p_n$ is the number of eigenvalues smaller than λ. When this process is carried out in floating-point arithmetic, it is common, even for matrices of quite modest order, for values of the later $p_i(\lambda)$ to fall outside the range of permissible numbers. The possibility of such an occurrence may be reduced by replacing the sequence $p_i(\lambda), i = 1, \ldots, n$ by a sequence $q_i(\lambda), i = 1, \ldots, n$ which is derived from the $p_i(\lambda)$ [7].

Parallelisation of the eigenvalue computation

The parallel implementation comprises three separate phases: setup, isolation and extraction. Each of these phases is now described in more detail.

Setup

In this phase, various arrays are assigned initial values and an interval which contains all of the eigenvalues is determined. From Gerschgorin's theorem the eigenvalues lie within the union of the n intervals $\alpha_i \pm (|\beta_i| + |\beta_{i+1}|)$ with $\beta_1 = \beta_{n+1} = 0$. Hence, initial upper and lower bounds for the eigenvalues are $\max\{\alpha_i + (|\beta_i| + |\beta_{i+1}|)\}$ and $\min\{\alpha_i - (|\beta_i| + |\beta_{i+1}|)\}$, respectively.

The interval containing all of the eigenvalues is divided into $k \cdot p$ subintervals, where p is the number of processors and k is a factor which is specified below. Multisection [5] of order $k \cdot p - 1$ is applied to produce a large collection of subintervals each of which may contain either zero, one or many eigenvalues. The value of the factor k determines the number of subintervals that are generated. With a large value of k, say 20 or 30, there is a greater probability that more of the subintervals will contain only a single eigenvalue: that is, the eigenvalue is isolated within the interval. The execution time of the setup stage is small compared with those of the other stages.

In the evaluation of the Sturm sequences, the function $\mathcal{N}(\epsilon, q_i(\lambda))$ (used to count the number of negative signs) defined to be $+\epsilon$ if $q_i(\lambda) \geq 0$ and $-\epsilon$ otherwise, is implemented using the FORTRAN SIGN-function. This is much more efficient than using IF-statements to test if $q_i(\lambda) = 0$ and then replacing its value with ϵ. The algorithm used to compute the number $\nu(\lambda)$ of the eigenvalues which are less than λ is given below.

Algorithm 1 *Counting eigenvalues using Sturm sequences*

$\nu(\lambda) = 0$
   $q(\lambda) = \alpha_1 - \lambda$
   $\nu(\lambda) = \frac{1}{2} - \mathcal{N}(\frac{1}{2}, q(\lambda))$
for $j = 2, \ldots, n$ **do**
   $q(\lambda) = \alpha_j - \lambda - \beta_j^2 / (q(\lambda) + \mathcal{N}(\epsilon, q(\lambda))$
   $\nu(\lambda) = \nu(\lambda) + (\frac{1}{2} - \mathcal{N}(\frac{1}{2}, q(\lambda)))$
end do

Isolation of eigenvalues

In the isolation phase, the intervals which contain more than one eigenvalue are distributed, in a cyclical fashion, over all of the processors. Thus, the number of intervals that any processor has to deal with differs by at most one from that of any other. Each processor operates on the group of intervals which has been assigned to it. This initial distribution is the only time that interprocessor movement of data occurs in this phase. Thus, after a bisection step has been completed, the groups of intervals remaining are not gathered and redistributed. Although this may result in load imbalance, it is preferable to accept this modest penalty rather than pay the price of the communication overhead associated with gathering and redistributing the data in order to avoid it. Furthermore, by choosing a large value of k in the setup stage, the probability of acute imbalance is reduced.

Each processor takes the first interval of its allocated group and performs bisection until all of the eigenvalues within this interval have been isolated. It then processes the next interval (if any) in its assigned group in the same way, and so on. This process is performed by all of the processors simultaneously until all of the eigenvalues in all of the groups have been isolated. The intervals which contain exactly one eigenvalue are then gathered together preparatory to the extraction phase.

Extraction of eigenvalues

In the extraction phase the eigenvalues are determined to the required accuracy from those intervals which contain exactly one eigenvalue. Here, the n intervals are distributed over the p available processors using a cyclic distribution. If $n \bmod p = r$ ($\neq 0$), then the first r processors receive an extra interval. Each processor operates on its allocated group of intervals and the entire phase is executed in parallel. When all of the eigenvalues have been determined to a predefined accuracy, the extraction phase terminates.

EIGENVECTORS OF THE TRIDIAGONAL MATRIX

The eigenvectors of the tridiagonal matrix are computed by inverse iteration. The iteration matrix is first reduced to a simpler form using LU decomposition. Thus, if T

is tridiagonal and has the form:

$$
T = \begin{pmatrix}
\alpha_1 & \beta_2 & & & & \\
\beta_2 & \alpha_2 & \beta_3 & & & \\
& \beta_3 & \ddots & \ddots & & \\
& & \ddots & \ddots & \beta_n & \\
& & & \beta_n & \alpha_n &
\end{pmatrix}
\tag{8}
$$

the LU decomposition of $(T - \lambda_j I)$, where λ_j is a computed eigenvalue of T, produces the matrices

$$
L = \begin{pmatrix}
1 & & & & \\
l_2 & 1 & & & \\
& l_3 & \ddots & & \\
& & \ddots & \ddots & \\
& & & l_n & 1
\end{pmatrix}, \qquad
U = \begin{pmatrix}
u_1 & \beta_2 & & & \\
& u_2 & \beta_3 & & \\
& & \ddots & \ddots & \\
& & & \ddots & \beta_n \\
& & & & u_n
\end{pmatrix}
\tag{9}
$$

where

$$
\begin{aligned}
l_i &= \beta_i / u_{i-1} \\
u_i &= (\alpha_i - \lambda_j) - l_i \beta_i \quad i = 2, \ldots, n
\end{aligned}
$$

Inverse iteration to determine the eigenvectors associated with λ_j necessitates the repeated solution of equations of the form $LU x_j = \tilde{x}_j$. This is achieved using forward and backward substitutions.

Parallelisation of the tridiagonal eigenvector computation

In this phase parallelism is exploited by distributing the eigenvalues equally over all of the processors. Then the processors compute, simultaneously, the eigenvectors of T associated with the sets of eigenvalues which have been assigned to them.

COMPUTING THE EIGENVECTORS OF THE TARGET MATRIX

Having determined Householder transformations P_j, $j = 1, \ldots, n-2$, such that

$$
A_{n-2} = (P_1 P_2 \ldots P_{n-2})^T A (P_1 P_2 \ldots P_{n-2}) = Q^T A Q
\tag{10}
$$

where $Q \in \Re^{n \times n}$ is orthogonal and A_{n-2} is tridiagonal, it is necessary to pre-multiply the eigenvectors of the tridiagonal matrix by Q in order to recover the eigenvectors of the matrix A. Thus, for each Householder transformation, P_j, it is required to compute $x_j' = P_j x_j$, where x_j is an eigenvector of the tridiagonal matrix. Note that, $x_j' = x_j - \beta(u^T x_j) u$. Thus, to compute x_j', two Level 1 BLAS subroutines may be employed as follows:

- compute $\delta = \beta u^T x_j$ (an *sdot* operation)

- compute $x_j' = x_j - \delta u$ (a *saxpy* operation)

It would appear that the use of Level 2 BLAS subroutines is not possible here. However, rather than premultiply each eigenvector in turn by the Householder transformations,

Table 1. Timings (in seconds) for a dense matrix of order 1024

	Number of processors						
	1	2	4	8	16	32	64
Reduction	61.04	31.32	17.09	9.69	5.85	4.61	3.71
Eigenvalue	55.09	28.43	15.42	8.48	4.88	2.66	1.61
Eigenvector	52.90	26.77	13.95	7.25	3.95	2.32	1.59
Total Time	169.09	86.53	46.46	25.41	14.69	9.58	6.91
Speedup	1.00	1.95	3.64	6.65	11.51	17.65	24.47
Efficiency	100%	98%	91%	83%	72%	55%	38%

Table 2. Timings (in seconds) for a dense matrix of order 2048

	Number of processors						
	2	4	8	16	32	64	128
Reduction	224.97	114.99	62.07	34.38	22.77	16.05	15.42
Eigenvalue	111.13	58.63	33.36	19.80	11.92	7.35	4.31
Eigenvector	227.30	104.45	54.46	28.22	15.34	8.95	5.93
Total Time	563.39	278.07	149.90	82.40	50.03	32.35	25.65
Speedup	1.00	2.02	3.76	6.84	11.26	17.42	21.96
Efficiency	100%	101%	94%	86%	70%	54%	34%

the k eigenvectors stored on each processor may be combined into a $n \times k$ matrix so that the vector updates may be achieved as matrix updates, viz:-

$$X' = X - b(u^T X)u = X - (b^T X^T u)^T u \qquad (11)$$

where $X \in \Re^{n \times k}$ and $b \in \Re^n$ has $2/u_j^T u_j$ as its j-th element. With this reorganisation, Level 2 BLAS subroutines sgemv and sger may be employed.

Parallelisation of the computation of the target matrix eigenvectors

As in the parallelisation of the computation of the eigenvectors of the tridiagonal matrix, the parallelism is based on the distribution of the eigenvectors over all of the processors. In this phase, each processor premultiplies the set of eigenvectors of the tridiagonal matrix which it has been assigned by the Householder transformations used in the construction of Q. To economise on storage, the vectors from which the Householder transformation may be generated are distributed cyclically over all of the processors. Thus, it is required that the processor which holds the vector used in the construction of a particular Householder transformation broadcasts this vector to all of the other processors.

THE IMPLEMENTATION

The parallel implementation has been developed using the MPI (Message Passing Interface [8] standard. MPI comprises a collection of library routines which are efficiently implementable on a wide range of computers. All of the communications necessary for the efficient implementation of the algorithm discussed here are taken from the library of collective communication. Three library routines are used viz. MPI_ALLREDUCE{}, MPI_BROADCAST{}, MPI_ALLGATHERV{}, which, respectively, combine, broadcast and gather information.

Table 3. Timings (in seconds) for a dense matrix of order 4096

	Number of processors					
	8	16	32	64	128	256
Reduction	440.79	232.63	135.79	86.20	69.09	58.82
Eigenvalue	127.66	78.81	52.17	31.05	22.19	15.16
Eigenvector	463.46	236.37	122.20	65.72	37.73	25.32
Total Time	1031.91	547.81	310.15	182.97	129.01	99.31
Speedup	1.00	1.88	3.33	5.64	8.00	10.39
Efficiency	100%	94%	83%	71%	50%	32%

Table 4. Timings (in seconds) for a dense matrix of order 5120

	Number of processors				
	16	32	64	128	256
Reduction	441.27	252.08	152.27	115.59	101.56
Eigenvalue	123.01	80.45	58.26	31.98	23.83
Eigenvector	442.51	227.45	120.61	67.70	43.78
Total Time	1006.79	559.99	331.15	215.27	169.17
Speedup	1.00	1.80	3.04	4.68	5.95
Efficiency	100%	90%	76%	59%	37%

THE EXPERIMENTS

Numerical experiments have been conducted for randomly generated matrices of orders ranging from 1024 to 8192. Tables 1–5 show some performance results in which the complete execution times have been broken down to show the proportions devoted to reduction, eigenvalue calculation and eigenvector calculation. The times for eigenvector computation include the times to compute both the eigenvectors of the tridiagonal matrix and from these the eigenvectors of the target matrix. The speedups in all cases are measured with respect to the execution time on the smallest number of processors necessary to perform the computation. Thus, for example, in the experiment reported in Table 5, the smallest number of processors used is 64 and the speedup is measured accordingly. Tables 6–7 show some performance results for routine PSSYEVX from the ScaLAPACK library which is included in the default library libsci.a on the Cray T3D and for routine PDSPEV from PeIGS [9], a collection of commonly used linear algebra subroutines for computing eigensolutions. The test data for these tables were generated by a program which assembles a pseudo-Fock matrix which is then made

Table 5. Timings (in seconds) for a dense matrix of order 8192

	Number of processors		
	64	128	256
Reduction	540.04	367.33	276.05
Eigenvalue	154.95	110.97	78.90
Eigenvector	568.37	302.59	175.04
Total Time	1263.37	780.89	529.99
Speedup	1.00	1.62	2.38
Efficiency	100%	81%	60%

Table 6. Timings (in seconds) of routine PSSYEVX from Cray T3D libsci.a

Order of Matrix	Number of processors						
	4	8	16	32	64	128	256
256	2.41	2.17					
512	8.68	6.82	5.04				
1024	51.4	29.0	17.9				
2048	299.0	144.0	76.7	52.3	35.2		
4096			410.4	243.7	176.6	110.3	
5120					233.3	169.7	
8192						491.3	

Table 7. Timings (in seconds) of PeIGS2.1 routine PDSPEV on the Cray T3D

Order of Matrix	Number of processors						
	4	8	16	32	64	128	256
64	0.3398	0.4606	0.5208		1.444		
128		0.8613	0.7955		2.050		
256		2.666	2.176		3.532		
512		11.44	8.347	8.236	8.038		
1024		75.92	52.67	37.52	32.18		
2048			383.1	253.4	195.9	173.9	
4096					2195.0		

diagonally dominant.

Tables 8–9 show the speedups of the ScaLAPACK and PeIGS routines, respectively. The bracketed figures are the corresponding speedups for matrices of the same orders when executed using the parallel subroutine discussed here. It will be observed that for most of the test cases the parallel subroutine discussed here is competitive with the ScaLAPACK routine and is much more efficient than PeIGS. For example, from Table 2 and Table 7 the parallel routine discussed here is over 6.7 times faster than PeIGS for a matrix of order 2048 when executed on 128 processors. Since the ScaLA-PACK routine is based on a block algorithm the more efficient BLAS 2 subroutines may be employed to greater advantage. This accounts for its superior performance as the order of the target matrix increases. It would appear, however, from Tables 8–9 that the parallel subroutine discussed here scales better. Although the same test data has not been used it seems that, from the results presented (and particularly for matrices of large order) that this routine can, within a certain range of matrix orders, outperform the ScaLAPACK routine. Another advantage of the subroutine discussed here is that its memory requirements are less than those of both ScaLAPACK and PeIGS.

Table 8. Speedup of routine PSSYEVX from Cray T3D libsci.a

Order of Matrix	Number of processors						
	4	8	16	32	64	128	256
256	1.00	1.11(1.13)					
512	1.00	1.27(1.70)	1.72(2.50)				
1024	1.00	1.77(1.83)	2.87(3.16)				
2048	1.00	2.08(1.86)	3.90(3.37)	5.72(5.56)	8.49(8.60)		
4096			1.00	1.68(1.77)	2.32(2.99)	3.72(4.25)	
5120					1.00	1.37(1.54)	

Table 9. Speedup of PeIGS2.1 routine PDSPEV on the Cray T3D

Order of Matrix	Number of processors						
	4	8	16	32	64	128	256
512		1.00	1.37(1.47)	1.39(2.14)			
1024		1.00	1.44(1.73)	2.02(2.65)	2.36(3.68)		
2048			1.00	1.51(1.65)	1.96(2.55)	2.20(3.21)	

For a matrix of order n the routine requires $2n^2/p + 12n$ words of storage. Thus, if 512 processors are available, it can solve systems of order ≈ 45000 without the necessity to use backing storage. In contrast, the equivalent ScaLAPACK routine with 512 processors requires approximately $5n^2/p + 5n$ words of storage and can solve systems of order ≈ 29000 while PeIGS requires approximately $4n^2/p$ words of storage and can solve systems of order ≈ 32000.

The subroutine utilises calls to certain BLAS subroutines and functions. Of these four are to BLAS level 1 subroutines and two are to BLAS level 2 subroutines.

The performance on the Cray T3D of the parallel implementation of this routine is impressive and indicates that other routines in the Harwell Library (and there are a number) which are built from the same kernel operations as Subroutine EA06 are prime candidates for re-engineering for efficient execution on this machine.

ACKNOWLEDGEMENTS

This work was supported by IRTU (Industrial Research and Technology Unit) and was carried out using the facilities of The University of Edinburgh Parallel Computing Centre.

REFERENCES

1. *Cray Research MPP Software Guide*, Cray Research Inc. (1993) order number SG-2508.
2. S.P. Booth, J. Fisher, P.H. MacCallum and A.D. Simpson. *Introduction to the Cray T3D at EPCC*, University of Edinburgh (1996).
3. Anon. *Harwell Subroutine Library (Release 11)* Theoretical Studies Dept. AEA Technology, Didcot, Oxfordshire (1993).
4. K. Murphy, M. Clint, R.H. Perrott, D. McLaughlin and I. Carberry, Re-engineering mathematical library software for execution on high performance computers, in: *High Performance Computing 1997*, Proceedings of the 1997 Simulation Multiconference, Atlanta, Georgia, A. Tentner ed., SCS, pp 207-212 (1997).
5. T.L. Freeman and C. Phillips. *Parallel Numerical Algorithms*, Prentice Hall (1992).
6. J.H. Wilkinson. *The Algebraic Eigenvalue Problem*, Clarendon Press, Oxford (1965).
7. J.H. Wilkinson and C. Reinsch. *Handbook for Automatic Computation, Volume II, Linear Algebra*, Springer-Verlag, Berlin (1971).
8. Message Passing Interface Forum, MPI: a message-passing interface standard (version 1.1), Revision of article to appear in the *International Journal of Supercomputing Applications*, 8(3/4), 157–416 (1994, June 1995).
9. D. Elwood, G. Fann and R. Littlefield. *Parallel Eigensolver System User Manual*, available from *anonymous@ftp://pnl.gov*.

PARASOL
AN INTEGRATED PROGRAMMING ENVIRONMENT FOR
PARALLEL SPARSE MATRIX SOLVERS

Patrick Amestoy[1], Iain Duff[2], Jean Yves L'Excellent[3]
and Petr Plecháč[4]

[1]ENSEEIHT-IRIT, Toulouse, France
[2]Rutherford Appleton Laboratory, UK and CERFACS, Toulouse, France
[3]CERFACS, Toulouse, France
[4]Rutherford Appleton Laboratory, UK

ABSTRACT

PARASOL is an ESPRIT IV Long Term Research Project whose main goal is to build and test a portable library for solving large sparse systems of equations on distributed memory systems. There are twelve partners in five countries, five of whom are code developers and five end users. The software is written in Fortran 90 and uses MPI for message passing. There are routines for both direct and iterative solution of symmetric and unsymmetric systems. The final library will be in the public domain.

We will discuss the PARASOL Project with particular emphasis on the algorithms and software for direct solution that are being developed by RAL and CERFACS in collaboration with ENSEEIHT-IRIT in Toulouse. The underlying algorithm is a multifrontal one with a switch to ScaLAPACK processing towards the end of the factorization (and solution). We will discuss the algorithms, the interface, and their current status and illustrate the performance of the direct solver on a range of problems from the PARASOL end users.

INTRODUCTION

PARASOL is a long term research (LTR) ESPRIT IV Project (No 20160) for "An Integrated Environment for Parallel Sparse Matrix Solvers". This Project started on January 1st, 1996, and its aim is to develop a parallel scalable library of sparse matrix solvers using Fortran 90 and MPI. At the end of the Project, the codes will be made available in the public domain.

The PARASOL Consortium is managed by PALLAS in Germany and consists of

- leading European research organizations with internationally recognized experience and an established track record in the development of parallel solvers

(CERFACS, GMD-SCAI, ONERA, Rutherford Appleton Laboratory (RAL), University of Bergen);

- industrial code developers who define the requirements for PARASOL, are providing test cases generated by their finite-element packages, and will use the developed software in production mode (Apex Technologies, Det Norske Veritas (DNV), INPRO, MacNeal-Schwendler (MSC), Polyflow);

- two leading European HPC software companies who will exploit the project results and are providing state-of-the-art programming development tools (GENIAS, PALLAS).

For more information see the project web site at *http://www.genias.de/parasol*.

The codes in the Library include direct methods, domain decomposition techniques, and multigrid algorithms. Within this project, RAL and CERFACS are involved in the development of direct solvers and are working in this context in close collaboration with ENSEEIHT-IRIT, Toulouse, France. The codes are implemented as portable prototypes integrated into the PARASOL Library. The four different solvers and their implementation within the PARASOL Library are depicted in Figure 1.

The Library provides its own communication and data exchange routines (the PARASOL Interface) as a higher level message passing protocol based on MPI and designed for specific data structures that arise in the finite-element approximation of partial differential equations and operations with sparse matrices. In this short paper, we focus on the sparse direct solver (MUMPS) developed by RAL and CERFACS in collaboration with ENSEEIHT-IRIT (see [2] for more details). We first give a brief overview of the PARASOL Interface Library [6] emphasizing aspects relevant to the

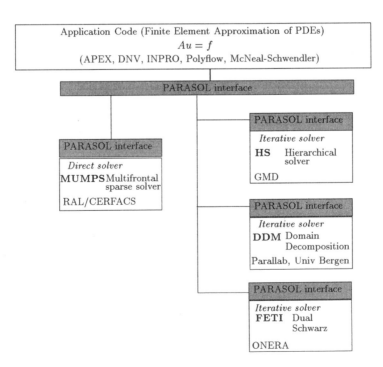

Figure 1. PARASOL Library

implementation of MUMPS. We then briefly describe the algorithm behind the MUMPS package.

The PARASOL project contains several examples which have been established as a test bed for the solvers designed and implemented by the developers. The test cases are provided by the industrial partners and range from small to medium size examples. We consider the performance of the MUMPS package on some of these examples in the last section.

THE PARASOL INTERFACE

Several very different solvers and algorithms for the solution of large sparse linear systems are integrated within the PARASOL Library. The Project addresses the issue of a higher level message passing protocol in order to overcome difficulties arising from the different data structures typically used by the different algorithms. These difficulties are exacerbated by a distributed memory environment. The package uses the MPI 1.1 message passing interface [8] as the basis for its own interface standard. The idea of the PARASOL Interface allows users (and developers of solvers) to exchange data between different modules of the package in a precisely defined way which is independent of the specific data structures needed internally in the user's codes or solvers.

The PARASOL Interface accommodates three modes of parallel execution:

- Host-node - one process inputs the problem definition and the others exchange data with it and compute the solution.

- Hybrid-host - one process inputs the problem definition and exchanges data with the others, but then helps the others to compute the solution.

- Hybrid-node - all processes input the problem definition and compute the solution.

although only the first one is currently efficiently implemented by the MUMPS solver.

It is responsibility of the user to define, configure and terminate the parallel environment in addition to controlling I/O resources. However, the PARASOL Library offers tools for controlling synchronization of data exchanges. The basic process within the PARASOL environment is called *an instance* and it is described by its descriptor, which also encodes an MPI communicator used for data exchanges related to a given instance. Every PARASOL routine is passed an instance descriptor as an argument and then performs operations only on data associated with this instance. The PARASOL Library allows the user to run several PARASOL instances simultaneously. The instance operates on its own set of private data structures, although it can initialize and use another instance. This feature provides an easy mechanism for integrating different solvers from the Library into a new solver. For example the domain decomposition solver can call the sparse direct solver. This new solver will use all parallel features of both solvers, without any changes of data structures.

The PARASOL Interface guide [6] gives a detailed description of the data exchange protocol and data structures defined for PARASOL routines. Here we restrict ourselves only to data structures relevant for the current version of MUMPS, that is the one for assembled sparse matrices. We will not discuss the implementation of data protocols related to structures like the description of finite-element meshes, element matrix format etc.

Sparse matrices are represented within the PARASOL package in the compressed column storage format, that is, for an $m \times n$ sparse matrix $A = (a_{ij})$ we have a descriptor which contains the data shown in Table 1.

Table 1. Sparse matrix descriptor

PSL_TYPE	MPI_DOUBLEPRECISION ...	MPI data type
PSL_NAME	PSL_MATRIX	data name
PSL_ATTR	PSL_SYMMETRIC,PSL_HERMITIAN,PSL_UNSYMMETRIC	matrix structure
PSL_FORM	PSL_SPARSEMAT	sparse matrix format
PSL_NROW	INTEGER $m > 0$	number of rows
PSL_NCOL	INTEGER $n > 0$	number of columns
PSL_NVAL	INTEGER $ne > 0$	number of entries

The matrix is stored in associated arrays:

col - stores pointers to the start of the column data (entries in a column) in row and val. The last entry of col points to the first free entry in row, val. The length of col is equal to PSL_NCOL + 1.

row - stores row indices of entries for each column. The length of row is equal to PSL_NVAL + 1.

val - stores values of entries for each column. The length of val is equal to PSL_NVAL + 1.

The right-hand side(s) are entered either in the sparse vector format, which has the same descriptor as a sparse matrix with the following fields modified: PSL_NAME = PSL_RHSIDE, PSL_SOLUTION, PSL_ATTR = PSL_LEFT, PSL_RIGHT (vector position), PSL_FORM = PSL_SPARSEVEC, or in the dense vector format (see Table 2). In the case of dense vectors, the arrays col, row are not used and the vectors are stored in val(1:PSL_LDIM,1:PSL_NVEC).

The PARASOL Library provides the user with easy access to the solvers via control routines for each solver and data exchange routines related to the transfer of necessary data structures relevant to the given solver.

We illustrate the use of the PARASOL Library for the MUMPS solver implemented within a simple code which reads a sparse matrix from an external file, maps the solver onto a given number of processors, and performs the solution. The skeleton for this is shown in Figure 2.

The generic calls to psl_init and the other psl_ calls are interpreted by the PARASOL Interface, through the defined configuration name, to be calls to the PARASOL control routines for the MUMPS solver.

Unlike the exchange routines, the control routines make heavy use of the internal structure of the MUMPS code. Their implementation is based on appropriate calls to the subroutine PSL_MUMPS. We use the name "MUMPS code" to refer to the whole set of routines for which the routine PSL_MUMPS serves as a driver.

Table 2. Dense vector descriptor

PSL_TYPE	MPI_DOUBLEPRECISION	MPI data type
PSL_NAME	PSL_RHSIDE,PSL_SOLUTION,...	data name
PSL_ATTR	PSL_LEFT,PSL_RIGHT	vector position
PSL_FORM	PSL_DENSEVEC	dense vector format
PSL_NROW	INTEGER > 0	number of rows
PSL_NVEC	INTEGER > 0	number of vectors
PSL_LDIM	INTEGER > 0	leading dimension of val

- `psl_mumps_init` - this routine initializes the MUMPS instance:

  1. creating a description of nodes, the current version efficiently supports PSL_MODE = PSL_HOSTNODE and the user has reserved one process (`rank = 0`),

  2. allocating private data structures for the instance, and

  3. calling the initialization phase of the MUMPS code,

- `psl_mumps_end` - this routine terminates the instance:

  1. deallocating the resources,

- `psl_mumps_map` - the mapping routine performs the analysis and factorization phases of the MUMPS code:

  1. setting some output/diagnostics parameters,

  2. receiving the sparse structure of the matrix (`n,ne,col,row`) on the master,

  3. performing the analysis phase of the MUMPS code,

  4. receiving numerical values of matrix entries (`val`) on the master, and

  5. numerical factorization,

- `psl_mumps_solve` - the solution routine calls the solution phase of the MUMPS code:

  1. receiving the right-hand side(s), and

  2. solving,

- `psl_mumps_endsolve` - the termination routine sends the solution to the user and deallocates the memory.

The data exchange routines of the form `host_` in Figure 2 must be coded by the user and should include calls to the following data exchange routines that support data communication needed between the user's code and the MUMPS solver:

- `psl_mumps_contract` - establishing a data exchange session,

- `psl_mumps_what2send` - send an inquiry,

- `psl_mumps sendIdata` - send INTEGER data (`col, row`),

- `psl_mumps_sendDdata` - send DOUBLE PRECISION data (`val`),

- `psl_mumps_need2recv` - request data,

- `psl_mumps_recvIdata` - receive INTEGER data, and

- `psl_mumps_recvDdata` - receive DOUBLE PRECISION data.

These routines are used only for data exchange between a user's application (program) and the MUMPS solver (package). The basic data communication for MUMPS requires: sending a matrix (PSL_NAME = PSL_MATRIX), and a right-hand side vector (PSL_NAME = PSL_RHSIDE), and receiving a solution vector (PSL_NAME = PSL_SOLUTION).

The current version (Version 2.0) of the MUMPS integration into the PARASOL package supports the PSL_HOSTNODE and PSL_HYBRIDHOST execution models, although

```
        PROGRAM example
! Include files
        INCLUDE 'mpif.h'        ! MPI definitions
        INCLUDE 'pslf.h'        ! PARASOL definitions
! Local data
        INTEGER self            ! MPI rank of the current process
        INTEGER id(PSL_IDSIZE) ! PARASOL instance descriptor
        INTEGER rc              ! return code

        CALL MPI_INIT(rc)       ! initialize MPI
        CALL MPI_COMM_RANK(MPI_COMM_WORLD,self,rc)    ! get MPI rank

        id = 0                          ! clear the instance desc.
        id(PSL_COMM) = MPI_COMM_WORLD ! use MPI communicator
        id(PSL_CONF) = PSL_MUMPS      ! name the configuration
        id(PSL_MODE) = PSL_HOSTNODE   ! select execution mode

        CALL psl_init(id,rc)    ! initialize PARASOL instance

        CALL psl_map(id,rc)     ! nodes compute mapping/reordering
        IF( self.EQ.0) THEN
          CALL host_serves_mapping_data_ex(....id,rc) ! reading data from a file
        END IF
        CALL psl_solve(id,rc)   ! nodes solve the system
        IF( self.EQ.0 ) THEN
          CALL host_serves_solution_data_ex(....id,rc) ! collecting the solution
          CALL host_outputs_data(....id,rc)  ! printing the solution
        END IF

        CALL psl_end(id,rc)     ! terminate PARASOL instance
        CALL MPI_FINALIZE(rc)   ! terminate MPI
        END
```

Figure 2. Skeleton of PARASOL Test Driver for MUMPS code

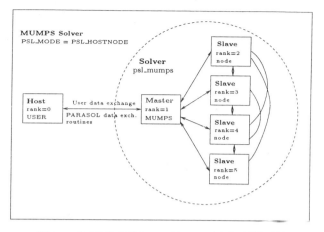

Figure 3. HOST_NODE execution model for MUMPS

only the former one fully benefits from the parallel features of the MUMPS Version 2.0 code. The execution mode PSL_HOSTNODE is schematically described in Figure 3. In this mode, one processor is dedicated to serve as a master process for PARASOL and the remaining processors are used for the parallel implementation of the solver.

MUMPS - MULTIFRONTAL MASSIVELY PARALLEL SOLVER

MUMPS is a parallel sparse direct solver for distributed memory architectures using a multifrontal method, which is a direct method based on the LU factorization of the matrix. We refer the reader to our earlier papers [1, 4, 5] for full details of this technique. The currently implemented version of MUMPS (Version 2.0) solves the system

$$\mathbf{Ax} = \mathbf{b},$$

where $\mathbf{A}$ is unsymmetric.

The structure of the matrix is first *analysed* to determine an ordering that, in the absence of any numerical pivoting, will preserve sparsity in the factors. In Version 2.0 of MUMPS, an approximate minimum degree ordering strategy is used on the symmetrized pattern $\mathbf{A} + \mathbf{A}^T$, and this analysis phase produces both an ordering and an assembly tree. The assembly tree is then used to drive the subsequent numerical factorization and solution phases. At each node of the tree, a dense submatrix (called a *frontal matrix*) is assembled using data from the original matrix and from the sons of the node. Pivots can be chosen from within a submatrix of the frontal matrix (called the *pivot block*) and eliminations performed. The resulting factors are stored for use in the solution phase and the Schur complement (the *contribution block*) is passed to the father node for assembly at that node. In the numerical factorization phase, the tree is processed from the leaf nodes to the root (if the matrix is reducible, we have a forest, and each component tree of the forest will be treated similarly and independently). The subsequent forward and backward substitutions during the solution phase process the tree from the leaves to the root and from the root to the leaves, respectively. A crucial aspect of the assembly tree is that it defines only a partial order for the factorization since the only requirement is that a son must complete its elimination operations before the father can be fully processed. It is this freedom that enables us to exploit parallelism in the tree (*tree parallelism*).

85

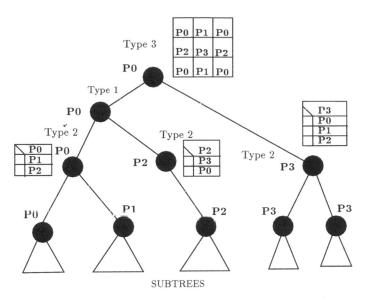

Figure 4. Distribution of the computations of a multifrontal tree

In the unsymmetric case, handled by Version 2.0 of MUMPS, threshold pivoting is used to maintain numerical stability so that it is possible that the pivots selected at the analysis phase are unsuitable. In the numerical factorization phase, we are at liberty to choose pivots from anywhere within the pivot block (including off-diagonal pivots) but it still may be impossible to eliminate all variables from this block. The result is that the Schur complement that is passed to the father node may be larger than anticipated by the analysis phase and so our data structures may be different from those forecast by the analysis. This implies that we need to allow dynamic scheduling during numerical factorization, in contrast to the symmetric positive definite case where only static scheduling is required.

A version of the multifrontal code for shared memory computers was developed by Amestoy and Duff [1] and was included in Release 12 of the Harwell Subroutine Library as code `MA41`. This was the basis for Version 1.0 of MUMPS that was released in May 1997.

In the current version of MUMPS (Version 2.0), both tree and node parallelism are exploited, and we distribute the pool of work among the processors, but our model still requires an identified host node to perform the analysis phase, distribute the incoming matrix, collect the solution, and generally oversee the computation. All routines called by the user for the different steps are SPMD, and the distinction between the host and the other processors is made by the MUMPS code. The code is organized with a designated host node and other processors as follows (notice that the following steps are easily implemented within the controlling strategy of the PARASOL Library):

1. **Analysis.** The host performs an approximate minimum degree algorithm based on the symmetrized pattern $\mathbf{A} + \mathbf{A}^T$, and carries out symbolic factorization. A mapping of the multifrontal tree is then computed, and symbolic information is transferred from the host to the other processors. Using this information, the processors estimate the memory necessary for factorization/solution.

2. **Factorization**. The host sends appropriate entries of the original matrix to the other processors, that are responsible for the numerical factorization. The numerical factorization on each frontal matrix is conducted by a *master* processor (determined by the analysis phase) and one or more *slave* processors (determined dynamically) as discussed later in this section. Each processor allocates an array for contribution blocks and factors; the latter should be kept for the solution.

3. **Solution**. The right-hand side is broadcast from the host to the other processors. These processors compute the solution using the (distributed) factors computed during Step 2, and the solution is assembled on the host.

For an efficient and more scalable parallelism on general matrices, the elimination of frontal matrices near the root of the tree has to be parallelized. Version 2.0 of MUMPS exploits both tree parallelism and node parallelism; this is done by introducing Type 2 and Type 3 nodes, as defined below.

We consider the assembly tree of Figure 4 where, instead of single nodes as the leaves, there are subtrees whose constituent nodes have frontal matrices of small order. Each subtree is processed by a single processor, to avoid communication at that stage. This mapping of subtrees to processors is performed by the analysis phase. For large problems, there will be more subtrees than processors which will aid in the overall load balancing of the computation.

Above the subtrees, there can still be some nodes processed by only one processor. These nodes (as well as nodes inside the subtrees) are called nodes of Type 1.

Consider a typical frontal matrix in the tree in which there are NPIV pivots to eliminate (that is, the pivot block has order NPIV) and NCB rows to update (that is, the order of the frontal matrix is NPIV+NCB). A node is of Type 2 if NCB is large enough. The partial factorization process is then parallelized with the first NPIV rows on one processor, called the *master of the node* and the NCB rows distributed among other processors (called the *slaves of the node*). For instance, in the Type 2 node on the right of Figure 4, P3 is the master, and P0, P1, and P2 are the slaves.

A pipelined factorization is used, and updates to the contribution blocks are performed in parallel. In our implementation, the assembly process is also fully parallel.

At the root node, a full LU factorization is performed. If the size of the root node is deemed large enough, the root node is said to be of Type 3, and is factorized using ScaLAPACK [3]. The assembly of the root node is directly distributed in a 2D cyclic grid and is completely parallel.

RESULTS

In this last section, we present some results to demonstrate the performance of the MUMPS code on a few test cases from the PARASOL set of test examples. We show the performance of MUMPS with different levels of parallelism as well as the speedup obtained on an IBM SP2.

The test cases are summarized in Table 3 where N, NE denote the size of the matrix and the number of entries, respectively. The size of the LU factors is reported in the column "LU-Fac", and the number of floating-point operations to factorize the matrix in the last column.

The results in Table 4 show that there is often good speedup on the test examples, indeed sometimes the performance is superlinear. This is caused by the reduction in memory requirements on individual processors with a consequent reduction in paging overheads.

Table 3. Some of the PARASOL test problems.

Problem	Origin	Type	N	NE $\times 10^6$	LU-Fac $\times 10^6$	No. flops $\times 10^9$
EXTRUSION-1	POLYFLOW	U	30412	1.79	49.7	36
OILPAN	INPRO	S	73752	1.84	20.6	8
MIXING-TANK	POLYFLOW	U	29957	1.99	62.7	142
CRANKSEG1	MSC	S	52804	10.6	80.1	101
BMW7ST_1	MSC	S	141347	3.74	54.0	31
WANG3	RBSMC	U	26064	0.2	11.5	11

Table 4. Times for Analysis and Factorization in seconds. Results obtained on the IBM SP2 at GMD, Bonn

	Analysis	LU Factorization				
Number processors	1	2	4	8	16	32
Test case						
EXTRUSION-1	5.2		536.7	179.1	72.7	69.1
OILPAN	4.4	133.7	25.9	22.0	21.0	
MIXING-TANK	4.4			607.6	83.8	80.1
CRANKSEG1	36.0				626.0	
CRANKSEG1 [a]	9.1			1251.4	1022.8	899.6
Number processors	1	8	12	16	20	24
BMW7ST_1	9.7	118.9	60.8	43.8	44.1	38.7

[a]Computed on SGI Origin at Parallab, Bergen

Table 5 compares MUMPS with the symmetric code by Gupta [7]. As MUMPS is an unsymmetric solver, the symmetric matrix is expanded into full memory storage format and subsequent operations are performed on this expanded form. Nonetheless, it exhibits a comparable performance with the well-tuned symmetric solver.

The importance of different levels of parallelism is demonstrated on an example of an unsymmetric matrix, WANG3 from the Rutherford-Boeing Sparse Matrix Collection. CPU times for the numerical factorization are shown in Table 6. The times for only using tree parallelism are in the column headed by L1, while L2 denotes that nodes of Type 2 are treated in parallel. With L3, the root node is assembled and factorized in parallel using ScaLAPACK.

The tuning of the code and also the influence of different levels of parallelism can be traced and visualized using VAMPIR, which is a tool developed by one of the PARASOL project partners (PALLAS Gmbh). The two figures in Figure 5 show communication and load balancing, with the vertical lines indicating messages being passed and the dark shaded regions work being performed on processors. Although it

Table 5. Comparison with the Gupta's solver for the test example BMW7ST_1 on 16 processors of the IBM SP2 at GMD (thin nodes with 128 MB of physical memory and 512MB of virtual memory). Times are in seconds.

	Gupta code	MUMPS 2.0
Analysis	500	10.2
Matrix redistribution		11.8
Factorization	22.1	43.8
Triangular solution	1.14	1.6

Table 6. Comparison of different levels of parallelism

No. procs	L1	L1 + L2	L1 + L2 + L3	
			Time	Speed-up
Seq CPU Time	71.0	71.0	71.0	1
2	61.3	89.3		
3	96.6	79.5		
4	46.9	65.7	77.9	
5	49.1	33.3	23.7	3.0
6	46.4	31.9	22.4	3.2
7	45.4	30.1	20.5	3.5
8	44.1	27.9	20.7	3 4

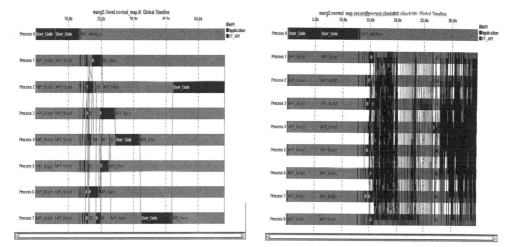

Figure 5. Output from the VAMPIR log of the MUMPS run: with only tree parallelism (L1) in the left figure, and with all levels of parallelism (L1+L2+L3) in the right figure.

may be difficult to see too much from these figures without prior experience with the tool, the poor parallelism from using only tree parallelism is seen in the left-hand figure while the right-hand figure shows more communication due to the L2 and L3 levels, the L3 level being the denser part at the right of the figure. The elapsed time is of course much reduced as can been seen from the times along the top edges of the figures.

REFERENCES

1.   P.R. Amestoy and I.S. Duff, Vectorization of a multiprocessor multifrontal code, *Int. J. of Supercomputer Applics.*, 3:41-59 (1989).
2.   P.R. Amestoy, I.S. Duff, and J.-Y. L'Excellent, *MUMPS MUltifrontal Massively Parallel Solver, Version 2.0*, Technical Report TR/PA/98/02 (1998).
3.   L.S. Blackford, J. Choi, A. Cleary, E. D'Azevedo, J. Demmel, I. Dhillon, J. Dongarra, S. Hammarling, G. Henry, A. Petitet, K. Stanley, D. Walker, and R.C. Whaley, *ScaLAPACK Users' Guide*, SIAM Press (1997).
4.   I.S. Duff and J.K. Reid, The multifrontal solution of indefinite sparse symmetric linear systems, *ACM Trans. Math. Softw.*, 9:302-325 (1983).
5.   I.S. Duff and J.K. Reid, The multifrontal solution of unsymmetric sets of linear systems, *SIAM J. Scientific and Statistical Computing*, 5:633-641 (1984).
6.   A. Supalov (editor), *PARASOL Interface Specification, Version 2.1*, (January 9th 1998).

7.   A. Gupta, M. Joshi, and V. Kumar. *WSSMP: Watson Symmetric Sparse Matrix Package. Users Manual: Version 2.0β*, Technical Report RC 20923 (92669), IBM T.J. Watson Research Centre, P.O. Box 218, Yorktown Heights, NY 10598 (1997).

8.   Message Passing Interface Forum, *MPI: A Message Passing Interface Standard Version 1.1*, Technical report (1995).

COMPUTATIONAL MODELLING OF MULTI-PHYSICS PROCESSES ON HIGH PERFORMANCE PARALLEL COMPUTER SYSTEMS

M. Cross, K. McManus, S. P. Johnson, C. S. Ierotheou, C. Walshaw,
C. Bailey and K.A.Pericleous

Centre for Numerical Modelling and Process Analysis
University of Greenwich
Wellington Street
London SE18 6PF UK.
email : m.cross.gre.ac.uk

INTRODUCTION

The demands of the process of engineering design, particularly for structural integrity, have exploited computational modelling techniques and software tools for decades. Frequently, the shape of structural components or assemblies is determined to optimise the flow distribution or heat transfer characteristics, and to ensure that the structural performance in service is adequate. From the perspective of computational modelling these activities are typically separated into:

- fluid flow and the associated heat transfer analysis (possibly with chemical reactions), based upon Computational Fluid Dynamics (CFD) technology

- structural analysis again possibly with heat transfer, based upon finite element analysis (FEA) techniques.

Until recently, little serious attention has been given to the coupled dynamic fluid-structure interaction problems in the design for operation context. Such problems are conventionally addressed by focusing on one phenomena with the effects of the other represented crudely. The CAE community has tended to focus its attention on either flow or structural mechanics phenomena. From a computational perspective this is not surprising, the Navier Stokes and related equations characterising the flow of fluids have conventionally been solved by finite volume techniques with segregated iterative solvers. Whereas, the stress-strain equations are almost exclusively solved using finite element methods traditionally with direct solvers. Despite the fact that in the last couple of years, a number of workers[1] have shown that the formal mathematical distinctions between finite volume and finite element methods are marginal, these classes of approach have, in fact, led to modelling software tools which are entirely distinct. This is clear from the leading CFD software tools (eg. PHOENICS, FLUENT, CFX4) and solid mechanics products (eg. ANSYS, NASTRAN, ABAQUS). The focus upon their physical domain of interest, and optimisation of their software tools in that context means that the interactions of fluids and solids phenomena are difficult to represent with any degree of detail. Whilst this focus into distinct CFD and FEA software tools has served the needs of much of the design function adequately enough, it will not remain so as design engineers become ever more demanding in the analysis of the performance environment. Moreover, this constraint to represent either fluid or solid behaviour has been less than adequate in the modelling and analysis of manufacturing processes for many years. One key reason for this is that most manufacturing processes involve material exhibiting both "fluid" and "solid" mechanics behaviour. Some processes, such as forging and super-elastic forming, which involve both 'flow' and 'stress' characteristics can be solved by FEA software using a large deformation formulation and sophisticated contact analysis to represent the surface interactions (see, for example, the papers in the NUMIFORM conference series[2]). However, many forming processes involve a change from liquid to solid state via solidification processes. One such process is twin roll casting for the production of metallic thin strip (typically steel or aluminium). Models of this process have either assumed the whole system to be a solid (where the liquid is a very soft solid) and then modelled it using a elasto-viscoplastic large deformation FE analysis[3]. Alternatively,the model has assumed the system to be liquid (where the solid is an extremely viscous liquid) and then modelled it assuming a non-Newtonian plastic approximation based upon FV methods[4].

Neither of these approaches is adequate to represent all the main phenomena present and their interactions. This is also true of the shape casting of metals, plastics moulding, composite shape manufacture and the control of soldering in electronics manufacture. To model such processes adequately requires software tools which facilitate the interactions of a range of physical phenomena, which includes fluid, solids and heat transfer as well as electromagnetic fields. There is also a demand emerging to analyse (what may be more generally called) the multi-physics aspects of the operational performance of engineering equipment in service, an example being aircraft flutter. It is also difficult to solve such mathematically coupled problems by combining phenomena specific software tools, so it can be argued that the next generation of design and manufacturing engineers will need computational modelling software tools which facilitate the analysis of multi-physics processes (i.e. the interactions amongst phenomena).

Finally, given that multi-physics modelling problems generally involve a large number of continuum variables (typically 10–20), a complex unstructured mesh (10^5–10^6 + nodes),a substantial temporal resolution (often 10^4–10^4 time steps) this leads to very large coupled sets of highly non-linear equations. Therefore, the numerical solution is extremely challenging. As such, the multi-physics software tools need to be implemented on high performance parallel computers.

In this paper we will focus upon the following key issues:- the design and implementation of a multi-physics software modelling toolkit; the exploitation of an SPMD paradigm based upon mesh partitioning for the parallelisation of computational mechanics software; the tools to i) aid the process of parallelisation and ii) partition the mesh dynamically at runtime.

KEY ISSUES IN THE DESIGN OF MULTI-PHYSICS MODELLING TOOLS

If modelling tools are to facilitate the analysis of interacting physical phenomena then a single software framework is required. This implies that a single solution strategy has to be employed which exploits the same mesh, formulation approach and style of solution procedures. Conventionally, finite volume (FV) discretisation techniques are used for the highly non-linear flow problems, especially for free surface, two-phase or otherwise physically complex scenarios. Either finite volume or finite element (FE) procedures can be used for heat transfer and electromagnetic processes, but for solid mechanics, finite element methods are used almost exclusively. As such, the choice of discretisation technique (FE or FV), although essential, is not straightforward, as the basis for a single software toolkit for multi-physics modelling. One effort to produce multi-physics tools has been led by Hughes of Stanford University and CENTRIC Corporation. The software product, SPECTRUM[5], is based upon a finite element discretisation procedure and facilitates fluid flow, solid mechanics and heat transfer with reasonable measures of coupling. Another effort, has been underway at Greenwich for some years. The resulting toolkit PHYSICA is based upon finite volume discretisation procedures and enables fluid flow, heat transfer and non-linear solid mechanics with a strong degree of coupling. In the following we briefly outline some of the key features of PHYSICA.

Finite Volume-Unstructured Mesh Context

The main reason for choosing FV procedures over their FE counterparts was because a) they generally involve segregated iterative solvers for the separate variables which are then coupled (very effective for highly non-linear equations) and b) their natural conservation properties at the cell or element level. Given that it is now well established and straightforward to generate a non overlapping control volume for any kind of mesh, then it is useful to exploit unstructured meshes for the accurate representation of geometrical features. Given that the key phenomena of interest can be expressed in a single form

$$\frac{\partial \int_V \rho A \phi dV}{\partial t} = \int_s \Gamma_\phi \nabla \cdot \phi \underline{n} ds + \int_V Q_V dV - \int_s Q_s \underline{n} ds \qquad (1)$$

where Table 1 provides a summary of the main continuum equations describing fluid flow, heat transfer, electromagnetics and solid mechanics, then it should be quite possible to extend established FV methods FV methods on single and multi-block structured meshes are well established for Navier Stokes fluid flow and heat transfer[6,7]. Their extension to turbulent compressible flow on unstructured meshes has been achieved without difficulty[8], as have procedures for solidification/melting phase change[8], free surfaces[10] and magnetohydrodynamic[11] systems. Essentially, whatever has been achieved in a structured context can be extended to unstructured meshes! The key gap in the FV context used to be solid mechanics, however, in the last five years a number of groups have worked on FV procedures for most non-linear solid mechanics problem classes[12-17]. They have been extended straightforwardly to unstructured meshes and are as accurate as their FE equivalents. At this stage we have demonstrated that there are now a range of FV-UM solution procedures that solve fluid flow, heat transfer, electromagnetics and solid mechanics processes. These procedures provide the basis for the design and implementation of a multi-physics modelling software tool.

Design of the multi-physics modelling software framework, PHYSICA

The core of PHYSICA is a three-dimensional code structure which provides an unstructured mesh framework for the solution of any set of coupled partial differential equations up to second order[18,19]. The design concept is as object oriented as possible,

Table 1. The key arguments for the generic partial differential equations characterising continuum phenomena.

Phenomenon	ϕ	A	Γ_ϕ	Q_V	Q_s
Continuity	1	1	0	S_{mass}	$\rho\underline{v}$
Velocity	$\underline{v}$	1	$\Gamma_{\underline{v}}$	$(S+\underline{J}\times\underline{B}-\nabla p)$	$\rho\underline{v}.\underline{v}$
Heat Transfer	h	1	$\frac{k}{c}$	S_h	$\rho\underline{v}h$
Electromagnetic Field	$\underline{B}$	1	η	$(\underline{B}\nabla)\underline{v}$	$\underline{u}.\underline{B}$
Solid Mechanics	$\underline{u}$	$\frac{\partial}{\partial t}$	μ	$\rho\underline{f_b}$	$\mu(\nabla\cdot\underline{u})^T + \lambda(\nabla\underline{u}-(2\mu+3\lambda)\alpha T)\underline{I}$

within the constraints of FORTRAN77. The challenge has been to build a multi-level toolkit which enables the modeller to focus upon the high level process of model implementation and assessment, and to simultaneously exert maximum direct control over all aspects of the numerical discretisation and solution procedures.

The object orientation is essentially achieved through the concept of the mesh as constructed of a hierarchy of objects - nodes, edges, faces, volumes which comprise the mesh, see Figure 1. Once the memory manager has been designed as an object hierarchy, then all other aspects of the discretisation and solution procedures can be related to these objects. This enables the software to be structured in a highly modular fashion, and leads to four levels of abstraction. The **Model** level - where the User implements the multi-physics models; the **Control** level - which provides a generic equation (for exploitation by the User) and solution control strategies; the **Algorithm** level - a whole set of tools for discretisation, interpolation, source construction, managing convection and diffusion, properties, system matrix construction, linear solvers, etc; the **Utility** level - file input-output tools for interaction with CAD software, memory manager, database manager, etc.

With the abstraction framework it is quite possible to implement discretisation and solution procedures to analyse distinct continuum phenomena in a consistent, compatible manner that particularly facilitates interactions. The initial version of PHYSICA has a) tetrahedral, wedge and hexahedral cell/element shapes, b) full adaptivity implemented in the data structures which are consistent for refinement/coarsening and c) a range of linear solvers. It has the following core models:

- single phase transient compressible Navier-Stokes flow with a variety of turbulence models

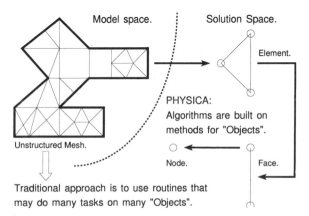

Figure 1. PHYSICA – Mesh abstraction to yield object oriented design approach

- convection-conduction heat transfer with solidification-phase change and simple reaction kinetics

- elastoviscoplastic solid mechanics

and their interactions. Work is currently at an advanced stage to include free surface flow, contact mechanics and electromagnetics. The PHYSICA toolkit outlined above and its prototypes are currently being used in the modelling of a wide range of engineering processes, including shape casting, twin roll casting, flutter and soldering.

THE STRATEGIES AND TOOLS USED TO EXPLOIT HIGH PERFORMANCE PARALLEL COMPUTERS

In the last decade there has been a substantial amount of research on strategies and tools for the parallelisation of computational mechanics (CM) codes. Much of this work has been charted by a number of conference series (see, for example, the SIAM Parallel Processing for Scientific Computing[20], Parallel CFD[21] and Domain Decomposition Proceedings series[22]). The predominant paradigm for the parallelisation of CM codes is known as Single Program Multiple Data (SPMD), typically based upon the partition of the geometric mesh. Here, whilst every processor essentially runs the same program, each one only operates on the data associated with its component of the partitioned mesh. On this basis, an unstructured CM code is parallelised using a number of key steps. These include **partitioning** the mesh into a number of submeshes which have the same number of nodes/elements (assuming, of course, the work done at each node is similar), in such a way as to minimise the number of cells along the interfaces between the submeshes. Modify the CM code so that it only declares a **reduced array space** for the data it operates on plus the halo/overlap nodes/cells at submesh interfaces. The insertion of **masks** to ensure functions only operate on the processor where the appropriate data are stored. The introduction of **synchronisation points** and **communication calls** whereby the program on each processor stops and exchanges, with its neighbours new data to be stored in the halo surfaces. In fact, the parallelisation is carried out in a generic fashion so that the number of processors and the mesh are both specified and the latter partitioned at run-time. Finally, the target code parallelisation task itself is performed by hand ie. there are no effective parallelising compilers. The software engineer carries out the above tasks and enables the data/control exchange through the use of message passing calls, typically, provided by PVM, MPI or machine specific libraries.

Mesh Partitioning Tools

There are now a range of techniques and tools for mesh partitioning. They are all based on the representation of the mesh by a graph, its reduction, an initial crude partition followed by optimisation heuristics to yield a high quality partition. At Greenwich we have developed a procedure, embedded in the software tool, JOSTLE, which is based upon a number of techniques[23]. These include a graph reduction technique to reduce the problem size; use of the Greedy algorithm to give a very rapid but crude initial partition; optimisation algorithms to ensure a minimum length interface and the submesh connection topology matches that of the parallel processors; recovery of the full graph subsequently optimised.

Recently, this technique has been extended to cope with dynamic load balancing and has been implemented in parallel[28]. JOSTLE is an order of magnitude faster than other commonly used methods, such as MRSB, for static partitions. In addition, JOSTLE maps the mesh partition onto the processor topology, dynamically modifies the partition as dictated by the physics with a computational cost that is about 10% of the initial partition and a movement of less than 1% of the total data. Since JOSTLE can be executed in parallel, once the data has been partitioned then the movement of data across the processor network is marginal.

Some parallel performance results

In Figure 2 we show the partitions of a 2D mesh with its mapping onto various topologies[26]. This mesh was used in a 2D prototype of PHYSICA to solve a problem involving a solidifying domain involving flow, heat transfer and solid mechanics simultaneously.

This was run in parallel on a TRANSTECH system where each compute node has an i860XP processor with 16/32 Mb of RAM plus a T805 transputer to facilitate the interprocessor communications.Figure 3(a) shows the speed-up results for a 60000 triangular mesh for a variety of mappings onto the processor topology, but all using synchronous communications. Notice that when the communications are made asynchronous where possible, the parallel efficiency rises to nearly 90% (see Figure 3(b)). The PHYSICA code has been configured to model the shape casting of metals[26]. In Figure 4 we show some results from this multi-physics model showing the residual convection in the liquid before solidification, the von Mises stress distribution when the shape is solid and its deformation by illustrating the gap formation along the centre section. This code has been tested on a number of parallel systems including a CRAY T3D, an IBM-SP2 and the DEC 4100 cluster at the University of Greenwich[27]. Obviously, the mesh was partitioned using the JOSTLE software. The speed-up curves for the CRAY T3D in Figure 5 show how the use of differing communications on the same platform can affect performance. Although the IBM SP2 speed-up appears poor the run times are better than the CRAY T3D as the individual node performance is faster. Clearly however, the IBM SP2 will not scale to larger numbers of processors as

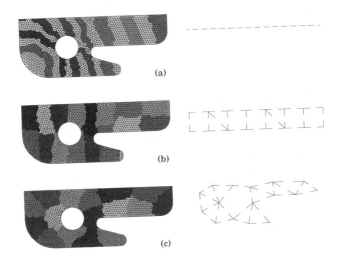

(a)

(b)

(c)

Figure 2. Partitions of a 2D mesh in (a) 1D pipeline (b) 2D grid and (c) a uniform topology with corresponding submesh connectivity graphs

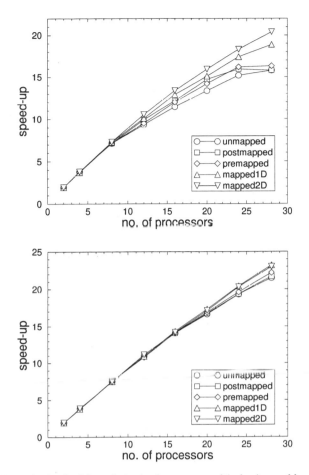

Figure 3. Speed-up obtained with optimised solvers on a multi-physics problem with a range of partition strategies using a 6000S triangle mesh with (a) synchronous and (b) asynchronous communications

the speed-up is less than ideal for only 14 processors. Although it is hard to predict the scalability of the DEC 4100 cluster beyond the available 12 nodes, the run time performance of both the CRAY T3D and the IBM SP2 is eclipsed by the DEC 4100 cluster. This comparison illustrates how far hardware has developed over a period of around 2 years.

Automatic Parallelisation Tools

Tools to facilitate interprocessor communications are now well established. However, whilst the user of parallel architectures might expect that compiler technology should be able to compile a standard FORTRAN code onto a parallel system there are many reasons why this is not the case. Practically, the best that can be hoped for are pre-processing tools that will restructure the scalar FORTRAN code as described above so that it may be compiled by standard compilers together with the interprocessor communication calls. Such pre-processing tools are available for optimising superscalar compilers. However, the task of pre-processing tools to aid the parallelisation of FORTRAN CM codes is really very complex. One programme of research to develop tools to automate the transformation of scalar FORTRAN codes into parallel form has been

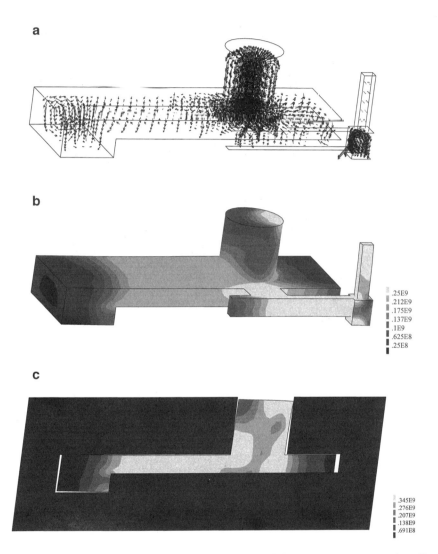

a

b

.25E9
.212E9
.175E9
.137E9
.1E9
.625E8
.25E8

c

.345E9
.276E9
.207E9
.138E9
.691E8

Figure 4. Some results from the multi-physics shape casting model. **a.** Thermally induced flow distribution in the melt after 30 secs cooling **b.** von Mises stress profile from the deformation analysis **c.** profile of the gap formation along the centre section of the casting.

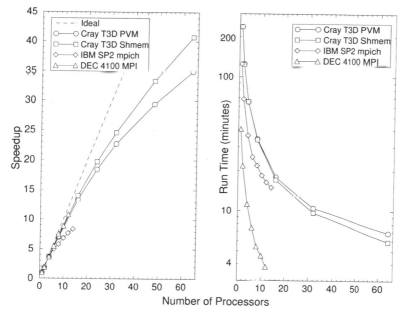

Figure 5. Parallel speed-up and run-time measurements for PHYSICA on a range of systems

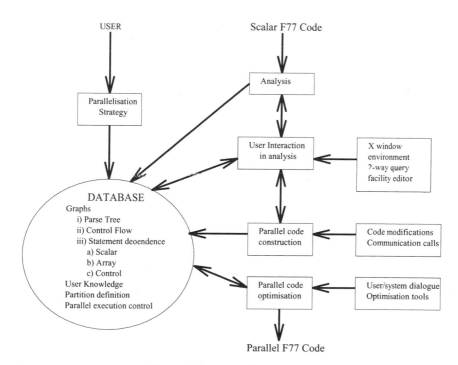

Figure 6. Structure of CAPTools.

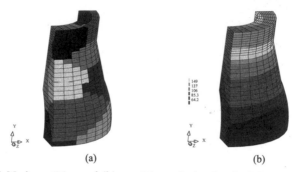

(a)   (b)

Figure 7. (a) Mesh partition and (b) von Mises solution for the FE structural analysis code parallelised automatically

underway at Greenwich for some years. The results of this research have been embedded into the software toolkit, CAPTools, to automate the parallelisation of mesh based (ie CM) FORTRAN codes[28]. The structure of CAPTools is illustrated in Figure 6. At the heart of the toolkit is a very sophisticated, accurate and comprehensive dependence analysis[29]. This is used to identify which loops and sections of code in subroutines are essentially scalar. The analysis is interprocedural so that dependencies can be identified throughout the code. An essential component of the parallelisation process is the user interaction phase. No matter how sophisticated the analysis, there will be code fragments where it is not possible to decide whether or not they are scalar without extra information from the user or code author. Moreover, it may be that some simple code fragment transformations could make a significantly greater proportion of the code parallelisable. Hence, some automatic analysis and code transformation could be beneficial. Once the code has been thoroughly analysed then the parallelisation of the code (ie. changing loop limits, reducing array storage, installing masks, and communications, etc), can proceed automatically. The initial version of CAPTools was limited to structured mesh codes. However, it has recently been extended to cope with unstructured mesh codes[30]. Figure 7 shows the partition (using JOSTLE) of a 3D finite element code for elastic structural analysis using hexahedral elements where the linear equation solver employs a preconditioned conjugate gradient technique. Figure

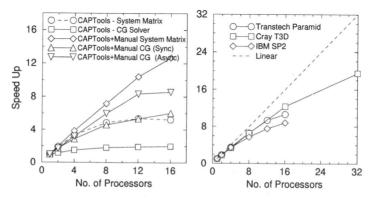

Figure 8. Parallel results (a) on the TRANSTECH system with automatic parallelisation and hand optimation plus (b) the optimised code speed-ups on a range of parallel systems

8 shows the performance of the FE code which has been automatically parallelised by CAPTools on a variety of parallel systems. The parallel performance is as good as a full manual parallelisation, when a few minor hand optimisations have been included. The key issue here is that the time complete the parallelisation process using CAPTools is a fraction of that required to complete the task by hand.

CONCLUSIONS

It is asserted that CM software tools of the next generation will need to focus upon interacting phenomena (ie. multi-physics). Because the modelling of such multi physics processes makes huge computational demands, then the emerging CM software tools will need to be ported to parallel machines to enable simulation runs to be completed in a reasonable time. In this contribution we have outlined the development of the multi-physics software framework, PHYSICA, which has been designed and implemented in parallel with high efficiencies on state-of-the-art parallel computer systems. One of the key tools which has facilitated the parallel performance of PHYSICA, is the mesh partitioning tool JOSTLE. This provides high quality mesh partitions very rapidly that a) map onto the processor topology to retain nearest neighbour communications and b) enable dynamic load balancing in parallel. The other tool of significance is CAPTools which automates much of the parallelisation process and yields parallel versions of CM codes that have very high parallel efficiencies, in a fraction of the time it would take to achieve manually. As such, the computational modelling of multi-physics processes on high performance parallel computers is not just viable - it in the future!

REFERENCES

1. S R Idelsohn and E Onate, Int J Numerical Methods in Engg, 37, 3323-3341 (1994).
2. NUMIFORM series - Numerical Methods in Industrial Forming Processes, Balkema, Netherlands, (1982-1995).
3. P J Jarrey et al, Light Metals, 1, 905-911 (1996).
4. P J Bradbury and J D Hunt, in Modeling of Casting, Welding and Advanced Solidification Processes VII (Eds M Cross and J Campbell) pub TMS, 739-746 (1995).
5. SPECTRUM, Centric Corporation, Palo Alto, CA, USA.
6. S V Patankar, Numerical Heat Transfer and Fluid Flow, Hemisphere, Washington, USA (1980).
7. H K Versteeg and W Malalasekera, An Introduction to Computational Fluid Dynamics - The Finite Volume Method, Longman, Harlow, UK (1995).
8. N Croft, K Pericleous and M Cross, in Numerical Methods in Laminar and Turbulent Flow (Eds C Taylor et al) Vol IX, Pineridge Press, Swansea, UK, 1269-1280 (1995).
9. P Chow and M Cross, Int J Num Methods in Engg, 35, 1849-1870 (1992).
10. K Pericleous, M Cross, G Moran, P Chow and K S Chan, Adv Comp Meths (in press, 1997).
11. K Pericleous, M Hughes, M Cross and D Cook, in Modeling of Casting, Welding and Advanced Solidification Processes VII (Eds M Cross and J Campbell), TMS, Pa, USA, 213-221 (1995).
12. Y D Fryer, C Bailey, M Cross and C-H Lai, App Math Modelling, 15, 639-645 (1991)
13. C Bailey and M Cross, Int J Num Methods in Engg, 38, 1757-1776 (1995).
14. G A Taylor, C Bailey and M Cross, Appl Math Modelling, 19, 746-770 (1995).
15. I Demirdzic and D Martinovic, Comp Meth Appl Mech Eng, 109, 331-349 (1993).
16. J H Hattel and P N Hanen, Appl Math Modelling, 19, 210-243 (1995.
17. E Onate et al, Int J Num Methods Engg, 37, 181-201 (1994).
18. M Cross, IMA J Math Appl Bus Ind, 7, 3-21 (1995).
19. M Cross, Jnl Comp Aided Materials Design, 3, 100-116 (1996).
20. Parallel Processing for Scientific Computing Proceedings, SIAM, Pa, USA.
21. Parallel Computational Fluid Dynamics Proceedings, North-Holland, Amsterdam, Netherlands.
22. Domain Decomposition Methods for Partial Differential Equations Proceedings, SIAM, Pa, USA.
23. C Walshaw, M Cross and M G Everett, Int J Supercomput Appl, 9, 280-295 (1995).

24. C Walshaw, M Cross and M G Everett, in Parallel Processing for Scientific Computing (Eds M Heath et al), SIAM, Pa (in press 1997).
25. K McManus, C Walshaw, M Cross, P Leggett and S Johnson, in Parallel Computational Fluid Dynamics - Implementation and Results using Parallel Computers (Eds A Ecer et al), North-Holland, Netherlands, 673-680 (1996)
26. P Chow, C Bailey, M Cross and K Pericleous, in Modeling of Casting, Welding and Advanced Solidification Processes VII (Eds M Cross and J Campbell) TMS, Pa, USA 213-222 (1995).
27. K McManus, S Johnson, P Leggett and M Cross, in Proceedings of HPCN'97 (in press).
28. C S Ierotheou, S Johnson, M Cross and P Leggett, Parallel Computing, 22, 163-195 (1996).
29. S Johnson, M Cross and M G Everett, Parallel Computing, 22, 197-226 (1996).
30. S Johnson, C S Ierotheou and M Cross, Computer Aided Parallelisation of Unstructured Mesh Codes,. University of Greenwich Report 15pp (1996).

PORTING INDUSTRIAL CODES TO MPP SYSTEMS USING HPF

L.M. Delves

N.A Software Ltd., 62 Roscoe Street, Liverpool L1 9DW UK
E-mail:*delves@nasoftware.co.uk*

ABSTRACT

There has been substantial interest in HPF as a high level parallel programming language for scientific and engineering codes; but as yet, little experience of its use for production work. Within the Esprit Pharos project, four industrial simulation codes were ported to HPF. We describe here the porting process, and results obtained on the Meiko CS2 and the IBM SP2.

INTRODUCTION

High Performance Fortran

HPF is a de facto standard set of extensions to Fortran95 intended to provide a high level programming language for scientific and engineering applications which can be efficiently compiled for both shared and distributed memory parallel architectures. By shielding the programmer from the architectural details, HPF makes it much easier to write, and to maintain, parallel codes; and to provide a single source for both serial and parallel architectures.

The first version of the HPF standard, HPF1, was completed in 1993 [1]; an extended version, HPF2, in 1997 [2]. Interest has been very substantial; and HPF compilers are now available from several vendors including DEC, IBM, PGI and NASL for a wide range of systems.

But as with all new languages, the use of HPF for production as opposed to experimentation, has been limited to date. Available compilers have implemented only (differing) subsets of the language; and there has been (healthy) scepticism as to their runtime efficiency, as well as (unjustified) scepticism about the capability of HPF to describe parallel constructs sufficiently fluently.

However, compilers have been developing fast. In 1995 the ESPRIT Pharos project was planned, to test the capabilities of the language and a current compiler (ours) to handle industrial codes. In this paper, we report on the results obtained.

The Pharos Project

Pharos was an Esprit–funded project aimed at:

- Bringing together an HPF-oriented toolset from three vendors;

- Using this toolset to develop HPF versions of four industrial–strength codes owned by project partners.

Partners in Pharos included:

Tools Vendors :

N.A. Software . HPF Compiler and source level debugger (*HPFPlus*);

Simulog : Fortran77 to Fortran90 conversion tool (*FORESYS*);

Pallas : Performance and communications monitor (*VAMPIR*).

Code Owners :

SEMCAP : Electromagnetic Simulation code (*SEMC3D*);

debis : Linear Elastostatics code (*DBETSY3D*).

CISE : Finite Element Structural code (*CANT-SD*);

Matra BAe Dynamics : Compressible Flow code (*Aerolog*);

HPF Experts :

GMD : developers of the ADAPTOR research compiler

University of Vienna : developers of Vienna Fortran, a precursor to HPF; and of the Vienna Fortran Compilation System implementing it and (now) HPF.

The project ran over the period January 1996 – December 1997, and produced HPF ports of all four codes. We describe here the porting process, and outline the results obtained

THE PHAROS CODES

The *SEMC3D* Code

The SEMC3D package is an Electromagnetic code developed to simulate EMC (ElectroMagnetic Compatibility) phenomena. It has a wide range of applications. For example, in the automotive industry electric and electronic components and systems must comply with increasingly tougher regulations. They must resist hostile electromagnetic environments and show reduced emissions.

The code simulates the electromagnetic environment of complex structures made up of metal, dielectric, ferrite and multilayers or human tissues. These structures could be surrounded with ionised gas, laid on a realistic or perfect ground, or with antennas or cables. The electromagnetic environment should be taken as the knowledge of electric and magnetic fields inside and outside the analyzed structures. Many source models can be used: NEMP (Nuclear ElectroMagnetic Pulse), plane wave, current injection, lightning and all analytical and digital signals.

The original Fortran 77 source code comprises approximately 3500 lines. The kernel of SEMC3D code is built around the Leap-Frog scheme in time and space. The Leap-Frog scheme uses only the components located within semi space-time away with regard to the updated component.

The code is computationally demanding, and a message–passing version was developed in the early 90's. This used a domain decomposition with overlap. At each step of the computation, the calculations are made in each domain and then the quantities on the boundaries of domains are updated. This is a coarse grain parallelisation. A major advantage of HPF is its intrinsic ability to provide a fine grained parallelisation with modest development effort; and this is what we sought to produce.

The DBetsy-3D Code

DBETSY is used extensively by Mercedes Benz to calculate stresses and deformations on the surfaces of objects. It uses the theory of linear elastostatics, solving the large dense linear systems involved using the residual matrix/frontal technique: using forward substitution, each substructure matrix is reduced to the degrees of freedom it shares with other substructures. The residual matrix is then formed and solved directly, and finally backwards substitution on each substructure is used to recover the complete solution.

Even with this technique it is not possible to store a complete substructure matrix in core. Hence, i/o to backing store is significant.

The CANT-SD Code

CANT-SD is a finite element code for static and dynamic analysis of 3D structures, and is used extensively by CISE under contract to ENEL-CRIS, mainly to perform seismic analysis of dams.

Most of the computational effort is spent in the solvers. There are two solvers in the package, for symmetric and for non-symmetric problems; because of the size of problems handled, both are out-of-core and like DBetsy-3D use a direct frontal method. Thus, the complete matrix is never constructed; but i/o can be significant.

Within Pharos, only the symmetric solver was parallelised; this as is usual with symmetric solvers provides more difficulties than the corresponding unsymmetric solver because of the less regular data structures used in the serial version to economise on storage.

A message–passing version of the code exists which implements coarse grain domain decomposition but provides little speedup. However, CISE is interested less in speedup than in memory availability and the ability to run larger problems than is possible on a single processor. Thus, their aim in developing an HPF version was to retain this capability without significant efficiency penalties, while obtaining a code version which is easier to maintain and develop further than the message–passing code.

The Aerolog Code

Aerolog is a CFD code developed by Matra BAe Defense and devoted to the study of compressible fluid flows around complex geometries. It has a modular implementation to allow easy use of various fluid models, including Inviscid Euler; viscous Navier–Stokes; turbulence modelling; etc.

The numerical solver is based on a multi–domain approach where the global mesh is composed of an assembly of locally structured mesh blocks. The basic numerical scheme is a second order explicit finite volume scheme with convergence acceleration techniques including a time implicit scheme and a multigrid scheme.

The version ported within Pharos includes approximately 11500 lines of code, and implements only the Euler inviscid fluid model with explicit solver (55 subroutines). A coarse grain parallelisation was sought for HPF, using a domain decomposition. The major difficulty was expected to be that provided by the boundary condition treatment, which makes heavy use of indirect addressing.

OUTLINE PORTING PROCEDURE

The porting procedure had the same steps for each code; these steps are those foreseen as appropriate at project start:

Code Cleaning : The original Fortran77 code was tidied up to remove language features which were either:

- non-standard Fortran77 (some of these remain in many codes either as legacies from earlier Fortran versions or because they have been widely supported by compilers);

- or can be replaced even within Fortran77 by better constructs.

This stage proved to be well-supported by the Simulog *FORESYS* tool. However, some work was necessary by hand.

Translation to Fortran90 : This provides a better base for HPF. The *FOREST90* tool (part of the *FORESYS* suite) proved to give good support for this, providing an automatic translation which at the least formed a tidy basis for further work.

Improving the Fortran90 version : to make the source better suited to HPF by utilising array syntax where possible, and by removing further features (storage association, sequence association, assumed size and explicit shape array arguments) which cannot directly be parallelised in HPF. Static storage allocation has been replaced by dynamic storage allocation, as part of this process, and COMMON blocks replaced by the use of modules. All array arguments to procedures were declared assumed shape, and the INTENT of all arguments declared.

This stage made use of the FORESYS vectorising tool; but much of the work had to be done manually.

Inserting HPF Directives : was the final stage. It involved also some further code restructuring:

- loops were restructured to be INDEPENDENT or to use FORALL;

- Procedures were marked as HPF_SERIAL where appropriate; see later.

- Some i/o statements were moved to improve parallelisation.

- Parallel i/o was implemented for those codes needing it.

- For the Aerolog code, quite extensive restructuring of the boundary condition treatment was undertaken to overcome inefficiencies stemming from the extensive use of indirect addressing; this restructuring made use of the HPF_LOCAL construct.

Major Parallel Constructs Used

The major sources of parallelism in the HPF codes are:

Use of Array Expressions : Many of the Fortran77 DO loops in the codes translate naturally to Fortran90 array assignments (and *FORESYS* carries out many of these translations automatically). These are very efficiently handled by the NASL *HPFPlus* compiler.

FORALL and INDEPENDENT DO loops : Computationally expensive code sections which can not readily be expressed using array syntax, often appear in loops in which the iterations may be carried out in any order (sometimes after loop restructuring). These loops can be labelled INDEPENDENT in HPF, or (in suitable cases) replaced by FORALL statements or blocks. Our compiler implements FORALL very efficiently, but currently ignores the INDEPENDENT directive. However, a very efficient work-around exists: INDEPENDENT loop bodies were wrapped in HPF_SERIAL procedures. Within a loop, the NASL *HPFPlus* compiler places the calls intelligently by analysing the processor residency of the actual arguments, and the resulting loops run in parallel with very low overheads.

Note that this approach can be used for both fine grain and coarse grain parallelism; in this application the granularity is low.

Parallel I/O : for the CANT-SD and DBetsy-3D codes, i/o to backing store can dominate. HPF does not define standard facilities for parallel i/o, because hardware/system support varies greatly and there is no agreed portable approach. However, the IBM SP2 supports a single filestore across the whole machine, and on this system parallel i/o can be implemented straightforwardly within an HPF program by using HPF_SERIAL or HPF_LOCAL routines.

HPF_LOCAL : This construct provides a low–level paradigm within HPF in which each processor sees only its local data: HPF_LOCAL procedures, like HPF_SERIAL procedures, therefore run very efficiently.

One major problem found in porting codes, is the use of indirect addressing, which is extremely hard for an HPF compiler to optimise when distributed arrays are involved (historically, this has always been hard for vector processing too). We used HPF_LOCAL procedures to handle the indirect addressing associated with the boundary conditions in the Aerolog code.

In other places we found the restructuring involved was not necessary. Some boundary conditions used in SEMC3D can only be handled using indirection. In the Fortran77 version, these conditions are located with an integer pointer array. In the initial translation we replicated these arrays on all processors (giving satisfactory serial efficiency but no speedup for that part of the code), but at a later stage we recoded the one dimensional pointer array to reflect the 3-D nature of the problem, and used a masked WHERE to retain parallelism:

```
integer  iptstr(1:nstr)
do i=1,nstr
   e(ipstr(i)) = 0
enddo
```

became

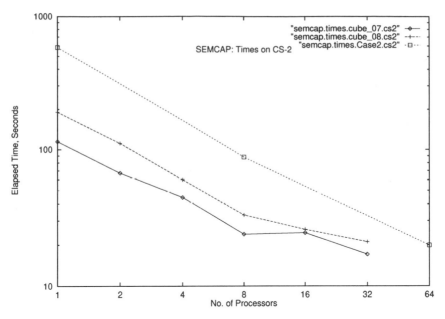

Figure 1: SEMC3D Benchmarks on the Meiko CS2

```
      real,    dimension(1:,1:,1:), intent(inout) :: e
      logical, dimension(1:,1:,1:), intent(in)    :: iptstr
!HPF$ DISTRIBUTE e*(BLOCK,BLOCK,BLOCK)
!HPF$ ALIGN       iptstr(I,J,K) WITH *e(I,J,K)
      WHERE(iptstr(:,:,:))
        e(:,:,:) = 0
      ENDWHERE
```

BENCHMARKS

SEMC3D

We give benchmarks on the Vienna Meiko CS2 for this code, using two datasets:

cube_09 : An 80*80*80 test cube, with primary data requirements around 20MB. This provides a test problem small enough to be run on a single processor of the Meiko CS2, which has only limited storage (48MB user–accessible per processor).

CAR_1,2 : Industrial models of a complete car, with much larger primary data requirements. These represent typical production problems for the code.

Figure 1 gives the benchmarked results for the cube_09 dataset, for the HPF and for the message-passing (MPI) versions of the code. The speedups achieved are very good; the code clearly scales well. However, it takes between three and four processors to overcome the poor single node performance of Fortran90 on the CS2, for the model problems.

The MPI times are roughly a factor 3 better than those achieved with HPF; this is wholly attributable to the lack of optimisation of the backend Fortran90 compiler.

Figure 2 gives results for the two CAR benchmarks. These could not be run on the CS2 using the serial code. The results for CAR2 show superlinear speedup initially, due to thrashing.

108

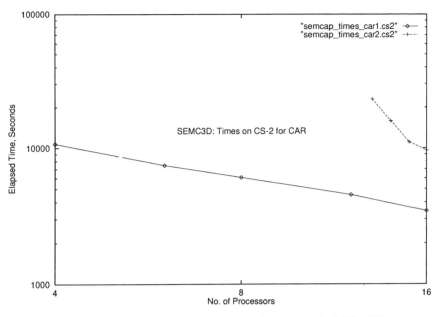

Figure 2. Figure 2: SEMC3D CAR Benchmarks on the Meiko CS2

DBetsy-3D

The target architectures for DBetsy-3D include a small IBM SP2 and a workstation cluster. Such systems are well suited to the coarse grain port carried out, in which the parallelism is limited by the number of subdomains into which the datasets are split.

In production use, the subdomains are chosen for geometric convenience, and (depending on the structure being modelled) there are likely to be up to 10 subdomains, but these may be of very different sizes. Hence, although scalability to many processors is not an issue (4-8 processors is the target system size) load balancing is of some importance.

In the HPF code, the subdomains are farmed out via the use of a loop with HPF_SERIAL body; each iteration runs on a processor defined by the ownership of the arguments passed to the hpf_serial routine. Load balancing can be achieved by aligning these arguments appropriately; in the implementation produced, this was automatically achieved from a knowledge of the size of each subdomain.

We benchmarked the code using two commercially significant models:

KWREN : 10 subdomains, with data sizes ranging from 0.3 to 14.1 MBytes

LEVER : 6 subdomains, with data sizes ranging from 5 to 54 MBytes.

Figure 3 shows the results obtained (the times shown for LEVER have been divided by 10 for convenience in plotting). On the SP2, there is little difference between the serial F77 and F90 times. The overheads on one processor due to the use of HPF are also very low (negligible for the KWREN dataset; less than 15% for the LEVER dataset). The speedup achieved is very satisfactory, being as high as can be expected given the existence of relatively small subdomains; the effectiveness of the load-balancing is demonstrated by the smooth decrease in time taken as the number of processors is increased.

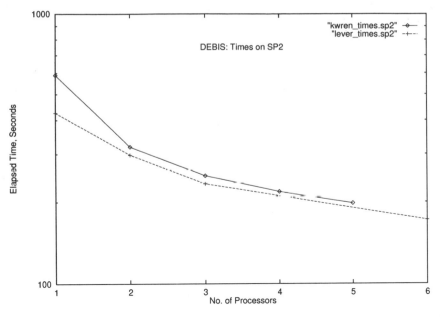

Figure 3. Figure 3: DBETSY-3D Benchmarks on the IBM SP2

CANT-SD

As noted above, the primary aim in the port of CANT-SD was to demonstrate that HPF provided easy access to the running of problems larger than was practicable on a single processor: memory, rather than processor, utilisation was most important. However, we certainly hoped to see some speedup.

CISE has a four–processor IBM SP2, so we benchmarked the code on the GMD SP2 using a test case representing a dam with its foundation, and comprising five substructures: 3 for the dam, and 2 for the foundations. There are 37692 variables in this test case, with a maximum frontal matrix size of 1905.

The benchmarking showed that (on both the SP2 and the CS2) there was essentially no difference between F90 and F77 serial times; and that HPF on one processor ran around 10% faster than the serial F90 code *. The parallel times are shown in Figure 4. We see that modest speedups are achieved; and certainly, with our compiler it is possible to access the memory on a large number of processors without disastrous loss of efficiency.

Aerolog

The Aerolog port was similar to that for DBetsy-3D in depending on coarse grain parallelism expressed through substructuring. However, it presented more difficulties because of the irregular nature of the boundary condition treatment. Handling these involved substantial restructuring of the boundary condition parts of the code, and the use of HPF_LOCAL to retain efficiency in the indirect addressing. The restructuring was carried out by Thomas Brandes and Falk Zimmermann of GMD.

Figure 5 shows the results obtained on the IBM SP2

---

*We observe this quite often with our HPFPlus compiler, on codes which make good use of array syntax

110

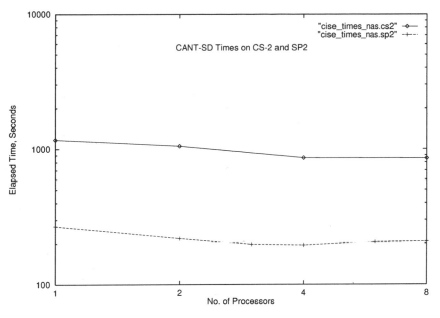

Figure 4. Figure 4: CANT-SD Benchmarks on the Meiko CS2

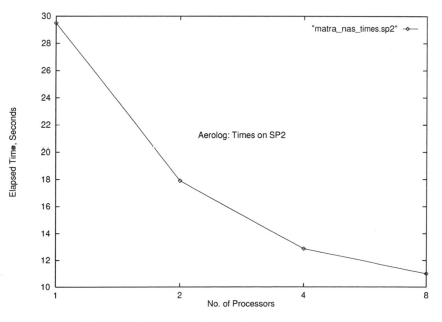

Figure 5. Figure 5: Aerolog Benchmarks on the IBM SP2

ACKNOWLEDGEMENTS

This work was supported in part via Esprit Project No P20162, Pharos. The work of producing HPF versions of the codes was carried out primarily by the code owners and by staff of GMD and the University of Vienna. We are indebted to our colleagues in this project for their work and for many helpful discussions, and aid in running on a variety of systems. Especial thanks are due to Christian Borel, Thomas Brandes, Roberto Guandalini, Kadri Krause, Junxian Liu, Ruth Lovely, John Merlin, Dave Watson and Falk Zimmermann.

REFERENCES

1.   High Performance Fortran, language definition published in *Scientific Programming* (1993)
2.   *High Performance Fortran Language Specification*, HPF Forum (August 1996)

DECOMPOSITION INDEPENDENCE IN PARALLEL PROGRAMS

S. Booth

EPCC
The University of Edinburgh
Edinburgh
EH9 3JZ

INTRODUCTION

It is very common for parallel programs to give slightly different results when running on different numbers of processors or even when using the same number of processors but different data decompositions. Though this variation is usually not at a significant level it can mask programming errors by providing an alternative explanation for varying results. The obvious methods for solving this problem rely on performing the affected sections of the code in serial. This approach may require very large amounts of memory on the processor performing the serial calculation and may significantly extend the run-time of the program to an extent that is unacceptable even for debugging purposes.

The purpose of this report is to present algorithms that will give bit-identical results on different processor decompositions while still scaling in terms of processor time and memory use per processor.

We examine two common operations:

1. Random number generation

2. Global summation

These two operations are sufficient to remove all decomposition dependence from many problems that use regular domain decomposition. We present results for the UKQCD collaboration's Generalised Hybrid Monte-Carlo simulation code (GHMC) on the 512 processor Cray T3D at EPCC.

GHMC

The GHMC algorithm is a Monte-Carlo algorithm that uses a modified form of the Metropolis algorithm to perform a global Monte-Carlo update. Each configuration variable is treated as a generalised coordinate of a classical mechanical system. Each

MC step consists of initialising this system with random "momenta"; iterating the dynamics forward for a fixed period of "time" using a time-reversible integration scheme; accepting or rejecting this new trial configuration depending on the change in energy.

In the code under consideration the variables are arranged in a 4-dimensional regular lattice that is distributed over a 4-dimensional grid of processors using domain decomposition. Each dimension of the lattice is decomposed separately, though the size of local hyper-cubic sub-lattice need not be the same on all processors.

This code requires two types of random numbers:

1. Distributed random numbers generated in large batches (proportional to the number of lattice sites). These are used to initialise the random momenta.

2. Single global random numbers for the global accept reject step.

Global summation operations are required as part of a conjugate gradient iterative solver used in each time-step and to sum the energies associated with each lattice site for the accept/reject step.

ALGORITHMS

In both cases we first solve the much simpler problem of developing a scalable and decomposition independent algorithm for 1-dimensional block data decompositions. The general problem is then solved by adding communication steps to re-map arbitrary data decompositions to and from this form.

Random Number Generation

The normal solution to implementing a random number generator (RNG) in parallel is to take a standard sequential algorithm and replicate it across processors. Each processor is initialised to a different starting state and each processor then uses the local generator to produce any random numbers it requires. Each processor therefore generates its own sequence of random numbers. For a given set of starting states, different data decompositions will result in different mappings between these sequences and the random numbers used by the application, which in turn gives rise to decomposition dependent results.

We take the approach of taking a single RNG sequence (equivalent to the output of a single sequential generator) and decompose generation of this sequence across processors. To implement this in parallel we require the RNG sequence to be divided into batches. Each batch is generated in parallel as a collective operation with each processor generating a different subset of the sequence.

It is essential to have some mechanism to efficiently advance the RNG algorithm for long distances in its sequence (*stepping* the generator). If the generator is based on a linear update rule it is usually possible to construct linear operations that step the generator n steps in constant time with the construction of the operator taking $log(n)$ time[1].

There are two possible methods of generating the batch in parallel:

1. Generate numbers on the processors that require them.

   Those parts of the sequence not required by the local processor are stepped over. This does not require any inter-processor communication though it may require a large number of stepping operations with some data decompositions.

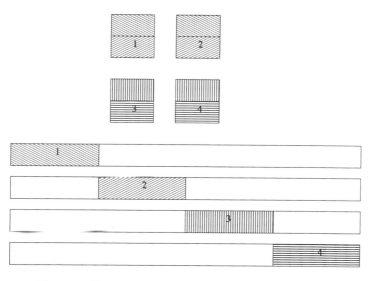

Figure 1. Decomposing the RNG sequence across processors

2. Split the batch into N consecutive strips and generate one on each processor (see figure 1.

   This only requires one stepping operation per processor (two if we wish to re-synchronise the generators) but inter-processor communication is required to re-distribute the results to the correct processor.

The T3D has a very good communication system so we use the second approach. A second stepping operations is used to re-synchronise the RNG states after each batch is generated. This allows the global random numbers to be generated with the normal RNG calls. Both stepping operators are saved to avoid the need to recalculate the same stepping operator if the same batch size is used multiple times. As can be seen from figure 2 this algorithm scales reasonably well with a moderate performance reduction compared to the replicated generator case.

The communication phase is required to map the random data from the 1-D block decomposition to the decomposition required by the application. Normally this would require the communication phase to be re-implemented for each application.

The single-sided communication model of the SHMEM library simplifies matters considerably and allows us to implement a general purpose RNG library. There are two kinds of RNG call:

1. A collective call executed by all processors. This generate the next batch of random numbers and stores the locally generated subsequence in a local data structure.

2. The RNG function call. This appears to be a conventional RNG call except with an additional parameter corresponding to the sequence number of the desired value. This is then decoded into the appropriate PE number and offset and the value retrieved using **shmem_get** calls.

As it is also possible to implement the normal replicated generator strategy using the same interface (the sequence number is just ignored in this case), the decomposition independence can be switched on and off without changing the application source code.

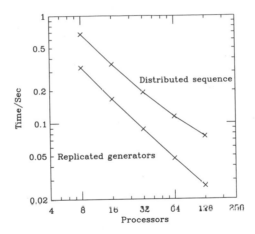

Figure 2. Performance of the sequence distributed generator compared to replicated generators

Global Summation

Global summation is another common source of decomposition dependent results. The summation of any set of numbers can be expressed as a binary tree with the final result at the root of the tree, the elements of the set on the leaves and intermediate results at the nodes. Floating point addition and subtraction are only an approximation to real number arithmetic. Unlike real numbers floating point arithmetic is not associative so it is possible to obtain different results by using a different summation tree.

The simplest implementation of a sum is to loop over the values adding them to a single running total; this results in a maximally tall tree with only a single addition at each level. The normal way of implementing a parallel sum is to perform a local sum over all the elements local to a processor and then perform a parallel sum of the results (figure 3). In this case the summation tree and hence the final result depends on the data distribution used.

To remove this dependence we have to coerce all data distributions to use the same summation tree. For any given summation tree only those Nodes at the same level of the tree can be calculated in parallel. We therefore choose the widest possible tree (figure 4) to introduce as much parallelism as possible.

The implementation of the global summation is similar to the random number case. As before there are 2 operations:

- A sum function that takes a single operand and a sequence number. The operand is inserted into a block distributed array using shmem_put calls.

- A collective summation operation that sums the distributed array.

As this implementation requires communication in both the initial redistribution of the data and in the final summation it runs about six times slower than the standard implementation (see figure 5).

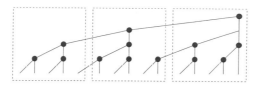

Figure 3. Summation tree for a local sum followed by a sum over processors

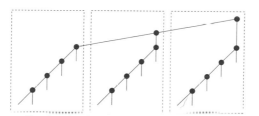

Figure 4. Summation tree for a decomposition independent sum

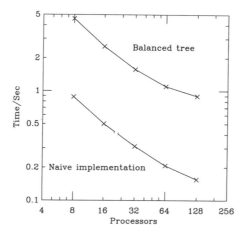

Figure 5. Performance of global sum. This figure shows the scaling performance of the sum routine used in the QCD code. The lower curve shows the performance of a DO loop sum followed by a sum over processors.

CONCLUSION

For the GHMC code it has been shown that decomposition independent versions of the random number generation and global summation routines are sufficient to give bit identical results on different processor numbers and topologies.

Decomposition independence imposes a performance penalty on parallel programs. However for the two operations discussed in this paper we have demonstrated that it is possible to construct scalable algorithms that give the same results on all data decompositions. In many codes the time spent in these operations is sufficiently small that these algorithms are suitable for use in production code. Even when this is not the case a check for decomposition independence provides a very strong test for the validity of parallel codes and the good scalability of these algorithms make it feasible to perform this test using realistic problem sizes and data decompositions.

REFERENCES

1. S.P. Booth *Parallel Random Number Generators* EPCC technical report (EPCC-TR96-04) University of Edinburgh (1996)

SOFTWARE PORTABILITY AND MAINTENANCE

Kenji Takeda, Ivan Wolton and Denis Nicole

Southampton HPCI Centre,
Department of Electronics and Computer Science,
University of Southampton,
Southampton SO17 1BJ

ABSTRACT

Topics related to software portability and maintenance are often overlooked in favour of code optimisation and other shorter-term programming issues. However, in the long run these can be crucial in ensuring continuity of work, for example when personnel change or a new machine becomes available.

We have gained some experience in working with HPCI consortia and on the HPC'97 procurement in these areas. This paper highlights some of the potential difficulties which have arisen, and describes techniques which may be used to minimise the amount of work required to port codes to different machines and maintain multi-developer projects.

INTRODUCTION: IS IT WORTH IT?

Good software management can reap real benefits for developers and users, especially on large projects. It is often thought that this is a significant overhead on the product development cycle, but if implemented sensibly it can provide significant savings in time and effort.

Portability means flexibility for project development. The ability to run programs on a variety of platforms enables development to be performed on different machines. This can be important for projects involving members at different institutions and those which span over many years. By using consistent programming procedures, and documenting them (automatically where possible), problems of handing over code to new developers can be overcome. This is also of great use when producing software packages for distribution to other users.

Portability also means flexibility for production runs. Robust programs can be rapidly moved onto machines as they become available. This enables more efficient use of resources at local and national levels. However, optimal performance need not be sacrificed. Both machine portable and vendor-specific optimisations can be managed readily to ensure that the best performance can be obtained on different systems.

Rapid code availability for benchmarking can be critical in choosing the right machine for specific, large applications. When procuring a new machine, be it a workstation or supercomputer, reliance on sanitized benchmarks can be deceptive. If real applications can be run on prospective systems then a significantly more realistic comparison can be obtained. Also, when such systems become available the ability to utilise new hardware quickly can prove very productive, as it is frequently associated with substantial leaps in performance. It has been found that as new services come online queues can be more accessible as the majority of users are expending a significant amount of time porting their code to the new machine.

The subjects of portability, code maintenance and the potential conflict with optimisation requires more space then is available in this paper, so we have attempted to summarise what we regard as best practice. For more detail refer to our web pages:

```
http://www.hpcc.ecs.soton.ac.uk
```

MAINTENANCE OF LARGE PROJECTS

Clearly, good programming practice is crucial, regardless of the size of project. Several tools and techniques are available to help in managing large projects while minimising the disruption to the programmer.

Programming Style

Every programmer has their own particular style of coding, but consistency in large projects should always be maintained. In some cases a project programming manual may be necessary setting out exact procedures to be followed. A few sensible guidelines may, however, be adequate for smaller projects. In our experience a few simple procedures can help considerably in reducing the program development cycle.

Write clear code, minimise clutter and ensure that it is easy to follow. This includes using visual cues, such as lines of asterisks and blank lines, to segregate portions of code, as well as consistent indentation, comments and subroutine/function headers. Consistent ordering of declarations, as regards arguments, COMMON variables and local variables, makes it easier to deduce the intended scope of a variable. The use of IMPLICIT NONE, while strictly outside the FORTRAN77 standard, is invaluable in picking up typos and avoiding inadvertent mistyping of variables. Preprocessor directives are useful for selectively including machine specific sections of code, to allow common source to be used in portable projects. In particular, machine specific optimisations should be required to be included explicitly by compiler directives and should have a portable alternative as the default.

File management is also an important issue. Use separate files for each function/subroutine to reduce hunting around in files. Include files should be used to keep track of common parameters and data structures. This cannot only be easier to manage but makes source code files more concise, such as when large common blocks are shared between subroutines. By putting subroutine and data definitions in separate files a standard annotation system can be implemented which helps greatly in working with complex programs.

Structure files in a sensible directory structure where appropriate. For example where different versions of a program are constructed using a makefile, specific routines should be placed in separate directories. The appropriate versions can then be linked in through the Makefile, perhaps as a library archive. Date and time functions are obvious candidates for this treatment.

120

Version Control

One of the keys of successful software management is proper version control. By keeping an audit trail of changes it is possible to develop along simultaneous threads and merge them at a later date. It also allows backtracking to help trace bugs which are created in a new version.

Southampton HPCI Centre uses the freely available, CVS package [1] for managing HPCI Consortia codes. Details for most platforms are available on our website together with source code and binaries for many systems:

http://www.hpcc.ecs.soton.ac.uk/hpci/tools/cvs .

CVS has many attractive features which make it suitable for HPCI projects. A central repository holds the most recent versions of files. From this repository authorised users can check out files and copy them to their own development environment easily. Support for shared, remote repositories is an important requirement for teams located at different sites, particularly for managing consortia codes.

Checking out copies of files rather than relying on file-locking (as in RCS [2]) enables multiple developers to work on files simultaneously. The last user to check a file back in must ensure correct merging with repository code. This encourages users to check out code for shorter periods of time.

Version stamping groups of files is based on diff rather than saving multiple copies of files. This keeps the size of the repository down. A semi-automatic annotation mechanism, including datestamping, encourages users to document changes. When checking files back in, a text editor automatically opens into which the programmer enters an appropriate comment. When the editor is closed the file is amended and committed to the repository.

Development Strategy

For HPCI projects a three-tier strategy is implemented, with CVS helping us to maintain coherency at each level:

1. A central CVS repository is held on a Southampton HPCI Centre server from which developers can check out any version of the held codes. This not only allows the latest versions to be centrally released, but also enables regression to previous versions. This is a critical and often overlooked need in such software development projects.

2. Developers check out programs for local development on their own workstations. This allows them to use tools which are not available on large, centrally administered systems. Good program editors and utilities such as FTNCHEK [3], FORGE Explorer, bounds-checking C compilers [4], Digital Visual FORTRAN and Visual C/C++ can be utilised. Test execution and performance simulation of programs in a controllable, local environment can also aid development greatly.

3. Copies, both source and binary-form, are held at remote HPC machines for execution. Synchronisation with developer versions is done automatically with awk/ftp scripts. Make is used to smooth out differences between machine environments. The C preprocessor can be used to aid portability but users should be cautious to consider the different lexical structure of C and FORTRAN when preprocessing source files.

SOFTWARE PORTABILITY: ADHERING TO STANDARDS

In order to make code truly portable adherence to language standards [5,6,7] is necessary. While this may seem restrictive, simultaneous maintenance of hardware-specific versions is relatively easy. Relaxation of the FORTRAN77 standard to allow INCLUDE statements and names up to 31 characters long (but still unique in the first 6 characters) is not likely to cause problems on modern HPC platforms. FORTRAN77 codes are more likely to have a long history of development on one class of architecture. Historic limitations on memory have often led to bad programming practices on such codes. The use of Fortran 90 constructs allows many of the problems outlined below to be circumvented. In order to check for non-standard code segments tools such as FTNCHEK [3] and lint can be used. Most compilers also have different levels of checking, so the most pedantic setting should highlight potential problem areas. Common problems encountered in scientific codes are outlined below.

Data Types

Equivalencing is best avoided in portable codes but if used it must be implemented with extreme care. Equivalencing CHARACTER and numeric data can be particularly troublesome as the FORTRAN77 standard specifically forbids this and defines no relation between the length of numerical storage units and character storage units.

Most users are aware that the length of a numeric storage unit varies between machines to the extent that, on the Cray T3D and T3E, a REAL is represented by 8 bytes whilst on many other machines it is represented by 4 bytes. Many programs require 8 byte REALS to achieve a desired accuracy. The dilemma as to whether to use REAL or DOUBLE PRECISION in a portable program can be partially solved by the use of compiler directives but this can cause problems if REALs are equivalenced to INTEGERs or LOGICALs as the storage relations between them may be changed in the process. The use of REAL*8 complicates the issue further and should be avoided in portable programs.

More complications can arise if mixed Fortran and C programing is used, as this may require further assumptions about the relative sizes of INTEGERs and REALS and their C equivalents. A more sophisticated approach may use a Macro preprocessor, such as the UNIX M4 or the C preprocessor, to achieve the appropriate declarations for a given system, but if possible such equivalencing should be avoided or eliminated. The consequence of getting it wrong is usually run time problems or numerical errors, but severe performance problems may also result. This frequently occurs when different data types are mixed in COMMON or EQUIVALENCE statements and the actual data lengths selected to achieve the required accuracy for REALs causes a data misalignment. In order to improve memory access performance, computer caches typically access several words of data at a time. Misaligned data can force the computer to access data a word at a time resulting in serious performance degradation. Some machines even refuse to access misaligned data. Usually misaligned data is detected at compile time and a warning issued. These warnings should not be ignored.

A more serious misuse of data types is a mismatch between the types of actual and dummy arguments of procedures. This is less commonly found but has caused problems in major production codes. It typically occurs when a large array is used as work space and parts of it are passed as actual arguments to a procedure. Within the procedure definition, the dummy arguments may be declared as of a different type to the actual array. This may work on some architectures but may fail totally on others as INTEGERs and REALs may be passed in different registers. A failure to match the numbers of dummy and actual arguments is another violation of the standard which may not cause problems on some systems but most definitely will on others.

Message Passing Programming

A major source of problems in message passing programs is inadequate consideration of how much buffering, if any, is required by the code and how this varies with problem size and the number of processors. PVM does not usually offer the user much control over the amount of memory required by message buffers or what happens when the available memory is exceeded. As a result programs may often hang or fail inexplicably due to large memory requirements for PVM buffers. Often the problem does not manifest itself until the code is moved from a development platform to a much larger production environment. Some control of resources may be offered by environment variables in the operating system but this can still be a bit of a black art to get right.

For this reason MPI is preferable as it allows much more control of how much buffering is used, and whether it is used at all. Despite this, many programs are written which are critically dependent on buffering being supplied by the MPI implementation (perhaps because they have been "translated" directly from PVM). We have experienced particular problems with IBM SP2 systems, where buffered message passing is provided up to a certain message size length, but messaging becomes synchronous when the message length is exceeded. Hence, with codes which assume system buffering is available, small test cases may run without problem but larger production runs deadlock when the buffer size is exceeded. The MPI standard [9,10] does not specify buffering as standard; it is therefore crucial for program portability that buffering is either specified explicitly, or a safe message-passing strategy is implemented.

The range of message tags in MPI is also an issue to be aware of. The MPI standard only guarantees tag values up to 32767 [9,10]. Vendors may implement a more generous range of values but are not obliged to do so. Seemingly spurious MPI errors can result from message tags being too large and can be difficult to trace, so adhering to the standard is advised.

Miscellaneous

It is often important to know how the memory requirements of a code vary with the problem parameters and the number of processors used, particularly when porting between machines with different configurations. It can be difficult to trace errors due to lack of memory so *a priori* knowledge of array requirements can be extremely useful. Tools such as the UNIX `size` (or `mppsize` on the T3D) which give information on the memory requirements of the executable file can be a useful, although these only report statically allocated memory.

Aliasing occurs when two procedure arguments or a procedure argument and a COMMON variable refer to the same memory location. This is permissible if none of the aliased variables are written to, but otherwise the value stored in the memory location may vary between systems or even between different runs on the same system.

Mixed compilation of C and FORTRAN can be a pain. Basic datatypes are generally handled well. Some systems force FORTRAN names to upper case, some to lower case; handling this needs either a tool or lists of name translations in the header files. Pre- and appended underscores can, however, be handled by a preprocessor with care. CHARACTER and COMPLEX datatypes can be problematic and are best avoided where possible; otherwise use an auto-tool such as `Cfortran` [8].

File processing is an area where care must be taken when moving between platforms. For examples, Windows NT systems do not distinguish between file.F and file.f, and UNIX's `'/dev/null'` is equivalent to Windows NT's `'NUL'`.

Unformatted datafiles are usually machine-specific and require special treatment. Some operating systems may provide assistance in converting from one vendor's format to another (e.g. the `assign` command on Cray systems). Portability with regard to unformatted data may be maximised by using IEEE standard formats but account may still have to be taken of differences between big-endian or little-endian representations and the specification of record lengths.

PROGRAM OPTIMISATION

In order for compilers to optimise efficiently they assume that users have adhered to language standards. This, in part, explains why some non-standard programs can cause errors to occur when optimisations are applied. Therefore, by sticking to the standards, more aggressive optimisations may be possible resulting in faster code.

Platform Independent Optimisations

General optimisations can yield performance gains on most systems, although the amount of speedup is architecture (and compiler) dependent. Modern optimising compilers are continuously being improved and knowledge of the available optimisation flags is important in obtaining maximum application performance. However, reducing the number of necessary operations and giving the compiler hints in the landscape can result in faster code. Some basic, but effective measures are outlined below.

Reordering and removal of conditional statements, especially inside loops will help keep the instruction pipeline flowing. Reordering, and removal where possible, of expensive operations such as divisions, square-roots, exponentials and trigonometric functions can be advantageous. Many vendors provide fast mathematics libraries for users able to sacrifice some accuracy. These can give excellent speedups where their use is mathematically justifiable.

Loop unrolling to expose more computations is a technique which can bring significant benefits. However, it can be a complex procedure and is best left to the compiler in most cases. Many compilers even allow the programmer to specify how many levels of unrolling the optimiser should attempt. On some modern compilers best performance is achieved by keeping loops small and simple so that the compiler can understand them fully. Loop interchange to keep array memory strides down to one can be particularly effective in increasing cache utilisation.

It should be noted that by rearranging the order of calculations the results may change as computational arithmetic is not necessarily associative due to rounding errors. It is very important that stringent checks on the accuracy of optimised code be carried out before being committed. Optimisation can often reduce the readability of code and the original version should always be retained (in the same source file) so it can be reverted to at any time. Here the use of preprocessor directives can be of great help.

Many compilers, and vectorising preprocessors, look for BLAS constructs and replace them with highly optimised routines. Therefore these can provide good performance on a wide variety of systems as most ship with some level of BLAS support.

Vendor-Specific Routines

Most vendors supply their own optimised library routines. Obviously this is a problem in terms of portability, but this can be easily overcome. Use optimised routines where possible to get the most out of a given machine. Portability can be maintained by using preprocessor directives to separate platform-independent function calls. Alternatively special wrappers/interfaces can be used to hide vendor-specific calls.

To maintain portability ensure there is a machine-independent equivalent for each function. Although this may not be as efficient, it at least allows the application to be run on different machines which can be tuned later. Also check for substitution of standard routines by vendor-specific ones using tools such as nm which creates an object map file.

CONCLUSIONS

Experience gained by the Southampton HPCI Centre regarding software portability and maintenance issues have been presented. It is hoped that developers will be able to utilise some of the techniques described here to reduce product development cycle times.

Maintainable and portable code is essential in projects of all sizes and need not be a barrier to rapid software development. In the long term it saves time, both in terms of code performance and reduced debugging time. In particular large projects can benefit greatly by ensuring continuity of work and the ability to make the most of available resources.

REFERENCES

1. Per Cederqvist et al., *Concurrent Versions System (CVS) reference, Version Management with CVS for CVS 1.9*, Signum Support AB, currently obtainable from
   http://www.loria.fr/cgi-bin/molli/wilma.cgi/doc.847210383.html
   Additional information is available at,
   http://www.hpcc.ecs.soton.ac.uk/hpci/tools/cvs/index.html
2. W.F. Tichy, RCS – A System for Version Control, *Software-Practice & Experience* 15:637-54 (1985). Currently available online at
   http://www.de.freebsd.org/de/doc/psd/13.rcs/paper.html
3. FTNCHEK, currently available at, http://netlib.cs.utk.edu/fortran/index.html
4. R. Jones and P. Kelly, *Bounds checking C compiler*, currently available at,
   http://www-ala.doc.ic.ac.uk/~phjk/BoundsChecking.html
5. *FORTRAN 77 Standard, ANSI X3J3/90.4.* Currently available online at,
   http://www.fortran.com/fortran/F77_std/rjcnf0001.html
6. *FORTRAN 90 Standard, ISO/IEC 1539 :1991*, Information technology-Programming languages-Fortran.
7. American National Standards Institute, *American National Standard for Information Systems -- Programming Language -- C_, ANSI X3.159-1989*
8. Cfortran, information currently available at
   http://www.physics.ohio-state.edu/~nylander/physics/cfortran.html
9. Message Passing Interface Forum, MPI: A message passing interface standard, *International Journal of Supercomputer Applications*, 8(3/4), special issue on MPI (1994).
10. M. Snir, S. Otto, S. Huss-Lederman, D. Walker and J.J. Dongarra, *MPI: The Complete Reference*, MIT Press, ISBN 0-262-69184-1 (1996).

A DESIGN ENVIRONMENT FOR STRUCTURED MAPPING OF SIGNAL PROCESSING APPLICATIONS ON PARALLEL PROCESSORS

Moe Razaz

University of East Anglia
School of Information Systems
Norwich, U.K.

ABSTRACT

We present an integrated software environment called Taurus, which is capable of structured mapping of signal processing applications on parallel computers. An application is first converted into a directed graph representation which is then turned into a multiprocessor code with the help of a code generator. Given the hardware interconnection topology and specification, a scheduler determines in what order the multiprocessor code should be mapped onto the individual processors in the hardware platform. The parallel implementation is shown in the form of a Gannt chart so that the user can see graphically the speed-up and processor utilisation. The major advantages of this environment are: (i) an intensive signal processing application can be easily implemented on a parallel platform (ii) processor specification and interconnection topology are user definable so that the same software can be used for implementation on different hardware platforms and (iii) a user can interact with the environment to enhance the performance of the parallel implementation. The design philosophy and the organisation of the integrated environment are presented and discussed.

STRUCTURED MULTIPROCESSOR IMPLEMENTATION

Digital signal/image processing algorithms and applications have in the past been implemented on single DSP chips. If we consider a conventional cycle of single DSP implementation, it starts with developing a design that meets the required specification. This design is then captured and verified by simulation before implementation on a DSP hardware. The implementation-testing step may have to be iterated several times until testing is successful. If the speed and memory specifications are too high, then single DSPs are unlikely to meet such requirements. The shortcomings of this conventional approach include: (i) a long cycle from design capture to the final product development,

(ii) hardware dependence and hence lack of portability to different DSP platforms, and (iii) lack of exploitation of parallelism.

The speed and complexity requirements of many intensive and real-time signal processing applications for example in control, communication, speech synthesis/ recognition and image processing, are too high and cannot be met by single DSPs. A possible solution would be to exploit parallelism on multiple DSP chips. This type of parallel implementation is often difficult, very time consuming and more importantly is hardware dependent. Another possibility is to implement the application in hardware, for example using a fast VLSI technology, so that the speed requirement can be met. However the latter is very expensive and does not have flexibility of programmable DSPs. An elegant solution to this problem would be to develop a structured methodology which can generate automatically multiprocessor code (from a single processor application software). A scheduler then can be used to map automatically the generated code onto a specific multiprocessor hardware platform. This solution offers many advantages over VLSI implementation including programmability, cost-effectiveness, short time-to-market and independence from hardware platform. At the University of East Anglia we have been developing such an integrated design environment, called Taurus, for structured mapping of signal processing applications on parallel processors[13-24]. This paper presents briefly the principles behind and building blocks of this environment.

DESIGN REPRESENTATION

In order to map a DSP-based design onto a multiprocessor platform in a systematic fashion it is necessary to represent an effective and structured model of the design. Various methods of design representations were considered such as Block Diagram Languages, Petri nets, reduced dependence graphs (RDG), signal flow graphs (SFG) and large grain data flow(LGDF) graphs [1-8]. Block Diagram Languages(BLD) are very suitable means for expressing DSP algorithms but cannot represent asynchronous systems and systems with multiple sample rates. Similarly, RDG and SFG have difficulty in modelling multirate systems. Although Petri nets do not suffer from these disadvantages they require expensive dynamic control for implementation which implies dynamic use of resources such as memory and CPU time, both of which are scarce commodity in real-time DSP applications. LGDF graphs are similar to BLD in their ability to express with ease algorithms in block diagrams, and can model multiple sample rates and asynchronous systems. They have also coarse granularity which enhances modularisation and decreases graph complexity, and thus improves design readability. Therefore a LGDF graph approach was chosen for design representation as it retains all the advantages of block diagrams without suffering from the limitations associated with the other approaches.

Briefly a typical acyclic large grain data flow graph consists of nodes and directed edges, as shown in Figure 1. A functional block or a task, is represented by a node. This can be a basic operation like add and multiply or a more complex functional block such as FFT, convolution and so on. The run time of a node is indicated by a number inside the node. The flow of information from one node to another and therefore their interdependencies is represented by a directed edge. A node is data driven in that it fires when sufficient tokens(input samples) are available to perform a task. The data transmitted and received by nodes in the form of tokens are shown on directed edges. The type of LGDF graphs we are concerned with are acyclic i.e. they do not contain any loops. This is a necessary condition for the graphs to be statically schedulable. No conditional nodes are also allowed as the flow of control in a LGDF is assumed to be data independent.

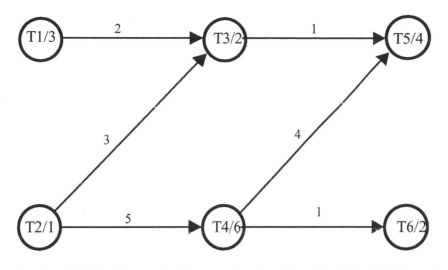

Figure 1.   A typical data flow graph; the processing time of a task is indicated inside each node and the number of tokens transmitted or received is shown on each edge.

THE MULTIPROCESSOR DESIGN ENVIRONMENT

The organisation of our design environment, Taurus, is shown Figure 2. The major building blocks are the Converter, Scheduler, Schedule Verifier, Schedule Editor, Precompiler (or Code generator) and Performance Analyser. Each of these blocks is a major software module. A supervisory shell glues these modules together and controls their interaction and operation, and an underlying database system provides access to data files. A typical signal processing application is first captured using the front-end SPW software [27], which has software facilities for design capture, simulation, and code generation.

The captured design is then translated into a large grain data flow (LGDF) graph representation   by the Converter which comprises three main building block modules. The first module parses the design netlist and generates a data structure for representing the LGDF graph; it also identifies any unconnected inputs. Parsing of the design C listing from the front-end SPW is performed by the second module which also   generates a text file identifying various sections of the code related to each node in the graph. The data obtained from modules 1 and 2 is combined by the third module to produce a graph description file in textual form. This file contains information about interconnection between nodes, and specific functions or instances and variables associated with each node. The graph description file is then used by the Converter to generate the necessary information in the form of files for the Scheduling system and multiprocessor code generator (Precompiler). Scheduling and precompilation are subsequently performed on this graph description file.

The Performance Analyser provides an indication of how good the multiprocessor implementation schedule is and how to interact with the system in a closed-loop iterative fashion in order to modify the schedule and hence improve throughput and efficiency.

In Taurus processor specification and interconnection topology are externally specified by the user. This independence of the multiprocessor hardware platform helps with portability and longevity of the software environment. Two types of parallel processing hardware platforms have been used in Taurus, the Transputer and TI TMS320C40 platforms. The Transputer platform, known as Meiko [27], is a network of reconfigurable processors for which a programming environment is provided by the Computing Surface Network (CSN) tools. The TI platform consists of a network of reconfigurable C40 DSP processors [25], with the programming environment provided by the 3L Parallel C tools [30]. The latter, like CSN for Meiko, provides the capability to configure, compile, link and execute a task code within its hardware environment.

Scheduling

The major function of task co-ordination and scheduling in Taurus is handled by the Scheduling system which consists of the Scheduler, Schedule Verifier and Graphical Schedule Editor (GSEdit). The Scheduler [16,19] assigns systematically the functional blocks (nodes in the LGDF graph) to various processors in the hardware platform. Different types of scheduling algorithms were considered [9,10,11,12,16] but static scheduling was chosen as it is performed at compile time and resource requirement in terms of memory and dynamic time is not demanding and hence is ideally suited for DSP applications. The Schedule Verifier checks if a schedule is permissible i.e. it can be executed to completion without a deadlock or livelock.

The Scheduler incorporates three main modules. Module 1 generates the precedence graph using the information provided by the Converte r. Module 2 derives Hu labels [10] for the precedence graph and passes this information via a Task Scheduling Window to Module 3, i.e. the main Scheduler. The latter creates permissible schedules and percentage utilisation for different processors, using the information provided by the precedence graph, its Hu labels and the characteristics of the hardware multiprocessor environment (namely the form of interconnection topology and individual processor specifications). The process of scheduling in Module 3 is controlled and influenced by the Task Scheduling Window, interprocessor communication times and homogeneous or heterogeneous nature of processors in the hardware platform.

The Graphical Schedule Editor (GSEdit) in Taurus provides the user with a central coherent interface for the checking and editing of multiprocessor schedules. Editing a schedule is allowed as long as the changes result in a permissible new schedule. When GSEdit is first executed a Main Window is displayed containing the complete Gantt chart for the first schedule to be operated on. The Gantt chart can be manipulated to zoom in on a region of the schedule, and hence displaying greater details. This region can then be moved up and down the Gantt chart to display different parts of the schedule. The form of display can also be changed to group similar tasks by colour and to display the intercommunications occurring. It is also possible to create new views onto the schedule independent of the view displayed in the Main Window, for example, processor utilisation and speed-up graphs can be shown concurrently with the Gantt chart.

Multiprocessor Code Generation

The major task of multiprocessor code generation in Taurus is performed by the Precompiler as shown in Figure 3. It uses the information from the current schedule description file together with the LGDF and precedence graphs to generate the necessary C source programs and control files for the compiler and linker in the target multiprocessor platform. The resulting executable programs implement the DSP application captured by the front end SPW package. Special use is made of sets of code segments called skeletons which provide a consistent abstracted interface to the hardware platform for the application code. By choosing certain skeletons the execution of the application on the hardware platform can be traced. A multiprocessor code generated by the Precompiler can be tuned to express additional functionality as required by the designer or the hardware platform.

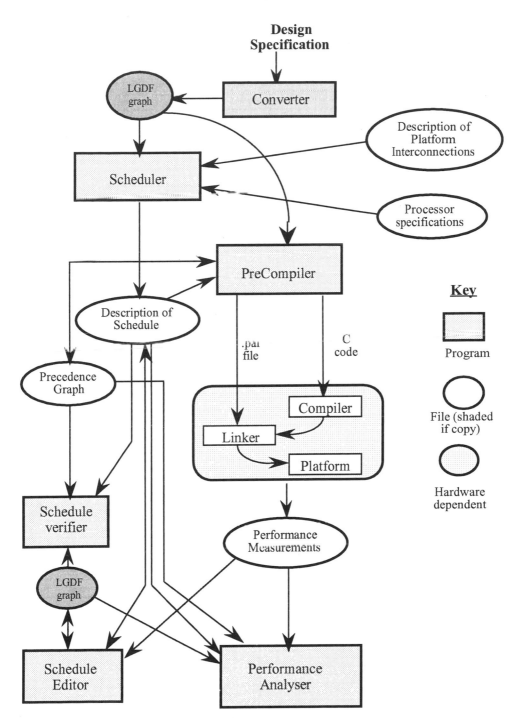

Figure 2. Detailed organisation of the design environment

The concept of using such a Precompiler for multiprocessor implementation is novel. It is more usual in similar design environments to use primitives or blocks internal to the system. For example applications are represented as a graph of interconnected stars in 28,29, where each star can either simulate or implement the functionality it represents, depending upon whether the application as a whole is to be simulated in software or implemented on a hardware platform. Such stars are internal and an integral part of the system, whereas for the Precompiler they are concerned with code generated externally and therefore have no part in the initial creation of the code.

Once the Precompiler has generated code for each processor, the code must be compiled and executed on the parallel processor hardware platform. The connectivity of processor code files must also be set up in order to represent the connection topology of the schedule. A task code must go through three stages to prepare it into a down-loadable form. Firstly, code files must be individually compiled to produce a series of *object files*. These are then linked with relevant libraries to produce a series of *task image files*. Once image files for each processor have been produced, they are configured by means of a configuration file, to produce an application file, which can then be down-loaded onto the hardware and executed.

To automate these three stages, three script files are produced by the Precompiler after code generation has taken place. The first of these is a Configuration file, declaring tasks and their connectivity to the *config* utility. The second script file is a Makefile, which is used to create the application by compiling, linking and configuring processor code. The final script file is for down-loading the application onto the hardware platform.

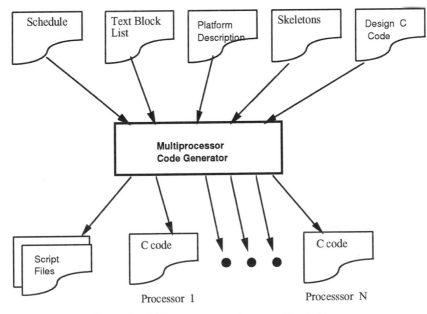

Figure 3. Multiprocessor code generation in Taurus

RESULTS AND DISCUSSION

We have used two types of parallel processing platforms to prove the concept of automatic code generation by Taurus., namely Transputers and C40 DSP processors. We started with the Transputer platform consisting of some 50 reconfigurable processors. A number of signal processing applications were successfully used for parallel implementation on the Transputer platform and different interconnection topologies [31] were also experimented for a given application. used. Typical examples implemented include: a 32 kbps ADPCM (Adaptive Differential PCM) system, a 9.6 kbps digital speech signal transmission and a quadrature amplitude modulation (QAM) digital radio link simulation. These experiments have clearly shown that Taurus can be used as an engineering environment for fast prototyping complex signal processing applications on a multiprocessor hardware platform.. A major advantage of Taurus is that it is independent of the hardware platform, so that the same application code can be easily implemented on different processors.

We have recently set up a reconfigurable hardware platform consisting of seven C40 processors. This is a multi-user Sun Sparc-based system in which the C40 processors communicate with a 6U VME CPU master board via a VME slave TIM-40 carrier board. The Sun host and the CPU board talk to each other via an ethernet. The transition from the Transputer platform to C40 was straightforward because of the ease with which a new hardware platform can be defined in Taurus (using special skeletons provided in the Precompiler). Work is currently being carried out to implement various applications on this platform.

Although the emphasis in this paper has been on signal processing applications and the use of special hardware platforms, Taurus can be easily used for other applications such as control, system design and image processing using different parallel processing hardware platforms.

ACKNOWLEDGEMENT

This project was partly supported by the BT, EPSRC and an innovation grant from the University of East Anglia. A number of people have contributed to the success of the project, and in particular Keith Marlow's contribution is acknowledged.

REFERENCES

1. W.B. Ackerman, Data flow languages, *IEEE Comput.* vol. 15 (1982).
2. T. Agerwala, Putting Petri Nets to work, *IEEE Compt.* 12:85 (1979).
3. R.E. Crochiere, and A.V. Oppenheim, Analysis of linear digital networks, *Proc. IEEE* 63:581-95 (1975).
4. A.L. Davis, and R.M Keller, Data flow program graphs, *IEEE Comput. vol.* 15 (1982).
5. R.M. Karp et al., The organisation of computations for uniform recurrence equations, *J. ACM* 14:563-90 (1967).
6. G. Korn, High speed block diagram languages for microprocessors, miniomputers in instrumentation, control and simulation, *Comput. Elec.Eng.* 4:143-59 (1977).
7. D.G. Messerschmit, A tool for structured functional simulation, *IEEE J.Comm.* vol. SAC-2 (1984).
8. J.L. Peterson, Petri Nets, *Comput. Survey* vol. 9, no. 3 (1967).
9. S. French, *Sequencing and Scheduling*, Ellis Horwood (1982).

10. T.C. Hu, Parallel sequencing and assembly line problems, *Oper. Res.* pp841-48 (1961).

11. N.F. Chen and C.L. Liu, *Proc. Sagamore Comp. Con. on Parallel Processing*, pp1-16, Springer Verlag, N.Y. (1974).

12. T.L. Adam et al., *Comm. ACM* 17:685-90 (1974).

13. M. Razaz, *Structured DSP Design and Development in a Multiprocessing System*, BT Research Report, pp1-74, (March 1993).

14. M. Razaz and K.A. Marlow, Toward an automatic mapping of DSP algorithms onto parallel processors, *Applications of Supercomputers in Eng.* 3:463-477 (1993).

15. M. Razaz and K.A. Marlow, An integrated multiprocessor system for DSP Design and Development, in: *Int. Symp. on DSP for Communication Systems* (Sept. 1992).

16. M. Razaz and K.A. Marlow, Scheduling DSP algorithms for parallel multiprocessor environment, in: *3rd IMA Conf. Maths. in Signal Processing.* pp10-27 (Dec. 1992).

17. K.A. Marlow and M. Razaz, A new precompiler for mapping DSP applications to multiprocessing systems, *Transputer Applications and Systems*, 1:296-311 (1993).

18. K.A. Marlow and M. Razaz, Visualization and analysis of multiprocessor DSP design implementations, in: *Advances in Parallel & Vector Processing*, B. Topping and M. Papadrakakis eds., pp245-9 (1994).

19. M. Razaz and K.A. Marlow, A multiprocessor algorithm scheduler for signal and image processing, Signal Processing VII: Theories and Applications (Holt, Cowan, Grant and Sandham eds.) 3:1605-8 (1994).

20. M. Razaz and K.A. Marlow, Performance comparison of two scheduling algorithms for parallel architectures, Advances in Parallel and Vector Processing, B. Topping and M. Papadrakakis eds., pp59-68 (1994).

21. M. Razaz and K.A. Marlow, A transputer-based parallel DSP environment, in: *Proc. IEE Colloq. on High Performance Applications of Parallel Architectures*, pp8.1-8.6, London, (Feb. 1994).

22. M. Razaz and K.A. Marlow, Designing DSP-based systems using multiple processor hardware platforms, *Proc. IEEE ISCAS*, pp109-12 (1994).

23. M. Razaz, T. Spendiff and K.A. Marlow, Code generator for parallel implementation of intensive algorithms on multiple DSP chips, in: *Conf. on DSP Chips in Real Time Measurement and Control*, UK, pp8.1-8.10 (Sept. 1997).

24. M. Razaz and K.A. Marlow, Taurus: A multiprocessor DSP prototyping environment, *IEEE SiPS*, pp263-72 (1997).

25. Texas Instruments, *TMS320C40 Parallel-processing DSPs, Product Bulletin* (1993).

26. Meiko Ltd., *Computing Surface, C for CS TOOLS*, Technical User Manual (1991).

27. *Signal Processing Worksystem (SPW)*, Comdisco Systems, Inc., Foster City, CA, (1991).

28. J.C. Bier et al., Gabriel , *IEEE Micro.* pp28-45 (Oct. 1990).

29. *The Almagest: Manual for Ptolemy Version 0.3.1*, Department of EECS, University of California, Berkeley, USA, (Jan. 1992).

30. 3L Ltd., *C4x Parallel C Version 2.0.2*, User manual (1995).

31. J.L. Hennessey and D.A. Patterson, *Computer Architecture A Quantitative Approach*, Morgan Kaufmann, San Francisco, pp378 (1995).

MATERIALS CHEMISTRY AND SIMULATION

NEW VISTAS FOR FIRST-PRINCIPLES SIMULATION

G. Ackland[1], D. Bird[2], P. Bristowe[3], M. Finnis[4], M.J. Gillan[5],
N.M. Harrison[6], V. Heine[7], P.A. Madden[8], M.C. Payne[7], A.P. Sutton[9]

[1]Physics Dept., Univ. of Edinburgh, Edinburgh EH9 3JZ
[2]Physics Dept., Univ. of Bath, Bath BA2 7AY
[3]Dept. of Materials, Univ. of Cambridge, Cambridge CB2 3QZ
[4]School of Mathematics and Physics, Queen's Univ., Belfast BT7 1NN
[5]Physics Dept., Keele University, Staffordshire ST5 5BG
[6]DCI, CCLRC Daresbury Laboratory, Warrington WA4 4AD
[7]Cavendish Laboratory, Univ. of Cambridge, Cambridge CB3 0IIE
[8]Physical Chemistry Laboratory, Univ. of Oxford, Oxford OX1 3QZ
[9]Dept. of Materials, Univ. of Oxford, Oxford OX1 3PH

ABSTRACT

We review recent progress in the first-principles treatment of condensed matter, with particular attention to the work of the UKCP consortium on the Cray T3D at EPCC. It is shown that HPC facilities are allowing first-principles methods to be innovatively applied in a great variety of fields, including surface science, catalytic chemistry, materials science and the earth sciences. We describe examples of UKCP work on molecular processes on metal and oxide surfaces, nanoindentation, acid-base reactions in zeolites, complex defects in oxides and semiconductors, liquid iron under Earth's core conditions, and liquid crystals. Recent progress in the development of linear-scaling first-principles techniques is also outlined. The vital role of HPC for future progress is stressed.

INTRODUCTION

First-principles calculations are revolutionising the understanding of matter on the atomic scale. The ability to do accurate quantum-based calculations on large assemblies of atoms is opening up completely new vistas in condensed-matter physics and chemistry, materials science, earth science, and even molecular biology. High-performance computing (HPC) is absolutely vital to this whole enterprise, and there is now a world-wide effort to apply HPC power to first-principles work. In Britain, a major role has been played by the U.K. Car-Parrinello consortium (UKCP), a collaboration of 10 research groups led by the present authors. In this overview of work done by the UKCP groups on the Cray T3D at Edinburgh, we will show how first-principles methods originally developed by condensed-matter physicists are being used in an ever expanding range of disciplines.

We want to emphasise science here, so we will say only a few words about the methods we are using. All our first-principles work is based on density-functional theory (DFT), the pseudopotential method, and plane-wave basis sets, which together give an accurate procedure for calculating the energy of the electronic ground state and the forces on atoms for any set of atomic positions. This information can then be used in many ways: to explore the energetics of the system; to search for equilibrium structures; or to simulate dynamical processes. This general approach is now very widely used to study condensed matter, and there are many reviews (see e.g. Refs. (1, 2)). Most of the work of UKCP is being done using a single code, known as CETEP (Cambridge-Edinburgh Total Energy Package)[3], which is being used collaboratively by the members of the consortium.

In order to show the broad range of first-principles research achieved on the Cray T3D, we will give brief summaries of our work in a variety of areas, including surface physics and chemistry, catalysis, defects in materials, and liquids; and we also mention the development of novel methods for treating very large systems.

OVERVIEW OF FIRST-PRINCIPLES WORK ON THE CRAY T3D

Some of the greatest opportunities for first-principles methods lie in surface science – an area of huge significance because of its relevance to economically important processes including corrosion and catalysis. We will describe projects carried out on the Cray T3D concerning molecular processes on metal and oxide surfaces, as well as a project on nanoindentation. This is followed by work relevant to catalysis in zeolites. Projects on extended defects in semiconductors and oxides come next. Very different from this is work on liquid metals and liquid crystals. Our review of UKCP work concludes by outlining current efforts to develop methods that are efficient for very large systems.

Molecular Processes on Metal Surfaces

Many important processes are catalysed by metal surfaces, but in most cases the detailed mechanisms are only guessed at. First-principles methods are opening the way to detailed and accurate calculations of multi-dimensional energy surfaces describing how the energy of the system depends on the positions of the atoms.

The group of Bird has put a major effort into understanding the interaction of hydrogen and other molecules with different metal surfaces[4, 5, 6, 7, 8, 9] – hydrogen is an important prototype for surface adsorption because of the wealth of experimental data available. One of the most important quantities measured experimentally is the sticking coefficient, i.e. the probability that a molecule arriving from the gas phase sticks to the surface, rather than bouncing off. The sticking coefficient depends on the energy of the incident molecule in quite different ways for different surfaces, and the reasons for this have been controversial. Bird's group have used CETEP calculations on the T3D to construct detailed energy surfaces for the H_2 molecule on various metal surfaces, and have found striking qualitative differences. For example, it has been found that the adsorption of H_2 on the Cu (001) surface involves an energy barrier, while on W (001) there is no barrier. In collaboration with the group of Holloway at the Liverpool IRC on Surface Science, the calculations have been used to give a detailed interpretation of the energy dependent sticking coefficient.

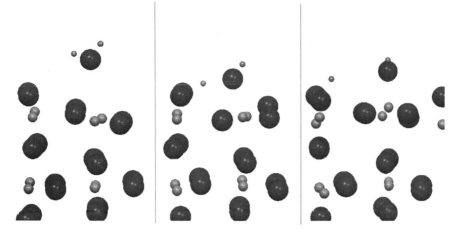

Figure 1. Three snap-shots from a dynamical simulation of an H_2O molecule being dissociatively adsorbed on the TiO_2O (110) surface, from Ref. (16). Large, medium and small spheres represent O, Ti and H atoms respectively. Time proceeds from left to right, with an interval of *ca.* 0.2 ps between successive frames.

Independently, Alavi and co-workers in the group of Finnis have been using first-principles T3D calculations to elucidate the oxidation of carbon monoxide (CO) on the surface of platinum[10]. This is one of the most widely studied reactions in surface science, and is hugely important because it is a key reaction in car exhaust catalytic converters. In any catalytic reaction like this, two of the main questions are: what is the transition state, and what is the origin of the reaction barrier? These questions are central in seeking ways to improve the efficiency of the catalyst. The calculations have revealed that competition of C and O atoms for bonding with the metal surface is crucial and the predominant energy barrier arises from the strength of the oxygen-metal bond. They have also yielded a simple explanation for much of the observed behaviour in CO oxidation.

Oxide Surfaces

The surfaces of oxide materials are important for many reasons: they are widely used as catalysts and gas sensors, and they play a key role in corrosion; in the earth sciences, processes at oxide surfaces are crucial in the formation and weathering of rocks and in the transport of pollutants in the environment. The equilibrium structure of these surfaces has been studied for many years by both quantum-chemistry and first-principles methods, and recent detailed comparisons with experiment leave no doubt that the theoretical predictions are realistic. With the increasing power of HPC, attention is turning increasingly to the complex processes involved in molecular adsorption and dissociation.

The Keele and Daresbury groups of the UKCP consortium have worked on the T3D to investigate the adsorpion of water, methanol, oxygen and other molecules on the surfaces of titania (TiO_2)[15, 16, 17, 18] and magnesia (MgO)[11, 12, 13, 14]. As an illustration of what has been achieved, we point to our work on the adsorption of water on the TiO_2 (110) surface. A controversial question in this area is whether water is adsorbed in molecular or dissociated form. To investigate this, we did dynamical simulations[16] in which the H_2O molecule comes onto the surface from the gas phase. The simulations show that when the coverage is not too high the molecule spontaneously dissociates (see Fig. 1). But our more detailed dynamical and static simulations[18] show that the balance is a delicate one, and at high coverage water seems to exist in a mixture of molecular and dissociated states.

In the past year, this work has been extended to a number of other molecules on the TiO_2 (110) surface, in order to help understand the systematics of acid-base processes on oxide surfaces. We believe that current advances of HPC power will make these methods a routine tool for investigating complex chemical reactions on oxide surfaces.

Nanoindentation and Atomic Force Microscopy

The effort to understand, characterise and manipulate matter on the nanoscale is becoming one of the dominant themes of condensed-matter science. Scanning-probe techniques, such as scanning tunelling microscopy (STM) and atomic force microscopy (AFM) are playing a leading role in the experimental effort. But accurate calculations are also essential in guiding and interpreting the experiments. It is clear that in the years to come first-principles methods will be in the forefront of the computational effort.

In order to demonstrate what can already be done, Payne's group have used the T3D to carry out CETEP simulations[19] of the atomic processes involved in nanoindentation – the process in which an atomic-scale tip is pressed into a surface; the group have also used similar techniques to investigate the imaging mechanisms in non-contact AFM[20].

In the nanoindentation study, a rigid tip made of aluminium atoms was pressed into the (111) surface of silicon, which was modelled as a slab of material 8 layers thick, with each individual layer having a repeating unit of 30 atoms. The tip was pressed slowly into the surface – slowly enough so that the atoms remained in equilibrium at all times. The calculations showed that there is an initial stage where the surface responds elastically, followed by plastic deformation in which the bonds between atoms are irreversibly broken. These events cause jumps in the energy, and the associated redistribution of electrons can be followed in the simulations. An important conclusion from this work is that the plastic deformation accompanying nanoindentation actually consists of two separate processes: the flow of sub-surface atoms to interstitial sites, and the extrusion of surface atoms around the walls of the tip. Insights such as these are crucial for interpreting nanoindentation experiments.

With the more powerful HPC facilities now becoming available, it is planned that this work will soon be extended to other materials and more complex processes on the nanoscale.

Catalysis by Zeolites

The work on metal and oxide surfaces outlined above strongly suggests that first-principles work will make an increasing impact on the understanding of catalysis. This message is reinforced by work of Payne's group on molecular processes in zeolite materials[21, 22, 23, 24, 25, 26]. Zeolites have a very open structure based on silica (SiO_2), with some of the silicon atoms replaced by aluminium. The open spaces in the structure form a regular network of pores connected by channels. Zeolites come in many different forms, but typically the pores are large enough to accommodate sizeable guest molecules, and the channels have diameters of $3-8$ Å, which is large enough to let some molecules diffuse through the crystal. Their atomic-scale porosity gives zeolites a large area of internal surface, and this makes them very effective catalysts. A well-known

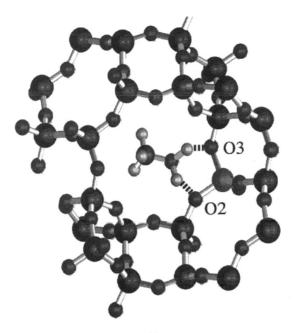

Figure 2. First-principles simulation of Shah *et al.*[21] shows the equilibrium geometry of methanol in the zeolite structure known as chabazite.

example is the conversion of methanol to gasoline using the zeolite known as ZSM-5. Some countries rely on this process for a large part of their gasoline production.

Most of the steps in the conversion of methanol (CH_3OH) to gasoline are still unknown, but the first step is believed to be proton transfer from the acid zeolite cage to form the reactive intermediate $CH_3OH_2^+$. The energetics of this process and its dependence on the zeolite structure have aroused vigorous controversy. In an effort to shed light on this, Payne's group, in collaboration with the group of Julian Gale at Imperial College, have used CETEP calculations[21, 22] on the T3D to look at this problem for methanol in chabazite and sodalite, two of the simplest zeolites. The simulations showed that the $CH_3OH_2^+$ ion is spontaneously formed in chabazite, which has large 8-membered rings of Si–O units separating the pores, but is energetically unfavourable in sodalite, which has smaller 6-membered rings (see Fig. 2).

This work has been very important, because it showed for the first time the key role of zeolite structure in influencing the key process of proton transfer. It augurs well for the ability of first-principles calculations on HPC machines to unravel the entire methanol $\rightarrow$ gasoline pathway and the pathways of other catalytic processes in zeolites in the near future.

Dislocations in Semiconductors

The group of Sutton in the Materials Department at Oxford University have used CETEP calculations to investigate dislocations in silicon. (Dislocations are extended line defects which control the response of the material to stress, and also influence its electrical properties.) The particular problem studied on the T3D concerns the structure and energies of kinks on the 90° partial dislocation. Dislocations move by nucleating and expanding double kinks. Since their motion causes plastic deformation, a good understanding of these kinks is very important.

Because of the complex geometry of the kinks, the CETEP calculations needed to be done on a very large repeating system of 512 atoms, and the availability of the T3D was crucial. Work on this problem had already been done using an empirical interaction model due to Tersoff, but of all the kink types obtained using this model, only one turned out to the stable in the first-principles calculations. Besides allowing the clear identification of the type of kink responsible for the motion of the 90° dislocation, the calculations have also given the first reliable values for the energies of formation and migration of the kink – crucial parameters for understanding the processes involved during the plastic deformation of silicon.

Grain Boundaries in Oxides

A further example of UKCP's first-principles work in the field of materials comes from the work of Bristowe's group on grain boundaries in oxides. In semiconducting materials, grain boundaries are very important because they are responsible for forming states in the electronic band gap which cause the trapping, scattering and recombination of charge carriers. As a result, polycrystalline oxide materials often show strongly non-linear current-voltage characteristics (varistor behaviour), which can be exploited in voltage or surge protector devices.

In an effort to understand this in more detail, Bristowe's group has carried out a CETEP investigation[27, 28, 29] on grain boundaries in TiO_2, an important oxide semiconductor. Full relaxation calculations on systems of over 100 atoms were needed for this work, and the power of the Cray T3D was crucial. The work showed that a high-angle tilt boundary in TiO_2 does indeed exhibit states in the band gap, but these states are not deep enough for the grain boundary to be electrically active. Building on this work, the next stage of the project will focus on the effect of impurities at grain boundaries in TiO_2, since these are known to promote varistor behaviour.

Liquid Metals

Dynamical first-principles simulation allows one to study solids and liquids in thermal equilibrium by using the time evolution to do statistical mechanics. This was first achieved by Štich et al.[30] in their pioneering work on liquid Si, and many other liquids (including water) have since been studied by others. The Keele group has used the Cray T3D to simulate alloys of liquid metals[31, 32, 33], which have long been intensively studied by experimentalists because of the dramatic composition dependence of their electrical properties. A fascinating example is the Ag/Se system, which turns from a good metal to a semi-ionic compound to a wide-gap semiconductor as the composition goes from pure Ag to pure Se. Detailed comparisons with very recent neutron-scattering measurements at the composition Ag_2Se demonstrated the excellent realism of the simulations[33]. The simulations also showed a dramatic change in the structure of the liquid on further increase of the Se content, with Se atoms spontaneously forming chains of increasing length as the pure Se compositions is approached. This allowed a detailed interpretation of the composition-dependent electrical properties.

A new and exciting area for these liquid-metal simulations is a project on the properties of iron under the conditions of the Earth's core. First-principles techniques are becoming increasingly important in the earth sciences, because they can give information which is very hard to obtain in the lab owing to the extreme conditions needed. Two of the great unknowns concerning Fe in the core are the viscosity of the liquid outer core, and the temperature distribution in the core. First-principles calculations can help with both. The Keele group, working in collaboration with the group of David Price at University College London, have recently reported T3D calculations[34, 35] on solid and liquid Fe which reproduce very well the known properties at high pressures and temperatures. The simulations also give an estimate for the viscosity of the liquid in the core which is much lower than some previous estimates would suggest. This has important implications for the nature of convection in the core. First-principles work is also in hand to calculate the melting temperature of Fe at the pressure of the boundary between the solid inner core and the liquid outer core. This will give a crucial constraint on the temperature distribution in the Earth's core.

Liquid Crystals

Liquid crystals have the properties of both solids and liquids. They consist of molecules whose *orientations* are aligned as they would be in a solid, but still the *positions* of the molecules are disordered as they would be in a liquid. They are extremely important technologically because they are widely used in displays – for example in the screen of your laptop computer. Ackland's group has begun a major initiative in the application of first-principles simulation to liquid crystals[36, 37], in order to find out things like:

- the intrinsic properties of the molecules, including electric multipole moments and polarisability, equilibrium shape, flexibility, etc;

- the interaction between the molecules, and the way this determines their behaviour in the liquid crystal;

- the mode and strength of attachment of the molecules to the surfaces of materials like silica used in devices.

Using the CETEP code on the T3D, the group has reported results[36] for the liquid-crystal molecule 4-4'-pentyl-cyanobiphenyl (5CB), for which there are good experimental data. The calculations produced excellent results for the known vibrational frequencies, dipole moment, equilibrium conformation, and the way the energy varies when the two phenyl groups are rotated relative to each other. The work has already made a substantial impact in the liquid crystal community, and a collaboration has been set up with SHARP(UK), which has led to CETEP studies on a ferroelectric liquid-crystal molecule used in practical displays.

This work is important because it is opening up completely new routes to determining the properties which govern the performance of real liquid-crystal devices.

Very Large Systems

Current HPC power is enough to do first-principles calculations on systems of a few hundred atoms. But for many problems this is not nearly enough. Problems involving complex defects in materials, surface nanostructures, or biomolecules – to name but a few examples – demand systems containing thousands of atoms. There is a fundamental problem here, because with current first-principles algorithms the number of computer operations (the op-count) goes at least as N^2, the square of the number of atoms, and this means that increases of HPC power are being poorly exploited. To break this bottleneck, we think it is very important to develop methods in which the op-count is proportional to N – so-called $O(N)$ or linear-scaling methods.

The UKCP groups have pursued two different approaches to this challenge. In the first approach, we aim to keep the full accuracy and generality of current plane-wave methods, while invoking the locality of the density matrix to achieve $O(N)$ behaviour. In the second, some accuracy and generality is sacrificed to allow the total energy to be expressed explicitly in terms of the electron density – this is known as the 'orbital-free' approach.

The first approach, also being followed by groups elsewhere in the world, has been implemented in the CONQUEST code[38, 39, 40], written specifically for parallel computers. The excellent linear-scaling properties of this code have been demonstrated on systems of up to ~ 6,000 atoms[39], but practical results for real-life problems have not been reported yet. The orbital-free approach has been implemented by Madden's group in the OF-AIMD code[41, 42] (the acronym stands for Orbital Free Ab Initio Molecular Dynamics). This code, running on the T3D, has achieved impressive simulations of grain boundaries in solid sodium using systems of ~ 6,000 atoms. The simulations have been used to study the thermal motion of a particular grain boundary in sodium at a temperature just below the melting point. This is the first time that dynamical first-principles simulations have been reported on systems of anything like this size.

OUTLOOK

The availability of HPC has enabled first-principles simulation to make a quantum leap forward. The work of the UKCP consortium illustrates how first-principles calculations are now tackling problems that before were completely inaccessible. Examples of this are our studies of nanoindentation, catalytic reactions in zeolites, and complex extended defects in semiconductors. In other areas, problems that needed a herculean effort can now be treated much more routinely, so that scientific progress is far more rapid. The projects on liquid metals and liquid crystals, and on surface reactions on metals and oxides are in this class. The use of *dynamical* first-principles simulation to study surface reactions is an interesting case. This was first done on an earlier parallel machine by the groups of Gillan and Payne[43], but each simulation at that time needed a sustained effort over several months. The Cray T3D machine has made such calculations almost routine, and they have been exploited throughout our recent work on oxide surfaces.

Another striking feature is the way HPC power is allowing first-principles methods to break completely new ground. Our projects on nanoindentation, catalytic reactions in zeolites, grain boundaries in oxides, and liquid crystals are all in areas where first-principles calculations had never been attempted before. Linear-scaling first-principles methods, often discussed before, have now actually been realised in practice, thanks to the stimulus of HPC.

What of the future? With the new machines now becoming available, first-principles methods are poised for major advances on several fronts. First is chemical reactions on surfaces. The results we have already reported give a foretaste of things to come, but the new machines will make it possible to trace out complete reaction pathways of relevance to catalysis, and to evaluate the influence of co-adsorbates, surface defects etc. A second front will certainly be chemical reactions in liquids, particularly water. Our recent work on liquid metals is but one example of what can be achieved in liquids, and the work of Parrinello, Sprik[44] and others has demonstrated the possibility of studying aqueous solutions from first principles. The next few years will certainly see these techniques used to study hydration structure of complex molecules, and to determine the pathways of reactions in solution. Thirdly, we expect rapid progress in the study of materials under extreme conditions. The work we have already done on liquid iron at Earth's core conditions shows the possibilities, and the new machines will greatly speed up work in this area. Finally, enhanced HPC power will enable linear-scaling first-principles methods to achieve their full potential, and these will find application to a variety of areas including surface nanostructures, complex defects in materials, and biomolecular problems.

ACKNOWLEDGEMENTS

We gratefully acknowledge the support of EPSRC through the High Performance Computing Initiative and through grants to individual groups. Technical support from EPCC and from the HPCI centre at CLRC Daresbury Laboratory has been vital to the work of the UKCP consortium.

REFERENCES

1. M.C. Payne, M.P. Teter, D.C. Allan, T.A. Arias and J.D. Joannopoulos, *Rev. Mod. Phys.*, 64:1045 (1992).
2. M.J. Gillan, *Contemp. Phys.*, 38:115 (1997).
3. L.J. Clarke, I. Štich and M.C. Payne, *Comput. Phys. Commun.*, 72:14 (1992).
4. J.A. White, D.M. Bird, M.C. Payne and I. Štich, *Phys. Rev. Lett.*, 73:1404 (1994).
5. M. Kay, G.R. Darling, S. Holloway, J.A. White and D.M. Bird, *Chem. Phys. Lett.*, 245:311 (1995).
6. J.A. White, D.M. Bird and M.C. Payne, *Phys. Rev. B*, 53:1667 (1996).
7. P.A. Gravil, J.A. White and D.M. Bird, *Surf. Sci.*, 352-354 (1996).
8. P.A. Gravil, D.M. Bird and J.A. White, *Phys. Rev. Lett.*, 77:3933 (1996).
9. D.M. Bird and P.A. Gravil, *Surf. Sci.*, 377-9:555 (1997).
10. A. Alavi, P. Hu, T. Deutsch, P.L. Silvistrelli and J. Hutter, *Phys. Rev. Lett.*, submitted.
11. L.N. Kantorovich, J.M. Holender and M.J. Gillan, *Surf. Sci.*, 343:221 (1995).
12. L.N. Kantorovich, M.J. Gillan and J.A. White, *J. Chem. Soc. Faraday Trans.*, 92:2075 (1996).
13. L.N. Kantorovich and M.J. Gillan, *Surf. Sci.*, 374:373 (1997).
14. L.N. Kantorovich and M.J. Gillan, *Surf. Sci.*, 376:169 (1997).
15. P.J.D. Lindan, N.M. Harrison, J.M. Holender, M.J. Gillan and M.C. Payne, *Surf. Sci.*, 364:431 (1996).
16. P.J.D. Lindan, N.M. Harrison, J.M. Holender and M.J. Gillan, *Chem. Phys. Lett.*, 261:246 (1996).
17. P.J.D. Lindan, N.M. Harrison, J.A. White and M.J. Gillan, *Phys. Rev. B*, 55:15919 (1997).
18. P.J.D. Lindan, N.M. Harrison and M.J. Gillan, *Phys. Rev. Lett.*, 80:762 (1998).
19. R. Pérez, M.C. Payne and A.D. Simpson, *Phys. Rev. Lett.*, 75:4748 (1995).
20. R. Pérez, M.C. Payne, I. Štich and K. Terakura, *Phys. Rev. Lett.*, 78:678 (1997).
21. R. Shah, M.C. Payne, M.-H. Lee and J.D. Gale, *Science*, 271:1395 (1996).
22. R. Shah, J.D. Gale and M.C. Payne, *J. Phys. Chem.*, 100:11688 (1996).
23. R. Shah, J.D. Gale and M.C. Payne, *Int. J. Quantum Chem.*, 61:393 (1997).
24. R. Shah, J.D. Gale and M.C. Payne, *Chem. Comm.*, 1:131 (1997).
25. R. Shah, J.D. Gale and M.C. Payne, *Phase Trans.*, 61:67 (1997).
26. R. Shah, J.D. Gale and M.C. Payne, *J. Phys. Chem. B*, 101:4787 (1997).
27. I. Dawson, P.D. Bristowe, M.-H. Lee, M.C. Payne, M.D. Segall and J.A. White, *Phys. Rev. B*, 54:13727 (1996).
28. I. Dawson, P.D. Bristowe, M.C. Payne and M.-H. Lee, *Mat. Res. Soc. Proc.*, 408:271 (1996).
29. I. Dawson, P.D. Bristowe, J.A. White and M.C. Payne, *Mat. Res. Soc. Proc.*, 409:453 (1997).
30. I. Štich, R. Car and M. Parrinello, *Phys. Rev. Lett.*, 63:2240 (1989).
31. J.M. Holender, M.J. Gillan, M.C. Payne and A.D. Simpson, *Phys. Rev. B*, 52:967 (1995).
32. F. Kirchhoff, J.M. Holender and M.J. Gillan, *Europhys. Lett.*, 33:605 (1996).

33. F. Kirchhoff, J.M. Holender and M.J. Gillan, *Phys. Rev. B*, 54:190 (1996).

34. L. Vočadlo, G. de Wijs, G. Kresse, M.J. Gillan and G.D. Price, *Faraday Disc.*, 106:205 (1997).

35. G.A. de Wijs, G. Kresse, L. Vočadlo, D. Dobson, D. Alfè, M.J. Gillan and G.D. Price, *Nature*, in press.

36. S.J. Clark, C.J. Adam, J.A. White, G.J. Ackland and J. Crain, *Liquid Crystals*, 22:469 (1997).

37. C.J. Adam, S.J. Clark, G.J. Ackland and J. Crain, *Phys. Rev. E*, 55:5641 (1997).

38. E. Hernández, M.J. Gillan and C.M. Goringe, *Phys. Rev. B*, 53:7147 (1996).

39. C.M. Goringe, E. Hernández, C.M. Goringe and I.J. Bush, *Comput. Phys. Commun.*, 102:1 (1997).

40. M.J. Gillan, D. Bowler, C. Goringe and E. Hernández, *Proc. Symposium on Complex Liquids, Nagoya, 1997*, T. Fujiwara *et al.* eds., in press.

41. B.J. Jesson, M. Foley and P.A. Madden, *Phys. Rev. B*, 55:4941 (1997).

42. S. Watson, B.J. Jesson, E.A. Carter and P.A. Madden, *Europhys. Lett.*, accepted.

43. A. De Vita, I. Štich, M.J. Gillan, M.C. Payne and L.J. Clarke, *Phys. Rev. Lett.*, 71:1276 (1993).

44. K. Laasonen, M. Sprik, M. Parrinello and R. Car, *J. Chem. Phys.*, 99:9080 (1993)

ON THE QUASI-PARTICLE SPECTRA OF YBA$_2$CU$_3$O$_7$

W.M. Temmerman,[1] B.L. Gyorffy,[2] Z. Szotek,[1] O.K. Andersen,[3] and
O. Jepsen[3]

[1]Daresbury Laboratory, Warrington, WA4 4AD, UK
[2]H H Wills Physics Laboratory, University of Bristol,
  Tyndall Avenue, Bristol, BS8 1TL, UK
[3]Max-Planck-Institut für Festkörperforschung,
  Postfach 800665, D-70506, Stuttgart, Federal Republic of Germany

INTRODUCTION

As is well known, the phenomenon of superconductivity arises when in a metal
electrons pair up and occupy a single quantum state. Since electrons normally repel
each other, one of the principal questions in the case of any superconductor is "why
do such Cooper pairs form?". Whilst for the conventional superconductors the answer
is that the attraction is due to the electron–phonon coupling, in the case of the new,
high temperature, superconductors (HTSC)[1], in spite of the unprecedented effort of
the past few years, the physical cause of the pairing remains a mystery[2]. Rather than
speculating on the microscopic nature of the pairing, in this paper, we implement a
semi-phenomenological strategy, whose aim is to determine which local orbitals the electrons occupy when they experience the attraction. Of course, because our description
of pairing is semi-phenomenological, we shed new light on the physical mechanism of
pairing only indirectly. As will be shown presently, progress towards such a goal is made
possible, in principle, by a particularly efficient representation of the electron-electron
interaction afforded by the Density Functional Theory (DFT) of Superconductivity[3, 4]
and, in practice, by the development of powerful numerical methods for solving the
corresponding Kohn–Sham–Bogoliubov–de Gennes (KS-BdG) equation[5, 6]. A central
feature of this theory is an *electron-electron interaction kernel* $K(\mathbf{r}_1, \mathbf{r}'_1; \mathbf{r}_2, \mathbf{r}'_2)$ which,
when corresponding to an attractive interaction, leads to superconductivity. This at-
tractive interaction is parametrized by a set of interaction constants $K_{RL,R'L'}$, where
R, R' and L, L' refer to the positions and orbital character, respectively, of the two
electrons.

Given the success of the Linear Muffin-Tin Orbital (LMTO) method in describing
the electronic structures of the HTSCs in the normal state without adjustable parameters, and yet, in material specific quantitative detail[8, 9], the above approach can be
said to be built on solid foundations. In particular we expect that, even though we
make a local approximation and treat the expansion coefficients $K_{RL,R'L'}$ as the ad-

High Performance Computing
Edited by R. J. Allan *et al.*, Kluwer Academic / Plenum Publishers, New York, 1999

justable parameters of the theory, they remain physically meaningful. Namely, if a specific coefficient $K_{RL,R'L'}$ with all the others set equal to zero, can be adjusted to give a good quantitative account of the quasi-particle spectrum in the superconducting state of a particular superconductor, then we shall conclude with appropriate caution, that the attraction operates between electrons in orbitals RL and $R'L'$. In this application to $YBa_2Cu_3O_7$ we shall use the so-called eight-band orthonormal tight-binding Hamiltonian[10] which provides quantitative insights to the band structure.

DENSITY FUNCTIONAL THEORY FOR SUPERCONDUCTORS

This is a relatively new formulation of the theory of superconductivity[3, 4, 11]. Although in its simplest version it deals only with the case of the instantaneous electron-electron interaction, with this restriction it is in principle exact and is fully equivalent to possible strong-coupling theories based on canonical perturbation theory[12]. However, the conventional, retarded electron-phonon mechanism of attraction can only be treated to the extent that it can be represented by a static, effective electron-electron interaction-potential. On the other hand the theory is fully applicable to pairing mechanisms due to electron-electron correlations such as spin fluctuations[13].

The DFT for superconductors is a close analogue of the very successful Spin-Density-Functional theory of magnetism[14]. Indeed one merely has to replace the magnetization density $m(r)$ by the superconducting order parameter, the *pairing amplitude*,

$$\chi(\mathbf{r}, \mathbf{r}') = \psi_\uparrow(\mathbf{r}), \psi_\downarrow(\mathbf{r}') , \tag{1}$$

where $\psi_\uparrow(\mathbf{r})$ and $\psi_\downarrow(\mathbf{r}')$ are electron annihilation operators for respectively a spin-up electron at $\mathbf{r}$ and a spin-down electron at $\mathbf{r}'$, and the theory of an equilibrium state with spontaneously broken gauge symmetry unfolds with minor modifications. In short, one introduces the grand potential functional $\Omega[n, \chi]$ of the electronic density $n(\mathbf{r})$ and pairing amplitude $\chi(\mathbf{r}, \mathbf{r}')$, and finds that it is minimized by

$$n(\mathbf{r}) = 2\sum_j [1 - f(E_j)]|v_j(\mathbf{r})|^2 + f(E_j)|u_j(\mathbf{r})|^2 \tag{2}$$

and

$$\chi(\mathbf{r}, \mathbf{r}') = \sum_j [1 - f(E_j)]u_j(\mathbf{r})v_j^*(\mathbf{r}') - f(E_j)u_j(\mathbf{r}')v_j^*(\mathbf{r}) \tag{3}$$

where $u_j(\mathbf{r})$ and $v_j(\mathbf{r})$ are respectively the electron- and hole-components of a Kohn-Sham like eigenvalue problem with the corresponding eigenvalues E_j. Moreover, $f(E) \equiv [1 + \exp(E/k_B T)]^{-1}$ is the Fermi function. The normalization of the eigenfunctions is of course such that

$$\int |u_j(\mathbf{r})|^2 d^3r + \int |v_j(\mathbf{r})|^2 d^3r = 1. \tag{4}$$

Remarkably, this eigenvalue problem is of the Bogoliubov-de Gennes (BdG) form[3, 4],

$$\left(-\frac{1}{2}\nabla^2 + V(\mathbf{r}) - \mu\right)u_j(\mathbf{r}) + \int \Delta(\mathbf{r}, \mathbf{r}')v_j(\mathbf{r}')d^3r' = E_j u_j(\mathbf{r})$$

$$-\left(-\frac{1}{2}\nabla^2 + V(\mathbf{r}) - \mu\right)v_j(\mathbf{r}) + \int \Delta^*(\mathbf{r}, \mathbf{r}')u_j(\mathbf{r}')d^3r' = E_j v_j(\mathbf{r}), \tag{5}$$

with the *effective pairing potential* $\Delta(\mathbf{r}, \mathbf{r}')$ being the functional derivative with respect to the pairing amplitude of the exchange-correlation contribution to the grand potential functional $\Omega_{xc}[n, \chi]$, i.e.,

$$\Delta(\mathbf{r}, \mathbf{r}') = \frac{\delta\Omega_{xc}[n, \chi]}{\delta\chi(\mathbf{r}, \mathbf{r}')}. \tag{6}$$

In the BdG equation (5), μ is the chemical potential,

$$V(\mathbf{r}) = V_{\text{ext}}(\mathbf{r}) + \int \frac{n(\mathbf{r}')}{|\mathbf{r} - \mathbf{r}'|} d^3 r' + \frac{\delta \Omega_{\text{xc}}[n, \chi]}{\delta n(\mathbf{r})} \tag{7}$$

is the *effective single-electron potential*, which is assumed independent of spin, and $V_{\text{ext}}(\mathbf{r})$ is the external potential, e.g. the Coulomb attraction from the protons. Since expressions (6) and (7) show that the effective potentials in the BdG equation (5) depend on the density and pairing amplitude, these coupled single-particle equations must be solved self-consistently using Eqs. (2)-(3) and (6)-(7). The effective single-particle spectrum E_j of the BdG equation may be thought of as the normal-state single-electron spectrum ε_i, doubled up by folding around the chemical potential μ, and subsequently split by the pairing potential Δ. Evidently, whilst exact, the above theory is useless until an explicit, approximate form for the functional $\Omega_{\text{xc}}[n, \chi]$ has been selected. To make such selection we rewrite without loss of generality[5] Eq. (6) as

$$\Delta(\mathbf{r}_1, \mathbf{r}_1') = \int \int K(\mathbf{r}_1, \mathbf{r}_1'; \mathbf{r}_2, \mathbf{r}_2') \, \chi(\mathbf{r}_2, \mathbf{r}_2') \, d^3 r_2 \, d^3 r_2', \tag{8}$$

where

$$K(\mathbf{r}_1, \mathbf{r}_1'; \mathbf{r}_2, \mathbf{r}_2') \equiv \frac{\delta^2 \Omega_{\text{xc}}[n, \chi]}{\delta \chi(\mathbf{r}_1, \mathbf{r}_1') \, \delta \chi(\mathbf{r}_2, \mathbf{r}_2')}.$$

The pair-interaction kernel $K(\mathbf{r}_1, \mathbf{r}_1'; \mathbf{r}_2, \mathbf{r}_2')$ describes the scattering of an $(\uparrow, \downarrow)$-pair at $(\mathbf{r}_2, \mathbf{r}_2')$ into an $(\uparrow, \downarrow)$-pair at $(\mathbf{r}_1, \mathbf{r}_1')$, and also the scattering of a $(\downarrow, \uparrow)$-pair at $(\mathbf{r}_2, \mathbf{r}_2')$ into a $(\downarrow, \uparrow)$-pair at $(\mathbf{r}_1, \mathbf{r}_1')$. This kernel is a functional of $n(\mathbf{r})$ and $\chi(\mathbf{r}, \mathbf{r}')$. In what follows we have parameterized K and employed the usual Local Density Approximation (LDA)[14] for $\delta \Omega_{\text{xc}}[n, \chi] / \delta n(\mathbf{r})$

EIGHT-BAND MODEL

Although the BdG equation, with the parametrized kernel K, can be solved self-consistently for Nb, using the $\mathbf{k}-$ and band-dependent formulation of the formalism and the LMTO method[5], the interesting cases of the HTSCs still represent a challenge with currently available computers. Thus, to explore the potentials of the formalism, for YBa$_2$Cu$_3$O$_7$ we adopted a computationally more tractable, albeit less accurate approach; we used the so-called eight-band model[10] for the LDA energy bands $\varepsilon_{\mathbf{k}i}$ and wave functions $\psi_{\mathbf{k}i}(\mathbf{r})$.

The eight-band model[15] gives a good quantitative description of the first principles LDA band structure of the superconducting cuprates, in the normal state, within an eV of the Fermi energy ε_F. In the 'eight-band model', orbitals located on non-generic structural elements separating the CuO$_2$ (bi)layers, such as the chain in YBa$_2$Cu$_3$O$_7$, have been deleted. As a consequence, the 'eight-band model' is two-dimensional*. As mentioned above, the attractive interaction is parametrized by a set of interaction constants $K_{RL,R'L'}$. In previous work[6, 7] we have investigated several cases where all but one of the coupling constants, $K_{RL,R'L'}$, were set equal to zero, and the other parameters of the model were those appropriate to a CuO$_2$ bilayer of a YBa$_2$Cu$_3$O$_7$ crystal. Here we concentrate on one of these scenarios of $d-$wave symmetry corresponding to the intra-layer, nearest neighbour, Cu $d_{x^2-y^2}-$Cu $d_{x^2-y^2}$ interaction. The coupling constant, $K_{dd}(a)$, was determined by the requirement that T_c agrees with experiments. For the

---

*The strong k_z-dispersion found in LDA calculations for YBa$_2$Cu$_3$O$_7$ and which is caused by hopping across the chain, has found little experimental support.

case of the intra-layer, nearest neighbour, Cu $d_{x^2-y^2}$−Cu $d_{x^2-y^2}$ interaction, the coupling constant, $K_{dd}(a)$, which gave T_c=92 K, worked out to be 0.68 eV[7].

The 'eight-band model' for a CuO$_2$ layer features eight orthonormal local orbitals φ_L (L=1 to 8). The first four of these are the σ-orbitals: O2 p_x, Cu $d_{x^2-y^2}$, O3 p_y, and Cu s. The others are the π-orbitals: O2 p_z, Cu d_{xz}, O3 p_z, and Cu d_{yz}. The centres of the Cu orbitals form a slightly orthorhombic planar lattice, and the oxygen atoms occupy slightly out of the plane positions, giving rise to a structure usually referred to as dimpled. A bilayer is a pair of such CuO$_2$ layers arranged as mirror images of each other with respect to the plane where yttrium atoms would be in a real material. So, the full unit cell of a bilayer contains in total 2×8 orbitals.

To calculate the quasi-particle spectrum for the 'eight-band model', we have solved the tight-binding Bogoliubov-de Gennes equation:

$$\sum_L \left[\begin{array}{cc} H^{\mathbf{k}}_{L'L} - \left(\mu + E^{\mathbf{k}}_j\right)\delta_{L'L} & \Delta^{\mathbf{k}}_{L'L} \\ \Delta^{\mathbf{k}\,*}_{LL'} & -H^{\mathbf{k}}_{L'L} + \left(\mu - E^{\mathbf{k}}_j\right)\delta_{L'L} \end{array} \right] \left[\begin{array}{c} u^{\mathbf{k}}_{L,j} \\ v^{\mathbf{k}}_{L,j} \end{array} \right] = \left[\begin{array}{c} 0 \\ 0 \end{array} \right] \qquad (9)$$

for the particle and hole amplitudes $u^{\mathbf{k}}_{L,j}$ and $v^{\mathbf{k}}_{L,j}$, respectively, where $\mathbf{k}$ is the Bloch vector, j is the band index, and L refers to the orbitals of the model. Here $H^{\mathbf{k}}_{L'L}$ is the lattice Fourier transform of the Hamiltonian matrix in the local orbitals representation, and the pairing potential matrix $\Delta^{\mathbf{k}}_{L'L}$ is related to the pairing amplitude $\chi^{\mathbf{k}}_{LL'}$, and the lattice Fourier transform of the interaction matrix $K^{\mathbf{k}}_{LL'}$. Namely,

$$\Delta^{\mathbf{k}}_{LL'} = \sum_{\mathbf{k}'} K^{\mathbf{k}-\mathbf{k}'}_{LL'} \chi^{\mathbf{k}'}_{LL'} \qquad (10)$$

and

$$\chi^{\mathbf{k}}_{LL'} = \sum_j \left[1 - 2f\left(E^{\mathbf{k}}_j\right)\right]\left(u^{\mathbf{k}}_{L,j} v^{\mathbf{k}\,*}_{L',j} + u^{\mathbf{k}}_{L',j} v^{\mathbf{k}\,*}_{L,j}\right)/2. \qquad (11)$$

Since these are the Kohn-Sham equations of a DFT, they are to be solved fully self-consistently with respect to $H^{\mathbf{k}}_{L'L}$ and $\Delta^{\mathbf{k}}_{L'L}$. However, to simplify matters we have taken $H^{\mathbf{k}}_{L'L}$ to be the same as was determined by a normal state calculation[7], and adjusted only $\Delta^{\mathbf{k}}_{L'L}$ and the chemical potential μ, to maintain the prescribed band filling, after each iteration.

RESULTS

Here we discuss our results for the gap anisotropy, low temperature specific heat and T_c vs. doping, and compare them to the relevant experiments.

For the attractive interaction operating between intra-layer, nearest neighbour, Cu $d_{x^2-y^2}$−Cu $d_{x^2-y^2}$ orbitals, we found the gap, on the Fermi surface shown in Fig. 1, to have the expected d−wave symmetry and to agree, quantitatively, with the values deduced from photoemission experiments[7, 16]. To highlight this, in Fig. 2 we compare our calculated gaps for the nearest-neighbour Cu $d_{x^2-y^2}$− Cu $d_{x^2-y^2}$ coupling, with the one deduced from the high-resolution experiments of Schabel et al.[16]. Considering the fact, that we **did not** fit our *one* free parameter, $K_{dd}(a)$, to any feature of the gap measured by Schabel et al.[16], but only to obtain T_c of 92K, the near quantitative agreement between theory and experiment for $\Delta^{\mathbf{k}}(0)$ (see Fig. 2) can be taken as evidence that the relation between the attractive force, represented by $K_{dd}(a)$, and the gap in the quasi-particle spectrum is correctly described by our BdG equation with the eight-band model.

150

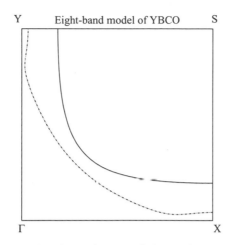

Figure 1. The odd (chain curve) and even (full curve) sheets of the Fermi surface of a CuO$_2$ bilayer of YBa$_2$Cu$_3$O$_7$ in the irreducible part of the two-dimensional Brillouin zone. It is this part of the electronic structure which the eight-band model is designed to reproduce accurately.

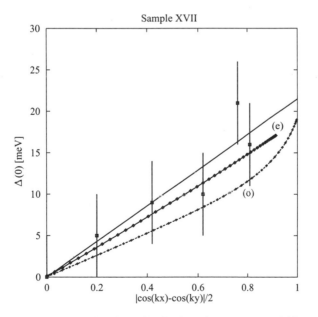

Figure 2. A comparison of the calculated gap for the intra-layer nearest-neighbour Cu $d_{x^2-y^2}$−Cu $d_{x^2-y^2}$, scenario with K_{dd} (a)=0.68 eV for the even (e) and odd (o) sheets of the Fermi surface, with the experimental data deduced by Schabel et al. 16, from photoemission measurements on their sample XVII.

There are two features of the specific heat $C_v^S(T)$ in the superconducting state that are of general interest: One is the size and nature of the jump, $\left[C_v^S(T_c) - C_v(T_c)\right]/C_v(T_c)$, and the other is the temperature dependence as T goes to zero. Here C_v^S and C_v stand for specific heats of superconducting and normal states, respectively. Unfortunately our eight-band model calculations became very difficult near T_c where the gap is very small. Also, fluctuation effects, which are large for high T_c superconductors, on account of the short coherence length, $\xi_0 \sim a$, make the interpretation of the experiments in terms of a jump problematic[17]. Therefore we did not pursue our computations of $C_v^S(T)$ near T_c. The second noteworthy feature of $C_v^S(T)$ is its approach to zero at low temperatures. In the usual 's-wave' case this is an exponential decay whilst for 'd-wave' superconductors is a power law. In our, two-dimensional calculations, the latter is T^2. Although the interpretation of the low temperature specific heat measurements is fraught with difficulties[18, 19], nevertheless, Moler et al.[20] were able to extract a T^2 contribution from their measurements on very high quality YBCO crystals and found $C_v^S(T)^{\text{exp}} = 0.95\,(T/T_c)^2$ (J/mole K). This is remarkably close to our result, $0.93\,(T/T_c)^2$ (J/mole K) for the Cu $d_{x^2-y^2}-$ Cu $d_{x^2-y^2}$ intra-layer nearest-neighbour scenario. Namely, the coefficient of the $(T/T_c)^2$ contributions to the low temperature specific heat worked out to be within a percent or so of the experimental value[7, 20].

One of the interesting, generic features of the first-principles calculations of the electronic structure of the HTSC materials, seems to be a set of bifurcated saddle points near the Fermi energy ε_F. We discovered[21] that the evolution of the relative position of ε_F, with respect to this Van Hove singularity, with doping can give a good quantitative account of the experimentally observed rise and fall of T_c accompanying such changes (see Fig. 3). We identify the characteristic saddle points of the 'eight-band model' as the cause of this behaviour. Such evidence in support of the Van Hove scenario may turn out to be a significant step towards solving the central problem of HTSC's, namely identifying the mechanism of pairing. Another point of interest is the width of the T_c versus $-\delta n$ curve. As indicated in Fig. 3, our calculations for the Cu $d_{x^2-y^2}-$Cu $d_{x^2-y^2}$ scenario imply a width at half maximum, $\delta n = 0.14$. This is remarkably close to the universal T_c versus $-\delta n$ curve deduced from experiments by Markiewicz[22]. In particular, he finds $\delta n = 0.15$.

NUMERICAL IMPLEMENTATION

The calculations are excess of 40,000 times more time consuming than normal bandstructure calculations. Since the Hamiltonian is doubled in size the diagonalization is eight times more expensive. The formalism involves two self-consistency cycles: first with respect to the normal and anomalous charges, and the second with respect to the chemical potential. On average about five guesses of the chemical potential were sufficient to find the correct one, making this part of the calculation five times more expensive. The self-consistency with respect to the charges is strongly dependent on the temperature at which the BdG equations are solved. At low temperatures usually about 100 iterations suffice, which is just slightly more than for a standard bandstructure calculations. However at temperatures closer to T_c thousands of iterations are required. Moreover, the **k** mesh for self-consistency requires in the excess of 65,500 points, namely a **k** mesh of 256 by 256 for the two dimensional Brillouin zone integral. In a standard bandstructure calculation an 8 by 8 mesh is usually sufficient. Futhermore, even the division by 256 is not sufficient to calculate the low temperature specific heat and this number needed to be increased by an order of magnitude. Therefore the

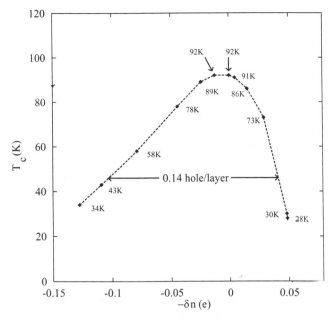

Figure 3. T_c versus deviation of the number of holes from that at optimal doping for the CuO_2 bilayer of $YBa_2Cu_3O_7$ for the intra-layer nearest-neighbour Cu $d_{x^2-y^2}$–Cu $d_{x^2-y^2}$ scenario. On the curve T_c's corresponding to different hole concentrations are marked.

consideration of the eight-band model where one diagonalizes a 32 by 32 Hamiltonian (due to two different sheets of the Fermi surface), becomes a problem which can only be tackled on large parallel supercomputers. Fortunately, this problem lends itself to an extremely efficient parallelization strategy which involves the parallelization over the k-points of the diagonalization for which we used standard LAPACK routines. The parallellization was accomplished trivially by looping over the number of k-points in the two dimensional Brillouin zone in steps of the number of processors. A global summation involving all processors was carried out to obtain the anomalous charge density for the next iteration. The loop over k-points already occurred in the main subroutine of the program ensuring that no compute cycles were wasted before and after the parallellization loop. Converging a calculation for a particular pairing interaction at a given, low, temperature requires around 32 SP2 processor hours; for temperatures close to a critical temperature T_c it requires of the order of 200 SP2 processor hours per temperature.

CONCLUSIONS

We have presented a semi-phenomenological approach to interpreting experiments which probe the electronic structure of $YBa_2Cu_3O_7$ in the superconducting state. Its principle virtue is that it is able to make contact with the experimental data in quantitative detail whilst avoiding the proliferation of adjustable parameters. It is based on an effective parameterization of the electron-electron interaction with the Density-Functional description of the superconducting state and the Tight-Binding-LMTO method for solving the Kohn-Sham-Bogoliubov-de Gennes equations. We have concluded that the experimental data from photoemission and specific heat measurements favours an electron-electron interaction which is predominantly between electrons with opposite spins sitting on nearest-neighbour Cu-sites of the CuO_2 layers. It should be stressed that d-wave pairing is not assumed, but is the direct consequence of this particular interaction and the electronic structure described by the eight-band model. Moreover, the fact that T_c, the quasi-particle spectra, the low temperature specific heat, and the T_c versus number of holes are all described quantitatively using only one adjustable parameter $K_{dd}(a)$, suggest that irrespective of the nature of the pairing interaction, these properties are related to each other as dictated by the structure of the simplest BdG equations.

REFERENCES

1. J.G. Bednorz and K.A. Müller, *Z. Phys. B*, 64:189 (1986).
2. J.F. Annett, N. Goldenfeld, and A.J. Leggett, in: *Physical Properties of High Temperature Superconductors*, Vol. 5, D.M. Ginsberg ed., World Scientific, Singapore (1996).
3. L.N. Oliveira, E.K.U. Gross, and W. Kohn, *Phys. Rev. Lett.*, 60:2430 (1988).
4. W. Kohn, E.K.U. Gross, and L.N. Oliveira, *J. Phys. (Paris)*, 50:2601 (1989).
5. M.B. Suvasini, W.M. Temmerman, and B.L. Gyorffy, *Phys. Rev. B*, 48:1202 (1993).
6. W.M. Temmerman, Z. Szotek, B.L. Gyorffy, O.K. Andersen, and O. Jepsen, *Phys. Rev. Lett.*, 76:307 (1996).
7. B.L. Gyorffy, Z. Szotek, W.M. Temmerman, O.K. Andersen, and O. Jepsen, (1997) submitted.
8. O.K. Andersen, A.I. Liechtenstein, C.O. Rodriguez, I.I. Mazin, O. Jepsen, V.P. Antropov, O. Gunnarsson, S. Gopalan, *Physica C*, 185-189:147 (1991).
9. O.K. Andersen, A.I. Liechtenstein, O. Jepsen, and F. Paulsen, *J. Phys. Chem. Solids*, 56:1573 (1995).
10. O.K. Andersen, O. Jepsen, A.I. Liechtenstein, and I.I. Mazin, *Phys. Rev. B*, 49:4145 (1994).
11. E.K.U. Gross, Stefan Kurth, Klaus Capelle, and Martin Lüders, in: *Density Functional Theory*, Vol. 337 of NATO ASI Series B, E.K.U. Gross and R.M. Dreitler eds., Plenum Press, New York (1995).
12. A.L. Fetter and J.D. Walecka, *Quantum Theory of Many-Particle Systems*, McGraw Hill (1971); A.A. Abrikosov, L.P. Gorkov, and I.Ye. Dzaloshinskii, *Quantum Field Theoretical Methods in Statistical Physics*, Pergamon Press (1965).
13. P. Monthoux, A. Balatsky, and D. Pines, *Phys. Rev. Lett.*, 62:961 (1989); D. Pines and P.J. Monthoux, *J. Phys. Chem. Solids*, 56:1651 (1995).
14. R.M. Dreizler and E.K.U. Gross, *Density Functional Theory* Springer-Verlag, Berlin-New York (1990).
15. O.K. Andersen, O. Jepsen, A.I. Liechtenstein, and I.I. Mazin, *Phys. Rev. B*, 49:4145 (1994); O.K. Andersen, A.I. Liechtenstein, O. Jepsen, and F. Paulsen, *J. Phys. Chem. Solids*, 56:1573 (1995).
16. M.C. Schabel, C.H. Park, A. Matsuura, Z-X Shen, D.A. Bonn Ruixing Liang, and W.N. Hardy, *Phys. Rev. B*, 55:2796 (1997).
17. A. Junod, in: *Physical properties of high temperature superconductors II*, D.M. Ginsburg ed., p13, World Scientific, Singapore, (1990).
18. J.W. Redcliffe, J.W. Loram, J.M. Wade, G. Witschek, and J.L. Talon, T. *Low. Temp. Phys.*, 105:903 (1996); A. Junod, in: *Studies in HTC*, A.V. Naklian ed., Ch. 15, Nova Scientific, New York, (1996).
19. Z.-X. Shen and D.S. Dessau, *Physics Reports*, 253:1 (1995).
20. K.A. Moler, D.J. Baar, J.S. Urbach, Ruixing Liang, W.N. Hardy, and A. Kapitulnik, *Phys. Rev. Lett.*, 73:2744 (1994).
21. Z. Szotek, B.L. Gyorffy, W.M. Temmerman, and O.K. Andersen, (1997) submitted.
22. R.S. Markiewicz, *J. Phys. Chem. Solids*, 58:1173-1310 (1997).

AB INITIO STUDIES OF HYDROGEN MOLECULES IN SILICON

B. Hourahine[1], R. Jones[1], S. Öberg[2], R. C. Newman[3], P. R. Briddon [4]
and E. Roduner[5]

[1]Department of Physics,
   The University of Exeter, Exeter, EX4 4QL
[2]Department of Mathematics,
   University of Luleå, Luleå, S-97187, Sweden
[3]Semiconductor IRC, Imperial College,
   Prince Consort Road, London, SW7 2BZ, UK
[4]Department of Physics,
   The University of Newcastle upon Tyne,
   Newcastle upon Tyne, NE1 7RU, UK
[5]Institut für Physikalische Chemie,
   Pfaffenwaldring 55, D-70569, Stuttgart, Germany

ABSTRACT

The results of first principles calculations of the structure and vibrational modes of H_2 molecules in Si located at i) isolated interstitial sites, ii) bound to oxygen impurities in the silicon and iii) trapped in voids in the lattice are reported. These results are compared with recent experimental investigations. The isolated molecule is found to lie at a T_d interstitial site, oriented along [011] and is infra-red active. The rotational barrier is at least 0.17 eV. The molecular frequency is a sensitive function of cage size and increases to the gas value for cages about 40% larger than the T_d site. We find that it is possible for H_2 molecules to cause an upward shift in the antisymmetric stretch mode of O_i when H is replaced by D, which could explain the anomalous shift in the 1075 cm^{-1} O-H related local vibrational mode. It is suggested that Raman active modes around 4158 cm^{-1} are due to molecules within voids.

INTRODUCTION

It has long been expected that hydrogen molecules exist as dark matter inside silicon[1, 2] but a major problem has been in devising an experiment which detects them. Very recently, experimental observations on two distinct types of molecular hydrogen within silicon have been reported[3, 6].

The experimental group at Imperial College Group observe an infra-red vibrational spectrum from hydrogen sorbed into silicon by soaking in an atmosphere of hydrogen.

High Performance Computing
Edited by R. J. Allan *et al.*, Kluwer Academic / Plenum Publishers, New York, 1999

Of great interest are a number of absorption bands detected around 3600 cm^{-1}. Our ab initio modelling studies, performed within the E6 HPCI consortium, and carried out on the T3D and T3E have revealed that these modes are due to hydrogen molecules both free and trapped by oxygen. The molecule is peculiar in a) being unable to rotate and hence without ortho and para molecular forms, and b) possessing a stretch frequency which is red-shifted from the gas value by an extraordinary large amount due to the screening of the intra-molecular electrons by those from the host, and c) being infra-red active.

By considering the effect of the size of the surrounding lattice cages, we can also show that the Raman active modes observed around 4158 cm^{-1} in material treated with hydrogen plasma are due to molecules trapped within voids in the silicon.

The detection and modelling of H$_2$ within silicon is a major step forward in understanding the behaviour of hydrogen in an extremely important technological material.

Calculational Methods

All of the reported structural and vibrational calculation were performed using the AIMPRO ab initio modelling program. The key features of this code are:

- **Density functional calculations** with either local spin or generalised gradient approximations

- **Real or reciprocal space calculations**

- **Fully self–consistent calculations**

- **All electron or norm–conserving pseudopotential calculations**

- **Cartesian Gaussian basis set**

- **Structure optimisation using a conjugate gradient method** with analytic evaluation of forces

- **Efficient parallel scaling** using native PBLAS and BLACS libraries

THREE TRAPS FOR H$_2$ IN SILICON

Interstitial H$_2$

Recently, a vibrational band at 1075 cm^{-1} due to an oxygen-hydrogen defect has been detected by Fourier transform infra-red spectroscopy in Czochralski grown Si which had been soaked in hydrogen at high temperatures[3]. The presence of hydrogen in the defect was confirmed by an upward shift of 1 cm^{-1} when H was replaced by D. Subsequent work [4] showed that the 1075 cm^{-1} band was composed of a line at 1075.1 cm^{-1} and a shoulder at 1075.8 cm^{-1}. Satellite bands were detected below the main peak, and are analogous with similar satellites seen for interstitial oxygen and attributed to ^{29}Si-^{16}O-^{28}Si and ^{30}Si-^{16}O-^{28}Si defects. This strongly implies that oxygen is also involved and in its normal bond centred location. Studies using mixtures of H and D showed that two hydrogen atoms were involved in the defect.

The infra-red measurements also revealed sharp H-related modes labelled ν_1, ν_2 and ν_3 at 3789, 3731 and 3618 cm^{-1} which shifted downwards by approximately a factor of $\sqrt{2}$ in the D case. In the mixed H-D case, an additional mode lay mid-way between the pure isotopic ones for each ν_i, implying that the hydrogen atoms are bonded together. The integrated intensities of ν_1 and ν_2 were correlated with the oxygen related 1075 cm^{-1} band implying that the molecule is bound near oxygen in these defects. Recently, studies have been made of the affect of annealing on the 1075 cm^{-1} and ν_3 bands[5]. On heating to between 50 and 130°C the intensities of ν_1, ν_2 and the 1075 cm^{-1} band decreased and there was a correlate d increase in the intensity of the ν_3 band. These are reversible changes so that re-heating the samples again at a lower temperature results in a decrease in the intensity of ν_3 and a recovery of the other bands. This can be understood if the molecule in the ν_3 site is less stable than those in the ν_1 and ν_2 centres but has a greater entropy, and lower free energy at 130°C arising from a greater number of available sites. The annealing study then showed that the number of sites for ν_3 was equal, to within a factor of about two, to the number of interstitial T_d sites. This strongly suggests that ν_3 is associated with the isolated molecule and is consistent with the observation of the same mode in float zone Si which had been soaked in hydrogen[4].

This assignment, however, causes several problems. Firstly, there is no splitting of the ν_3 band due to ortho- and para-forms of the molecule. Secondly, the infra-red activity of the molecule is unexpected, and thirdly the position of the ν_3 band is ~ 540 cm^{-1} below the gas value, and inconsistent with the recent assignment of Raman active modes around 4158 cm^{-1} to molecules at interstitial sites[6].

The observation of a single stretch mode of the isolated molecule is in contrast with Raman scattering studies of H_2 molecules, introduced from a plasma, in GaAs[7]. Here, *two* H-H stretch modes, separated by 8 cm^{-1}, attributed to para- and ortho- H_2 molecules occupying the lowest energy rotational states $J = 0$ and $J = 1$ respectively, were detected in the expected intensity ratio of 1:3. Although the energy difference between these states corresponds to 170K, they are both occupied, even at cryogenic temperatures, because of the absence of strong nuclear spin-flip processes. It seems likely that this is also true in Si. In this case, if the molecule was freely rotating, then two H-H modes would be expected, whereas only one is detected. This implies, contrary to the case of GaAs, that molecular rotation is prevented by the crystalline field.

The effects of a crystal field on molecular rotation can be understood in terms of a model whereby rotating the molecule by θ from its equilibrium axis is accompanied by an increase in energy equal to $V(1 - cos\ 2\theta)\left(\frac{r_0}{r}\right)^2$ where r_0 is the molecular bond length. The molecular Schrödinger equation is then separable and the solutions in the variables θ and ϕ have definite parity. If $VI/\hbar^2$ is small, then the lowest energy states are the rotational levels $J = 0$ and $J = 1$. Here I is the moment of inertia of the molecule. On the other hand, if $VI/\hbar^2$ is large, then they are symmetric and asymmetric combinations of harmonic oscillator states bound in the potential wells around $\theta = 0$ and $\theta = \pi$. We denote the symmetric and asymmetric states by '+' and '−' respectively . The Pauli exclusion principle tells us that H_2 molecules with parallel nuclear spins must occupy the '−' state and those with anti-parallel spins occupy the '+' state. In the case of D_2, those with symmetric spin functions occupy the '+' state while those with asymmetric spin functions occupy the '−' state. For the molecule in Si, we believe $VI/\hbar^2$ is large and the energy difference between the '±' states, $\mu_+ - \mu_-$, is very small. This difference leads to distinct centrifugal potentials,

$$\mu_\pm \left(\frac{r_0}{r}\right)^2,$$

in the two cases. For free-rotation, $V = 0$, and $\mu_- - \mu_+ = \frac{1}{2}\hbar^2 J(J + 1)/I$. The centrifugal potentials lead to a H-H stretch mode for the $J = 1$ state lying 5.9 cm^{-1}

Table 1. Calculated and experimental frequencies, cm^{-1}, of H_2 molecules in Si with different alignments

Alignment	H-H	H-D	D-D
[110]	3708.4	3217.1	2622.2
[111]	3713.0	3221.4, 3221.6	2625.5
[100]	3606.8	3128.6	2559.1
Exp.[4]	3618.3	3264.8	2642.5

below that of $J = 0^8$. However, as V increases, the separation between the two stretch modes diminishes. If we assume that the splitting in ν_3 is in fact less than 0.1 cm^{-1}, and has not been so far resolved, then $(\mu_- - \mu_+)$ must be less than $0.02\hbar^2/I$ and this imposes a lower limit on the rotation barrier. Using the expressions for μ_+ and μ_- given by Stern[9], we find the barrier to be at least 0.17 eV. This means that the molecular rotation is frozen below about 70K.

Thus at low temperatures, the molecules are aligned along one or more axes. The observation of only one mode in the H-D case means that either the hydrogen atoms are equivalent, and the centre of the molecule lies on a mirror plane or a C_2 axis, or, if they are inequivalent, then the two H-D modes must be separated by at least 20 cm^{-1} and only one is thermally populated at 10K.

To investigate the structure and vibrational properties of the isolated molecule, we have carried out density functional cluster calculations using a 148 atom H-terminated cluster centred on the tetrahedral interstitial site. The wave function basis consisted of independent s and p Gaussian orbitals, with eight different exponents, sited at each Si site and three at each H atom of the molecule. A fixed linear combination of two Gaussian orbitals was sited on the terminating H atoms. In addition three Gaussian s and p orbitals were placed at each Si-Si bond centre. The charge density was fitted with eight independent Gaussian functions with different widths on each Si atom, and four (three) on the central (terminating) H atoms. Three were placed at each bond centre. All atoms, except the terminating H ones, were allowed to relax by a conjugate gradient method. The second derivatives of the energy were found between the two central H atoms. Further details of the method can be found in ref. 10.

We found that the molecule is stable when oriented along [100], [111] and [110]. These three configurations are essentially degenerate in energy. The rotational barrier is also very small but the uncertainty in the energies can easily exceed 0.1 eV. This is similar to the situation in GaAs[11]. The vibrational modes are given in table 1. These are all close together and lie within 90 cm^{-1} of the ν_3 band. They are slightly lower than that found for the molecule trapped near oxygen (3855.6 cm^{-1}). Stretch modes around 3561 cm^{-1} have also been found by a previous calculation[12] and other first-principles methods [13, 14], although the [110] orientation was not considered. These calculations also find a low barrier to reorientation of the molecule. We find the H-H length to be 0.785 Å for the [110] and [111] structures and 0.788 Å for the [100] alignment. These are slightly longer than the gas value (0.74 Å). It appears that the energy calculations are not sufficiently accurate to resolve the reorientation energy or the stable alignment. However, the vibrational modes do present us with a method of discriminating between the structures.

Molecules lying along [100], with D_{2d} symmetry, would be infra-red inactive while those along [111] possesses inequivalent H atoms. The latter defect produces two distinct H-D modes shown in Table 1. As these are separated by 0.25 cm^{-1}, they would

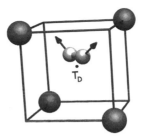

Figure 1. Schematic figure of the stable form of the molecule at a T_d site aligned along [110]. Arrows show an exaggerated movement of the atoms in the stretch mode leading to a transition dipole along [001]

give two bands with intensities corresponding to the different populations of the defects, contrary to the observations. The ν_3 band could not arise from these defects if the H-D splitting was as large as this. The [110] orientation1 is infra-red active with equivalent H atoms. Only one mode is then expected in the H-D case. We identify the ν_3 band with a molecule lying along [110], although a [111] alignment cannot be completely ruled out. The stretch mode is about 90 cm^{-1} above that of the ν_3 band but considerably lower than that calculated for an isolated molecule. With the present method, this occurs at 4424.8 cm^{-1}: about 10% higher than the observed value, although anharmonicity must play a significant role in this overestimate.

The transition dipole moment of the molecule can be calculated directly. This is done by finding the change in the dipole moment of the cluster when the atoms are displaced according to their normal coordinates. The induced dipole lies along [001] and arises as there is a slight displacement of the H atoms along this direction. The effective charge of the induced dipole is 0.1e and comparable with that of the the molecule trapped by oxygen. For the [111] alignment, the effective charge is very similar at 0.08e.

The molecule is not stable at the centre of the puckered hexagon (H site) but this site lies on a diffusion trajectory linking different T_d sites. The energy of a molecule constrained to lie at the H site is 0.72 eV greater than that at a T_d site and this then is an estimate of the migration energy of the molecule. It is appreciably greater than the diffusion energy for H$^+$ or H^0 around 0.43 eV[15].

The low frequency of the molecule is due to to the leaking of charge from the surrounding bonds screening the intra-molecular proton-electron attraction.

The Molecule Bound To An Oxygen Impurity

There are several obvious locations for the molecule close to O$_i$ in the silicon lattice but since the binding between the molecule and the O atom is so weak, these are close in energy and difficult to distinguish. Furthermore, it seems that there are a large

Table 2. Local vibrational modes, cm^{-1}, of the H$_2$-O$_i$ complex in silicon

	H-H, ^{16}O	H-D, ^{16}O	D-H, ^{16}O	D-D, ^{16}O	H-H, ^{18}O
Calculated	3855.6	3339.5	3350.1	2726.7	3855.6
	1129.8	1129.7	1129.6	1129.5	1078.1
	733.3	685.6	690.6	678.6	731.4
	696.2	663.2	654.6	653.7	685.3
	655.3	650.1	645.0	630.9	649.0
	643.6	627.6	627.4	536.3	638.9
	630.6	539.5	542.6	533.6	622.6
	576.7	536.0	536.1	532.8	576.1
Exp.[4]	3788.9	3304.3		2775.4	
	3730.8	3285.3		2716.0	
	3618 3	3264.8		2714.9	
	1075.1			1076.6	

number of configurations differing in the orientation of the molecule and all of these have to be considered as candidate structures.

We suppose that the molecule resides near a T_d cage site and oxygen decorates one of the nearby Si-Si bonds. We first consider the case when the O atom bridges a [11$\bar{1}$] Si-Si bond forming part of the cage surrounding the T_d site as shown in Fig. 2. The molecule in the relaxed configuration then is oriented almost perpendicular to the Si-O-Si bond and nearly along [1$\bar{1}$0]. The H-H bond length is 0.77 Å and each H atom is about 2.37 Å from O. It has been pushed slightly away from the T_d site as the nearest two Si atoms are 2.19 Å but these are not bonded to O. The vibrational modes of the cluster are given in table 2.

The H$_2$ stretch mode at 3855 cm^{-1} is again lower than that calculated for an isolated molecule. The molecular frequency depends on the size of the surrounding cage. The 3855 cm^{-1} mode lies close to the experimental modes around 3750 cm^{-1} associated with oxygen. The two H atoms are almost equivalent and this explains why

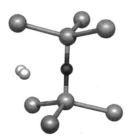

Figure 2. Schematic illustration of O$_i$H$_2$ defect

only one H-D mode is present although the calculations show a 10 cm^{-1} splitting caused by deviations from ideal σ_h symmetry.

The mode at 1129.8 cm^{-1} is clearly due to O_i as it shifts downwards by only 0.3 cm^{-1} when D replaces H but by 52 cm^{-1} with ^{18}O. This is comparable with the 51 cm^{-1} shift observed in the 1136 cm^{-1} mode of O_i when ^{18}O replaces ^{16}O. The small downward shifts with D demonstrate that there is very little direct coupling between O_i and the molecule and this is essential to our arguments, based on anharmonicity, if an upward shift is to be seen in the deuterated case. It is not clear why the experimental O mode is shifted downwards by as much as 65 cm^{-1} from that of isolated O_i.

The modes at 733 to 576 all involve the movement of H and none has been detected so far. They represent H_2 bend and librational modes. Their shifts with ^{18}O given in table 2 demonstrate that many of them involve the movement of O.

Of particular interest here is the mode at 577 cm^{-1}. This represents a librational mode as the two H atoms are displaced almost parallel to their bond. We now argue that the anomalous frequency shift of the 1075 cm^{-1} band is to be understood through an anharmonic coupling between an overtone, or combination band, of these low frequency modes and the O mode resulting in a Fermi resonance. Let $|n_{Ox} >$ be the n'th oscillator state for the vibrations of the oxygen atom whose fundamental occurs at $\nu_O = 1129.8$ cm^{-1}, and

$|m_H >$ be the m' th state for a mode whose frequency, ν_H, is about half that of the O_i mode, ie. either the modes in the region of 550 cm^{-1}. The states of the coupled system are then described by linear combinations of $|n_{Ox}, m_H >$. The effect of anharmonicity, V, is to couple together these states and second order perturbation theory gives the shift in the energy of the $|1_{Ox}, 0_H >$ state to be dominated by

$$\frac{< 1_{Ox}, 0_H |V| 0_{Ox}, 2_H >^2}{(\nu_O - 2\nu_H)}.$$

This follows as the energy denominator is particularly small for these modes. The perturbation is negative for the 577 cm^{-1} mode and lowers the energy of the $|1_{Ox}, 0_H >$ state, and hence that of the fundamental transition. On the other hand, when H is replaced by D, as the frequency of this mode drops below 575 cm^{-1}, then the perturbation acts to raises the energy of the state. Thus the two cases reinforce the tendency to depress the O mode in the H case below that of D. Another ways of describing the effect is an anti-crossing between the O mode and an overtone of the librational mode (as shown in fig. 3). If this mechanism is correct, there has to be unreported modes in the 550 cm^{-1} region.

We have also investigated other positions for the molecule. A second and almost degenerate configuration occurs when the molecule lies in the mirror plane. This has similar modes to the first and could account for the second defect which is observed. A third possibility is that the molecule lies along the Si-O-Si axis near a T_d site. This site is stable but the molecule is then close to the Si neighbour of O_i and this results in an O related mode that is strongly coupled with H in conflict with the experiment. However, the energy of this structure appears to be lower than that shown in Fig. 2 by 0.5 eV. This may be a due to the proximity of O_i to the surface of the cluster and further investigations are needed to clarify the most stable configuration.

H_2 Trapped Inside Voids

Since the drop in the stretch frequency of the molecule from the gas–phase value is due to screening from the surrounding lattice, then if the cage size was increased, the

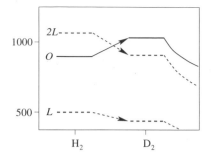

Figure 3. Schematic of the anti-crossing of the oxygen mode and an overtone of a librational mode

charge density at the centre of the cage would decrease and the molecular frequency increase.

This is seen in Fig 4 where a relaxed molecule is placed in a cage of increasing size by scaling the surrounding Si-Si bonds. It is seen that the gas value is reached for voids about 1.4 times the actual cage. This shows that the large red-shift in the molecular stretch frequency is due to the size of the surrounding cage and is not an artifact of the calculational method. It is also consistent with the result that the crystal field effects are sufficiently strong to freeze molecular rotation.

These results strongly suggest that Raman active modes at 4158 cm^{-1} found in H plasma treated Si are due to molecules in voids where the electron density arising from the lattice is small and are not due to isolated molecules at interstitial sites. A realistic model of the void must have surface dangling bonds saturated by H[16] and a correlation of the Raman signals due to these Si-H bonds and the 4158 cm^{-1} band is then to be expected. Such a correlation has already been detected[17].

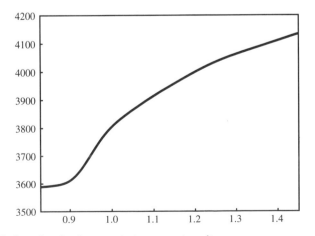

Figure 4. Variation of molecular stretch frequency (cm^{-1}) verses multiples of the equilibrium interstitial cage size

CONCLUSIONS

In conclusion, we have shown that hydrogen molecules at interstitial sites in Si are not rotating and do not exist in ortho- and para-forms. The molecule is either aligned along [110], with C_{2v} symmetry, or along [111] if the H-D splitting is smaller than calculated here. The stretch mode is some 540 cm^{-1} red-shifted from the gas value and is infra-red active. This frequency is very sensitive to the cage size and increases rapidly to the gas value for cages about 40% larger than that at a T_d site. The anomalous upward shift of the O-mode with D can be explained through a Fermi resonance interaction with librational modes of the molecule around 530 cm^{-1}. Such modes have not been reported and experimental investigations in this region are called for.Modes reported around the gas value are due to molecules in voids created by the plasma. The migration energy of the molecule through the solid is found to be 0.72 eV.

REFERENCES

1.   A. Mainwood, and A. M. Stoneham, *Physica*, 116B:101 (1983).
2.   J. W. Corbett, S.N. Sahu, T.S. Shi, and L.C. Snyder, *Phys. Letts.*, A 93:303 (1983).
3.   V. P. Markevich, M. Suezawa, and K. Sumino, *Mater. Sci. Forum*, 196-201:915 (1995).
4.   R. E. Pritchard, M. J. Ashwin, R. C. Newman, J. H. Tucker, E. C. Lightowlers, M. J. Binns, R. Falster, and S. A. McQuaid, *Phys. Rev. B*. in press.
5.   R. E. Pritchard, M. J. Ashwin, J. H. Tucker, and R. C. Newman, submitted for publication.
6.   K. Murakami, N. Fukata, S. Sasaki, K. Ishioka, M. Kitajima, S. Fujimura, J. Kikuchi, and H. Haneda, *Phys. Rev. Lett.*, 77:3161 (1996).
7.   J. Vetterhöffer, J. Wagner, and J. Weber, *Phys. Rev. Lett.*, 77:5409 (1996).
8.   L. Pauling, and E. B. Wilson, *Introduction to Quantum Mechanics*, McGraw Hill, New York (1935).
9.   T. E. Stern, *Proc. Roy. Soc.*, A 130:551 (1931).
10. R. Jones, and P. R. Briddon, Identification of defects in semiconductors, in: *Semiconductors and Semimetals*, M. Stavola ed., R. K. Willardson, A. C. Beer, and E. R. Weber treatise eds., Academic Press (1997) in press.
11. S. J. Breuer, R. Jones, P. R. Briddon, S. Öberg, *Phys. Rev. B.*, 53:16289 (1996).
12. R. Jones, *Physica B*, 170:181 (1991).
13. Y. Okamoto, M. Saito, and A. Oshiyama, *Phys. Rev. B*, (1997).
14. C. G. Van de Walle, unpublished.
15. Gorelkinski, N. N. Nevinnyi, *Mater. Sci. Eng. B*, 36:133 (1996).
16. S. Muto, S. Takeda, and M. Hirata, *Phil. Mag. A*, 72:1057 (1995).
17. A. W. R. Leitch, V. Alex, and J. Weber, *Mater. Sci. Forum*, in press.

QUANTUM MONTE CARLO SIMULATIONS OF REAL SOLIDS

W. M. C. Foulkes[1], M. Nekovee[1], R. L. Gaudoin[1],
M. L. Stedman[1], R. J. Needs[2], R. Q. Hood[2],
G. Rajagopal[2], M. D. Towler[2], P. R. C. Kent[2],
Y. Lee[2], W.-K. Leung[2], A. R. Porter[2], and S. J. Breuer[3]

[1]Blackett Laboratory, Imperial College
[2]Cavendish Laboratory, Cambridge University
[3]Edinburgh Parallel Computing Centre, University of Edinburgh

INTRODUCTION

Computer simulation has become a standard tool in almost all areas of science, engineering and technology, and its pervasive importance was one of the clearest themes to have emerged from the recent U.K. Technology Foresight Exercise. The Foresight Steering Group Report emphasised that the impressive recent advances in our understanding of complex systems are largely attributable to the use of computer simulation, and identified the field as a main priority for the future.

Although computer simulation is used in many different contexts, the underlying questions are often surprisingly similar. What happens when many simple objects come together and interact? How does the complex behaviour of the whole emerge from the simple laws obeyed by its parts? The constituent objects may range from the electrons and nuclei making up a crystal of silicon to the cables and girders holding up the Millennium Dome, but the common aim is to predict the complex large scale behaviour from the simpler small scale behaviour. In essence, computers are used to build bridges between different length scales.

Figure 1 shows a wide selection of different areas of physics, chemistry and engineering linked by arrows corresponding to changes of length scale. Computer simulations are already being used to help climb most of the arrows, and a large part of the work being discussed at this conference fits somewhere on the diagram. At the top level of the tree, the technological and economic importance of computer simulation is large and growing rapidly. Aircraft, cars, and bridges are already designed using computers, and new drugs and materials soon will be. Our work, being at the bottom of the tree, may appear far from the wealth creating activities at the top, but it provides the scientific understanding needed to underpin progress at higher levels.

To understand why work at the bottom of the tree is interesting and important, think about the problems facing someone higher up. Imagine, for example, that you're a materials scientist trying to use computer simulations to assess the effects of different manufacturing processes on the properties of steel. Since the mechanical behaviour of steel is governed by the dislocations, impurities, grain boundaries, and other defects introduced during manufacture, your job is to model a complicated system of inter-acting defects. Your starting point is a thorough quantitative understanding of the comparatively simple properties of single defects and interacting pairs of defects, but where is this understanding going to come from? Some of the information you need can probably be measured, but experiments are difficult, expensive and limited. The rest can only be obtained from the results of simulations one level lower on the tree. You can't begin to model steel without first modelling its constituent defects.

Almost everywhere you look on the tree, some of the data required for simulations at one level can only be obtained from the results of simulations on the level below. The descent continues right down to the quantum mechanical atomic/molecular level, where the many-electron Schrödinger equation at last provides a concise and accurate universal law of nature. There are quantum field theoretical levels below it, of course, but the Schrödinger equation requires no input from these and is accurate enough to explain almost everything higher up the tree. Since the quantum mechanical length scale at the bottom of Figure 1 is the largest at which we can write down a reliable "grand unified theory", it serves as a natural root for the tree above it.

The "Consortium for Computational Quantum Many-Body Theory" was one of the most sharply focused of the consortia set up under the High Performance Computing Initiative (HPCI). Our goal is the development of a deeper conceptual understanding of the role of quantum mechanics in determining the electronic structure and properties of real materials. In pursuit of this goal, we're developing accurate and reliable methods for climbing from the lowest box in Figure 1 to the levels above. Our quantum mechanical calculations start from "first principles" and require only the atomic numbers

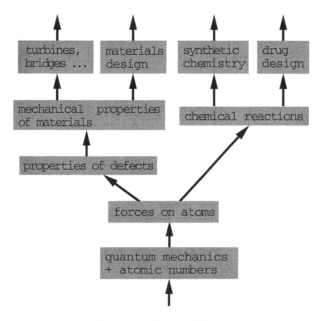

Figure 1. The simulation

of the atoms involved as input; if they live up to expectations, they should provide a firm base on which the higher levels of the simulation tree can rest.

Although the methods we use are very different, our goals are strikingly similar to those of the larger UKCP Consortium. In the longer term, we believe that the many-electron approach we have adopted[1] will supersede the one-electron approach[2] they favour, but this is a matter of conjecture of course. The good news is that both our Consortium and the UKCP Consortium are world leaders in their fields, and that the United Kingdom is well placed to seize the great opportunities offered by first principles materials simulation over the next twenty years. Within that time scale, we are confident that quantum mechanical simulations will be used to design drugs, catalysts, chemical syntheses, advanced materials, and electronic devices.

THE CHALLENGE

Our "grand unified theory", the many-electron Schrödinger equation, is generally believed capable of predicting and describing almost everything experienced in everyday life. It doesn't matter whether you're interested in the biochemical reactions taking place in the human body, the flow of electrons through a transistor, or the strength of steel — they're all in there somewhere.

Given its enormous multitude of complicated ramifications, the equation itself is surprisingly simple,

$$
\left(-\frac{1}{2} \sum_{i=1}^{N} \nabla_i^2 + \sum_{i=1}^{N} v(\mathbf{r}_i) + \frac{1}{2} \sum_{i=1}^{N} \sum_{\substack{j=1 \\ (j \neq i)}}^{N} \frac{1}{|\mathbf{r}_i - \mathbf{r}_j|} \right) \Psi(\mathbf{r}_1, \mathbf{r}_2, \ldots, \mathbf{r}_N) = E \Psi(\mathbf{r}_1, \mathbf{r}_2, \ldots, \mathbf{r}_N) \quad .
$$

It's a linear second-order complex partial differential equation of a well-known type, the only significant computational problem being that the sums over i and j include terms for every electron in the system. An object weighing a gram or two contains something like 10^{23} electrons, so the many-electron Schrödinger equation is a partial differential equation in roughly 10^{23} variables. Most scientific and engineering problems involve partial differential equations in 3 or at most 6 variables, so dealing with 10^{23} is quite a challenge.

The traditional response to this challenge has been to stand back in horror and start approximating. Instead of considering all the electrons together, the idea has been to look at the electrons one by one, replacing the complicated fluctuating forces due to the others by an average force known as a mean field. This trick reduces the 10^{23} dimensional many-electron Schrödinger equation to a much simpler three-dimensional equation for each electron. The mean field depends on the electron density, which isn't known until the one-electron equation has been solved to find all the one-electron quantum states, so it's necessary to use some sort of iterative procedure to home in on a consistent solution.

The mean-field theory most often used in solids is density functional theory.[2] Although exact in principle, density functional theory involves an unknown quantity called the exchange-correlation energy functional which has to be approximated in practice. Several rival approximations are in widespread use, but the simplest and best known is the local density approximation (LDA). The LDA and its alternatives work surprisingly well in many cases, and density functional calculations are already being used to investigate questions of genuine industrial and economic importance. This is the province of the UKCP Consortium, the members of which apply density functional theory to a wide range of topics in condensed matter physics, materials science, geophysics and biochemistry. Unfortunately, the LDA is by no means universally accurate or reliable. When it works, as in simple metals and semiconductors, it seems to work very well indeed; but when it doesn't, as in transition metal oxides, high temperature superconductors, and hydrogen-bonded materials such as water, there isn't much one can do. There's no simple systematic way to improve the standard density functional approach and one is forced to go back to the full many-electron Schrödinger equation instead.

QUANTUM MONTE CARLO

The main approaches used to solve the many-electron Schrödinger equation in real solids are quantum Monte Carlo[1] (QMC) and the *GW* method.[3] The two are complementary — QMC simulations tell you about ground state properties while the *GW* method gives information about excitations — and we use both. In this article we'll concentrate on our QMC work since it's slightly easier to explain and particularly straightforward to implement on massively parallel computers. The generic term "quantum Monte Carlo" covers several different methods, but the ones we use are called variational QMC and diffusion QMC.

Variational QMC

According to quantum mechanics, the probability that a measurement of the positions of all N electrons in a solid finds them at r_1, r_2, ..., r_N is proportional to $|\Psi(r_1, r_2, \ldots, r_N)|^2$, where $\Psi(r_1, r_2, \ldots, r_N)$ is the many-electron wavefunction. The idea behind variational QMC is to use a computer to generate sets of random positions distributed in exactly the same way as the results of this idealised measurement, and to average the outcomes of many such computer experiments to obtain quantum mechanical expectation values. Given the form of the many-electron wavefunction one can generate the required samples using the well-known Metropolis algorithm, but the exact many-electron wavefunction is an unknown function of an enormously large number of variables and has to be approximated.

The first approximation is to replace the macroscopic piece of solid containing roughly 10^{23} electrons by a small model system containing no more than a few thousand. This sounds drastic, but in fact it works very well, especially when we apply periodic boundary conditions (which means that an electron leaving one face of the model system immediately re-enters through the opposite face) to get rid of any surfaces. The replacement of an effectively infinite solid by a small model system subject to periodic boundary conditions is still an approximation, of course, and the associated finite-size errors caused us significant problems at the beginning. Luckily, however, our attempts to understand these errors have been very successful[4] and we have been able to reduce them to the point that they can safely be ignored for most purposes. We consider this an important advance and expect the techniques we have developed to become standard practice in the field.

The second approximation is more problematic and harder to improve. Since we don't know the exact many-electron wavefunction (even for our small model system) we have to guess it. At first glance this looks like a hopeless task, but a surprisingly large fraction of those few quantum many-body problems that have ever been "solved" have in fact been solved by guessing the wavefunction (think of the BCS theory of superconductivity and Laughlin's theory of the fractional quantum Hall effect). Most of our calculations are done using trial functions of the Slater-Jastrow type,

$$\Psi_T = D^\uparrow D^\downarrow \exp\left[-\sum_{i>j} u(\mathbf{r}_i, \mathbf{r}_j) + \sum_i \chi(\mathbf{r}_i)\right] \quad ,$$

where $D^\uparrow$ and $D^\downarrow$ are Slater determinants of spin-up and spin-down single-particle orbitals obtained from Hartree-Fock or LDA calculations, the function $u(\mathbf{r}_i, \mathbf{r}_j)$ correlates the motion of pairs of electrons, and $\chi(\mathbf{r}_i)$ is a one-body function as introduced by Fahy.[5] The Slater determinants build in the antisymmetry required by the Pauli principle, and the u and χ functions are adjusted to minimise the total energy (or more precisely[6] the variance of the total energy) in accordance with the variational principle.

This simple trial wavefunction, first applied to real solids at the end of the eighties,[5] has proved astonishingly accurate for the weakly correlated solids studied so far. Following careful optimisation of the u and χ functions, total and cohesive energies are accurate to within 0.2eV per atom, approximately five times better than the best LDA calculations.[7] The full range of solids for which this trial wavefunction is accurate is still not known, but so far it has exceeded all reasonable expectations.

Diffusion QMC

The major limitation of the variational QMC method is obvious: what happens when the assumed trial wavefunction isn't accurate enough? The direct approach is to add more variational parameters and resort to brute force optimisation, but this limits one to an assumed functional form which may not be adequate, particularly in strongly correlated systems where the quantum state is markedly different from the Fermi liquid state found in weakly correlated solids.

An alternative and much better approach is diffusion QMC, which is based on the imaginary-time Schrödinger equation,

$$\left(-\frac{1}{2}\nabla_{\mathbf{R}}^2 + V(\mathbf{R})\right)\Psi(\mathbf{R},\tau) = -\frac{\partial\Psi(\mathbf{R},\tau)}{\partial\tau} \quad ,$$

where we have adopted a very condensed notation in which $\mathbf{R} = (\mathbf{r}_1, \mathbf{r}_2, \ldots, \mathbf{r}_N)$ is a $3N$-dimensional vector containing all the electron positions, $\nabla_{\mathbf{R}}^2$ is shorthand for $\sum_i \nabla_i^2$, and $V(\mathbf{R})$ is the sum of all the potential energy terms appearing in the ordinary many-electron Schrödinger equation. The imaginary-time Schrödinger equation reduces to the usual time-dependent Schrödinger equation when τ is replaced by the imaginary variable it, but we will be interested only in real τ from now on.

It's straightforward to show that as long as the starting state, $\Psi(\mathbf{R}, \tau = 0)$, isn't orthogonal to the ground state energy eigenfunction, the solution of the imaginary-time Schrödinger equation becomes proportional to the ground state in the limit as $\tau \to \infty$. The imaginary-time development is just a mathematical trick used to convert an arbitrary starting state into the ground state without assuming any particular functional form.

The clever part is that there's a simple algorithm that can be used to carry out the imaginary-time development in systems containing hundreds or thousands of electrons. If the $V(\mathbf{R})$ term is ignored, the imaginary-time Schrödinger equation is just a diffusion equation describing a population of "walkers" diffusing in a $3N$-dimensional "configuration space"; if the $\nabla_{\mathbf{R}}^2$ term is ignored, it's just a rate equation analogous to the equation describing radioactive decay. Putting these two things together, it can be seen that the imaginary-time Schrödinger equation describes walkers diffusing in $3N$-dimensional space while dying out or multiplying at a rate determined by the value of $V(\mathbf{R})$. This simple physical picture leads directly to a simulation method in which a collection of walkers multiply and reproduce as they diffuse randomly around in a $3N$-dimensional vector space. After a long time τ, the density of diffusing walkers becomes proportional to the ground state wavefunction $\Psi_0(\mathbf{R})$ (not $|\Psi_0(\mathbf{R})|^2$ as you might expect; it's $\Psi_0(\mathbf{R})$ that plays the role of a probability density here), and one can start accumulating ground state expectation values as in variational QMC.

There's one major problem with the algorithm just described. The diffusion QMC method finds the overall ground state of the system, and for the many-electron Schrödinger equation this happens to be a totally symmetric function of the particle coordinates. Electrons are fermions, however, and according to the Pauli principle many-fermion states have to be totally antisymmetric. The simple diffusion QMC algorithm gives a many-boson state of no physical interest.

To ensure that the diffusion QMC simulation produces an antisymmetric state we're forced to make the so-called fixed node approximation, in which we guess the shapes of the regions of configuration space within which the wavefunction is positive and negative and solve the imaginary-time Schrödinger equation to find the lowest energy state consistent with that guess. Fixed-node diffusion QMC is perhaps best regarded as a variational method in which, instead of assuming a trial form for the whole wavefunction, we assume only a trial form for the nodal surface (the $3N-1$ dimensional configuration-space surface on which the wavefunction is zero). We use fairly simple guesses based on mean-field wavefunctions and our experience so far suggests that these give very good results, accurate to considerably better than 0.1eV per atom.[7] The fixed-node approximation is the only significant approximation in our calculations.

COMPUTATIONAL ISSUES

Like most Monte Carlo methods, our QMC calculations are ideally suited to massively parallel architectures. We simply put one walker (or several walkers) on each node and run different simulations on each. Some communication between the many parallel simulations is necessary, but the communications requirements are quite small and the performance increases linearly with the number of nodes used.

One of the comments we received in response to our first HPCI grant proposal was that our work was so "embarrassingly parallel" that it wasn't worth running on the Cray T3D; it wouldn't provide enough of a technical challenge! In our view this is nonsense. Our use of the T3D and T3E has allowed us to do world-beating physics which would not have been possible on lesser machines or networks of workstations. Yes, our algorithms are easy to parallelise, but surely this is a strength rather than a weakness? It means that we can get our programs running quickly and efficiently and concentrate our attention on the science instead of on its computational implementation. The T3D and T3E have enabled us to study larger systems and achieve higher accuracies than any other group in the world, and to complete ambitious scientific projects that nobody else could have attempted. Our successes have demonstrated the tremendous potential of the combination of quantum Monte Carlo methods and massively parallel computation. This, we believe, is the sort of work the T3D and T3E were intended for.

We have, of course, put considerable effort into optimising our codes, and thanks to the expert help we have received from Edinburgh Parallel Computing Centre (EPCC) they are now amongst the most efficient on the T3D and T3E. We have found it very useful to have a named EPCC postdoc (currently Stephen Breuer) assigned to look after our Consortium, and hope that the EPSRC will provide funds to continue this system in future.

SOME EXAMPLE APPLICATIONS

Our consortium has been involved in a wide variety of different projects, leading to 25 publications in refereed international journals, 6 of which were in Physical Review Letters. Our access to the T3D and T3E has allowed us to establish a position as one of the world's two leading groups in our field, and we have recently been invited to write major review articles in both "Reports on Progress in Physics" and "Advances in Chemical Physics". We plan to use these to explain the techniques we have developed in the hope that this will help to demystify the subject and encourage its further growth. We can't possibly cover all our QMC work in this short paper, so we've picked four representative examples.

Total Energy Calculations

Our first few projects were concerned with testing the accuracy and reliability of QMC calculations of ground state energies.[7] We used QMC to calculate the cohesive energies of a number of different solids and compared the results against mean-field values obtained using density functional theory within the LDA. Typical results are summarised in Table 1.

Finite-size errors in QMC simulations are typically a significant fraction of an eV, so highly accurate calculations such as that in Table 1 rely on the techniques we have developed for eliminating these errors;[4] it's only after the finite-size errors have been eliminated that the full accuracy of the QMC approach becomes apparent. For weakly correlated solids like Ge, it's now clear that diffusion QMC calculations of total energy differences (and hence interatomic forces) are at least an order of magnitude more accurate than calculations using any alternative method.

Relativistic Electron Gas

Close to the nuclei of heavy atoms, where the electron density is very high, the electrons move at a significant fraction of the speed of light and relativistic effects become important. These are not included in the many-electron Schrödinger equation, but can be incorporated via perturbation theory. We have been studying the relativistic effects in various systems to order $1/c^2$ using QMC methods. After a starter project in which we studied the relativistic effects in various atoms,[8] we concentrated on the unpolarised relativistic electron gas. Our results enabled us to check the relativistic version of the LDA used in density functional calculations of solids and molecules containing heavy atoms. We found that the contact term is well described by the LDA, but that the effects of retardation are very poorly described and require a non-local functional. A short description of this work[9] appeared in Physical Review Letters, and we are currently writing up our new results on the polarised relativistic homogeneous electron gas.

Table 1. The Cohesive Energy of Ge Obtained Using Three Different Methods

Method Used	Cohesive Energy (eV/atom)
LDA Calculation	4.59
Diffusion QMC Calculation	3.85
Experiment	3.85

Silicon Quasiparticle Bandstructure

This project demonstrated the power of diffusion QMC by calculating the band structure of silicon, obtaining excellent agreement with experiment.[10] Although similar calculations had been done in the past,[11] ours were much more complete and precise than their predecessors and were the first to demonstrate the very high accuracy of the QMC approach. The excited state issues that can be addressed using QMC are somewhat limited, but in those cases where QMC can be applied it works very well and gives a good account of the many-body effects missed by mean-field methods. Unlike perturbative methods for studying excited states, QMC can also be used to study strongly correlated systems, and hence provides a unified framework for studying ground states and excitations throughout the periodic table. No other approach shows such promise.

Another result of this work was a better understanding of the theory underlying QMC calculations of excitation energies. A widely believed "folk theorem" states that the fixed-node diffusion QMC energy is always greater than or equal to the energy of the ground state eigenfunction with the same symmetry as the trial state used to define the nodal surface. In fact, we have recently proved that this theorem is correct only when the relevant ground state eigenfunction is non-degenerate, and have devised a simple analytic example to demonstrate our point.

Exact Density Functional Theory

In order to understand the physics behind mean-field approximations such as the LDA, it helps to think about one particular electron (let's call it the labelled electron) moving through the sea of nuclei and other electrons making up a solid. The labelled electron is attracted to the positively charged nuclei, which are so massive that they can be treated as immobile, and repelled by the other electrons, which are negatively charged.

As a rough approximation, it seems sensible to replace the fluctuating forces due to the other electrons by the static electrical (Coulomb) force due to the *average* electronic charge density. This simple mean-field approximation, known as the Hartree approximation, helps keep the labelled electron away from regions where there are lots of other electrons on average, which is a good start, but misses something important. As the labelled electron moves around, the others stay out of its way; you can think of the labelled electron carrying round a little "exclusion zone", usually known as the exchange-correlation (XC) hole, within which other electrons rarely venture. The electron density near the labelled electron is therefore less than the average density, and the Hartree approximation doesn't take this into account. Other mean-field approaches such as Hartree-Fock theory attempt to build in the effects of the XC hole in an approximate way, but these approximations aren't particularly accurate.

Density functional theory is based on a remarkable theorem, first proved almost 35 years ago,[12] which states that it is in principle possible to devise an exact mean-field theory. In other words, the mean field can be chosen in such a way that the energies and electron densities obtained by solving the one-electron equations come out exactly right. There's even a prescription for constructing the exact mean field given the detailed shape of the XC hole.[13] Practical applications of density functional theory have had to rely on approximations such as the LDA only because no available method has been able to calculate the shapes of XC holes in real materials.

Quantum Monte Carlo methods can calculate accurate XC holes, however, and Figure 2 shows the result of our calculation of the shape of the hole around a labelled

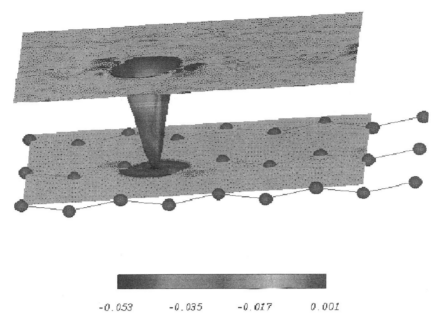

-0.053   -0.035   -0.017   0.001

Figure 2. The exchange-correlation hole around an electron near the centre of a bond in crystalline silicon

electron at a particular point in the (110) plane of silicon.[14] We've now assembled enough data to allow us to plot the shape of the XC hole anywhere in a piece of silicon, and have made a movie to show how the hole changes as the labelled electron moves around. The *exact* mean fields calculated from our XC hole data for silicon and other systems[15] are significantly different from the commonly used approximations, and we believe that these results will help the development of better mean-field methods in the future.

CONCLUSIONS

This article was intended to emphasise the importance of first principles materials simulation and to explain where QMC fits in. We hope we have convinced you that the impact of quantum mechanical materials simulation will continue to grow in the future, and that QMC will be the method of choice when high accuracy is required.

Mean-field methods such as density functional theory within the LDA are fine when the necessary accuracy is a few tenths of an eV per atom, but many problems require much higher accuracy than this. Chemical reaction rates, for example, are sensitive to energy differences of order $k_B T \simeq 0.025$ eV at room temperature, and many biochemical processes require even higher accuracy. There is no doubt that quantum mechanical materials simulation promises a technical revolution, with computers used to design drugs, chemical syntheses, and smart materials, but this will not happen until accuracies of order 0.01 eV per atom can be attained routinely. No method applicable to solids or large molecules can yet provide such precision, but quantum Monte Carlo is the closest and (in our opinion) the most likely to succeed.

Fortunately for us, it seems that the EPSRC concurs with this opinion. Computational quantum many-body theory, including both QMC calculations and methods for excited states such as the GW self-energy calculations that form the other main part of our work, were highlighted in the EPSRC "Review of Condensed Matter Physics" under "Future Growth Areas/Priorities", and were featured in the "Analysis of Community Questionnaire" as a "likely major growth area in CMP research over the next 5 years". The United Kingdom is among the world leaders in most areas of first principles materials simulation, and with the continuing support of the EPSRC there is no reason why we should not continue to remain at the forefront.

REFERENCES

1. See, e.g., B. L. Hammond, W. A. Lester, and P. J. Reynolds. "Monte Carlo Methods in Ab Initio Quantum Chemistry," World Scientific, Singapore, (1994); D. M. Ceperley and M. H. Kalos, *in*: "Monte Carlo Methods in Statistical Physics," K. Binder, ed., Springer-Verlag, Berlin (1984); K. E. Schmidt and M. H. Kalos, *in*; "Applications of the Monte Carlo Method in Statistical Physics," K. Binder, ed., Springer-Verlag, Berlin (1984); K. E. Schmidt, *in*: "Models and Methods in Few-Body Physics, Proceedings of the 8th Autumn School, Lisbon, Portugal, 13-18 Oct. 1986," Springer-Verlag, Berlin (1987).
2. See, e.g., R. G. Parr and W. Yang. "Density-Functional Theory of Atoms and Molecules," Oxford University Press, New York (1989); R. M. Dreizler and E. K. U. Gross. "Density Functional Theory, An Approach to the Quantum Many-Body Problem," Springer-Verlag, Berlin (1990).
3. See, e.g., M. S. Hybertsen and S. G. Louie, *Phys. Rev. B*. 34:5390 (1986); R. W. Godby, M. Schluter, and L. J. Sham, *Phys. Rev. B*. 56:10159 (1988); L. Hedin and S. Lundquist, *Solid State Physics, Advances in Research and Applications* 23:1 (1969).
4. L. M. Fraser, W. M. C. Foulkes, G. Rajagopal, R. J. Needs, S. D. Kenny, and A. J. Williamson, *Phys. Rev. B*. 53:1814 (1996); A. J. Williamson, G. Rajagopal, R. J. Needs, L. M. Fraser, W. M. C. Foulkes, Y. Wang, and M.-Y. Chou, *Phys. Rev. B*. 55:4851 (1997).
5. S. Fahy, X. W. Wang, and S. G. Louie, *Phys. Rev. Lett.* 61:1631 (1988).
6. A. J. Williamson, S. D. Kenny, G. Rajagopal, A. J. James, R. J. Needs, L. M. Fraser, W. M. C. Foulkes, and P. Maccallum, *Phys. Rev. B*. 53:9640 (1996).
7. G. Rajagopal, R. J. Needs, S. Kenny, W. M. C. Foulkes, and A. James, *Phys. Rev. Lett.* 73:1959 (1994); G. Rajagopal, R. J. Needs, A. James, S. D. Kenny, and W. M. C. Foulkes, *Phys. Rev. B*. 51:10591 (1995).
8. S. D. Kenny, G. Rajagopal, and R. J. Needs, *Phys. Rev. A*. 51:1898 (1995).
9. S. D. Kenny, G. Rajagopal, R. J. Needs, W.-K. Leung, M. J. Godfrey, A. J. Williamson, and W. M. C. Foulkes, *Phys. Rev. Lett.* 77:1099 (1996).
10. A. J. Williamson, R. Q. Hood, R. J. Needs, and G. Rajagopal, accepted for publication in *Phys. Rev. B*. (1998); P. R. C. Kent, R. Q. Hood, M. D. Towler, R. J. Needs, and G. Rajagopal, submitted to *Phys. Rev. B*. (1997).
11. L. Mitas and R. M. Martin, *Phys. Rev. Lett.* 72:2438 (1994).
12. P. Hohenberg and W. Kohn, *Phys. Rev.* 136:B864 (1964); W. Kohn and L. J. Sham, *Phys. Rev.* 140:A1133 (1965).
13. J. Harris and R. O. Jones, *J. Phys. F*. 4:1170 (1974); D. C. Langreth and J. P. Perdew, *Phys. Rev. B*. 21:5469 (1980); O. Gunnarson and B. I. Lundqvist, *Phys. Rev. B*. 13:4274 (1976).
14. R. Q. Hood, M.-Y. Chou, A. J. Williamson, G. Rajagopal, R. J. Needs, and W. M. C. Foulkes, *Phys. Rev. Lett.* 78:3350 (1997); R. Q. Hood, M.-Y. Chou, A. J. Williamson, G. Rajagopal, and R. J. Needs, accepted for publication in *Phys. Rev. B*. (1998).
15. M. Nekovee, W. M. C. Foulkes, A. J. Williamson, G. Rajagopal, and R. J. Needs, accepted for publication in *Advances in Quantum Chemistry* (1997).

AB INITIO INVESTIGATIONS OF THE DYNAMICAL PROPERTIES OF ICE

I. Morrison,[1] S. Jenkins,[1] J.C. Li,[2] and D.K. Ross[1]

[1]Department of Physics,
University of Salford,
Salford, M5 4WT
[2]Department of Physics,
UMIST,
Manchester, M60 100

ABSTRACT

In this paper we present a summary of results of *ab initio* simulations of the dynamical properties of various phases of ice. The calculations are performed using the *ab initio* pseudopotential method within the Generalized Gradient Approximation to Density Functional Theory. The dynamical properties have been assesses within the harmonic approximation via the numerical evaluation of dynamical matrices from atomic forces. Supercells are used to model the various ordered and disordered phases under consideration. Detailed analysis of the normal modes is performed via resolution into pure "intra" and "inter" molecular character. The nature and role of "intra" and "inter" mode coupling is discussed and its dependence on local geometry explained. The couplings are seen to explain anomalous splittings observed in incoherent inelastic neutron spectra both in the low frequency (translational optic) and in the high frequency (molecular stretching) range. In addition, calculations of the dynamical properties of various exotic high pressure ice phases are presented. Changes in the spectra of the high pressure phases are explained by sub-lattice coupling between the inter-penetrating tetrahedrally co-ordinated sub-lattices associated with these structures.

INTRODUCTION

Inelastic neutron scattering experiments provide a direct probe of excitations in condensed matter through a wide range of energies. The low end of the excitation spectra in insulators is dominated by the excitation of lattice vibrations or phonons. Such spectra contain a wealth of information regarding the microscopic nature of the inter-atomic bonds which bind atoms together in the condensed phase. In complex systems, such as the phases of ice investigated in this study, the complexity of the

High Performance Computing
Edited by R. J. Allan *et al.*, Kluwer Academic / Plenum Publishers, New York, 1999

spectra makes a direct microscopic interpretation of the spectra difficult. In such systems computational *ab initio* simulation of lattice vibrations provides a direct method for determining the microscopic nature of vibrational modes which dominate different regions of the observed spectra.

The computational studies presented here are motivated by a series of incoherent inelastic neutron spectra performed on ice structures across the whole of the phase diagram (Li, 1996). The complexity of the various structures, and consequently of the vibrational spectra, is due to the range of bond strengths present (ice is constructed from covalently bonded water molecules bonded together by hydrogen bonds) and the molecular orientational disorder characteristic of many of the stable phases. Of particular interest in these studies is the translational optic region of the spectra, this is the frequency range where the characteristic microscopic motion is dominated by distortions of hydrogen bonds such that neighbouring molecules move out of phase. The experimental spectra, for numerous ordered and disordered phases of ice, show a double peak structure in this region (Li and Ross, 1993; Li, 1996). This structure has led to the speculation of the existence of two hydrogen bond types, an empirical parameterization of microscopic bond strengths assuming two very different bond types randomly distributed throughout the lattice yields a spectra in agreement with experiment (Li and Ross, 1993). Another suggestion of the origin of the double peak structure relies on TO-LO splitting as observed in ionic crystals (Tse et al., 1991). Our studies here show the feature is in fact due to coupling between intra molecular (covalent bond) stretching modes and inter (hydrogen bond) stretching modes. This unexpected coupling is also seen to be responsible for unusual features in the spectra in other frequency ranges.

The low pressure phases of ice are tetrahedrally co-ordinated structures where oxygen atoms lie on either a diamond (cubic) or wurtzite (hexagonal) lattice (Hobbs, 1974). Hydrogen atoms then occupy the lattice according to the "ice rules" whereby a single hydrogen atom lies between each oxygen oxygen pair such that each oxygen atom is covalently bonded to two hydrogen atoms and hydrogen bonded to a further two. This arrangement of molecules results in four possible geometrical arrangements of molecules involved in a hydrogen bond as displayed in Figure 1.

We have performed simulations on numerous cubic and hexagonal structures, here we will describe simulations performed on four hexagonal structures modelled by eight molecule supercells. The different structures are characterised by different geometrical orientations of neighbouring water molecules. The four structures and numbers of hydrogen bond types in each structure are labelled F (4 B bonds, 4 D Bonds), G (4 A bonds, 4 D bonds), R (4 B Bonds, 1 C Bond, 3 D Bonds) and S (4 A Bonds, 1 C Bond, 3 D Bonds), the labelling is according to Howe (1987). We call these unit cells "supercells" as they are constructed to be as representative as possible of the orientational disorder present in ice Ih (the normal phase of ice). It should be noted that we find the principle features of the spectra, as described later, to be only weakly dependent on the structure. In addition, we present results concerning the high pressure phases ice VII and ice VIII, both structures consist of two inter-penetrating diamond ice lattices. In ice VII the hydrogens occupy disordered sites within the ice rules, in ice VIII an ordered structure is formed. Again the disorder is modeled by a supercell approach.

METHODS

The computations are performed using the *ab initio* pseudopotential method (Payne et al., 1992) using the MSI implementation of the CASTEP code. Here the many elec-

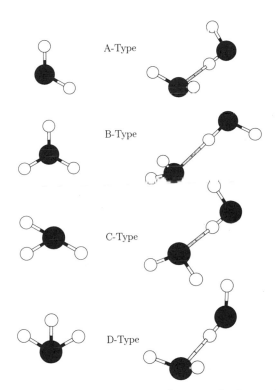

Figure 1. The four possible types local orientations of water molecules possible in hexagonal ice structures.

tron problem is treated using density functional theory via the solution of the Kohn Sham equations. *Ab initio* pseudopotentials are used to remove the oxygen "1s" electrons from the problem and the remaining single particle electron states are described in terms of a plane wave basis. The generalized gradient approximation (Perdew et al., 1992) (GGA) to the exact exchange-correlation functional is used to approximate the many electron nature of the problem (Perdew et al., 1992). The conjugate gradient minimization scheme is used to minimize the resulting energy functional with respect to single particle coefficients. The Hellman-Feynman theorem yields accurate atomic forces which can be used to minimize the total energy of the system of interacting electrons and atomic nuclei with respect to atomic positions in order to predict equilibrium structures. This method is well established and provides an accurate description of the structure and bonding in a wide range of materials. In ionic and covalently bonded materials typical structural parameters are predicted within 1% of their observed values. However, in hydrogen bonded systems the predictive accuracy is not as good, after relaxation of all structural parameters the method typically predicts hydrogen bond lengths that are approximately 2% too large. One important aspect of this approach is that covalent and hydrogen bonds are modeled on an equal footing, this feature is important if the subtle mixing between intra and inter molecular modes described later is to be correctly reproduced.

The dynamical properties of the ice structures are assessed within the harmonic approximation by numerical finite difference evaluation of the zone center dynamical matrix (Morrison et al., 1997; Ackland et al., 1997). Here rows of the dynamical matrix are evaluated from atomic forces obtained when atoms are displaced from their equilibrium positions by a small shift and the energy functional iteratively minimized again. Positive and negative shifts in all three Cartesian directions result in the cancellation of third order non-linear contributions. Subsequent diagonalisation of the dynamical matrix yields the frequency and eigenvectors of the normal modes of vibrations. More efficient, but more computationally complex, approaches exist which analytically evaluate the dynamical matrix via second order response functions (Gonze, 1997), but these methods are presently not applicable to the GGA functional.

RESULTS

The calculated vibrational spectra of the structures F, G, R and S are presented in Figure 2. These spectra are obtained by broadening the frequencies, obtained by diagonalising the dynamical matrix, with Gaussians of half width half maximum equal to 15 cm^{-1}. Because of the limited size of the "supercells" used here, and the resulting (very) limited integration of the Brillouin zone, direct comparison with experimental spectra should be made with care. However, essential features of the spectra are clearly visible and qualitative interpretation of the true spectra is possible. In order to facilitate interpretation of the spectra further projections of the spectra onto "pure" lattice and molecular modes is also shown. The projection is performed by weighting the Gaussian functions used to produce the spectra by dot products of the normal mode eigenvectors with "pure" mode eigenvectors. The "pure" modes in question can be divided into intra and inter molecular modes. The "pure" intra molecular modes consist of individual water molecules vibrating as in the isolated water molecule case, see Figure 3, and are characterized as Symmetric Stretch, Anti-Symmetric Stretch and Bending for obvious geometrical reasons. The pure inter molecular character is labeled Librational, pure modes consist of individual "rigid" molecules rotating about a fixed center of mass, and Vibrational, pure modes consist of translations of individual "rigid" water molecules.

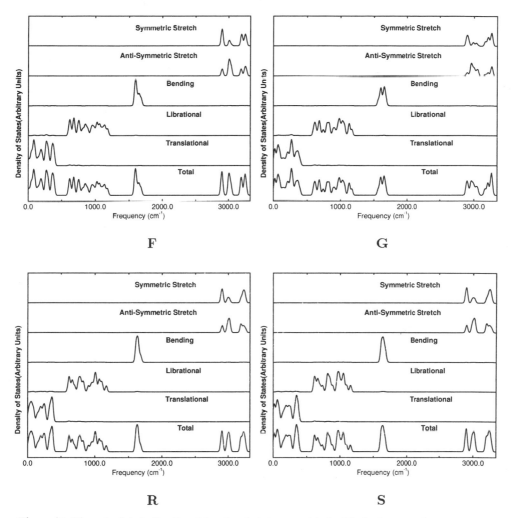

Figure 2. The calculated vibrational density of states associated with the hexagonal structures as described in the text.

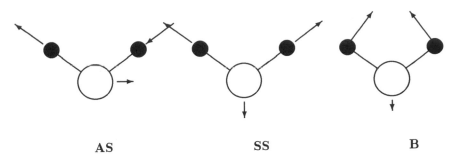

AS   **SS**   **B**

Figure 3. The normal modes of vibration of an isolated water molecule, AS refers to the anti-symmetric stretch mode (frequency 3754 cm^{-1}), SS refers to the symmetric stretch mode (frequency 3656 cm^{-1}) and B to the bending mode (frequency 1597 cm^{-1}.

It can be seen that each type of motion dominates a particular frequency range of the spectra. Close inspection of the calculated spectra reveals small, but important, mixing of character in parts of the frequency range. Since this mixing will be the subject of the discussion a magnification of the intra molecular stretch frequency range for the ice F structure is presented in Figure 4. In this figure the Librational and Translational contribution to the total spectra has been magnified in order to emphasize the mixing between intra and inter molecular character in this frequency range.

Finally in Figure 5 we present the calculated spectra for the ice VII and ice VIII structures considered in this study.

DISCUSSION

We begin the discussion by noting the overall similarity seen in the spectra for all of the "open" wurtzite structures modeled here. This similarity is also true for diamond structures and demonstrates that the principle features of the spectra are determined by the local tetrahedral co-ordination of water molecules present in both diamond and wurtzite structures. Furthermore, the discussion here will be limited to an explanation of features observed in the covalent bond stretch frequency range (2800–3700 cm^{-1}) and in the translational optic region (200–500 cm^{-1}).

In the covalent bond stretch region two distinct bands are observed, projection onto pure anti-symmetric and symmetric intra molecular modes shows these features are not simply the broadened molecular frequencies, both bands being resolved into a mixture of anti-symmetric and symmetric motion. The origin of these two bands can be seen by examining the magnified spectra in Figure 4. The higher frequency band is seen to have essentially zero contribution from pure molecular translational motion, the low frequency band is seen to have a small, but important, contribution from molecular translations. The splitting into the two high frequency bands results from this maximal/minimal mixing with the low frequency translational molecular modes, the maximal mixing case having the lower frequency. Closer examination of Figure 4 shows that each band is comprised of two further peaks, analysis of characteristic eigenvectors in each of the four regions provides us with a microscopic explanation of this maximal/minimal mixing. A schematic of characteristic coupled vibrations between neighbouring molecules in the four frequency ranges is presented in Figure 6. High frequency modes are characterized by neighbouring molecules both undergoing symmetric stretch type motion out of phase with each other (Figure 6a), this results

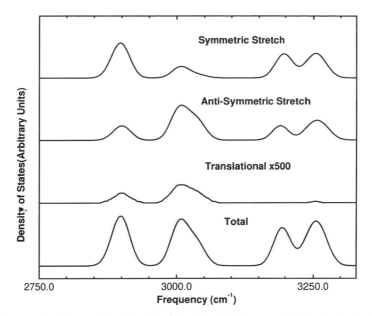

Figure 4. A magnified view of the vibrational density of states associated with the ice F structure, the projection onto pure translational modes has been magnified by 500.

in minimal distortion of the hydrogen bond and a subsequent high frequency. Low frequency modes in this region are characterized by neighbouring molecules undergoing symmetric stretch type motion in phase with each other (Figure 6d), this results in maximal distortion of the hydrogen bond and a subsequent low frequency. The intermediate frequency modes in this region are characterized by the symmetric/antisymmetric and antisymmetric/antisymmetric couplings as shown in Figure 6b and 6c. The general principle determining a particular frequency here is explained by the amount of distortion of the hydrogen bonds in a particular mode, these distortions result in mixing with the molecular optic modes and a correspondingly lower frequency.

We now turn our attention to the translational optic frequency range, in all structures considered two peaks are seen with frequencies 280 cm^{-1} and 370 cm^{-1}. A schematic of the characteristic motion at these two frequencies is shown in Figure 7. The explanation of the two peak structure is seen to be the converse of the maximal/minimal coupling observed in the high frequency range. The characteristic motion in the low frequency peak is essentially rigid molecule motion with minimal mixing with covalent bond stretch modes. In the higher frequency peak, at the upper edge of the translational optic region, the mixing with covalent bond stretch modes is maximal and consequently the frequency is maximal. A direct comparison with experimental spectra shows the study here to predict a splitting approximately 20% too large and the upper edge of the translational optic region to be about 80 cm^{-1} too high. These differences are primarily due to known difficulties with modeling hydrogen bonds within approximations to DFT, an improved functional would be expected to improve agreement with experiment but would not change the geometrical reasons for the various splittings discussed here.

Finally, we will make some comments regarding the translational optic region of the calculated spectra of ice VII and ice VIII as modeled here. We note that principle features of the spectra in this region are similar to the "open" structures spectra with a

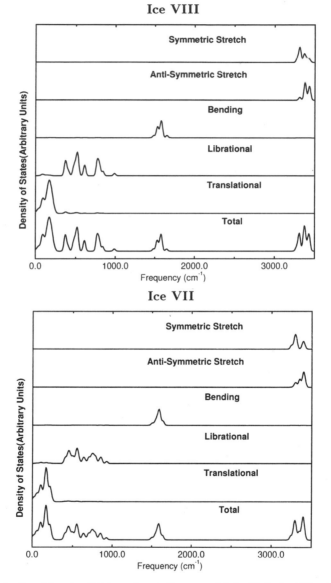

Figure 5. The vibrational density of states and projections associated with the ice VII and VIII structures considered in the text.

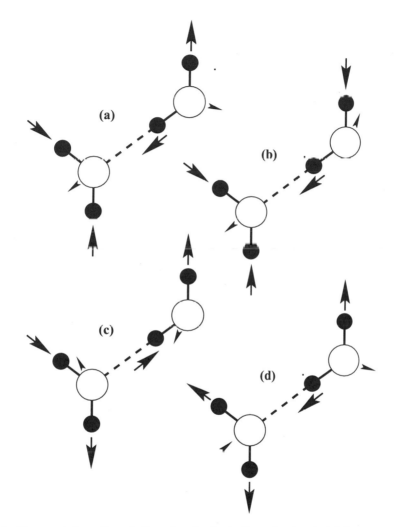

Figure 6. Characteristic motion of adjacent molecules in the molecular stretch frequency range shown in Figure 5. The four figures are representative of the motion in the four peaks seen in Figure 5 with (a) being the highest frequency and (d) the lowest.

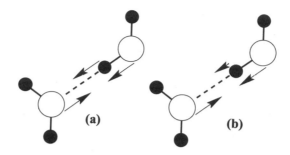

Figure 7. Characteristic motion of adjacent molecules in the molecular optic frequency range of the ice F structure. (a) is representative of motion in the low frequency peak at 280 cm^{-1} and (b) of motion in the high frequency peak at 370 cm^{-1}.

general reduction in frequency. This behaviour can be explained by couplings between the translational optic modes of a single "open" diamond sublattice and the weak interactions between the two sublattices which constitute the structures. In general a coupling to a weaker interaction will drive down the frequency.

ACKNOWLEDGEMENTS

We would like to thank the EPSRC for financial support for SJ and JCL and the EPSRC and Nuffield foundation for providing the computational facilities used in this study. Additionally we thank M.H. Lee for providing the GGA oxygen pseudopotential used in this study.

REFERENCES

Ackland, G.J., Warren, M.C., and Clark, S.J., 1997, Practical methods in *ab initio* lattice dynamics, *J. Phys.: Condens. Matter*, 9:7861.

Howe, R., 1987, The possible ordered structures of ice Ih, *Journal de Physique*, C1:599.

Li, J., and Ross, D.K., 1993, Evidence for two kinds of hydrogen bond in ice, *Nature*, 365:327.

Li, J., 1996, Inelastic neutron scattering studies of hydrogen bonding in ices, *J. Chem. Phys.*, 105:6733.

Morrison, I., Li, J., Jenkins,S., Xantheas, S.S., and Payne, M.C., 1997, *Ab initio* total energy studies of the static and dynamical properties of ice Ih, *J. Phys. Chem.*, 101:6146.

Klug, D.D., Tse, J.S., and Whalley, E., 1991, The longitudinal optic transverse optic mode splitting in ice Ih, *J. Chem. Phys.*, 95:7011.

Hobbs, P.V., 1974, *Ice Physics*, Clarenden Press, Oxford.

Gonze, X., 1997, First principles responses of solids to atomic displacements and homogeneous electric fields: implementation of a conjugate gradient algorithm, *Phys. Rev. B*, 55:10337.

Perdew, J.P., Chevary, J.A., Vosko, S.H., Jackson, K.A., Pederson, M.R., Singh, D.J., and Fiolhais, C., 1992, Atoms, molecules, solids and surfaces – applications of the generalized gradient approximation for exchange and correlation, *Phys. Rev. B*, 46:6671.

Payne, M.C., Teter, M.P., Allan, D.C., Arias, T.A., and Joannopolous, J.D., 1992, Iterative minimization techniques for *ab initio* total energy calculations – molecular dynamics and conjugate gradients, *Rev. Mod. Phys.*, 64:1045.

PHASE SEPARATION OF TWO IMMISCIBLE LIQUIDS

S.I. Jury, P. Bladon, S. Krishna and M.E. Cates

Department of Physics and Astronomy,
University of Edinburgh, JCMB, King's Buildings,
Mayfield Road, Edinburgh EH9 3JZ, U.K.

ABSTRACT

The separation of a binary fluid mixture into its constituent phases involves the interaction of numerous physical phenomena. Studying such a transition gives insight into the underlying mechanisms, their respective strengths and at what stage in the separation each becomes important or dominant. Large scale simulations of 3 dimensional spinodal decomposition in a binary fluid using the Dissipative Particle Dynamics method show the approach to a linear time dependence in domain coarsening. We present interface and velocity maps which clearly demonstrate the Siggia mechanism for domain growth.

INTRODUCTION

When a two component system is quenched into a thermodynamically unstable region it will phase separate, forming two phases with different compositions. The kinetics of this process has been the focus of much attention in recent years [1, 2, 3]. Computational studies have been performed using various simulation methods with varying degrees of success [9, 10, 11, 12]. Resolution of interface physics requires length and time scale to be studied over orders of magnitude. The relative simplicity of the system coupled to its non-equilibrium behaviour suggest simulations as an appealing method for examining phase separations. Depending on the depth of quench and initial composition, separation may occur via nucleation or spinodal decomposition. Nucleation occurs when the system is locally stable and requires fluctuations in concentration to exceed a given threshold. Spinodal decomposition occurs when the system is locally unstable: any fluctuation in concentration will lead to separation. Domains of a characteristic size $L(t)$ form a structure that is (statistically) independent of time when lengths are scaled by $L(t)$, implying that the size of the domain varies as a power law in time $L(t) \sim t^\alpha$. Spinodal decomposition is, then, an example of dynamic scaling [1].

Although dynamic scaling appears to be a universal feature of systems undergoing spinodal decomposition, different growth mechanisms give rise to different growth laws. Systems such as alloys, which coarsen by species interdiffusion, have a growth law

$L \sim t^{1/3}$. This can be described by the theory of Lifshitz and Slyozov [4], provided the volume fraction of the minority phase is sufficiently small. Coarsening driven by coalescence of diffusing droplets also leads to a $t^{1/3}$ growth law. Additional coarsening mechanisms are possible in fluids, where there is coupling of the domain structure to hydrodynamic transport mechanisms. By analysing model H [5], the appropriate dynamical model for a conserved order parameter coupled to a viscous incompressible fluid, three scaling regimes can be identified, which occur at successively longer times and larger lengthscales. Firstly, there is a $t^{1/3}$ due to order parameter diffusion: i.e. species interdiffusivity; secondly there is a regime where viscous hydrodynamic forces dominate transport, first predicted by Siggia [6], with a growth law of t^1; lastly, there is the $t^{2/3}$ regime predicted by Furukawa [2], where fluid inertia dominates transport. The scaling of the different regimes in terms of the physical parameters, the order parameter mobility λ, the fluid viscosity, η, and the surface tension, σ, can be summarised as [1]:

$$L(t) \sim \begin{cases} (\lambda \sigma t)^{1/3}, & L \ll (\lambda \eta)^{1/2}, & \text{(diffusive)} \\ \sigma t/\eta, & (\lambda \eta)^{1/2} \ll L \ll \eta^2/\rho\sigma, & \text{(viscous hydrodynamic)} \\ (\sigma t^2/\rho)^{\frac{1}{3}}, & L \gg \eta^2/\rho\sigma, & \text{(inertial hydrodynamic)} \end{cases}$$

The existence of scaling laws does not necessarily give any insight into the physical mechanisms responsible for their creation. Siggia proposed that the t^1 growth law occurs due to a Rayleigh - type instability in the fluid network [6]. The surface tension causes pressure differences across the curved fluid interfaces resulting in a squeezing of fluid from regions of high curvature into low. Regions of high curvature — necks — should pinch off, causing the structure to coarsen. The physics of this involves σ and η, but not ρ or λ, so that a simple dimensional analysis of the quantities involved results in the t^1 growth law.

Using neutron scattering and light scattering techniques, experiments have observed early time diffusive growth, as well as the late time viscous dominated regime [7,8]. Computer simulations of multi-phase flow are an appealing alternative to experiments, since they (in principle) allow selection of the desired flow regime. More importantly they allow visualisation of the phase separation process itself. Computer simulations have their limitations: the length and timescales accessible by simulation depend critically on the efficiency of the model, and most importantly, the amount of computer resources available. Many different techniques have been used to investigate spinodal decomposition in fluids. Evidence for a t^1 growth law has been claimed in simulations of model H [9], lattice Boltzmann simulations [10], molecular dynamics [11] and a 'Direct Simulation Monte Carlo' technique applied to binary gases [12]. One conclusion that can be drawn from these studies, is that observing an extended regime of t^1 growth is often difficult; finite size effects can be severe, and in many cases exponents have to be extracted by extrapolating effective exponents to infinite system sizes.

The late time inertia dominated regime predicted by Furukawa has not yet been observed experimentally [7,8,13], or by numerical methods in three dimensions, although lattice gas simulations in two dimensions (where there is no direct equivalent of the Siggia mechanism) do show the inertial growth exponent of 2/3 [14].

In this paper, we use Dissipative Particle Dynamics (DPD) to investigate the phase separation of a binary symmetric fluid in the viscous regime. In addition to obtaining scaling functions and growth laws, as previous authors have done, we obtain visualizations of the coarsening mechanism in action. This allows direct confirmation that the mechanism proposed by Siggia is indeed responsible for evolving the structure in the viscous dominated regime.

DISSIPATIVE PARTICLE DYNAMICS

DPD is a relatively new technique for simulating hydrodynamic behaviour at the mesoscopic level. In DPD, a set of particles in a continuum act as momentum carriers whose behaviour is governed by a set of stochastic differential equations [15]. Each particle interacts with other particles in its local environment with conservative, random and dissipative forces that conserve mass (i.e. particle number) and momentum. The emergence of isothermal Navier-Stokes hydrodynamics is thus guaranteed at large length and timescales. Note that momentum is exchanged in an isotropic and Galilean invariant fashion, which is not easily ensured in lattice based techniques. Modification of the conservative forces that act between fluid particles allows many different mesoscale models to be defined at the thermodynamic level. The ease with which a model system may be defined has led to a number of applications: model colloidal suspensions [16]; multi-component fluids [17, 18]; polymers [19] and polymer mixtures [20].

The set of stochastic differential equations describing particle motion are [15]

$$\frac{d\mathbf{r}_i}{dt} = \mathbf{p}_i, \quad \frac{d\mathbf{p}_i}{dt} = \sum_{ij} \Omega_{ij}\hat{\mathbf{r}}_{ij} \tag{1}$$

where the sum runs over all pairs of particles. Ω, which we can view as a force, is given by

$$\Omega_{ij} = w(r_{ij})\left(\alpha_{ij} + \sigma\theta_{ij} - w(r_{ij})\gamma(\mathbf{v}_{ij} \cdot \hat{\mathbf{r}}_{ij})\right) \tag{2}$$

The particle separation vector is denoted by $\mathbf{r}_{ij} = \mathbf{r}_i \quad \mathbf{r}_j$, $r_{ij} = |\mathbf{r}_{ij}|$, $\hat{\mathbf{r}}_{ij} = \mathbf{r}_{ij}/r_{ij}$, with the relative velocity $\mathbf{v}_{ij}$ defined similarly. $w(r)$ is a weight function which we choose to be [15]

$$w(r) = \begin{cases} \alpha_{ij}(1 - r/r_c) & (r < r_c) \\ 0 & (r \geq r_c) \end{cases} \tag{3}$$

r_c is a cutoff distance, which we choose as our unit of length. The first term in Ω acts as a conservative force, with maximum repulsion given by α_{ij}, which may be different for different pairs of particles. The second term corresponds to a random force acting between pairs of particles, with properties

$$\theta_{ij}(t) = 0, \quad \theta_{ij}(t)\theta_{kl}(t') = (\delta_{ij}\delta_{kl} + \delta_{il}\delta_{jk})\delta(t - t'). \tag{4}$$

The term involving γ describes a dissipative force that acts to decrease the relative velocity of the two particles. The random and dissipative forces act to drive the system towards an equilibrium state, with a temperature defined by

$$\sigma^2 = 2\gamma k_B T. \tag{5}$$

The stationary distribution of DPD, $f(\mathbf{r}_i^N, \mathbf{v}_i^N)$, is the canonical ensemble

$$f(\mathbf{r}_i^N, \mathbf{v}_i^N) \propto \exp\left(-H/k_B T\right) \tag{6}$$

$$H = \sum_i \frac{\mathbf{p}_i^2}{2m_i} + \sum_{ij} U(r_{ij}) \tag{7}$$

where $U(r)$ is the pair potential that generates the conservative forces.

The formulation of DPD in this manner allows any thermodynamic model, defined via the conservative forces, to be simulated with the correct canonical distribution. However, a key advantage of DPD over molecular dynamics techniques which simulate the canonical ensemble is that the viscosity, and hence the hydrodynamics, can be varied *independently* of the thermodynamics. An estimation of the dissipative contribution to the viscosity is given in [21] as,

$$\eta_D = \frac{2\pi\gamma\rho^2}{15} \int_0^\infty dr r^4 w^4(r) \qquad (8)$$

Altering γ, while keeping the thermodynamic parameters fixed, changes the dissipative contribution to the viscosity, and hence the total viscosity. The thermodynamics, determined by the conservative forces are unaffected, provided constant temperature is maintained.

To construct a binary fluid, we define two species A and B that interact with different conservative forces. By choosing $\alpha_{AA} = \alpha_{BB}$, $\alpha_{AB} > \alpha_{AA}$ for the AA, BB and AB interactions we obtain a symmetric binary fluid. Our binary fluid model is defined by the following choice of parameters: $\alpha_{AA} = \alpha_{BB} = 20$; $\alpha_{AB} = 100$; $\rho = 10$; $\gamma = 30$ and $k_B T = 1$. The viscosity of the homogeneous phase was measured using Lees-Edwards sliding periodic boundary conditions [22], and the surface tension by integrating the difference between the longitudinal and transverse stress through an AB interface in a slab geometry [23]. All results in this paper were obtained by using a conventional velocity Verlet algorithm to integrate the stochastic differential equations [20], with a timestep of 0.01, chosen so as to ensure that the measured temperature was within about 1% of that implied by eqn(5).

The parameters were chosen so as to access the viscous regime. This was made possible by choosing $\eta^2/\rho\sigma$ so that the crossover to the inertial $t^{2/3}$ regime occurred at lengthscales much larger than those accessible in the simulation.

PARALLEL D.P.D. CODE

A system consisting of 10^6 particles, with equal amounts of A and B, was simulated in a box of size $\sim 45r_c^3$ using a message-passing code developed for the Cray T3D at Edinburgh. Since the force falls to zero for $r > r_c$, only particles separated by r_c or less need be considered in the force calculation. This allows us to use a traditional spatial domain decomposition wherein each processor is allocated a region of space. At the beginning of each timestep neighbouring processors exchange information about particles lying a distance r_c from their boundaries. In normal molecular dynamics only the positions need be exchanged, however the D.P.D. algorithm requires that velocities are passed as well. The top layer of each processor is passed to the bottom of the processor above, this is repeated for the bottom layer and for the other two dimensions. A halo is thus created around each processor giving the processor enough information to perform the force calculation for all particles within its domain. Simulating two different phases is straightforward. A species label is attached to each particle allowing the processor to decide which value of α to use for every interaction.

An extra complication arises due to the stochastic term in the DPD force calculation. In molecular dynamics, the forces F_{ij} and F_{ji} may be calculated different processors, since they depend only on position they will be guaranteed t e equal and opposite. This will not be the case with DPD since the random comp nent will be different on the different processors. To overcome this difficulty a unique integer label is assigned to every particle, which we denote `particle-ident`, and interactions are only considered if `particle-ident(i)` is greater than `particle-ident(j)`. This is sufficient to ensure interactions are considered once on a single processor. The `particle-ident` label becomes a unique identifier which must remain with the particle at all times. An extra round of communication is required to pass the random forces back to the correct processor.

When extracting velocity fields and order parameter fields a degree of coarse graining was employed. This was done for two reasons. Firstly, each configuration consists of 6×10^6 real numbers (a minimum of 48MB per file) making storage difficult and analysis time consuming. Secondly, for the velocity field in particular, thermal noise would add an unwanted random component onto the underlying hydrodynamic flow field, averaging over time and space will remove this. So space was divided into a 32^3 grid and the fields defined at each gridpoint. This involves further communications as each processor sends information back to a *master* processor which does the coarse graining.

Typical simulations involved 10^6 particles running on 256 processors taking 1000 processor hours per 2000 timesteps. The graphics package A.V.S. was used to produce pictures and animations.

The inner loop:

```
exchange:  positions, velocities, species, particle-ident
                              ↓
                     calculate force
                              ↓
            exchange:  random force component
                              ↓
       add random force component to other forces
                              ↓
                  integrate total force
                              ↓
                    exchange particles
```

SCALING FUNCTIONS

The scaling hypothesis assumes the existence of a single length scale $L(t)$, which implies that the structure factor has a scaling form given by

$$S(k,t) = L(t)^d F(kL(t)).$$
(9)

Here we define a lengthscale $L(t) = 2\pi/k$, where

$$k = \frac{\int k S(k)_r dk}{\int S(k)_r dk}$$
(10)

and $S(k)_r$ is the radial average of the three dimensional mixed structure function as defined by

$$S(k) = [\rho_k^A - \rho_k^B].[\rho_{-k}^A - \rho_{-k}^B]$$
(11)

$$\rho_k^A = \int \exp(-i\mathbf{k}.\mathbf{r})\rho^A(\mathbf{r})d\mathbf{r}^3$$
(12)

with $\rho^A(\mathbf{r})$ the number density of particles of type A at $\mathbf{r}$. Figure (1) shows the behaviour of $L(t)$ vs t. The behaviour at very early times $t \leq 10$ arises from the finite size of the coarse-graining grid used to extract the interface and is not plotted. After time $t = 40$ the system appears to enter a scaling regime with $L(t) \sim t$.

189

One can use $L(t)$ in the scaling regime to collapse the data onto a single, universal, scaling function $F(x)$. The results are shown in figure (2). The collapse of the data onto a universal curve over nearly a decade in time is excellent, with the data showing an x^{-4} Porod tail, characteristic of scattering from the well defined interfaces within our sample. Other features of the curve resemble those reported by other studies [24].

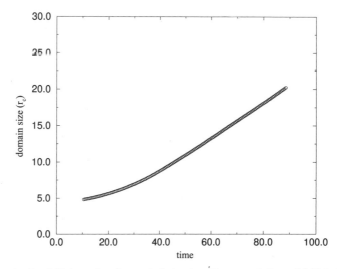

Figure 1. Domain size $L(t)$ in units of r_c vs t. Late stage linear variation of $L(t)$ indicates viscous dominated transport.

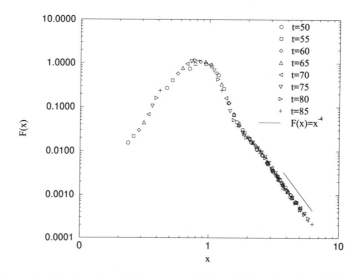

Figure 2. Universal scaling function $F(x)$ vs x , $x = k/k_m$, with k_m the value of k for which $S(k)$ is maximum. Curves are all for the late time viscous dominated separation regime.

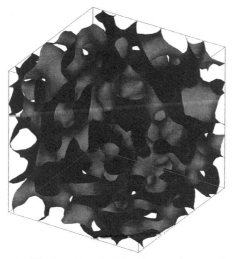

Figure 3. The interface dividing the two fluids at late times.

THE SIGGIA MECHANISM

The presence of a single characteristic length at late times that scales as t^1 is clear evidence that the system is in a viscous dominated hydrodynamic regime. However one can gain substantial physical insight into the coarsening process by observing the evolution of the interface, together with the corresponding velocity field, as a function of time. The velocity field was calculated by coarse graining the velocity of the DPD particles over 200 timesteps, during which time the interfaces remain (effectively) fixed. Figure (3) shows the system at a late stage. Only the interface between the two phases is shown. One can see that the fluid forms a percolating structure of fluid tubes with a well defined length scale — the width of the tubes. The mechanism by which this structure coarsens is shown in figure (4). A region joining two tubes - a 'neck' is shown during the breaking process. Surface tension causes a pressure gradient that drives material from the neck, causing it to pinch off, break and retract. This is a direct observation of fluid tube necking, the mechanism proposed by Siggia [6] to explain the observed t^1 growth law. More details of the velocity maps, and a quantitative discussion of the amplitude of the t^1 growth law, will appear elsewhere [18].

CONCLUSIONS

Using DPD to simulate a binary, symmetric, fluid in three dimensions, we have directly observed the 'necking' instability predicted by Siggia to arise in the viscous hydrodynamic regime, along with the corresponding t^1 growth law. A key factor in the success of this work was, within the DPD model, the ability to to tune the hydrodynamics *without* altering the thermodynamics, allowing us to access the desired flow regime.

ACKNOWLEDGEMENTS

We would like to thank EPSRC for support via the U.K. Colloidal Hydrodynamics Grand Challenge. We would like to thank the Edinburgh Parallel Computing Center

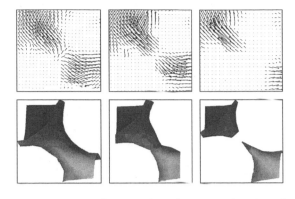

Figure 4. The lower figures show a selected neck at three successive times during breaking. The upper figures show the corresponding velocity fields for the fluid inside the neck.

for help developing the code. SIJ would like to thank Unilever PLC for a CASE Award. We would like to thank Patrick Warren for useful discussions.

REFERENCES

1.   A.J. Bray, *Advances in Physics* 43:357-459 (1994).
2.   H. Furukawa, *Advances in Physics* 34:703-50 (1986).
3.   J.D. Gunton et al., in: *Phase Transitions and Critical Phenomena*, C. Domb and J.L. Lebowitz eds. Academic Press, New York (1983) pp267-482.
4.   I.M. Lifshitz and V.V. Slyozov, *J. Phys. Chem. Solids* 19:35-50 (1961).
5.   P.M. Chaikin, and T.C. Lubensky, in: *Principles of Condensed Matter Physics*, Cambridge University Press (1995).
6.   E.D. Siggia, *Phy. Rev.* A20:595-605 (1979).
7.   S.H. Chen S.H. et al., *Progress in Colloid and Polymer Science*, 93:311-316 (1993).
8.   N. Wong and C.M. Knobler, *Phys. Rev.* A24:3205-211 (1981).
9.   T. Lookman, W. Yanan, J.A. Francis and S. Chen, *Phys. Rev* E53:5513-16 (1996).
10.   S. Chen and T. Lookman, *J. Stat. Phys.* 81:223-35 (1995).
11.   M. Laradji, S. Toxvaerd and O.G. Mouritsen, *Phys. Rev. Lett.* 772253-56 (1996).
12.   S. Bastea and J.L. Lebowitz, *Phys. Rev. Lett.* 78:3499-502 (1997).
13.   T. Hashimoto et al., *Physica A* 204:261-76 (1994).
14.   S. Bastea and J.L. Lebowitz, *Phys. Rev.* E52:3821-3826 (1995).
15.   P. Espagñol and P.B. Warren, *Europhys. Lett.* 30:191-196 (1995).
16.   P.J. Hoogerbrugge and J.M.V.A. Koelman, *Europhys. Lett.* 19:155-60 (1992); J.M.V.A. Koelman and P.J. Hoogerbrugge, *Europhys. Lett.* 21:363-368 (1993); E.S. Boek, P.V. Coveney and H.N.W. Lekkerkerker, *J. Phys. - Cond. Mat.* 8:9509-12 (1996); E.S. Boek, P.V. Coveney, H.N.W. Lekkerkerker and P.V.D. Schoot, *Phys. Rev.* E55:3124-38 (1997).
17.   P.V. Coveney and K.E. Novik, *Phys. Rev.* E54:5134-41 (1996); P.V. Coveney and K.E. Novik, *Phys. Rev.* E55:4831 (1997).
18.   S.I. Jury, P. Bladon, M.E. Cates and S. Krishna, in preparation.
19.   A.G. Schlijper, P.J. Hoogerbrugge and C.W. Manke, *J. Rheology.* 39:567-79 (1995); Y. Kong, C.W. Manke, W.G. Madden and A.G. Schlijper, *J. Chem. Phys.* 107:592-602 (1997).
20.   R.D. Groot and P.B. Warren, *J. Chem. Phys.* 107:4423-4435 (1997).
21.   C.A. Marsh, G. Backx and M.H. Ernst, *Phys. Rev.* E56:1676-1691 (1997).
22.   M.P. Allen and D.J. Tildesley, in: *Computer Simulation of Liquids*, Clarendon Press, Oxford (1987).
23.   J.S. Rowlinson and B. Widom, in: *Molecular Theory of Capillarity*, Clarendon Press, Oxford (1982).
24.   C. Yeung, *Phys. Rev. Lett.* 61:1135-8 (1988); H. Furukawa, *J. Phys. Soc. Jap.* 58:216-21 (1989); M. Takenaka and T. Hashimoto, *J. Chem. Phys.* 96:6177-90 (1992).

COMPUTER SIMULATION OF LIQUID CRYSTALS ON THE T3D/T3E

Mark R. Wilson,[1] Michael P. Allen,[2] Maureen P. Neal,[3]
Christopher M. Care, Douglas J. Cleaver,[4]

[1]Department of Chemistry, South Road, University of Durham, Durham,
DH1 3LE.
[2]University of Bristol, H. H. Wills Physics Laboratory, Royal Fort, Tyndall
Avenue, Bristol, BS8 1TL.
[3]School of Mathematical and Information Sciences, Coventry University,
Priory St., Coventry, CV1 5FB.
[4]Materials Research Institute, Sheffield Hallam University, Pond Street,
City Campus, Sheffield, S1 1WB.

INTRODUCTION

This article describes the recent simulation studies of the UK Complex Fluids
Consortium on the Cray T3D and T3E in Edinburgh. The common theme of our recent
work has been *thermotropic liquid crystal phases*. These materials occur sandwiched
between conventional liquid and solid phases. As their name implies, they can exhibit
both liquid-like and solid-like properties in the bulk.

In the simplest liquid crystal (LC), the *nematic* phase, molecules line up in a
preferred direction (the *director*), but the phase is liquid in all other respects. On
cooling a nematic, the molecules may form into layers, whilst still retaining normal
liquid properties such as flow and structural disorder within the layers: this is a *smectic*
liquid crystal phase. The molecular order within a liquid crystal gives rise to anisotropic
electrical and optical properties leading to important uses for thermotropic LC materials
in display technology.*

The aim of liquid crystal simulations is to complement laboratory experiments by
providing greater insight into the link between molecular shape and structure and the
technologically important bulk properties of liquid crystals. This should assist future
design and synthesis of new materials for display (and other) applications. Simulation
can also provide fundamental insights into the detailed structure and dynamics of liquid
crystal phases at the molecular level; and allows researchers to test a variety of statistical
mechanical theories that aim to explain the wealth of interesting physical phenomena
which occur in liquid crystal systems.

In this article we begin by describing the simulation models and methods most
currently used in our work. We then go on to describe some of the interesting results
which arise from the study of these models.

---

*e.g. Liquid Crystal Displays (LCDs)

High Performance Computing
Edited by R. J. Allan *et al.*, Kluwer Academic / Plenum Publishers, New York, 1999

SIMULATION MODELS AND ALGORITHMS

The Gay-Berne pair potential[1] has been a common feature in much of our simulation work. It has a similar form to the well-known Lennard-Jones potential, with a strong repulsion between particles at close separations and attraction between particles at longer distances. However, unlike the Lennard-Jones potential, both attractive and repulsive terms are orientationally dependent. Figure 1 shows the form of the Gay-Berne potential for two molecules plotted for three fixed relative orientations. The depth and position of the potential wells in Figure 1, and the molecular length to breadth ratio, can easily be adjusted by varying a number of parameters within the potential. This provides a series of ellipsoidal potentials which can be prolate or oblate in shape and can exhibit very different phase behaviour.

The phase behaviour of an ensemble of molecules interacting via the Gay-Berne potential is most easily studied by molecular dynamics (MD) or Monte Carlo simulation. The former is most easily adapted to massively parallel machines. The consortium has developed two families of parallel MD codes: one based on a replicated data algorithm, and most suitable for simulations of a few thousand molecules; the other based on a geometrical domain-decomposition algorithm, which allows simulation of several tens of thousands of molecules[3] (see table 1). The programs run very efficiently on the Cray T3D, typically using 64, 128 and 256 processors, and have been successfully ported to the IBM SP2 and Cray T3E. Figure 2 shows the linear scaling achieved for the domain decomposition algorithm (GBMEGA[4]) for simulations on the Cray T3D.

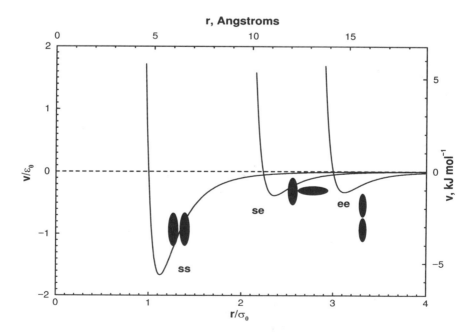

Figure 1. The Gay-Berne potential v for two molecules plotted for three fixed relative orientations. (The potential is plotted for the relative well-depths and shape anisotropies given in reference 2. The scales show the potential in reduced units and an approximate comparison with real units.)

Table 1. Gay-Berne simulation programs developed by members of the Complex Fluids Consortium.

Program	Description	Algorithm	no. of molecules
GBMESO	Single or multiple Gay-Berne sites	leap frog /RD[a]	500-10000
GBMOL	Gay-Berne + Lennard-Jones sites[b]	leap frog /RD[a]	250-10000
GBMEGA	Single Gay-Berne sites	leap frog /DD[c]	1000-250000

[a]Replicated data algorithm.
[b]Suitable for flexible molecules.
[c]Domain decomposition algorithm.

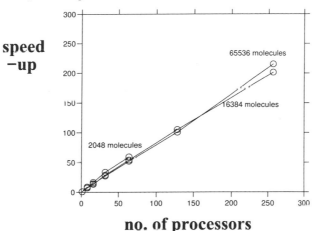

Figure 2. Simulation times for the domain decomposition Gay-Berne algorithm (GBMEGA) on the Cray T3D.

STUDIES OF RIGID MOLECULES

Elastic Constants in the Nematic Phase

The distortion of a nematic liquid crystal from a uniformly aligned state can be described by continuum elastic theory. At the heart of this theory are the Frank elastic constants K_1, K_2, K_3 which determine how susceptible the nematic is to splay, twist and bend deformations. The Frank constants influence the shape of disclinations defects in the nematic and play a critical role in describing how nematics respond to applied fields. They are therefore of key importance in display applications.

The Bristol group has developed a method for calculating the Frank elastic constants by monitoring equilibrium director fluctuations as a function of wavevector.[5, 6] This technique has been applied to the Gay-Berne potential using a system size of 8000 molecules.[7] A special director constraint technique has been used to simplify the data analysis and reduce statistical errors. K_1, K_2 and K_3 are seen to systematically increase with density at constant order parameter with $K_3 > K_1, K_2$ ($K_1 > K_2$ in most cases) as expected for elongated molecules.

The Direct Correlation Function near the Isotropic-Nematic Transition

The direct correlation function $c(1, 2)$ is the *Holy Graille* of liquid state structure theory. It is related to the pair correlation function $h(1, 2)$, which may be straightforwardly measured in diffraction experiments and computer simulations, and lies at the heart of modern density functional theories of liquids. The function $c(1, 2)$ is expected to have a shorter spatial range than that of $h(1, 2)$. This is especially interesting close to the ordering transition in a liquid crystal, when the range of $h(1, 2)$ becomes extremely large. The simplest description of the transition involves $c(1, 2)$ remaining very short ranged (of molecular dimensions) while orientational correlations represented by appropriate components of $h(1, 2)$ diverge.

The Bristol group (MPA) have calculated $c(1, 2)$ for the first time in the vicinity of the isotropic-nematic ordering transition,[8] for a simple molecular model using large system sizes with the GBMEGA program. The results are shown in Figure 3, and confirm that the form of $c(1, 2)$ is almost unaltered as the system is cooled on approach to the transition, whereas $h(1, 2)$ shows extremely long-ranged effects. The spatial integral of the second- and fourth-rank components of $c(1, 2)$ approaches the mechanical instability limit for the isotropic phase, as predicted by density functional theory. The simulations also showed that the orientational correlation lengths and correlation times are well described by a Landau-de Gennes theory. These results could not have been achieved without very long simulations of large systems on the Cray T3D.

Smectic-A* Twist-Grain-Boundary (TGB) Phase

By dissolving right-handed or left-handed *chiral* molecules in a liquid crystal, the director can be made to twist in a helical fashion uniformly through space. Chiral or twisted nematic phases are commonly used in display devices, and rely on the pitch of the helix being commensurate with the wavelength of visible light. On cooling such a system, a chiral smectic phase is formed, but it is no longer possible for the layered system to support a uniform twist. Instead, to relieve the strain energy, the phase splits into domains of untwisted smectic, separated by twist grain boundaries.

The Bristol group (MPA), in collaboration with the Durham group (MRW) has carried out the first simulation of the twist grain boundary phase, using the Cray T3D.[9] The typical smectic layers can be clearly seen in the Figure 4, and molecules are colour coded by orientation to highlight the formation of distinct domains. The overall twist from the front to the back of the simulation cell is maintained by specially developed periodic boundary conditions.

Within each twist grain boundary, the distortion is produced by an array of screw dislocations. A typical example is shown in the second part of Figure 4, which is a

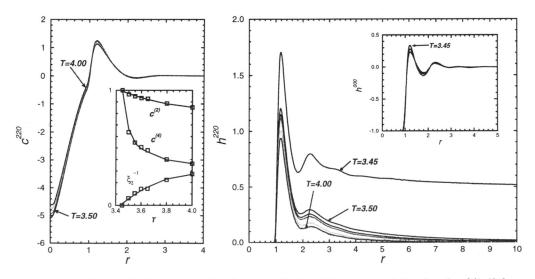

Figure 3. Plots of the direct correlation function $c(1, 2)$ and the pair correlation function $h(1, 2)$ for Gay-Berne particles in the liquid phase.

Table 2. Comparison with experimental data.

Property	Simulation $/\sigma_0$	Experiment[a] /nm
Smectic layer spacing d_0	4.3	4.1
Helix pitch p	320	500
Smectic domain size l_s	20	24
Dislocation line separation l_d	11	15

[a]K. J. Ihn, J. A. N. Zasadzinski, R. Pindak, A. J. Slaney, J. Goodby. *Science*, **258**, 275–278 (1992).

cylindrical core sample taken from the grain boundary region. In the figure, molecules are reduced in size to aid clarity. The molecular centres are arranged in a 'spiral staircase' around the screw dislocation line.

Very large systems are required in order to allow these structures to develop properly. The overall cell length from front to back determines the helical pitch, which must be several hundred nanometres for a realistic comparison with experiment; twisted structures of the order of the wavelength of visible light are of technological interest, because of their potential applications in optical device design. The transverse dimensions of the simulation cell must also be very large, otherwise the orientation of smectic layers is dictated by the requirement that they fit with the periodic boundary conditions, rather than being driven by the physics. It is also necessary to simulate the system for a relatively long time (of the order of nanoseconds) to allow these structures to develop spontaneously. The close comparison with experimental results is indicated in table 2.

The Influence of Steric and Electric Effects on Mesophase Structure

As part of the research programme in Coventry (MPN), two separate studies of Gay-Berne based mesogens have been undertaken to date: one study concentrating on steric factors and a second study concentrating on electronic effects.

The *steric study* comprised a comparison of two multi-site Gay-Berne systems: a *zigzag* and a *triangle model* formed from three rigidly joined Gay-Berne sites. This led to

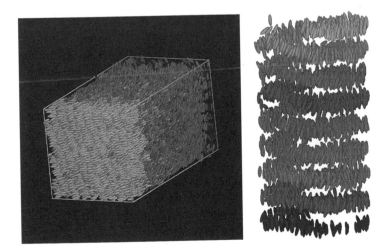

Figure 4. A snapshot from a molecular dynamics simulation showing the molecular order in a smectic-A twist grain boundary phase. Also shown is a close up view of molecules within a grain boundary showing a screw dislocation defect.

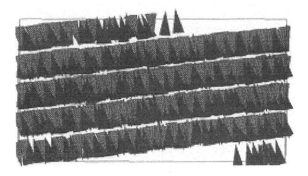

Figure 5. A typical configuration of a smectic layer for a anti-ferroelectric phase, viewed perpendicular to the layer

the first simulation of an anti-parallel phase[10] driven by steric effects. Snapshots of an anti-ferroelectric smectic phase exhibited by the triangle model are illustrated in Figure 5. This is in agreement with the behaviour of the achiral compound 1-propylbutyl4'-(4"-n-nonyloxphenylproprioloyloxy)biphenyl-carboxylate which forms an antiferroelectric-like phase on cooling.[11] This link between the steric shape of the molecule and long-range packing effects is of importance in the design of phenylpropriolates which exhibit a wide variety of ferro- and anti-ferroelectric chiral phases.

Simulation studies of large systems of dipolar single-site Gay-Berne molecules have been reported recently in the literature.[12] The magnitude and location of the dipole within a liquid crystal molecule have been shown to alter significantly the molecular organisation in smectic layers.

There has also been a growing interest in the molecular quadrupoles within liquid crystal molecules. Barbero and Durand[13] have shown that the tilt of the director is an intrinsic property of well ordered quadrupolar layered systems. In agreement with this Poniewierski and Sluckin[14] have shown that electric quadrupoles favour the formation of a smectic C phase, and that more generally this propensity is favoured by the presence of a (224)-term in the spherical harmonic expansion of the intermolecular potential.

The Coventry group have investigated the effect of the addition of a longitudinal and of a transverse electric quadrupole to a single-site Gay-Berne particle. The magnitude and direction of the quadrupole was shown to have a marked effect upon the onset and duration of the smectic phase.[15] In particular, the longitudinal quadrupolar Gay-Berne fluid was found to form a tilted smectic phase.[15] For this tilted phase, the best normal to the smectic plane was determined from the positions of the centres of mass of the quadrupolar Gay-Berne molecules employing a novel optimisation technique. Greater amplitude peaks were found in the longitudinal pair correlation function resolved parallel to the plane normal, rather than parallel to the director of the system, demonstrating a tilt between the layer normal and the director of the system. The tilt was found to be 8 degrees, somewhat less than the observed tilt in many experimental systems. The results demonstrate that the influence of the electric quadrupole upon the formation of tilted smectic phases is determined critically by both its magnitude and direction, thus informing the design of smectic-C mesogens. Recently the Southampton group have also observed a tilted smectic phase for a quadrupolar Gay-Berne fluid.[16]

Liquid Crystal Mixtures

As a result of collaboration between several groups in the consortium, a generalised form for the range parameter governing the interaction between Gay-Berne particles has been developed.[17] This has led to the simulation of bi-disperse liquid crystal mixtures.[18] This work uses a generalized version of the GBMEGA program which is suitable for use with Gay-Berne particles of different shapes. As seen experimentally for a binary mixture of nematogens, the simulated nematic phases are found to be miscible. In a mixture of two Gay-Berne particles with different length/breadth ratio, the nematic order parameter for the longer species is found to be larger than that of the shorter species.

Free Standing Films

The Sheffield-Hallam group have investigated free standing liquid crystalline films of Gay-Berne particles. The Gay-Berne parameterisation used in this work was determined from Gibbs ensemble simulations, which showed, (for a certain parameter set), bulk phase isotropic-vapour and nematic-vapour coexistence. The aim of this work was to investigate the effect of free surfaces upon phase coexistence and look at the progress of ordering across the slab through the isotropic-nematic transition. It is believed that the width of the film will have a marked effect upon the phase behaviour, thus making the simulation of different system sizes of particular interest.

Initial work focussed on a *small system* consisting of 1626 Gay-Berne particles in a 6:1:1 box, with the film occupying approximately half of the box. The system was cooled through the isotropic-nematic transition and the surfaces were seen to induce a preferred nematic direction upon the film, with the particles lining up normal to the interfaces. This can be see from the profile and snapshot in Figure 6. The profile shows the density (ρ), nematic (P_2) and z-direction (Q_{zz}) order parameters, with the high value of Q_{zz} in the film indicating alignment parallel to the z-axis (normal to the interfaces). This is confirmed by the snapshot. At temperatures slightly higher than those shown in Figure 6, orientational order was found only at the centre of the slab, indicating that for this system the nematic phase is seeded away from the liquid-vapour interface. Simulations of a 7168 particle system in a 22:1:1 box are currently under way on the Cray T3E to investigate the influence of system size on these results.

STUDIES OF FLEXIBLE MOLECULES

Liquid Crystal Dimers

The GBMOL[19] program has been written by the group in Durham to facilitate the simulation of flexible molecules composed of combinations of Gay-Berne and Lennard-Jones particles, employing traditional *molecular mechanics* force fields to handle intramolecular interactions. GBMOL was first used to simulate liquid crystal dimer molecules composed of two mesogenic units linked by a flexible chain.[20] Experimentally, these systems have been shown to exhibit large odd-even effects in transition temperatures for an homologous series of mesogens. The model systems show spontaneous growth of a smectic-A phase over a period of 6 ns as the system is cooled below the clearing point[20] (as shown in Figure 7). This is the first atomistic simulation where this has been achieved. At lower temperatures still, the molecules form a smectic-B phase in which the Gay-Berne sites exhibit hexagonal packing within a layer.

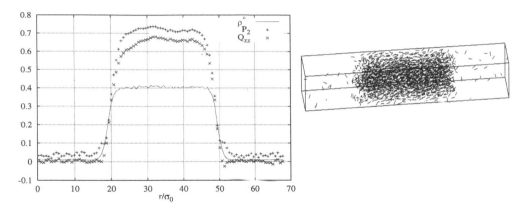

Figure 6. Nematic order parameter (S_2), order parameter along the long axis of the simulation box (Q_{zz}), and density profile (ρ^*) across the simulation box, for a system of 1626 Gay-Berne particles exhibiting nematic-vapour coexistence. Also shown is a snapshot illustrating the distribution of molecules at this state point.

Model mesogens are seen to remain flexible in the smectic-A phase and a large coupling occurs between internal molecular structure and molecular environment. The latter leads to an energy difference between *gauche* and *trans* dihedral angles in the smectic-A phase which is 2 kJ/mol higher for each of the *odd* dihedrals angles in the chain. Individual conformers in which the molecules remain linear in shape are seen to be preferentially chosen over conformers in which the molecules are forced to bend with respect to the director of the surrounding fluid.

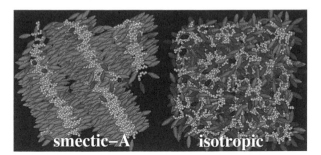

Figure 7. Snapshots showing liquid crystal dimer molecules in a) the nematic phase at 350 K, b) the isotropic liquid phase at 400 K. The Gay-Berne particles are colour coded with the colours red/green/blue indicating mutually orthogonal molecules.

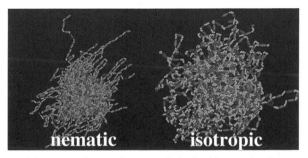

Figure 8. Nematic and isotropic phases of a main chain liquid crystal polymer with 6 methylene spacers. Colour coding as in figure 7.

Main Chain Liquid Crystalline Polymers

The Bristol group, in collaboration with Durham, and the group in St. Petersburg (A. V. Lyulin, I. Neelov) have carried out molecular dynamics simulations of a family of model main-chain liquid-crystalline polymers (LCPs).[21] Such a polymer is formed from Gay-Berne mesogenic units connected to each other through flexible methylene spacers. We have studied the effect of varying the spacer length, using 5, 6, 7 and 8 methylene units, and have examined the region of the isotropic - liquid crystalline transition. Our preliminary results indicate that liquid crystalline ordering may occur spontaneously on lowering the temperature, and that odd-even dependences of thermodynamic properties on spacer length occur, in agreement with existing experiments. Local orientational time correlation functions, and local translational mobility, have been studied both for the mesogenic elements and the bonds in the flexible spacer. The anisotropy of both orientational and translational local dynamical properties have been compared with theoretical predictions and with Brownian dynamics results for a freely-jointed chain in a liquid crystalline orienting field. Figure 8 shows the isotropic and nematic phases for the case of 6 methylene spacers.

ACKNOWLEDGEMENTS

The authors wish to thank the UK EPSRC for the grant of computer time on the Cray T3D and T3E in Edinburgh, that has made this work possible.

REFERENCES

1. J.G. Gay and B.J. Berne, Modification of the overlap potential to mimic a linear site-site potential, *J. Chem. Phys.*, 74:3316 (1981).
2. E. de Miguel, L.F. Rull, M.K. Chalam and K.E. Gubbins, Liquid-crystal phase-diagram of the Gay-Berne fluid, *Molec. Phys.*, 74:405, (1991).
3. M.R. Wilson, M.P. Allen, M.A. Warren, A. Sauron and W. Smith, Replicated data and domain decomposition molecular dynamics techniques for simulation of anisotropic potentials, *J. Comput. Chem.*, 18:478 (1997).
4. GBMEGA: A domain decomposition molecular dynamics program written for the HPCI consortium "Simulation and Statistical Mechanics of Complex Fluids". Authors: M.A. Warren, M.P. Allen, W. Smith, M.R. Wilson, A. Sauron.
5. M.P. Allen D. Frenkel, Calculation of liquid-crystal Frank constants by computer-simulation, *Phys. Rev. A.*, 37:1813 (1988).
6. M.P. Allen and D. Frenkel, Correction, *Phys. Rev. A.*, 42:3641 (1990).
7. M.P. Allen, M.A. Warren, M.R. Wilson, A. Sauron and W. Smith, Molecular-dynamics calculation of elastic-constants in Gay-Berne nematic liquid-crystals, *J. Chem. Phys.*, 105:2850 (1996).
8. M.P. Allen and M.A. Warren, Simulation of structure and dynamics near the isotropic-nematic transition, *Phys. Rev. Lett.*, 78:1291-1294 (1997).
9. M.P. Allen, M.A. Warren and M.R. Wilson, Molecular dynamics simulation of the twist grain boundary phase, *Phys. Rev. E*, (to appear May 1998).
10. M.P. Neal, A.J. Parker, and C.M. Care, A molecular dynamics study of a steric multipole model of liquid crystal molecular geometry, *Molec Phys*, 91:603 (1997).
11. J.W. Goodby, I. Nishiyama, A.J. Slaney, C.J. Booth and K.J. Toyne, Chirality in liquid-crystals – the remarkable phenylpropiolates, *Liq. Cryst.*, 14:37, (1993).
12. R. Berardi, S. Orlandi and C. Zannoni, Antiphase structures in polar smectic liquid-crystals and their molecular-origin, *Chem. Phys. Lett.*, 261:357 (1996).
13. G. Barbero and G. Durand, Order electricity and the nematic, smectic-A, smectic-C phase-transitions, *Molec. Cryst. Liq. Cryst.*, 179:57 (1990).

14. A. Poniewierski and T.J. Sluckin, Nematic, smectic-A, and smectic-C ordering in a system of parallel cylinders with quadrupolar interaction, *Molec. Phys.*, 73:1, 199 (1991).

15. M.P. Neal, A.J. Parker, *Chem. Phys. Letts.*, in preparation.

16. G.R. Luckhurst, private communication.

17. D.J. Cleaver, C.M. Care, M.P. Allen, and M.P. Neal, Extension and generalization of the Gay-Berne potential, *Phys Rev E*, 53:559 (1996).

18. R.A. Bemrose, C.M. Care, D.J. Cleaver and M.P. Neal, A molecular dynamics study of a bi-disperse liquid crystal mixture using a generalized Gay-Berne potential, *Molec Phys*, 90:625 (1997).

19. GBMOL: A replicated data molecular dynamics program to simulate combinations of Gay-Berne and Lennard-Jones sites. Author: M.R. Wilson, University of Durham, (1996).

20. M.R. Wilson, Molecular dynamics simulations of flexible liquid crystal molecules using a Gay-Berne/Lennard-Jones model, *J. Chem. Phys.*, 107:8654 (1997).

21. A.V. Lyulin, I. Neelov, M.P. Allen, M.S. Al-Barwani, M.R. Wilson, and N.K. Allsopp, Molecular dynamics simulation of main chain liquid crystalline polymers, *Macromolecules*, submitted.

202

A FIRST PRINCIPLES STUDY OF SUBSTITUTIONAL GOLD IN GERMANIUM

J. Coomer,[1] A. Resende,[1] P. R. Briddon,[2] S. Öberg,[3] R. Jones,[1]

[1]Department of Physics, Exeter University,
Stocker Rd., Exeter, EX4 4QL
[2]Department of Physics, University of Newcastle upon Tyne,
Newcastle upon Tyne, NE1 7RU
[3]Department of Mathematics, University of Luleå,
Luleå, S-97187, Sweden

ABSTRACT

Spin-polarised local density functional cluster techniques are employed to investigate the behaviour of substitutional gold in germanium. The electronic structure of the defect is then identified using a combination of Slater's transition argument and Janak's theorem. The three acceptor and single donor levels are identified and their energies are compared with experiment.

INTRODUCTION

The Ib metals, Cu, Ag and Au are of interest as they are known to introduce deep levels in the band gap which interfere with the electronic properties of devices. Treatment of such anomalies (e.g. via passivation) requires a detailed knowledge of the properties of the defect. A proper understanding of common impurities in germanium will become increasingly important as SiGe becomes integrated into mainstream semiconductor technology. Gold is especially important technologically as it is used in fast-switching devices to reduce minority carrier lifetimes and is also widely employed as a material for metallic contacts on semiconductor devices.

The levels introduced by impurity gold atoms in the germanium band-gap were first located using Hall effect measurements in the 1950's. They were found to be $E_c - 0.2\ eV$ and $E_v + 0.15\ eV$ [1] for the Au^{--} and Au^- states respectively. A donor level and a third acceptor level were later identified and determined to lie at $E_v + 0.04\ eV$ [?] and $E_c - 0.04\ eV$ [3]. A more recent experiment employing DLTS has corroborated all these values.

The substitutional gold impurity in silicon is believed to have tetragonally distorted C_{2v} symmetry with spin state $S = \frac{1}{2}$. Although this is not directly observed for the case of gold in silicon, possibly due to dynamical Jahn-Teller distortions, E.P.R studies

High Performance Computing
Edited by R. J. Allan *et al.*, Kluwer Academic / Plenum Publishers, New York, 1999

on the other Ib substitutional impurities confirm this configuration. The electronic configuration is then $b_2^{\uparrow\downarrow} b_1^{\uparrow} a_1^0$ (see Fazzio et al., 1985 [4] for a review of Ib metals in silicon). This configuration is expected to be unchanged for the case of substitutional gold in germanium and is readily accounted for by the vacancy model (Watkins, 1983). This model assumes that the substitutional transitional metal ions will behave like closed d-shells within a vacancy. The d-orbital of the substitutional gold atom is split under the T_d symmetry to form a t_2 and an e-level deep in the valence band. The levels that are introduced into the forbidden gap are mostly vacancy-like and are partially filled by the electrons from the gold atom. Under this scheme we understand the gold defect to be isoelectronic with the negative vacancy and this results in the spin $\frac{1}{2}$ state predicted.

THE CODE

AIMPRO (*Ab Initio* modelling program) [6] is a computer code developed over the past 10 years for modelling atomic clusters of over 1000 atoms in size. It has proven portability, being currently utilised on numerous scalar machines of differing architecture as well as having been optimised for differing parallel architectures including T3D, T3E and SP2 supercomputers. The overall parallel scalability of the code is excellent as the majority of the computing operations involve linear matrix operations employing optimised libraries such as ScaLAPACK.

Features of AIMPRO include:

- s,p and d Cartesian Gaussian basis;

- spin polarisation;

- Ceperley-Alderley Exchange correlation;

- Bachelet-Hamann-Schlüter norm-conserving pseudopotentials;

- conjugate gradient geometry optimisation;

- *conventional* but parallel diagonalisation;

- Slater's transition argument combined with Janak's theorem used to determine the many body levels.

METHOD

Tetrahedral H-terminated clusters of 131 and 297 atoms were used to find the structure and calculate the acceptor and donor levels of substitutional gold in germanium. A Cartesian Gaussian basis set was used siting N s, p orbitals on the germanium and hydrogen atoms and N independent d-orbitals on the gold atoms. The charge density was fitted to M Gaussian functions. The values for N and M used in this investigation were : Au(6,12), Si(4,5) and H(2,3). Pseudowavefunctions of Bachelet et al. [7] were expanded in independent combinations of the N Gaussian orbitals for the central shell of 5 atoms in each cluster. For the remaining atoms a fixed linear combination of the orbitals was used. The interatomic forces and self-consistent energies were calculated and the cluster was relaxed iteratively using a conjugate gradient algorithm.

Firstly, a neutral 131 cluster consisting of $Ge_{70}Au_1H_{60}$ was fully relaxed. The t_2 level in the gap contains 2/3 electrons in spin-up and 1/3 electrons in spin down

Table 1. Calculated Acceptor
and Donor levels compared
with experiment

Level	AIMPRO	EXPERIMENT
(-3/-2)	E_c-0.11	E_c-0.04
(-2/-)	E_c-0.26	E_c-0.20
(-/0)	E_c-0.39	E_c-0.65
(0/+)	E_v+0.33	E_c-0.04

thus yielding a spin density of a_1 symmetry. This procedure prevents Jahn-Teller distortion which is important for the elucidation of the acceptor and donor levels . The central shell of 17 atoms of the relaxed cluster was then inserted into a larger cluster, $Ge_{180}Au_1H_{116}$.

According to a previously described method [8] employing Slater's transition argument and Janak's theorem we can find the acceptor-donor levels by partially occupying the Kohn-sham levels. The energy difference between the highest occupied and lowest conduction band state when occupied by an additional half an electron each is equivalent to the acceptor level energy if the relaxation occurring is negligible.

To determine the acceptor energy from the relaxed cluster, half an electron was added to the t_2 level and half to the next unoccupied level taken to represent the conduction band bottom. The energy difference between these two partially occupied levels was then recalculated and scaled by an amount so that the energy gap is given correctly. This accounts for the band gap distortion arising from both D.F.T. and finite cluster size. A similar method was used to determine the other many-body levels.

The donor calculation was also carried out using a larger 501 atom cluster in order to compare with the 297 atom cluster results.

RESULTS

The gold-germanium bond length was found to be 2.88 Å as compared with 2.43 Å for the Ge–Ge bond length established by AIMPRO (c.f. experimental value, 2.44 Å). The band gap was calculated to be 2.85 eV giving a rescaling factor of 0.26 to bring the gap into agreement with the experimentally observed gap of 0.74 eV. The rescaled levels are compared against the most established experimental results in table 1. The calculated triple and double acceptor levels are in good agreement with experimental values. The single acceptor and donor levels, however are found to lie deeper in the gap than their experimental counterparts.

The donor level calculated from the 501 cluster with bond centres was located at E_v+0.33, confirming that the 297 atom cluster is sufficient to model such transition metal impurities.

CONCLUSION

Ab initio calculations have been performed on germanium clusters with a substitutional gold impurity. A combination of Slaters scheme and Janak's theorem was employed to calculate the many-body levels of the defect. The triple acceptor nature of the substitutional gold impurity is confirmed and the double and triple acceptor levels are located successfully by AIMPRO.

REFERENCES

1. W.C. Dunlap Jr., Gold as an acceptor in germanium, *Physical Review* 97:614 (1954)
2. Morton, Hahn and Schultz, *Atlantic City Photoconductivity Conference* John Wiley and Sons, Inc., New York (November 1954)
3. H.H. Woodbury and W.W. Tyler, Triple acceptors in Germanium, *Physical Review* 105:84 (1957)
4. A. Fazzio, M.J. Caldas, A. Zunger, Electronic structure of copper, silver and gold impurities in silicon, *Physical Review B* 32:934 (1985)
5. G.D. Watkins, *Physica* B117-B118:9 (1983)
6. R. Jones, and P. R. Briddon, Identification of defects in semiconductors, in *Semiconductors and Semimetals*, ed. M. Stavola, treatise editors, R.K. Willardson, A.C. Beer, and E.R. Weber, Academic Press (1997) in press.
7. G.B. Bachelet, D.R. Hamann and Shlüter, *Phys. Rev. B* 26:4199 (1982)
8. A. Resende, J. Goss, P.R. Briddon, S. Öberg and R. Jones, Theory of gold-hydrogen defects in silicon *Matter. Sci. Forum* 1 (1997)

APPLICATIONS OF SELF–INTERACTION CORRECTION TO LOCALIZED STATES IN SOLIDS

Z. Szotek,[1] W.M. Temmerman,[1] A. Svane,[2] H. Winter,[3] S.V. Beiden,[4] G.A. Gehring,[4] S.L. Dudarev,[5] and A.P. Sutton[5]

[1]Daresbury Laboratory, Warrington, WA4 4AD, UK
[2]Institute of Physics and Astronomy, University of Aarhus,
 DK-8000 Aarhus C, Denmark
[3]Forschungszentrum Karlsruhe, INFP, Postfach 3640,
 Karlsruhe, Germany
[4]Physics Department, University of Sheffield, Sheffield, UK
[5]Department of Materials, University of Oxford, Parks Road,
 Oxford OX1 3PH, UK

INTRODUCTION

Density Functional Theory provides an exact mapping of a many–body electron problem which occurs in solids onto a one–electron problem[1]. Instead of considering, for N interacting electrons in an external potential $v(\mathbf{r})$, the 3N-dimensional Schrödinger equation for the wavefunction $\Psi(\mathbf{r}_1, \mathbf{r}_2, \mathbf{r}_3, ..., \mathbf{r}_N)$, DFT expresses this many–body problem in terms of the electronic density distribution $n(\mathbf{r})$ and a universal exchange and correlation functional of the density, $E_{xc}[n(\mathbf{r})]$. The task of solving the many–body problem is then reduced to finding sufficiently accurate expressions for $E_{xc}[n(\mathbf{r})]$ and then solving the relevant one–electron Schrödinger equation with an effective potential of which the exchange–correlation potential is a prominent part. Surprisingly, DFT has turned out to be a powerful and extremely successful scheme of calculating electronic properties of solids. This is owing to a simple and practical approximation for $E_{xc}[n(\mathbf{r})]$, the local density approximation, where the exchange and correlation potential, $E_{xc}^{LDA}[n(\mathbf{r})] \equiv \int n(\mathbf{r})\epsilon_{xc}(n(\mathbf{r}))d\mathbf{r}$, is approximated by the exchange–correlation energy per particle, $\epsilon_{xc}(n)$, of a homogeneous electron gas of density n. This simple function of density can be very precisely calculated and allows for an accurate determination of the ground state energies and charge densities of any system[1].

The LDA seems to be a highly accurate approximation when the electron is delocalized and travels 'fast' through the solid. However, when the 'static' electron-electron interactions become so strong that the electron gets localized on an atomic site in the solid, one finds that the LDA, and even its gradient corrections, fail to describe the correct groundstate[2]. This obviously means that the $E_{xc}[n(\mathbf{r})]$ can not anymore be adequately represented by the LDA. For when the electron slows down upon localization it starts responding to a different potential than the effective LDA potential. To take this into account, it seems more appropriate to consider orbital–dependent functionals, $E_{xc}[\{\phi_i\}]$, where $\{\phi_i\}$ is a set of one–electron orbitals with the orbital index i. These functionals are a possible evolution on the LDA and can be considered as a basis of the orbital–dependent DFT[1]. They can be treated as the starting point for new approximations to DFT. The self–interaction–corrected LDA (SIC-LDA) is one such approximation[3]. It is based upon the observation that localized orbitals in the LDA give rise to an error due to a spurious self–interaction contained in the effective one–electron potential. This error increases the more localized the orbital becomes. For a 'fast' electron, however, this self–interaction is negligible or zero. Therefore a functional can be constructed where for the orbitals which contain the localized electrons one explicitly subtracts the self-interaction term from the LDA Coulomb and E_{xc} term, while for the orbitals referring to the delocalized electrons one uses the LDA. This functional forms the basis of the SIC–LDA scheme which has the property that when all the orbitals are delocalized it reduces to the LDA. It also allows for a much improved treatment of the electron–electron correlations, and is much more adequate, than LDA, for systems where both localized and delocalized electrons are present. This feature will become apparent while discussing its accuracy in the examples below.

Solids where both localized and delocalized electrons are present give rise to some of the most fascinating phenomena in solid state physics. They include the heavy fermion compounds and their complicated phase diagrams at low temperatures, the still unexplained phenomenon of the high T_c superconductivity, the observation of collosal–magneto resistance, with its potentially wide ranging technological applications. The theoretical investigations have usually been confined to either studies of generic model Hamiltonians which allow for a very accurate treatment of the electron–electron interactions, or materials specific LDA calculations with a not fully satisfactory treatment of the 'static' electron–electron correlations. Here we demonstrate the power of the SIC-LDA approach by applying it to elemental cerium, cerium pnictides, and NiO, of relevance to the high T_c superconductivity. To describe the correct physics of these systems special care has to be given to the best possible treatment of both the localized and itinerant electrons which participate to the same degree in defining the properties of these systems.

ELEMENTAL CE

Cerium metal has been widely studied over the years, both experimentally and theoretically, mostly due to its famous isostructural $\gamma \rightarrow \alpha$ phase transition, occurring at the pressure of ~ 8 kbar. This transition, that has a phase boundary terminating at a critical point– a feature which makes Ce unique among elemental metals– is accompanied by a volume collapse of 14.8% and quenching of the magnetic moment[4].

The LDA, in its spin-polarized extension (local spin density (LSD) approximation), was not able to describe the γ–phase and to explain this transition, leading only, in agreement with experiments, to the formation of a magnetic moment for the volumes in the neighbourhood of the γ-phase. The SIC-LSD, however, was able to provide a uniform description of this transition with the total quenching of the magnetic moment and volume collapse of 24 % at the transition pressure of -1 kbar, which compares favourably with the value of -7 kbar, extrapolated from the experimental phase diagram to T = 0K, at which the calculations were performed. The dramatic volume change can be seen in Fig. 1, where the calculated total energy of Ce as a

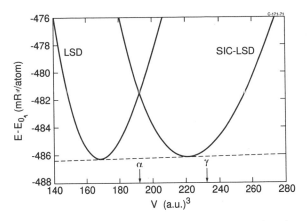

Figure 1. Cohesive energy of Ce (in mRy/atom) as a function of atomic volume (in a_0^3/atom). The curve marked 'SIC-LSD' corresponds to the calculation with one localized f–electron per Ce atom, while the curve marked 'LSD' corresponds to itinerant f–electrons. The common tangent marks the phase transition.

function of volume is shown[3]. The two different curves correspond, respectively, to the α–phase (curve marked LSD; SIC–LSD reduces to the LSD for delocalized electrons) and γ–phase (curve marked SIC–LSD). The LSD minimum of the total energy, corresponding to the α–phase, is located in the non–magnetic region at V = 168 (a.u.)^3, while the spin magnetic moment is m = $1.32\mu_B$ at the SIC–LSD total energy minimum, corresponding to the γ–phase. Within these *ab initio* calculations, the $\gamma \rightarrow \alpha$ transition can be viewed as the transition between the phase with fully localized f electrons and the phase, where the f electrons are fully delocalized.

In a relativistic generalization of the SIC–LSD, the orbital moment of the γ–phase of Ce can also be studied (Fig. 2)[5]. In Fig. 2 the orbital moment can be seen to extrapolate to the atomic value of 3 at large volumes. The relativistic and spin–polarized SIC scheme allows for each of the 14 possible f–states in Ce to become localized. One finds that these solutions have quite different orbital moments and correspond to different total energies. The solution with the lowest total energy is consistent with all three Hund's Rules, and the spin and orbital moments are anti–parallel aligned (see also Fig. 2). Therefore, one can say that the relativistic SIC–LSD scheme forms a bridge between the atomic and band pictures, and enables one to obtain a good description of both the localized and itinerant properties of a rare earth metal, within the relativistic spin-polarized band theory. The $\gamma \rightarrow \alpha$ transition in the framework of the relativistic band theory can be explained as a transition from a localized state with maximum spin and orbital moments to a delocalized state without spin and orbital moments.

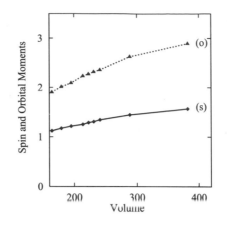

Figure 2. Spin (s) (in μ_B) and orbital (o) moments as a function of volume (in (a.u.)^3) within the relativistic SIC-LSD approximation.

CE MONOPNICTIDES

Similarly to pure cerium, cerium pnictides also undergo a variety of magnetic and structural phase transitions under pressure. These transitions can be well reproduced within the SIC-LSD approach which may be used to describe the localized as well as delocalized electron states. We have performed total energy calculations as a function of volume for CeN, CeP, CeAs, CeSb and CeBi for B1 and B2 phases, in the ferromagnetic (F) arrangement of Ce moments, and with the $f-$electron treated as either delocalized (LSD) or localized (SIC-LSD)[6, 3]. Our main results for the relevant transition pressures and volumes, together with the experimental data, are summarized in Table 1.

It can be seen that the calculations faithfully reproduce the pressure behaviour of the Ce pnictides. In particular, we find that only for CeN the f electron is delocalized and a transition takes place from a cubic B1 structure to a cubic B2 structure. For CeP two phase transitions are observed, first the delocalization transition and subsequently the structural transition from B1 to B2 structure. The two transitions merge into a single one for CeAs, namely a transition from a B1 structure with the f electron localized to a B2 structure with the f electron delocalized. For the two remaining systems, CeSb and CeBi, the structural transition occurs first and only at higher pressures does the f electron become delocalized.

NIO

In NiO, the SIC–LSD allows for a better description of the strong Coulomb correlations between electrons of the $3d$ shell of nickel ions[3, 7, 8]. In this case we compare our results to a recently proposed method combining the LSD approximation with a term describing the on–site Coulomb repulsion, U, between localized electrons, the so–called

Table 1. Calculated and experimental transition pressures for the electronic and structural phase transitions in the cerium pnictides.[a]

compound	transition	P_t (kbar)		V_h ($a_0{}^3$)		V_l ($a_0{}^3$)	
		theo.	expt.	theo.	expt.	theo	expt.
CeN	B1(d) → B2(d)	620	-	148	-	141	-
CeP	B1(l) → B1(d)	71	90,55	325	308	297	298
CeP	B1(d) → B2(d)	113	150(40)	288	285	246	247
CeAs	B1(l) → B2(d)	114	140(20)	332	315	265	274
CeSb	B1(l) → B2*(l)	70	85(25)	400	398	353	354
CeSb	B2*(l) → B2*(d)	252	-	311	-	295	-
CeBi	B1(l) → B2*(l)	88	90(40)	427	399	376	360
CeBi	B2*(l) → B2*(d)	370	-	317	-	304	-

[a]The errorbars in the quoted experimental transition pressures are estimates based on the hysteresis loop observed in the experimental PV curves. Also quoted are the specific volumes on the two sides of the transition. The notation (d) and (l) refers to calculations with delocalized or localized Ce f–electrons, i.e. tetravalent or trivalent Ce atoms. B2* denotes the distorted B2 structure.

LSDA+U scheme[9, 10]. In Fig. 3 we show the results of both schemes for the band gap, spin magnetic moment, and total energies, as functions of the lattice constant. Similarly to the LSDA+U, the application of SIC–LSDA to NiO leads to a substantial band gap of 3.15 eV, at the experimental volume, which compares favourably both with the LSDA+U and the observed gaps. The LSD, on the other hand, gives negligible band gap due to an inadequate treatment of the strong Coulomb correlations among 3d electrons. Also, the equilibrium lattice constant of 4.18 Å, calculated within the SIC–LSDA, is a substantial improvement on the LSD value. One can see in Fig. 3, that both the SIC–LSDA and the LSDA+U predict the energy minima at the same lattice constant. Concerning the magnetic moment, both schemes differ slightly in the magnitude, but give similar behaviour as a function of the lattice constant.

PARALLEL IMPLEMENTATION

Finally, we would like to remark on the implementation of this orbital-dependent SIC–LSD formalism. It is characterized by quantities which either need a real space representation or are most efficiently evaluated in k-space. The resulting computer codes involve successive transformations between real- and k–space. They are about ten times more CPU demanding than the usual DFT electronic structure codes due to the two step self–consistency involved. They demand energy functional minimization with respect to the orbitals, and are dealing with orbital dependent potentials. As a result for every band and k–point a different Hamiltonian matrix is diagonalized as opposed to the conventional band structure calculations where, for a particular k–point, all bands and corresponding eigenvectors are obtained from one matrix diagonalization. Additionally, the relativistic generalization leads to a doubling of the matrix sizes. Concerning the parallelization paradigm implemented here, it is two–fold, namely, with respect to bands in the orbital localization part, and with respect to k-points in the charge self–consistency part.

CONCLUSIONS

The local density approximation (LDA) to density functional theory (DFT) provides a simple and rather successful scheme for describing many interacting electrons in

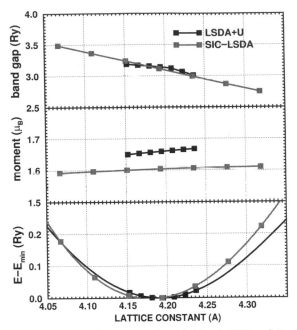

Figure 3. The bandgap, the magnetic moment of a nickel ion in NiO and the total energy of the antiferromagnetic unit cell calculated using LSDA+U and SIC–LSDA and plotted as a function of the lattice constant. The observed size of the bandgap equals 4.2 eV[11], the observed values of the magnetic moment of a nickel ion in NiO lie in the interval between 1.66 μ_B and 1.90 μ_B[12], and the equilibrium lattice constant equals 4.17 Å.

a solid. There exist, however, important cases where electron correlations are too strong to be properly treated within the LDA. Prominent examples are the high–temperature superconductors, the 3d transition metal oxides, and the rare earths such as cerium and its compounds. The self–interaction corrected local density approximation (SIC–LDA) provides a physically transparent and successful approach to improve the LDA description of these strongly correlated electron systems.

REFERENCES

1.   *Density Functional Theory*, E.K.U. Gross and R.M. Dreizler eds., NATO ASI series, Series B: Physics Vol. 337 Plenum Press, New York and London (1995).
2.   J. Yamashita and S. Asano, *J. Phys. Soc. Japan* 52:3514 (1983);
     K. Terakura, A.R. Williams, T. Oguchi, and J. Kübler, *Phys. Rev. B*, 30:4734 (1984).
3.   W.M. Temmerman, A. Svane, Z. Szotek, and H. Winter, *Electronic Density Functional Theory: Recent Progress and New Directions*, J.F. Dobson, G. Vignale and M.P. Das eds., Plenum Press (1997).
4.   A. Svane, *Phys. Rev. B*, 53:4275 (1996).
5.   S.V. Beiden, W.M. Temmerman, Z. Szotek and G.A. Gehring, *Phys. Rev. Lett.* 79:3970 (1997).
6.   A. Svane, Z. Szotek, W.M. Temmerman, and H. Winter, *Solid State Commun.*, 102:473 (1997).
7.   A. Svane and O. Gunnarsson, *Phys. Rev. Lett.*, 65:1148 (1990).
8.   Z. Szotek, W.M. Temmerman and H. Winter, *Phys. Rev. B*, 47:4029 (1993).
9.   A.I. Liechtenstein, V.I. Anisimov and J. Zaanen, *Phys. Rev. B*, 52:R5467 (1995).
10. S.L. Dudarev, G.A. Botton, S.Y. Savrasov, Z. Szotek, W.M. Temmerman, and A.P. Sutton, *Physica Status Solidi*, to be published.
11. G.A. Sawatzky and J.W. Allen, *Phys. Rev. Lett.*, 53:2339 (1984).
12. V.I. Anisimov, J. Zaanen and O.K. Andersen, *Phys. Rev. B*, 44:943 (1991).

COMPUTATIONAL CHEMISTRY

COMPUTATIONAL CHEMISTRY IN THE ENVIRONMENTAL MOLECULAR SCIENCES LABORATORY

David A. Dixon, Thom H. Dunning, Jr., Michel Dupuis, David Feller, Deborah Gracio, Robert J. Harrison, Donald R. Jones, Ricky A. Kendall, Jefferey A. Nichols, Karen Schuchardt and Tjerek Straatsma

William R. Wiley Environmental Molecular Sciences Laboratory, Pacific Northwest National Laboratory, P.O. Box 999, Richland WA, 99352, USA

INTRODUCTION

There are numerous serious environmental issues facing the world. Many of these have anthropogenic sources and are due to the production of materials for the consumer or for national defense and to energy production and consumption, e.g. global warming. For example, four decades of nuclear weapons production at Department of Energy facilities across the United States has resulted in the interim storage of millions of gallons of highly radioactive mixed wastes in hundreds of underground tanks, extensive contamination of the soil and groundwater at thousands of sites and hundreds of buildings that must be decontaminated and decommissioned.[1] The single most challenging environmental issue confronting the DOE and perhaps the United States, is the safe and cost-effective management of these wastes. Questions that must be addressed include "What is the physical and chemical form of the wastes in the tanks and in the ground?" ; "How can the radioactive wastes be safely processed?"; and "How can the processed waste be safely stored?"[2-4] The answers to these questions are, in general, unknown and the scientific basis required to meet these complex technological issues is not available today.[5] In order to address these issues, major new scientific advances are required. An important technique that can be used to solve such complex problems is computational science which often enables the replacement or curtailment of expensive experiments, especially experiments involving radioactive species. Computational science can be used to provide fundamental answers to the questions enumerated above, to provide the conceptual and numerical bridge for the extrapolation of experimental data available for the lanthanides to the actinides and to allow us to reliably extend the experimental data into other regions of parameter space.

There are a number of examples of how computational science can be used to address complex scientific and technical issues. Much of the tank waste is composed of radionuclides with half-lives on the order of 30 years; these include ^{90}Sr and ^{137}Cs. However, there is a significant amount of material which contains atomic species with very long half-lives, some of which may be on the order of 10^6 years; these include heavy elements such as the actinides and lighter transition metal atoms such as ^{99}Tc. Furthermore these metals can and do form highly toxic compounds. If metallic waste is released into the environment, it can be transported away from the release site by water if it is in a soluble chemical form. A key ingredient in the success as well as the cost-effectiveness of remediation efforts is an understanding of the chemical properties of the wastes, including thermodynamics and reactivity (kinetics). In order to effectively process tank waste, to design *in situ* remediation technologies and to understand fate and transport in the environment, the wastes need to be characterized and their chemical form (speciation) determined. For radioactive wastes, this is an especially costly process. One of the important interactions of metals (and their ions) in natural aqueous solutions involves dissolved carbonates that come from the dissolution of minerals such as calcite, $CaCO_3$, or the dissolution of CO_2 in H_2O. How do these metals interact with carbonates and how does this affect their speciation?[6] Another example involves the proposed use of in-situ phosphate barriers for the immobilization of actinides,[7] which requires basic chemical information related to actinide phosphates: their structures, thermochemical stability and solubility characteristics. For tank waste processing, the *de novo* design of highly-selective, efficient ligands for separation systems requires information about molecular structures, force fields and complexing energies for various metal ions and various ligands.[8-14] What are the best types of mono- and bi-dentate ligand sites that can selectively bind actinide ions in different oxidation states? How can the binding sites be optimally connected in order to control the steric interactions and use them to the best advantage?

A specific example of tank waste processing is associated with the Savannah River Site. The production of fissionable materials for the nuclear weapons program generated significant amounts of highly radioactive waste stored in underground tanks up to 1 million gallons in volume. Most of this alkaline waste exists in carbon-steel tanks at the Hanford and Savannah River Sites (SRS), with a combined volume exceeding 100 million gallons. These wastes include a sludge phase containing the actinides. Final disposal of this sludge involves vitrification and storage in a deep geologic repository such as Yucca Mountain.[15] Radioactivity in the remaining liquid waste, or supernatant, comes primarily from the soluble, short-lived fission products, Cs and Sr. Removing the Cs and Sr leaves a low-level waste form, allowing more economic disposal. At SRS, the "In-Tank Precipitation" (ITP) process removes soluble Cs and K from alkaline solutions by using an organic reagent, tetraphenylborate (BPh$_4^-$ where Ph = phenyl = C_6H_5, also abbreviated as TPB). A full-scale demonstration of the feasibility of the process began in 1983. Radioactive commissioning of the permanent facility was initiated by addition of 37,300 gal of NaTPB solution in September 1995. By December, 1995, personnel noted a high rate of TPB decomposition leading to the formation of significant quantities of benzene.[16] The formation of benzene poses a serious safety issue because of potential flammability in the vapor space. Similarly, radiolysis of the waste forms hydrogen. The transport of these flammable species within the operating vessels and piping defines the safety margin for operations. To better understand the chemistry, SRS proposed two Process Verification Tests (PVT). The Defense Nuclear Facilities Safety Board (DNFSB), which monitors safety of such operations, expressed reservations.[17,18] Although the DNFSB allowed the first test to proceed, the Board recommended no future tank additions until completion of fundamental chemistry studies.

"(1) Conduct of the planned test PVT-2 should not proceed without improved understanding of the benzene that it will generate and the amount and rate of release that may be encountered for that benzene."

The Board further recommended:

"(2) The additional investigative effort should include further work to (a) uncover the reason for the apparent decomposition of precipitated TPB in the anomalous experiment, (b) identify the important catalysts that will be encountered in the course of ITP and develop quantitative understanding of the action of these catalysts, (c) establish, convincingly the chemical and physical mechanisms that determine how and to what extent benzene is retained in the waste slurry, why it is released during mixing pump operation and any additional mechanisms that might lead to rapid release of benzene and (d) affirm the adequacy of existing safety measures or devise such additions as may be needed."

Thermochemical data is required to better understand the decomposition process and to define the safe operating regime for the In-Tank Precipitation process at the Savannah River Site. Such an effort in computational chemistry aimed at minimizing experimental work on radioactive elements is in parallel with the "Stockpile Stewardship" goals of DOE's Defense Production Program which seeks to develop, among other technologies, the computational tools needed for maintaining the nuclear weapons stockpile without weapons testing.

SOFTWARE DEVELOPMENT

In order to help address some of these extremely complex issues, one part of the work at the Environmental Molecular Sciences Laboratory (EMSL) has focused on developing software and new theoretical methods and applying these new methods and tools to solve complex environmental problems. An important area that we have focused on is that of computational chemistry.[19] Computational chemistry, as we practice it, is to find out information about the electronic structure of molecules by solving the Schrodinger equation (Eq. (1)) for the electronic motion in a molecule given only the Born-Oppenheimer (fixed nuclei) approximation. In Equation (1), H is the Hamiltonian and Ψ is the

$$H\psi = E\psi \tag{1}$$

wavefunction. The Hamiltonian for the electronic Schrodinger equation is actually quite simple as shown in Eq. (2). The first term is the kinetic energy of the electrons, the second

$$H = \sum_{\mu} \frac{-\nabla_{\mu}^2}{2} - \sum_{A,\mu} \frac{z_A}{r_{\mu} - r_A} + \sum_{\mu < v} \frac{1}{r_{\mu v}} \tag{2}$$

term is the attraction between the electron and the nuclei and the third term is the repulsion between the electrons. This equation which is simple to write down is not simple to solve. In fact, closed-form solutions do not exist for essentially any atomic or molecular systems other than the most simple such as the hydrogen atom or the H_2^+ molecule. There are two approaches to solving this equation from first principles (ab initio) in general use at this time, one based on molecular orbital (MO) theory and one based on density functional theory (DFT).[20] We use both approaches to address a variety of problems.

In order to solve the Schrodinger equation in the molecular orbital approach, one must have some representation of the wave function. The wave function is usually expanded in a set of functions (orbitals) centered on the atoms. This set of functions is known as the atomic orbitals or the basis set and needs to be as complete as possible. The generation and selection of the appropriate set of atomic orbitals is known as the "1-particle problem." The usual solution to Eq. (1) in the molecular orbital approach is to include a single Slater determinant in the equation. For a closed shell molecule this leads to the Hartree-Fock (HF) solution. The HF method scales formally as N^4 where N is the number of basis functions although in practice, with clever use of advanced numerical techniques, the scaling can be made to be in the range of $N^{2.5}$. The HF solution provides a description of the energy of an electron in the average field of the other (n-1) electrons, but electrons instanteously adjust to the positions of all of the other electrons; the resulting error is referred to as "the correlation energy problem." The solution of the correlation energy problem is known as solving the "n-particle problem" because it corrects for the interaction of the single electron with the remaining n electrons and is required to obtain quantitative accuracy.

Many methods have been developed to treat the n-particle problem, but there is a growing consensus that for systems based predominantly on a single configuration, (for example, those that do not contain transition metal atoms), one can treat the correlation problem adequately by using coupled cluster based methods. Until the last 10 years, it was believed that most of the problems in the accurate treatment of molecular systems were due to the difficulties in treating the n-particle problem. In fact, based on full configuration interaction benchmark studies, it has been found that many of the problems previously attributed to the solution of the n-particle problem were actually due to deficiencies in the solution of the 1-particle problem, the choice of the atomic orbital basis sets. One of the focus areas in the EMSL is on the development of highly accurate basis sets, the correlation-consistent basis sets.[21] Although, there is a good understanding today of the means to solve many problems to high accuracy, the computational cost can be very high. One must use very large extended basis sets and the higher level correlation treatments such as coupled cluster with all single and double excitations plus a correction for the connected triples (CCSD(T))[22] are computationally expensive and have approximately an N^7 scaling where N is the number of basis set functions.

The second way to solve the Schrodinger equation has its origin in the days when quantum mechanics was founded and is called density functional theory. In this case, rather than solving Equation (1) explicitly, one can use a theorem from Hohenberg and Kohn[23] which states that the exact energy is a functional of the density as shown in Eq. (3).

$$E[\rho] = T[\rho] + U[\rho] + E_{xc}[\rho] \qquad (3)$$

If one knows the exact form of each term in Eq. (3), then one could compute the energy exactly given an electron density. The first and second terms correspond to the kinetic energy of the electrons and to the classical Coulomb electrostatic energy, respectively. These two terms can be solved for in a straight-forward manner. However, it is the last term, E_{xc}, the exchange-correlation energy, that contains all of the many-body contributions to the energy. If the form of this term were known exactly, then one could solve for the total energy using Eq. (3). The exact form of this term is not known, however and it is necessary to use some approximation for it. The first level of approximation usually used is the local density approximation and this is usually taken as Slater exchange[24] together with the correlation energy of the uniform electron gas.[25] This term can be improved by incorporating gradient (nonlocal) corrections to the exchange and correlation energy. Common corrections are those provided by Perdew,[26] and Lee, Yang and Parr[27] for the correlation energy and by Becke[28] for the exchange energy giving the BP (Becke-Perdew) and BLYP (Becke and Lee, Yang, Parr) methods.

Use of approximate exchange-correlation functionals allows one to solve the density functional equation exactly for an approximate energy expression whereas molecular orbital methods solve an exact expression for the energy approximately. An advantage of DFT methods as described so far is that they formally scale as N^3 as compared to the N^4 scaling of Hartree-Fock theory and the N^5 - N^7 scaling of common correlation methods. Thus DFT methods, if derived without the inclusion of Hartree-Fock exchange, are computationally much cheaper than Hartree-Fock based methods and exhibit a much better scaling with the size of the molecule. In fact, one can show that the Hartree-Fock method is a special case of DFT with the exchange term computed exactly and the correlation term equal to zero. Such a relationship has been used by Becke to develop adiabatic correction methods which include some component of Hartree-Fock exchange. In the chemical community, DFT has not been practiced as long as HF based methods have been, so there is less experience with this method and its areas of applicability are not as well understood. We have implemented DFT by expanding the density in atomic orbital basis sets. Following MO theory, these basis sets are usually taken as Gaussians or sums of Gaussians.

As one studies new environmental systems by using computational techniques, one wants to include more of the system in the actual model. This leads to scaling problems that must be carefully considered. One has to consider three scales as shown in Figure 1. These correspond to the normal ranges of length and time for chemical systems and processes, which scale as nlogn to n^2 for the former and as n for the latter, where n is the number of particles. However, as noted above, we must also consider the accuracy of the simulation. This can vary from order n to n!, depending on the level of accuracy required and the level of theory, where n can be the number of particles or the number of basis functions in the calculation. Let's consider some of the reasons for striving for accurate predictions. For example, a factor of 2 - 4 in catalyst efficiency may determine whether a process is economically feasible or not. A factor of 10 in a rate constant at room temperature (25° C) corresponds to a change in the activation energy of 1.4 kcal/mol. As a second example, if one has a 50:50 starting mixture of two components, a change in ΔG of 2.8 kcal/mol leads to a change in the equilibrium constant of 100, leading to a 99:1 mixture at 25° C. Such accuracies mean that we would like to predict thermodynamic information to within 0.1 to 0.2 kcal/mol and rate constants to within ~15%, certainly very difficult by today's standards,

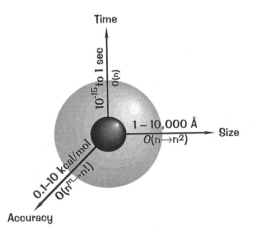

Figure 1. Scaling of computational complexity for molecular simulations

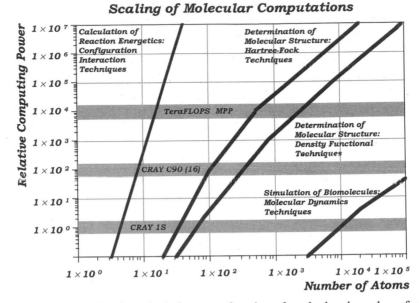

Scaling of Molecular Computations

Figure 2. Scaling of molecular calculations as a function of method and number of atoms in the simulation.

even for gas phase processes. Examples of the scaling of the various computational chemistry methods is shown in Figure 2.

In order to meet the computational challenges raised by the need for accuracy, the size and complexity of the problems and the required timeliness of the solution, one has to look beyond single processor computer architectures to massively parallel processing (MPP) computer architectures. In today's market, some flavor of MPP architectures provides the most performance for the dollar investment. For theoretical chemistry to have maximal impact on solving environmental problems, the computational power of MPP computers must be harnessed. In order to meet the computational challenge, the EMSL in its Molecular Sciences Computing Facility (MSCF) has invested in a new state-of-the-art massively parallel IBM SP computer system and in a Database/Archive system. The IBM SP system has 512 nodes, 67 Gbytes of memory and almost 3 terabytes of disk. The peak processing speed is almost 250 Gflop/s. However, in order to make optimal use of such a parallel computer, a new generation of software also must be created. The software must be scaleable in terms of the number of nodes and in terms of the problem size and portable to different computer architectures so that it can have broad applicability with minimal cost for porting. Finally such software must be high performance and take full advantage of modern theoretical and algorithmic advances. Such software, together with the tools developed to operate and utilize massively parallel computer systems as well as the software needed to handle and interpret the large amounts of data that will be generated, represent just as important a part of the computing revolution as do the high performance computers themselves.

A key component of any successful modern software development program for scientific applications is the use of teams of computer scientists, applied mathematicians, application developers and users to design and implement the software. The synergy of such efforts allows for the development of the highest performing software with the best algorithms and the longest in-use lifetime. Such teams help to minimize long-term development costs by developing software which is, to the maximum extent possible, portable and readily maintained and updated. This is especially true when tackling "Grand Challenge" computational problems where changes in computer architecture occur on a regular basis and new algorithms are constantly being developed.

The computational chemistry software being developed in the EMSL is contained in the program system NWChem,[30,31] a suite of programs running on advanced parallel processing computer systems. Version 1.0 of NWChem provides the following electronic structure capabilities:

- Hartree-Fock energies, gradients and second derivatives.
- Multiconfiguration self consistent field (MCSCF) energies and gradients.
- Density functional theory with a broad range of local and gradient-corrected functionals (with N^3 and N^4 formal scaling) energies and gradients.
- Many-body perturbation theory (MP2-MP4) energies plus MP2 gradients.
- Coupled cluster [CCSD and CCSD(T)] energies.
- Single and multi-reference configuration interaction energies.
- Segmented and generally contracted basis sets including the correlation-consistent basis sets under development at EMSL.
- Effective core potential energies, gradients and second derivatives.

NWChem 1.0 also supports molecular dynamics calculations with a variety of empirical (classical mechanical) force fields for the simulation of macromolecular and solution systems. Functionality includes:

- Energy minimization.
- Molecular dynamics simulation.
- Free energy calculation.

The above calculations can be performed by using multiconfiguration thermodynamic integration and multiple step thermodynamic perturbation theory. Interaction potentials may include first order and self consistent electronic polarization. Long range interaction can be approximated using a simple reaction field technique or particle mesh Ewald summation. The MD driver may also be used to perform quantum dynamics, in which all forces and energies are obtained from quantum mechanical gradients.

Much of the software development has been focused on providing the required capabilities and the appropriate scaling and portability. We provide three examples of the scaling of the software in Figures 3-5 for correlated calculations at the MP2 level, DFT and molecular dynamics. Excellent scaling over a large number of processors is found in all three examples.

NWChem can be accessed as a stand-alone module or via the Extensible Computational Chemistry Environment (ECCÉ) that is also being developed in the EMSL.[31] ECCÉ is a high level computational environment based on an object-oriented data model developed at the EMSL. ECCÉ combines automated metadata and database management, modern "intelligent" graphical user interfaces, automated calculation initiation and monitoring, scientific visualization, analysis tools and access to a hierarchical mass storage system. This interactive environment allows the user to have ready access to computational resources, both hardware and software, on parallel computing systems that are highly sophisticated. In order to optimize the use of the computational resources as well as minimize the user's investment of time, an advisory capability based on prior results and experimental validation is needed. Such a Computational Chemistry Advisor (CCA) can provide information such as what accuracy (e.g. energies or geometries) can be expected from a given basis set and particular treatment of the correlation energy and the computational cost associated with such a calculation. The CCA is currently under development and will be based on the EMSL Computational Results Database.[32] An example of such an advisor is the Basis Set Advisor which has been developed in the EMSL. As noted above, there are many different forms for the basis sets for treating the 1-particle problem. In order to make a wide range of basis sets, including the correlation-consistent

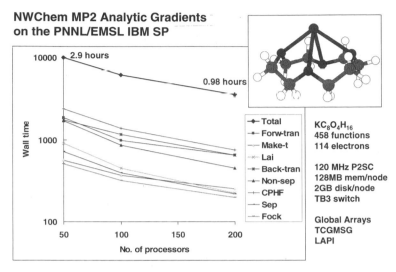

Figure 3. Scaling of MP2 analytical gradients with the NWChem program on the MSCF IPM SP. Wall-clock time as a function of the number of processors for various steps in the calculation. The molecule is K+ bonded to 12-crown-4 ether.

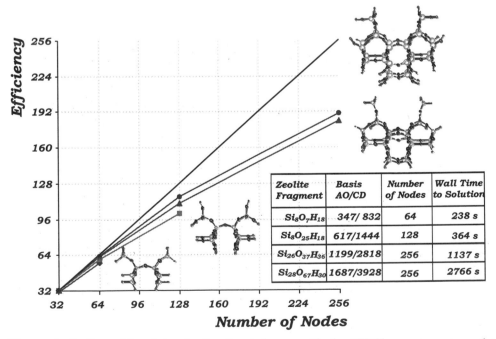

Zeolite Fragment	Basis AO/CD	Number of Nodes	Wall Time to Solution
$Si_8O_7H_{18}$	347/ 832	64	238 s
$Si_8O_{25}H_{18}$	617/1444	128	364 s
$Si_{26}O_{37}H_{36}$	1199/2818	256	1137 s
$Si_{28}O_{67}H_{30}$	1687/3928	256	2766 s

Figure 4. Scaling of local density functional theory with the NWChem program on the MSCF IPM SP. Efficiency as a function of the number of processors for different fragment models of a zeolite cluster. The solid line denotes linear speedup

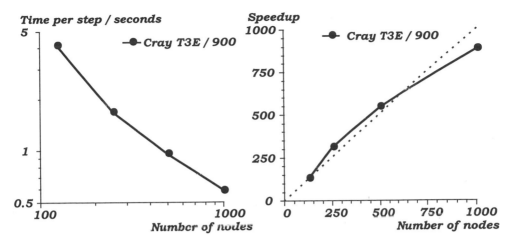

Figure 5. Scaling of molecular dynamics calculations with the NWChem program system on a Cray T3E/900. Time per step as function of the number of nodes and speedup as a function of the number of nodes. The dashed line denotes linear speedup. The calculations are of a 216,000 atom octanol simulation with the AMBER force field.

basis sets, generally available, a Basis Set Browser was developed.[33] The Basis Set Browser has made it much easier for a wide range of complex basis sets to be used in many different programs without transcription errors.

ADDRESSING AN ENVIRONMENTAL PROBLEM WITH COMPUTATIONAL CHEMISTRY

One of the examples given above is the formation of benzene from the degradation of tetraphenylborate in the In-tank Precipitation Process at the Savannah River Site. Extensive experimental work to measure various reaction rates and to identify possible catalysts is ongoing at the Savannah River Technical Center. An important set of data that was missing was thermodynamic data on the types of species that could beformed in the degradation process. We used computational chemistry methods to estimate the thermodynamics of a range of possible reactions. The geometries were optimized at the local density functional theory (LDFT) level with a polarized double zeta basis set (DZVP2).[34] Second derivative calculations were also done at this level to show that a minimum energy structure had been obtained and to calculate the vibrational frequencies for use in the thermodynamic analysis as well as for spectroscopic analysis. The final energies on which the heats of formation are based were calculated at the gradient-corrected DFT level (DFT/BP) with the DZVP2 basis set.

Bond exchange reactions were used to obtain the unknown heats of formation of the neutrals based on the heat of formation of $B(OH)_3$ reported in the JANAF Tables.[35] We first checked $\Delta H_f(B(OH)_3)$ by using the CBS-4 method and verified that the calculated value was within 2 kcal/mol of the experimental value. A difficulty due to experimental uncertainty is that there are different values for ΔH_f (B_g): the older JANAF value of 133.8 $\pm$ 2.9 kcal/mol and the newer value of 136.2 $\pm$ 0.2 kcal/mol (0 K), 137.7 (298 K).[36] Our calculated values for $DH_f(B(OH)_3)$ based on these two values are shown below where they are compared to experiment.

CBS-4 (JANAF) $\Delta H_f(B(OH)_3) =$ -239.1 ± 2.9 kcal/mol
CBS-4 (New) $\Delta H_f(B(OH)_3) =$ -235.3 ± 0.2 kcal/mol
Expt (JANAF) $\Delta H_f(B(OH)_3) =$ -237.1 ± 0.6 kcal/mol

An example of how the calculations were done for the heats of formation of the neutral phenylated compounds is shown below. The energetics (ΔE) of the bond exchange reaction (4) were calculated at the LDFT and DFT/BP levels. The first term in each of the energy expressions given below is the electronic energy difference for reaction (4) and the second term is the difference in zero point energies.

$$B(OH)_3 + Ph\text{-}H \rightarrow B(OH)_2Ph + H_2O \tag{4}$$

$\Delta E(LDFT//DZVP2) = 24.43 - 1.75 - 22.7$ kcal/mol
$\Delta E(DFT/BP//DZVP2) = 25.45 - 1.75 = 23.7$ kcal/mol

The LDFT and DFT/BP energy differences are essentially the same showing the consistency of our approach. The reaction energy $\Delta E(K)$ is to a good approximation $\Delta H(298K)$. Because the heats of formation of $B(OH)_3$, Ph-H (benzene) and H_2O are known, we can use Hess's Law to calculate $\Delta H_f(B(OH)_2Ph))$ and this value is given below for the two different levels of calculation.

$\Delta H_f(B(OH)_2Ph) = -135.9 + 0.8 = -135.1$ kcal/mol (LDFT//DZVP2)
$\Delta H_f(B(OH)_2Ph) = -136.9 + 0.8 = -136.1$ kcal/mol (DFT/BP//DZVP2)

We used another approach to obtain the heats of formation of the neutrals. We used the absolute fluoride affinity (FA) scale to calculate the heats of formation of the anions and to predict the fluoride affinity of $B(OH)_3$. In this way, we avoid direct calculations on small atomic anions. The absolute fluoride affinity scale has recently been recalibrated for FA(HF) = 45.8 ± 1.6 kcal/mol giving $FA(CF_2O)$ = 49.9 kcal/mol.[37] The fluoride affinity (FA) is defined as $-\Delta H$(rxn 5)

$$A + F^- \rightarrow AF^- \tag{5}$$

The value of $FA(B(OH)_3)$ was calculated at the DFT/BP//DZVP2 level with LDFT/DZVP2 geometries and frequencies relative to the well-established value for $FA(CF_2O)$ as shown in reaction (6).

$$CF_3O^- + B(OH)_3 \rightarrow CF_2O + B(OH)_3F^- \tag{6}$$
$$\Delta H(\text{rxn}) = 13.8 \text{ kcal/mol}$$

This gave $FA(BOH)_3$ = 36.1 ± 1.6 kcal/mol and $\Delta H_f(B(OH)_3F^-)$ = -332.8 kcal/mol. Bond exchange reactions as shown by reaction (7) were then used to calculate the heat of formation of the appropriate anion. This yields $\Delta H_f (B(OH)_3Ph^-)$ = -221.5 ± 2.9 kcal/mol.

$$B(OH)_3F^- + Ph\text{-}Ph \rightarrow B(OH)_3Ph^- + Ph\text{-}F \tag{7}$$
$$\Delta H(\text{rxn}) = 40.2 \text{ kcal/mol}$$

The calculated thermodynamic values can be used to calculate a variety of reaction energies. The global reactions responsible for benzene production are given in Table 1. They show that benzene will be produced exothermically by hydrolysis of the neutrals or the anions. These results suggest that understanding the kinetics is critical to controlling the benzene production in the ITP process. The global reactions responsible for phenol and biphenyl production are given in Table 2. They show that the oxidation/hydrolysis of the anions lead to the very exothermic formation of biphenyl and phenol. Both the phenol and the biphenyl are produced in a more exothermic reaction than production of benzene.

Table 1. Global Reactions for Benzene Production

Ion Hydrolysis	ΔH(kcal/mol)
$[BPh_4]^- + H_2O \rightarrow [BPh_3OH]^- + Ph\text{-}H$	-8.5
$[BPh_3OH]^- + H_2O \rightarrow [BPh_2(OH)_2]^- + Ph\text{-}H$	-13.9
$[BPh_2(OH)_2]- + H_2O \rightarrow [BPh(OH)_3]^- + Ph\text{-}H$	-14.1
$[BPh(OH)_3]^- + H_2O \rightarrow [B(OH)_4]^- + Ph\text{-}H$	-14.9
Neutral Hydrolysis	
$BPh_3 + H_2O \rightarrow BPh_2OH + Ph\text{-}H$	-24.9
$BPh_2OH + H_2O \rightarrow BPh(OH)_2 + Ph\text{-}H$	-24.3
$BPh(OH)_2 + H_2O \rightarrow B(OH)_3 + Ph\text{-}H$	-23.7

These results are consistent with experiments which show that phenol production is enhanced by addition of oxygen to the system and, with enough oxygen present, phenol production can be enhanced over benzene production consistent with the thermodynamic predictions. The results also show that biphenyl should be produced with excess oxygen, even more favorably than phenol. However, biphenyl is produced only at low levels under a broad range of conditions. This suggests that the paths for benzene and phenol production are coupled but that the pathway for biphenyl production is different.

The thermodyanmic data can be used to estimate the maximum electron affinity of $BPh_4^{\cdot}$ (Reaction (8)) from Reaction (9) which assumes that the binding of the Ph to the B in the radical $BPh_4\bullet$ is extremely weak. This is consistent with very preliminary calculations

Table 2. Global Reactions for Biphenyl and Phenol Production

Ion Oxidation/Hydrolysis	ΔH(kcal/mol)
$[BPh_4]^- + 1/2O_2 + H_2O \rightarrow [BPh_2(OH)_2]^- + Ph\text{-}Ph$	-76.3
$[BPh_3OH]^- + 1/2O_2 + H_2O \rightarrow [BPh(OH)_3]^- + Ph\text{-}Ph$	-82.0
$[BPh_2(OH)_2]^- + 1/2O_2 + H_2O \rightarrow [B(OH)_4]^- + Ph\text{-}Ph$	-82.9
Ion Oxidation/Hydrolysis	
$[BPh_4]^- + 1/2O_2 + H_2O \rightarrow [BPh_3OH]^- + PhOH$	-51.2
$[BPh_3OH]^- + 1/2O_2 + H_2O \rightarrow [BPh_2(OH)_2]^- + PhOH$	-56.7
$[BPh_2(OH)_2]^- + 1/2O_2 + H_2O \rightarrow [BPh(OH)_3]^- + PhOH$	-56.9
$[BPh(OH)_3]^- + 1/2O_2 + H_2O \rightarrow [B(OH)_4]^- + PhOH$	-57.7

on $BPh_4\bullet$ and more extensive ones on $B(OH)_4\bullet$.

$$BPh_4^- \rightarrow BPh_4 + e^- \tag{8}$$

$$BPh_4^- \rightarrow BPh_3 + Ph^- \rightarrow BPh_3 + Ph + e^- \tag{9}$$

The electron affinity is calculated as

$$EA = \Delta H_f(BPh_3) - \Delta H_f(BPh_4^-) + \Delta H_f(Ph) - \Delta H_f(Ph^-) = 101.9 \text{ kcal/mol } (4.42eV)$$

In the same way the maximum electron affinity of $B(OH)_4\bullet$ can be calculated from reaction (10).

$$B(OH)_4^- \rightarrow B(OH)_3 + OH^- \rightarrow B(OH)_3 + OH + e^- \tag{10}$$

This gives

$$EA = \Delta H_f(B(OH)_3) - \Delta H_f(B(OH)_4^-) + \Delta H_f(OH) - \Delta H_f(OH^-) = 86.1 \text{ kcal/mol } (3.73eV)$$

An exact calculation of the electron affinity (reaction (11)) can be done at the DFT/BP level as the radical is stable giving EA = 79.5 kcal/mol (3.45eV) based on a direct calculation. This suggests that the binding energy of an OH radical to $B(OH)_3$ is on the order of 0.3 eV or 7 kcal/mol.

$$B(OH)_4^- \rightarrow B(OH)_4 + e^- \tag{11}$$

CONCLUSIONS

The need to solve complex environmental problems in a timely fashion requires that the best scientific tools be employed. We have developed a broad range of computational chemistry software for advanced computer architectures in order to help address these problems. Our focus has been on developing tools for MPP architectures that have high performance as well as being portable and scaleable. In order to make these tools broadly available to both experts and non-experts, we have been developing a problem solving environment (PSE) for computational chemistry based on a fundamental understanding of the data models in this area. We are linking the PSE to a range of data bases to enable the computational chemist to make the best use of the available data and to optimally use his/her time and the available computer resources. Finally, we have begun using these new tools to address real environmental problems at DOE production sites.

ACKNOWLEDGMENT

This research was supported by the U. S. Department of Energy under Contract No. DE-AC06-76RLO 1830. We acknowledge the support of the the the Office of Biological and Environmental Research, the Division of Chemical Sciences (Office of Basic Energy Sciences), the Savannah River Technology Center and the Division of Mathematics, Information and Computational Sciences (Office of Computational and Technology Research). Some of this research was performed on the IBM SP computer in the Molecular Science Computing Facility in the Environmental Molecular Sciences Laboratory at PNNL. EMSL operations are supported by the DOE's Office of Biological and Environmental

Research. The Pacific Northwest National Laboratory is a multiprogram national laboratory operated by Battelle Memorial Institute.

REFERENCES

1. *U.S. DOE Complex Cleanup. The Environmental Legacy of Nuclear Weapons Production*, OTA-0-484, U.S. Department of Energy, Office of Technology Assessment, Washington, D.C. (1991).
2. R.E. Gephart and R.E. Lundgren, *Hanford Tank Cleanup: A Guide to Understanding the Technical Issues*, PNL-10773, Pacific Northwest National Laboratory, Richland WA (1995).
3. R.G. Riley and J.M. Zachara, *Chemical Contaminants on DOE Lands and Selection of Contaminant Mixtures for Subsurface Science Research*, DOE/ER-0547T, U.S. Department of Energy, Office of Energy Research, Washington, D.C. (1992).
4. G.E. Brown, Jr., R. Chianelli, L. Stock, B.R. Stults, S. Sutton and S. Traina, eds. *Molecular Environmental Science: Speciation, Reactivity and Mobility of Environmental Contaminants,* Report of DOE Molecular Environmental Science Workshop, July 5-8, 1995, Airlie Center, VA, U.S. Department of Energy, Office of Energy Research, Washington, D.C. (1996).
5. G.W. Gee and N.R. Wing, eds., *In-Situ Remediation: Scientific Basis for Current and Future Technologies*, Battelle Press, Columbus, OH (1994).
6. D.L. Clark, D.E. Hobart and M.P. Neu, *Chem. Rev.*, 95:25 (1995).
7. K.L. Nash, L.R. Morss, M.P. Jensen, E.H. Appelman and M. Schmidt, *Phosphate Mineralization of Actinides by Measured Addition of Precipitating Anions*, Annual Report (FY1995) for the ESP Program, U.S. Department of Energy, Office of Energy Management, Washington, D.C. (1995).
8. L. Rao and G.R. Choppin, A Survey of Separation Technologies of Interest for the U. S. DOE Complex, Chapter 18 in: *Separation Techniques in Nuclear Waste Management* (1994).
9. J.M. McKibben, *Radiochim. Acta.*, 36:3 (1984).
10. J.D. Navratil and W.W. Schulz, eds., *Actinide Separations*, ACS Symposium Series 117, American Chemical Society, Washington, D.C. (1980).
11. B.N. Laskorin, D.I. Skorovarov, E.A. Filippov and V.V. Yakshin, *Soviet Radiochemistry*, 27:140 (1985).
12. L. Cecille, M. Casarci and L. Pietrelli, eds., *New Separation Chemistry Techniques for Radioactive Waste and Other Specific Applications*, Elsevier Applied Science, London (1991)
13. K.L. Nash and G.R. Choppin, eds., *Separations of f Elements*, Plenum Press, New York (1995).
14. B.P. Hay and J.R. Rustad, *J. Am. Chem. Soc.*, 116:6316 (1994).
15. R.L. Murray *Understanding Radioactive Waste, 4th Ed.*, Battelle Press, Columbus, OH (1994)
16. D.D. Walker, M.J. Barnes, C.L. Crawford, R.F. Swingle, R.A. Peterson, M.S. Hay and S.D. Fink, *Decomposition of Tetraphenylborate in Tank 48H (U)*, WSRC-TR-96-0113, Rev. 0, May 10, 1996 Savannah River Technology Center, Savannah River GA (1996)
17. Defense Nuclear Facilities Safety Board Recommendation 96-1 to the Secretary of Energy, Aug. 14, U.S. Department of Energy, Washington, D.C. (1996).
18. Robinson, R. Sanders, J. Miyoshi, L. Fortenberry, K. Zavadoski, R. *Savannah River Site In-Tank Precipitation Facility Benzene Generation: Safety Implications*, DNFSB/Tech-14, Feb. 3, 1997 U.S. Department of Energy, Washington, D.C. (1997).
19. (a) D.M. Hirst, *A Computational Approach to Chemistry*, Blackwell Scientific, Oxford (1990);
    (b) G.H. Grant and W.G. Richards, *Computational Chemistry*, Oxford University Press, Oxford (1995);
    (c) W.J. Hehre, L. Radom, P.vR. Schleyer and J.A. Pople, *Ab Initio Molecular Orbital Theory*, John Wiley and Sons, New York (1986).
20. R.G. Parr and W. Yang, *Density Functional Theory of Atoms and Molecules*, (Oxford University Press, New York, 1989); (b) J. Labanowski and J. Andzelm, eds., *Density Functional Methods in Chemistry*, Springer Verlag, New York (1991).
21. (a) T.H. Dunning, Jr. *J. Chem. Phys.* 90:1007 (1989);
    (b) R.A. Kendall, T.H. Dunning, Jr. and R.J. Harrison, *J. Chem. Phys.* 96:6796 (1992);
    (c) D.E. Woon and T.H. Dunning, Jr. *J. Chem. Phys.* 99:1914 (1993);
    (d) K.A. Peterson, R.A. Kendall and T.H. Dunning, Jr. *J. Chem. Phys* 99:1930 (1993).
22. R.J. Bartlett *J. Phys. Chem.* 93:1697 (1989);
    (b) S.A. Kucharski and R.J. Bartlett *Adv. Quantum Chem.* 18:281 (1986);
    (c) R.J. Bartlett and J.F. Stanton, in *Reviews of Computational Chemistry, Vol. V*, K.B. Lipkowitz and D.B. Boyd, eds., Chapt. 2, p. 65, VCH Publishers: New York (1995).
23. P. Hohenberg and W. Kohn, *Phys. Rev. B*, 136:864 (1964).

24. J.C. Slater, *Adv. Quantum Chem.* 6:1 (1972).

25. S.H. Vosko, L. Wilk and M. Nusair *Can. J. Phys.* 58:1200 (1980).

26. J.P. Perdew, *Phys. Rev. B* 33:8822 (1986).

27 C. Lee, W. Yang and R.G. Parr,. *Phys. Rev. B*. 37:785 (1988).

28. (a) A.D. Becke, *Phys. Rev. A* , 38:3098 (1988);
(b) A.D. Becke, *Int. J. Quantum Chem. Symp.* 23:599 (1989);
(c) A.D. Becke, in *The Challenge of d and f Electrons: Theory and Computation*, D.R. Salahub and M.C. Zerner, eds., p166, ACS Symposium Series, No. 394, American Chemical Society, Washington, D.C., (1989).

29. M.F. Guest, E. Apra, D.E. Bernholdt, H.A. Fruechtl, R.J. Harrison, R.A. Kendall, R.A. Kutteh, X. Long, J.B. Nicholas, J.A. Nichols, H.L. Taylor, A.T. Wong, G.I. Fann, R.J. Littlefield and J. Niepolcha. *Future Generation Computer Systems*, 12:273 (1966).

30. D.F. Bernholdt, E. Apra, M.F. Fruechtl, R.A. Guest, R.J Harrison, R.A. Kendall, R.A. Kutteh, ; X. Long, J.B. Nicholas, J.A. Nichols, H.L. Taylor, A.T. Wong, G.I. Fann, R.J. Littlefield and J. Niepolcha *Int. J. Quantum Chem: Quantum Chem. Symp.* 29:475 (1995).

31. (a) E.N. Thornton, G. Black, T.L. Keller, K.L. Schuchardt, C.R Younkin and D.R. Jones, *Creating a Scientific Environment with AVS/Express*, in *Proceedings of the 1995 International Advanced Visual Systems Users and Developers Conference*, (April 1995);
(b) T.L. Keller and D.R. Jones, *Metadata: The Foundation of Effective Experiment Management*, in *Proceedings of the IEEE First Metadata Conference*, (April 1996).

32. D. Feller and K.A. Peterson, *J. Chem. Phys.*, 108:154 (1998).

33. The Browser is available on the World Wide Web at
http://www.emsl.pnl.gov:2080/forms/basis.form/html

34. N. Godbout, D.R. Salahub, J. Andzelm and E. Wimmer, *Can. J. Chem.*, 70:560 (1992).

35. M.W. Chase, Jr., C.A. Davies, J.R. Downey, Jr., D.J. Frurip, R.A. MacDonald and A.N. Syverud, *J. Phys. Chem. Ref. Data*, *14*, Supple. 1 (1985) (JANAF Tables).

36. (a) B. Ruscic, C.A. Mayhew and J. Berkowitz, *J. Chem. Phys.* 88:5580 (1988);
(b) E. Storms and B. Mueller, *J. Phys. Chem.* 81:318(1977);
(c) P.R. Rablen and J.F. Hartwig, *J. Am. Chem. Soc.* 118:4648 (1996).

37.(a) J.W. Larson and T.B. McMahon, *J. Am. Chem. Soc.* 105:2944 (1983);
(b) D.A. Dixon,and B.E. Smart, *J. Phys. Chem.* 93:7772 (1989);
(c) P.G. Wenthold and R.R. Squires, *J. Phys. Chem.* 99:2002 (1995);
(d) C.G. Krespan and D.A. Dixon, *J. Fluorine. Chem.* 77:117 (1996).

MACROMOLECULAR MODELLING ON THE CRAY T3D

Matthew D. Cooper,[1] Julia M. Goodfellow,[2] Ian H. Hillier,[1]
Christopher A. Reynolds,[3] W. Graham Richards,[4] Michael A. Robb,[5]
Paul Sherwood,[6] and Ian H. Williams[7]

[1]Chemistry Department, University of Manchester, Manchester, M13 9PL
[2]Crystallography Department, Birkbeck College, London, WC1E 7HX
[3]Biological Sciences Department, University of Essex, Colchester, CO4 3SQ
[4]New Chemistry Laboratory, University of Oxford, Oxford, OX1 3QT
[5]Chemistry Department, King's College, London, WC2R 2LS
[6]CCLRC Daresbury Laboratory, Warrington, WA4 4AD
[7]School of Chemistry, University of Bath, Bath, BA2 7AY

INTRODUCTION

The two basic methods of computational chemistry, namely electronic structure calculations (quantum mechanics, QM) and those based upon force fields (molecular mechanics, MM) are now widely used, in a routine fashion, to model many aspects of the structure and reactivity of macromolecular systems. Energy minimizations based upon quite simple representations of inter–atomic interactions via MM force fields can be used to predict the geometric structure of systems having many thousands of atoms, whilst their motion, particularly important in many biological problems can be followed using molecular dynamics (MD) simulations. These latter studies are particularly computationally intensive due to the quite long time scales that often need to be simulated.

To model situations where electronic effects are important, such as those involving bond breakage and formation as well as electronic polarisation accompanying interactions involving polar or charged groups, electronic structure calculations (QM) are usually employed. The poor scaling of such calculations with the size of the system means that the size of the system that can be treated is still a computational bottleneck. Furthermore, many modelling problems such as the calculation of potential energy surfaces and molecular dynamics or Monte Carlo (MC) simulations that include some degree of QM require the calculation of many molecular structures, which again is a computational bottleneck.

A particularly important area of research at present is the development and use of so-called hybrid methods that employ both QM and MM methods. These methods are based

High Performance Computing
Edited by R. J. Allan *et al.*, Kluwer Academic / Plenum Publishers, New York, 1999

upon the recognition that in many large molecular systems there is often a region where electronic rearrangement occurs, such as at an enzyme active site, which must be treated using QM, while the remainder of the system, the environment, where no significant electronic effects occur, can be modelled using MM. The QM region responds to the forces from the MM region, namely the electrostatic, van der Waals, and bonded interactions, which are all crucial in determining the behaviour of the QM region.

Both QM and MM methods, as well as hybrid QM/MM methods have been implemented on the massively parallel CRAY T3D and used by a number of groups to study a variety of problems in the area of Macromolecular Modelling. In this article we review the advances that have been made possible using this new computer hardware. We first summarize the software that has been implemented followed by examples of the macromolecular simulations that have been carried out.

In the area of calculations of the structure and motion of macromolecules, simulated using atomistic force fields, the code AMBER[1] is widely used both academically and industrially. In collaboration with the code developers and the support team at the Edinburgh Parallel Computing Centre (EPCC) a parallel version of AMBER was implemented on the CRAY T3D, with an efficiency of greater than 70% being achieved on up to 128 processors for medium to large calculations. The value of such a parallel implementation is evident since molecular dynamics simulations of a large macromolecular system is reduced from about one month on a typical workstation to under a day on the parallel computer.

A new approach to parallelising Monte Carlo free energy perturbation calculations which are widely used in macromolecular modelling has been developed by the Essex group.[2] The free energy perturbation method offers a powerful way of computing free energy differences for interactions between biomolecular species. It is ideally suited to parallel machines because it requires long simulation times. Moreover, in the case of large free energy differences, the simulations have to be repeated many times with each "window" representing a small portion of the overall change. The first window may represent 2% of the change from molecule A to B and so the overall change would require 50 simulations. In a break from the traditional approach of parallelising the non-bonded interactions, we have developed a version of the BOSS code[3] which runs all the windows simultaneously. Our strategy of implementing this approach carries an overhead of 50% due to increased equilibration times, but when implemented in a hybrid approach involving parallelisation of both the non-bonded interactions and of each window, offers the potential of truly massive parallelisation over thousands of processors since the communication problems are minimised. This new method has been tested on perturbations of nitroimidazoles to their radical anions, since this reduction is important in the activation of nitroimidazole bioreductive anti-cancer agents.

The Quantum Chemistry calculations utilised two codes, Gaussian,[4] and GAMESS(UK).[5] An initial serial implementation of Gaussian92 allowed individual calculations to be carried out on each node of the T3D. This proved especially useful when a large number of energy calculations of different configurations of the same system were required, as illustrated by a Monte Carlo simulation of solvation of structures of formamide using a hybrid QM/MM approach, to study the effect of bulk water on the barrier to rotation about the carbon-nitrogen bond in this molecule.[6] This is a prototype for the peptide (HNCO) bond, for which there have been several experimental and theoretical studies.

The efficient parallel implementation of the calculation of electronic wavefunctions at the self-consistent field (SCF) level on distributed memory computers[7] is well documented.

We have used such implementations of Gaussian94 and GAMESS(UK) in the studies described here. Use was also made of methods to treat electron correlation, important in many bond breaking and forming processes, using these two codes, with particular emphasis on density functional theory (DFT) methods within Gaussian and the second order Moller–Plesset (MP2) method within GAMESS(UK). Both these correlation methods scaled well on the T3D. Finally, these two QM codes were coupled with appropriate MM codes, AMBER in the case of Gaussian, and DL-POLY[8] in the case of GAMESS(UK), to allow hybrid QM/MM simulations to be carried out.

SCIENTIFIC RESULTS

Macromolecular Dynamics

We first give examples of atomistic simulations of macromolecular structure and dynamics carried out using the parallel AMBER code.

Membranes form the barrier between the cell cytoplasm and its surroundings, a barrier that many drugs and molecules must cross. The conventional model of a biological membrane is a lipid bilayer in which are embedded many proteins and sterols, all held together by non-covalent interactions. MD simulations can be used to study the motions and conformations of the lipid molecules and the effect of added solute molecules. The Oxford group have put cholesterol molecules into a computer-modelled hydrated bilayer of dimyristoyl phosphatidyl choline molecules and have used molecular dynamics simulations to characterise the effect of this important molecule on membrane structure and dynamics.[9] The main physical effects of cholesterol were successfully reproduced, notably an increase in motional order of those lipid atoms adjacent to the cholesterol and the important effect that cholesterol has on the ability of solutes to diffuse through the membranes. This latter result supports previous ideas on how cholesterol reduces the permeability of membranes.

The Birkbeck group have developed and applied new methods to model protein stability and folding/unfolding.[10-12] Such studies have relevance to many practical problems, such as those involving prions, the stability of both the protein P53 (implicated in many cancers) and of eye lens crystallins (relevant to cataracts). The particular system studied, the small enyzme barnase, has had its folding and unfolding pathways well-studied experimentally. The unfolding of this protein has been studied using both very high temperature MD and a novel technique to increase the speed and realism of unfolding using a solvent insertion algorithm at a realistic temperature.

The group at Essex carried out simulations on the activation of G-proteins by G-protein coupled receptors.[13] G-protein coupled receptors (GPCRs) constitute a family of several hundred receptors of immense medical importance - about 70% of all drugs bind to GPCRs. The receptors are involved in passing messages from small molecules, such as substance P or adrenaline across the cell wall to the nucleus. This process involves many steps, including the interaction between the receptor and the G-protein. From a therapeutic point of view, the GPCR represents the main target for intervention but since high affinity binding to the receptor requires the G-protein, it is important to study this interface. The crystal structure of the heterotrimeric G-protein is known but more recent work on the structure of the intracellular loops which couple to the G-protein has opened the way to study this interface. The simulations have confirmed the involvement of the C-terminus of the alpha subunit and have helped identify the role of particular residues. In addition, where alternative binding sites have been located, these have helped to expand novel ideas on the role of domain-swapped dimers in the activation of GPCRs.[13]

Macromolecular Reactivity

In the area of biochemistry, an understanding of the mode of action of enzymes in catalysing specific chemical transformations has been aided by molecular modelling methods, particularly those based upon hybrid QM/MM calculations.[14-17] Such studies are of potential value in drug discovery in view of the use of transition state analogues as pharmaceutically active enzyme inhibitors. In the QM/MM calculations the active site of the enzyme, including the substrate and important catalytic residues is treated by QM methods, while the remainder of the enzyme is modelled at a lower level by atomistic force field methods. The groups at Manchester and at Oxford have used hybrid QM/MM methods to understand enzyme catalysis at a molecular level.

At Manchester results have recently been obtained in studies of the enzyme papain.[18] Cysteine proteases as typified by the enzyme papain are a class of enzymes that have been widely studied over the years and have been shown to play important roles in degradation of muscular protein and in immunopharmacological responses. Defects in the regulation of cysteine protease activity have also been reported in connection with several disease processes, including cancer and muscular dystrophy. The overall principles of substrate recognition, catalysis and inhibition are now reasonably well documented. However, the mechanism by which cysteine proteases hydrolyse their substrates is still poorly defined at the molecular level so that the relationship between the molecular structure of the enzyme and of the substrate to the rate of the catalytic reaction is not clear. It is now widely accepted that the activity of cysteine proteases is dependent on the establishment of an active site thiolate-imidazolium ion pair between cysteine and histidine residues. Many other residues, both in and around the active site, are considered to be of importance for the proper functioning of the enzyme based on comparison of amino acid sequences and available X-ray structures.

For the enzyme-catalysed reaction the potential energy surface associated with the thiolate-imidazolium ion pair and with the subsequent amide hydrolysis were studied at the QM(AM1)/MM level. In the study of the ion pair, the complete substrate from the M M optimisation, Ace-Phe-Ser-Ile-Nme, was considered. In the amide hydrolysis study, the substrate chosen was N-methyl-acetamide, which was modelled at the QM level. This small substrate was used to permit the local geometry changes associated with the chemical reaction to occur; this would be difficult using a larger substrate. To quantify the effect of the enzyme environment on the potential energy surface, the energies of the optimised QM structure were also calculated in the absence of the MM atoms. In the absence of the enzyme the pair of neutral residues is strongly favoured. However, the protein environment has a large effect on proton transfer energetics. The potential energy surface for proton transfer with the protein environment present is extremely flat, the zwitterionic form being predicted to be stabilised over the structure having neutral residues by ≈ 3 kcal/mol.

The potential energy surface for amide hydrolysis was mapped out by minimising the substrate-active site structure for a range of S^{-}-C(O) and (N)H-N distances and showed conclusively that the reaction is a concerted one, proceeding via a transition state without the intervention of a tetrahedral intermediate. This structure was subsequently characterised as a transition state by calculation of the harmonic force constants. The predicted structure shows that during the reaction, the substrate moves towards the active site increasing the hydrogen bonding with the oxyanion hole (Gln-19, Cys-25) and with the imidazole proton involved in the reaction. The contribution of Gln-19 to the oxyanion hole which stabilises the transition state was modelled by repeating the calculation using an enzyme structure in which the residue is replaced by alanine. Such a mutation raises the calculated barrier to the reaction from 20.1 to 21.0 kcal/mol, close to the experimental increase of 2.4 kcal/mol. The transition state structure also shows reduced interaction of the substrate with the active site reducing its stabilization by the enzyme. Thus, these calculations have revealed important features of the reaction occurring at the active site, which are largely inaccessible by experiment.

The Oxford group completed a major investigation of acetyl-coenzyme A enolization in citrate synthase.[19] This study has gained in importance since it throws some light on a controversial aspect of enzyme mechanisms, that of the existence of low-barrier hydrogen bonds (LBHB). These are known to exist in the gas phase but their existence within enzyme active sites is disputed: see, for example, a recent article in *Science*.[20] This study does not support the existence of LBHB.

Bio-organic Reactivity and Structure

The groups at Bath and the Daresbury Laboratory have used both hybrid QM/MM, and *ab initio* QM methods to study aspects of bio-organic reactivity and the Manchester group have studied biologically important intermolecular interactions.

Equilibrium Isotope Effects for Hydrolysis of Methyl Glucoside in Vacuo and in Water.[21,22]   *Ab initio* gas-phase equilibrium isotope effects have been calculated for a range of isotopic substitutions for α and β anomers of methyl-glucoside using HF/4-31G optimized geometries and force constants computed with GAMESS(UK) on the T3D. These isotope effects were in accordance with AM1 gas-phase isotope effects and with HF/4-31G* values for smaller cut-down models for methyl-glucoside. Aqueous-phase equilibrium isotope effects have been calculated by means of hybrid QM/MM methods in which each glucoside is embedded within a 12Å-radius sphere of water molecules (TIP3P potential). Numerical hessian matrices, for displacements of solute atoms only, were computed using the AM1 QM/MM potential in CHARMM24b1. Approximate AM1 QM/MM hessians for the solutes, together with all waters within 4.0Å of it, were used to initiate *ab initio* HF/4-31G QM/MM optimizations using Chemshell/GAMESS(UK); only those waters for which there were hessian elements were allowed to move. HF/4-31G QM/MM numerical hessians, for displacements of solute atoms only, were then computed. However, equilibrium isotope effects from neither AM1 nor 4-31G QM/MM hessians were satisfactory. To obtain reasonable results it proved necessary to include all waters within 2.5Å of the solute in the hessian determinations.

Aldehyde Alkylation by Zinc Complexes.[23]   Many important metalloenzymes contain zinc as the key element within the active site.   As a preliminary to future theoretical investigations of mechanisms of zinc metalloenzymes, this project concerns a more modest but still very challenging goal: the catalytic and stereoselective mechanism of alkylation of aldehydes by dialkylzinc reagents. A recent theoretical investigation based on *ab initio* MP2=fc/6-31G(d)//HF/3-21G(d)   calculations   for   a   model   system   found   the stereodetermining and rate limiting step to be **5→6**; however, we are not convinced that this work contains the whole story since it omits the possibility of alternative reaction pathways and intermediates which we had proposed previously on the basis of semiempirical calculations. We are therefore re-investigating the reaction mechanism using better theoretical methods, and have so far performed the following density functional theory calculations using Gaussian.

- B3LYP/3-21+G geometry optimization for structures **1** to **6**.
- Comparison of B3LYP/3-21+G and BLYP/3-21+G optimised geometries of **3** shows very little difference but the B3LYP method takes twice as long.
- B3LYP/3-21+G optimization for transition structure **5** and B3LYP/6-311++G(d,p) optimizations for other species are in progress.

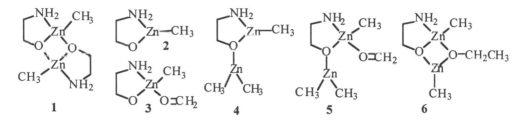

The Structure and Intermolecular Interactions of a Creatinine Designed–Receptor Complex.[24] Creatinine is an important end product of nitrogen metabolism in vertebrates and appears in the urine of healthy individuals. The search for a suitable host complexing this guest molecule has led to experimental studies of the mode of binding of creatinine to hosts ranging from quite small molecules such as pyrimidine derivatives to large polycyclic molecules. Of particular interest is the recent report of an artificial receptor which can possibly be used for creatinine assays.[25] Structural and energetic features of the intermolecular interactions of a creatinine designed-receptor complex were investigated using *ab initio* electronic structure methods.[24] Both the host and receptor can adopt different tautomeric forms and it was found that in the complex both molecules are considerably different from their gas-phase structures. A polar environment has an important role in determining the binding energy of the complex and may lead to proton transfer in the complex.

'On The Fly' MMVB Dynamics

The prediction of accurate potential energy surfaces for complex chemical reactions is the first essential step in the modelling of the dynamics of such reactions. The use of direct dynamics, where the energies and forces are calculated by QM methods "on-the-fly", rather than by the use of potential energy functions, is increasing and is, naturally extremely computationally demanding. Many dynamical processes important in biological systems, involve excitation energy transfer and hence involve more than one potential energy surface.

A central technical achievement of the King's College group has been the successful embedding of hybrid molecular mechanics/valence bond (MMVB) into a molecular dynamics simulation method.[26] In this implementation the MMVB energy and gradient are used to solve the equations of motion. The trajectories are propagated using a series of local quadratic approximations to the MMVB potential energy surface and the stepsize is determined by a trust radius. Initial conditions (i.e. the initial velocities) are determined by random sampling of each excited state normal mode within an energy threshold ΔE_{limit}, starting at the Franck-Condon geometry (S_0^*) on S_1. This generates an ensemble of trajectories - a "classical wavepacket". For each value of ΔE_{limit}, 256 trajectories are usually run, each for several hundred steps. Each trajectory is run on an individual processor of the T3D, thus ensuring very efficient load balancing. The surface-hopping algorithm of Tully and Preston is then used to allow excited state trajectories to transfer to the ground state in the conical intersection region, where strong nonadiabatic coupling effects appear. The difference in energy between S_0 and S_1 at the hop is then redistributed along the component of the momentum parallel to the nonadiabatic coupling vector. A simple Stoke's Law viscosity model has been implemented to simulate a loss of kinetic energy due to solvent interactions.

A prototype system, the decay of the S_1 excited state of azulene, has been investigated by this approach.[27] The rapid radiationless decay from the excited S_1 state to the ground S_0 state is found experimentally to occur in less than 1ps. MMVB direct dynamics showed that such ultrafast S_1 decay can by explained by relaxation through an unavoided S_1/S_0 crossing (i.e. a conical intersection). Trajectories, initially on the S_1 excited state curve, were followed and it was found that for the majority of such trajectories hopping from the $S_1 \rightarrow S_0$ surface occurred within ~10fs. Thus, these MD simulations suggest that fast intersystem crossing will approach 100% efficiency since the surface hop takes place within one vibrational period.

CONCLUSIONS

The full range of molecular modelling methods, including atomistic simulations, quantum chemistry calculations and appropriate schemes involving both approaches, are needed to tackle the many and complex problems of macromolecular structure and reactivity. Such problems are amongst the most challenging being addressed by computational chemists today and are important in understanding biological processes at a molecular level and in the associated area of drug discovery.

This article has highlighted how a start can be made on the solution to these problems aided by high performance parallel computers. There will undoubtedly be further rapid developments in macromolecular modelling in the near future involving both new modelling methods and their implementation on more powerful computer hardware.

ACKNOWLEDGMENTS

We thank EPSRC and BBSRC for support of this research.

REFERENCES

1. D.A. Pearlman, D.A. Case, J.C. Caldwell, G.L. Seibel, U.C. Singh, P. Weiner, and P.A. Kollman, *AMBER 4.0*, University of California, San Francisco (1992).
2. J.D. Wright, and C.A. Reynolds, Exploiting the parallelism inherent in the windowing approach to Monte Carlo free energy perturbation calculations, *J. Mol. Struct.* (Theochem), in press.
3. W.L. Jorgensen, *Boss Version 3.5*, Yale University, New Haven, CT (1994).
4. M.J. Frisch, G.W. Trucks, H.B. Schlegel, P.M.W. Gill, B.G. Johnson, M.A. Robb, J.R. Cheeseman, T.A. Keith, G.A. Peterson, J.A. Montgomery, K. Raghavachari, M.A. Al-Laham, V.G. Zakrezewski, J.V. Ortiz, J.B. Foresman, J. Cioslowski, B.B. Stefanov, A. Nanayakkara, M. Challacombe, C.Y. Peng, P.Y. Ayala, W. Chen, M.W. Wong, J.L. Andres, E.S. Replogle, R. Gomperts, R.L. Martin, D.J. Fox, J.S. Binkley, D.J. Defrees, J. Baker, J.P. Stewart, M. Head-Gordon, C. Gonzalez, and J.A. Pople, *Gaussian 94*, Revision A.1, Gaussian Inc., Pittsburgh, PA (1995).

5. M.F. Guest, J.H. van Lenthe, J. Kendrick, K. Schoeffel, P. Sherwood, R.D. Amos, R.J. Buenker, M. Dupuis, N.C. Handy, I.H. Hillier, P.J. Knowles, V. Bonacic-Koutecky, W. von Niessen, A.P. Rendell, V.R. Saunders, and A.J. Stone, *GAMESS(UK)*, CCLRC Daresbury Laboratory, Warrington (1996).

6. J.S. Craw, J.M. Guest, M.D. Cooper, N.A. Burton, and I.H. Hillier, Effect of hydration on the barrier to internal rotation in formamide. Quantum mechanical calculations including explicit solvent and continuum models, *J. Phys. Chem.* 100:6304 (1996).

7. M.D. Cooper, N.A. Burton, R.J. Hall, and I.H. Hillier, Combined Hartree–Fock and density functional theory: A distributed memory parallel implementation, *J. Mol. Struct. (Theochem)* 315:97 (1994).

8. W. Smith, and T.R. Forester, *DL POLY*, CCLRC Daresbury Laboratory, Warrington (1996).

9. A.J. Robinson, W.G. Richards, P.J. Thomas, and M.M. Hann, Behaviour of cholesterol and its effect on head group and chain conformation in lipid bilayers: A molecular dynamics study, *Biophys. J.* 68:164 (1995).

10. M. Knaggs, M. Williams, and J.M. Goodfellow, Protein hydration, stability and unfolding, *Biochem. Soc. Trans.* 711 (1995).

11. J.M. Goodfellow, M. Knaggs, M. Williams, and J.M. Thornton, Modelling protein unfolding: A solvent insertion protocol, *Faraday Discussions* 103:339 (1996).

12. M.A. Williams, J.M. Thornton, and J.M. Goodfellow, Modelling protein unfolding: Hen egg-white lysozyme, *Protein Engineer.* 10:895 (1997).

13. C.R. Gouldson, C.R. Snell, and C.A. Reynolds, A new approach to docking in the β(2)-adrenergic receptor which exploits the domain structure of G-protein coupled receptors, *J. Med. Chem.* 40:3871 (1997).

14. J. Aqvist, and A. Warshel, Simulation of enzyme reactions using valence-bond force-fields and other hybrid quantum-classical approaches, *Chem. Rev.* 93:2523 (1993).

15. U. C. Singh, and P.A. Kollman, A combined *ab initio* quantum-mechanical and molecular mechanical method for carrying out simulations on complex molecular systems: Application to the CH_3Cl and Cl^- exchange-reaction and gas phase protonation of polyethers, *J. Comp. Chem.* 7:718 (1986).

16. M.J. Field, P.A. Bash, and M. Karplus, A combined quantum mechanical and molecular mechanical potential for molecular dynamics simulation, *J. Comp. Chem.* 11:700 (1990).

17. B. Waszkowycz, I.H. Hillier, N. Gensmantel and D.W. Payling, Aspects of the mechanisms of catalysis in phospholipase A_2. A combined *ab initio* molecular orbital and molecular mechanics study, *J. Chem. Soc. Perkin Trans.* 2 1795 (1989).

18. M.J. Harrison, N.A. Burton, and I.H. Hillier, Catalytic mechanism of the enzyme Papain: Predictions using a hybrid quantum mechanical/molecular mechanical potential, *J. Amer. Chem. Soc.* 119:12285 (1997).

19. A.J. Mulholland, and W.G. Richards, Acetyl–Co A enolization in citrate synthase. A quantum mechanical/molecular mechanical (QM/MM) study, *Proteins: Struct. Funct. and Genetics* 27:9 (1997).

20. E.L. Ash, J.L. Sudmeier, E.C. de Fabo, and W.W. Bachovchi, A low-barrier hydrogen bond in the catalytic triad of serine proteases? Theory versus experiment, *Science* 278:1128 (1997).

21. A.J. Turner, P. Sherwood, J.A. Barnes, and I.H. Williams, Theoretical modelling of equilibrium isotope effects for the rate-determining step of acid-catalysed hydrolysis of alpha- and beta-methylglucopyranosides: Evaluation of cut-off models and solvation effects using hybrid quantum-mechanical/molecular mechanical methods, in preparation.

22. A.J. Turner, P. Sherwood, and I.H. Williams, Protium/deuterium fractionation factors for organic molecules in aqueous solution: Calculations using hybrid quantum-mechanical/molecular-mechanical methods, in preparation.

23. C.F. Rodriquez and I.H. Williams, Mechanism of aldehyde alkylation by dimethylzinc catalysed by a beta-aminoalcohol-zinc complex: A density functional study, in preparation.

24. J.S. Craw, M.D. Cooper, and I.H. Hillier, The structure and intermolecular interactions of a creatinine designed-receptor complex, studied by *ab initio* methods, *J. Chem. Soc. Perkin Trans.* 2 869 (1997).

25. T.W. Bell, Z. Hou, Y. Luo, M.G.B. Drew, E. Chapoteau, B.P. Czech, and A. Kumar, Detection of creatinine by a designed receptor, *Science*, 269:671 (1995).

26. B.R. Smith, M.J. Bearpark, M.A. Robb, F. Bernardi and M. Olivucci, "Classical wavepacket" dynamics through a conical intersection: application to the S_1/S_0 conical intersection in benzene, *Chem. Phys. Lett.* 242:27 (1995).

27. M.J. Bearpark, F. Bernardi, S. Clifford, M. Olivucci, M.A. Robb, B.R. Smith, and T. Vreven, The Azulene S1 state decays *via* a conical intersection. A CASSCF study with MMVB dynamics, *J. Amer. Chem. Soc.* 118:169 (1996).

ACCURATE CONFIGURATION INTERACTION COMPUTATIONS OF POTENTIAL ENERGY SURFACES USING MASSIVELY PARALLEL COMPUTERS

Abigail J. Dobbyn[1]* and Peter J. Knowles[2]

[1]HPCI Centre, CLRC Daresbury Laboratory,
Daresbury, Warrington WA4 4AD, U.K.
[2]School of Chemistry, University of Birmingham,
Edgbaston, Birmingham B15 2TT, U.K.

INTRODUCTION

A great deal of useful information, for example reaction rates, product state distributions and insights into reaction mechanisms can be obtained from the theoretical study of chemical reactions. Such research is of fundamental importance for the modelling of processes occurring in combustion and in the atmosphere. As the first step in such studies, it is often necessary to obtain highly accurate global potential-energy surfaces (PESs), or at least definitive information about the height of reaction barriers and reaction energies.

For dynamical calculations to be worthwhile, the information about the internuclear interaction energies must be of chemical accuracy. This means that a large proportion of the static, as well as the dynamic, electron correlation energy must be included. In the case of PESs, this must be done with uniform accuracy. Therefore, it is necessary to use multi-reference configuration interaction (MRCI) methods, which, in contrast to other methods that utilise only a single reference configuration, are capable of describing excited states, near-degeneracy effects and multiple open-shell systems, as well as the two-electron cusp. These methods are inevitably very costly in all aspects of computing resources. As well as involving large numbers of arithmetic operations, very large intermediate data structures must be manipulated, demanding large central memory and peripheral storage, with fast movement of data between them. A further consideration is that PESs must be global, so that they cover not only the interaction region, but extend well into both the entrance and exit channels. In order to achieve this, a very large number of points have to be calculated, of the order of a thousand for a triatomic system.

The increasing availability and performance of parallel computers [1,2] offers the possibility of at least partially satisfying the very heavy resource demands of MRCI

---

*Current address:
Department of Chemistry, University of Manchester, Oxford Road, Manchester M13 9PL, U.K.

High Performance Computing
Edited by R. J. Allan *et al.*, Kluwer Academic / Plenum Publishers, New York, 1999

computations. The use of parallel computers in quantum chemistry is not new [3, 4, 5], although the majority of work done in the past has concentrated on Self Consistent Field (SCF) programs, with more accurate methods being addressed more recently [3, 4, 5, 6, 7, 8]. There have been several previous presentations of parallel MRCI programs [4, 5, 6, 7], the most recent of which is that forming a part of the COLUMBUS program package [6, 7]. The parallelisation is there carried out across pairs of segments of the residual and coefficient vectors, which are dynamically assigned to the processors. This implementation has shown excellent scaling, and is also fully portable to many different machines. All of these parallel MRCI implementations unfortunately suffer from the use of an underlying ansatz which is much more computationally expensive than it needs to be. Following the suggestion of Meyer [9], the internally contracted MRCI (ICMRCI) approach [10, 11, 12, 13, 14] has emerged as the method of choice for MRCI computations, particularly where the number of reference configurations is large.

Our aim is to develop a scalable and portable parallel MRCI program. In balance with these aims, particularly the first, is a pragmatic approach which hopes to minimize the changes between the sequential and parallel codes. This should make the development quicker and easier, as well as reducing the chance of introducing errors into the code. Furthermore, it is important to realize that the MOLPRO program system [15] and the MRCI program within it are continually evolving, so that it is essential that the parallel version of the code should be easily maintained. Given the wide variety of different parallel computers that are routinely used at present it is highly desirable that the parallel code should be portable to a large number of different platforms (as the sequential version of MOLPRO already is). This is ensured by the use of the Global Array subroutine libraries [16], together with TCGMSG [17], which are portable software toolkits for the development of parallel programs.

CONTRACTED MULTIREFERENCE CONFIGURATION INTERACTION

In order to carry out contracted multireference CI computations within our program suite MOLPRO [15], four distinct tasks can be identified:

1. Calculation of 1- and 2-electron hamiltonian integrals, and their sorting into canonical order for use by the other program stages.

2. Self-consistent field calculation (SCF) to obtain molecular orbitals for use as starting guesses in the next program stage.

3. Multiconfiguration self-consistent field calculation (MCSCF), usually within the complete active space (CASSCF) scheme [18], which is carried out to optimize the molecular orbitals for that reference space [19, 20].

4. Iterative direct diagonalization of the MRCI hamiltonian matrix.

The evaluation and sorting of the two-electron integrals program stage has been parallelised, but this will not be described further here. Stages (ii) and (iii), the SCF and MCSCF, have not been parallelised. For the calculation of PESs, parallelisation of stage (ii) is not necessary as orbitals from an adjacent geometry can be used as starting guesses in the MCSCF program, which itself will converge rapidly because the neighbouring-geometry orbitals will be close to the solution. In the case when only a few important geometries are being investigated, for example at transition states or asymptotes, MCSCF orbitals can conveniently be obtained on a scalar computer, and so again parallelisation of the MCSCF program is not an issue.

The key computational stage is then the diagonalization of the hamiltonian matrix, which is typically of dimension 10^6. This matrix arises from the representation of the wavefunction as a linear combination of contracted configurations [11],

$$\Psi = \sum_Z c_Z \Psi_Z \tag{1}$$

and projection of Schrödinger's equation onto the basis yields the eigenvalue problem

$$\mathbf{Hc} = E\mathbf{c} \tag{2}$$

Solution is reached using an iterative method; a modified Davidson procedure is used here [21, 11]. An iterative method is used because it is not possible to hold the whole of the hamiltonian matrix in memory, as well as the fact that usually only the lowest one or two eigenvalues and eigenvectors are required. At each iteration the wavefunction is represented by a relatively small set of trial vectors, $\mathbf{c}$, which are generated by the operation of the hamiltonian on the wavefunction, $\mathbf{g} = \mathbf{Hc}$. Each iteration therefore consists of the calculation of the residual vector $\mathbf{g}$, followed by the update of $\mathbf{c}$, in preparation for the next iteration. This procedure is repeated until the wavefunction has converged and the trial vectors span the subspace in which the eigenvector lies.

The evaluation of $\mathbf{g}$, the most computationally intensive part of the MRCI program, can be represented schematically as

$$q_X = \sum_Y \sum_\lambda f_\lambda \alpha_\lambda(X, Y) c^Y + \text{other terms} \tag{3}$$

The equations will not be presented in detail, and for further information the reader is referred to Refs. [11, 12, 13]. The configuration basis, and therefore the vectors $\mathbf{c}, \mathbf{g}$, can be partitioned into different types, with different formulae for the hamiltonian matrix elements. f_λ is a set of two-electron integrals $(rs|tu)$, with the index λ running over 0, 1, 2, 3 or 4 internal molecular orbitals (MOs), according to case. The coupling coefficients $\alpha_\lambda(X, Y)$ arise from the topological structure of the contracted configurations. They are very numerous, and are therefore evaluated as needed from a minimal set of symmetric group representation matrices, which are set up at the start of the calculation [12] and held in local memory on each processor; the amount of memory required depends strongly on the case, but is typically between 10^5 and 10^6 words. The 'other terms' will be made up of one-electron integrals and the relevant coefficients and in some cases coupling coefficients; these terms are not computationally expensive and will not be discussed further.

From the above schematic overview of the equations involved in the MRCI program we can see that it is necessary to store the following quantities: one- and two-electron integrals $(N_{\text{basis}}^4/8)$, where N_{basis} is the number of atomic/molecular orbital (AO/MO) basis functions; c^Z and g_Z $(N_H$, the dimension of the hamiltonian matrix). These quantities are too large for all of them to be held simultaneously in the core memory. They are therefore stored as a set of scratch files. Each file is direct access and word addressable but is organized in records of any length. All input and output (i/o) to these scratch files is through a set of FORTRAN subroutines: *reserv*, which reserves space for a record of a specified length on the specified file; *lesw*, which reads a one-dimensional array of double precision data of particular length from an offset from the start of the record on a specified file; *sreibw*, which writes in an analogous fashion. These FORTRAN subroutines implement buffering, and perform raw read and write operations through a set of lower-level C subroutines. The files are generally scratch files, i.e., they are deleted at the end of a calculation, but can individually be made

Table 1. Generic batching algorithm for MRCI

```
check on amount of memory available
consideration of memory required to hold relevant gX, cY and fλ as well as work space
decision on batching scheme
do ibatchX = 1, nbatchX
        zero (if first contribution ) or read in gX for X in batch X1 : X2
        do ibatchY = 1, nbatchY
                read in cY for Y in batch Y1 : Y2
                do ibatchλ = 1, nbatchλ
                        read in fλ for λ in batch λ1 : λ2
                        calculate αλ(X, Y) for X1 to X2 and Y1 and Y2
                        for X in batch X1 : X2
```
$$g_X = g_X + \sum_{Y=Y_1}^{Y_2} \sum_{\lambda=\lambda_1}^{\lambda_2} f_\lambda \alpha_\lambda(X,Y) c^Y$$
```
                end do
        end do
        write gX for X in batch X1 : X2 to file
        end do
end do
```

permanent or 'named', so that quantities such as the molecular orbitals can be saved from one calculation to the next.

The normal algorithms for calculating **g** recognize firstly that the coupling coefficients $\alpha_\lambda(X,Y)$ are extremely numerous, and in some cases very sparse, and so they cannot conveniently be stored in full in memory; thus, for a given interaction, the calculation is driven by the sequential computation of $\alpha_\lambda(X,Y)$, combining with all other quantities which are held in full in memory. In some cases however, either when a very large calculation is being attempted, or if a more moderate calculation is carried out on a computer with only a relatively small amount of core memory, it is not possible to hold the whole of any one of the coefficient segments, c^Y or g^X, in core memory. Therefore the calculation of the residual has to be divided up in some way, so that it is essential to use algorithms such that all of g_X, c^Y and f_λ are read in in batches if necessary. The code will then have the general structure shown in Table 1 for the evaluation of g_X (for a single interaction type with λ going over just one number of internal MOs).

Notice that the inner part of the algorithm can generally be implemented very efficiently as a sequence of two vector-vector, matrix-vector or matrix-matrix operations. The order of the multiplications is not necessarily as shown, but will be chosen to minimize the operation count.

The order in which the loops in Table 1 are implemented will vary, as will the relative size of the batches $nbatch_X$, $nbatch_Y$ and $nbatch_\lambda$, depending partly on the relative sizes of g_X, c^Y and f_λ. In general, an attempt is made to hold the smallest quantities completely in memory and to batch over the larger quantities, preferably just once, so that each is read just once through. There are cases where there are quite different algorithms depending on the amount of memory available, and therefore the amount of batching required.

We have given above an overview of the memory management aspects of the MRCI program in its scalar version as it existed before this research began. The purpose of doing this rests on the recognition that the general model of paging to and from disk, adopted for the small-memory workstations of the last decade, maps quite closely to the situation prevailing with many modern parallel computers, where for each processor there is a relatively small amount of fast local memory, with slower access to storage on other processors. This non-uniform memory-access (NUMA) programming model unites sequential and parallel programming, and simplifies the design of scalable parallel algorithms. In the following section, we describe how these original paging algorithms can therefore be used in the parallel context for the effective distribution of data without wholesale redesign of a complicated code.

PARALLEL IMPLEMENTATION

The sequential code described above is seen to rely heavily on the use of scratch files, and therefore on the availability of fast i/o and relatively large amounts of disk space. It is clear that such reliance on i/o will be impossible when using massively parallel processing (MPP) machines, which to date have typically shown extremely poor i/o capabilities (of which the performance seldom scales with the number of processors), and often have relatively little hard disk available.

Therefore the first step necessary for the development of a scalable program to run on a MPP computer is the elimination of the i/o and the removal of the scratch files. This we do by replacing the reading and writing to scratch files by reading and writing to global memory, that is the sum of the memory of all the participating processors. So, for example, all the two-electron integrals will be held in memory, with a roughly equal number of them stored on each processor. This is straightforwardly implemented using the Global Array subroutine library [16].

Global Arrays

The Global Array (GA) subroutine library implement a portable "shared-memory" programming model for distributed-memory computers, developed specifically for use in parallel quantum-chemistry codes. In evaluating the advantages of using such a library and its underlying programming model, it is worth considering the two main alternatives: a shared-memory model, and a message-passing model. The main advantage of using a shared memory model is the ease of programming, as well as the fact that the one-sided communication does not require synchronization between processors, which facilitates load-balancing strategies. The disadvantage of such a model is, however, that it is not always obvious to the programmer whether or not they are using non-local data, i.e., data not held on that processor, so that the extra cost of obtaining this data is not fully accounted for. In contrast, the main advantage of a message-passing paradigm is that it is very clear when non-local memory is being accessed, which in fact leads to the main disadvantage, which is the difficulty of designing such programs, as well as the requirement for explicit cooperation between processors, and the synchronization which must necessarily occur (though the impact of this can be reduced to some extent by the use of asynchronous communication and buffering of data).

The Global Array subroutine libraries take the advantages of both these models: they allow for ease of programming, and lack of synchronization between processors, while acknowledging that the non-local data does take more time to access. Global arrays can be created and destroyed using the GA functions *ga_create* and *ga_destroy* respectively, while data can be read from, written and accumulated to a global array with the GA subroutines *ga_get*, *ga_put* and *ga_acc*.

The global arrays (GA) are implemented using the fastest native communication, and interoperate with MPI [22] and TCGMSG [17]. MPI and TCGMSG provide basic message passing subroutines, such as send and receive, as well as global operations, and TCGMSG includes a shared counter (*nxtval*), which is a useful tool for dynamic load balancing. The *nxtval* counter has been used extensively in the parallel implementation, and in a few cases the message passing and global operations have also been found useful.

GA within MOLPRO

In order to develop a completely replicated version of the code, i.e., where all processors do exactly the same work, the records on each files are replaced by global arrays. The substitution of the real i/o for reading and writing to the global arrays is simple to carry out, and can be done within the existing FORTRAN subroutines which manage the i/o, without any changes whatsoever to the rest of the code. Therefore, within the *reserv* subroutine there is now a call to the GA function *ga_create*; within *lesw* there is a call to *ga_get* and within *sreibw* there is a call to *ga_put*. In this initial replicated-data implementation all processors read the data they require, while the processors only write the data which they 'own' (which can be determined using the GA subroutine *ga_locate_region*), after which they synchronize. When we wish to save the scratch files, it is possible to 'name' them, in which case they are then treated as a real file and written to disk. In this case, only the zeroth processor reads and writes to the file, and broadcasts the data to the other processors. This facility is particularly useful for restarting the program from previously calculated molecular orbitals. This replicated-data mode of the code is denoted by the variable *mpp_state* having the value one.

Obviously, in order to develop a parallel version of the code, it is necessary to go beyond this replicated-data approach to a distributed-data approach, where processors work on complementary data, so that each processor needs to read and write distinct pieces of data. In this case each processor simply reads and writes the data on which it is working, without reference to any other processor. At present this type of i/o does not work with named, i.e., real, files. This distributed-data mode of the code is denoted by the variable *mpp_state* having the value two. At the start and end of each stage of the program, where there is a call to *sreibw* with *mpp_state=2*, the processors synchronize.

The parts of the code which have not been parallelised at all, e.g. the SCF, will run exclusively within the *mpp_state=1* mode. The MRCI part of the code will have within it both the *mpp_state=1* and the *mpp_state=2* modes. When it is reading and writing to the global arrays in the *mpp_state=1* mode, this does not necessarily mean that there has been no work done in parallel, but may mean that the work has been divided up across the tasks which are required to evaluate some length of data, rather than across the length of data itself, which will be the case in the distributed-data mode. To facilitate such a type of parallelism, a new FORTRAN i/o subroutine is introduced, *accw*, which will in these cases replace a call to *sreibw*. In this subroutine the data which is passed to it is first summed globally and then the completed data is written to the appropriate global array. In cases when it is not the first contribution to **g** that is being formed, there will more than just a change from *sreibw* to *accw* required. In order that the original value of **g** is not also summed, it is not read in at the start of the subroutine, but always zeroed, and the original value of **g** is read in and added to the new contribution after the latter has been summed. In fact, each processor does not need access to the whole of the completed data, but only the part which it 'owns'. Therefore, it is found to be more efficient here to use a message-passing routine such that each processor receives that data which it requires from all the other processors, and performs the necessary summation, before accumulating the data to the global array.

At the start of the calculation it is necessary to divide the total amount of allocated memory between that used for the global arrays, i.e., global memory, and that used as local memory. This division can be altered between program steps, which is particularly useful after the sort of the two-electron integrals which requires quite large amounts of global memory. In practice, the amount of global memory allocated during the MRCI program stage is typically double that necessary to hold the two-electron integrals. The GA subroutine libraries include a local memory management library called MA, which is used by the MPP version of MOLPRO.

The above constitutes a complete framework from which we can proceed to develop a truly parallel version of the code, in which the work, and where possible the memory requirements, of the program are divided up over the processors. In particular, a mechanism for the progressive introduction of parallel code has been set up. An initial replicated version of any code, using $mpp\_state=1$, can be gradually transformed into a parallel code through successive replacement of code segments, using $mpp\_state=2$ where necessary for supporting distributed data structures.

Analysis of Computation and Assignment of Tasks

As has been stated previously, the calculation of the residual vector **g** is the most computationally intensive part of the code. Often, it is possible to identify one, or a few, particular types of matrix element as being the dominant computational burden in a given calculation, and although this analysis is case-dependent, it allows the identification of the most important parts of the code to be parallelised first of all. Unfortunately, however, all the other interactions as well as the work involved in the initialization and the calculation of the density matrix can account for a significant fraction (e.g., 5%) of the total CPU time. Although this may not seem significant, a consideration of Amdahl's law makes it quite clear that this work must also be parallelised if reasonable speed ups are to be obtained using more than a small number of nodes. For example if a program running on 64 processors has 95% of its work parallelised, the best speed-up it can hope to obtain is 15.4, i.e., only 24% efficiency. It is the fact that the total time for the MRCI program step is not overwhelmingly dominated by any one interaction, or distinct set of localized tasks, which must limit our expectations of obtaining excellent speed-ups.

At this stage it is now necessary to consider how the work involved in these calculations can be divided up over the processors, i.e., the division of the work into separate tasks and the assignment of these tasks to individual processors.

As mentioned earlier in this paper, we hope normally to take advantage of the considerable work that has already gone into the paging strategies used throughout the program, so that for example, one might, in the scheme of Table 1, envisage each processor carrying out the work of a single batch, i.e., each task will be associated with a batch. Generally speaking, the work should be parallelised across just one set of batches, i.e., $nbatch_X$, $nbatch_Y$ or $nbatch_\lambda$, thus leading to the complete distribution of one of these quantities. It could be argued that the parallelisation is best carried out over $nbatch_X$, as this results in the distribution of the output, and thus avoiding the need for a summation of the result in *accw* over the processors.

There are several advantages in using the batching schemes of the sequential program. Firstly, the work of the development of the parallel code is greatly reduced as the mechanism to divide up the work, as well as the program infrastructure to deal with these batches, already exists in the code. Secondly, this approach simplifies the decision on how the work should be divided up, as a careful analysis has already been done in the sequential code as to the most efficient way to batch up the work. Thirdly, the changes between the sequential and parallel codes should by this means kept to a minimum. It should be noted however that in certain cases there may be other optimal ways to divide up the work in the parallel version of the program, which differ from the batching existing in the original sequential code.

In the design of any parallel algorithm, and in the setting up and allocation of tasks here, there are two main important factors which must be taken into account: Firstly, the minimization of overheads (latency and bandwidth) associated with communication between processors, which corresponds here to the mindful use of accesses to the global arrays, and where possible the maximization of use of local data. Secondly, the load balancing has to be optimized, so that each processor does an equivalent amount of work.

The above describes the general scheme through which effective parallelisation of the ICMRCI code has been achieved.

Table 2. Details of butadiene benchmark calculations. All calculations were carried out in the C_{2h} symmetry group and used the complete active space.

test case	A	B
Basis set	cc-pVDZ	cc-pVTZ [a]
Size of integral file	41 Mb	495 Mb
No. orbitals	$(32a_g,11a_u,32b_u,11b_g)$	$(61a_g,26a_u,61b_u,26b_g)$
Reference space	4 el. in $(1\text{-}3a_u,1\text{-}2b_g)$	
Frozen core	$(1\text{-}2a_g,1\text{-}2b_u)$	
No. ref. config.	28	
No. contracted CSFs	199183	817599
No. uncontracted CSFs	3933377	19375033

[a] not including the *d* functions on the hydrogen atoms.

Table 3. Details of butadiene benchmark calculations carried out on the SP2. All calculations were carried out in the C_{2h} symmetry group and used the complete active space.

test case	C	D
Basis set	cc-pVDZ	aug-cc-pVDZ [a]
Size of integral file	41 Mb	121 Mb
No. orbitals	$(32a_g,11a_u,32b_u,11b_g)$	$(44a_g,17a_u,44b_u,17b_g)$
Reference space	4 el. in $(1\text{-}3a_u,1\text{-}3b_g)$	6 el. in $(6\text{-}7a_g,1\text{-}3a_u,1\text{-}2b_g)$
Frozen core	$(1\text{-}2a_g,1\text{-}2b_u)$	
No. ref. config.	57	142
No. contracted CSFs	344595	1569335
No. uncontracted CSFs	8590550	46994460

[a] not including the diffuse functions on the hydrogen atoms

Table 4. Details of butadiene benchmark calculations carried out on the T3D/T3E. All calculations were carried out in the C_{2h} symmetry group and used the complete active space.

test case	E	F	G
Basis set	cc-pVTZ [a]	cc-pVTZ	
Size of integral file	495 Mb	909 Mb	
No. orbitals	$(61a_g,26a_u,61b_u,26b_g)$	$(70a_g,32a_u,70b_u,32b_g)$	
Reference space	4 el. in $(1\text{-}3a_u,1\text{-}3b_g)$	6 el. in $(7\text{-}8a_g1\text{-}3a_u,1\text{-}2b_g)$	
Frozen core	$(1\text{-}2a_g,1\text{-}2b_u)$		
No. ref. config.	57	142	
No. contracted CSFs	1254162	1679787	3294112
No. uncontracted CSFs	42408920	59263550	144758424

[a] not including the d functions on the hydrogen atoms

Table 5. Details of HOCO benchmark calculation. The calculation was carried out using the C_s symmetry group and used the complete active space.

test case	H
Basis set	aug-cc-pVTZ
Size of integral file	702 Mb
No. orbitals	$(106a',55a'')$
Reference space	11 el. in $(7\text{-}12a',1\text{-}3a'')$
Frozen core	$(1\text{-}3a')$
No. ref. config.	3048
No. contracted CSFs	10100454
No. uncontracted CSFs	1243997054

EXAMPLE APPLICATIONS AND DEMONSTRATION OF SCALING

We have carried out several different benchmark calculations, the details of which are presented in Tables 2, 3, 4 and 5, on both SP2 and Cray T3D/T3E MPP machines. Table 6 gives an approximate guide to the capabilities of the machines used. Of course, the processor speeds and bandwidths etc. relate to the theoretical peak performance and are unlikely to be achieved in any scientifically productive code. (It should be noted that while the characteristics of the T3D/T3E are quite universal, there are several different varieties of SP2s.) Not reported in this table is the memory bandwidth; this varies depending on the location of the data to be fetched, e.g. whether it is in the cache or not. The memory bandwidth of the SP2 is greater than that of T3D/T3E, so that it is easier to obtain a higher percentage of the peak performance on this machine than on the others. This then reinforces the difficulties associated with obtaining very good speed-ups on this machine; the large ratio of the processor speed to the

Table 6. Characterization of MPP computers used for benchmark calculations.

	No. of processors	Processor speed	Processor memory	bandwidth	latency
Daresbury SP2	26	480 Mflops/s	256 Mb	100 Mb/s	20 μs
Edinburgh T3D	512	150 Mflops/s	64 Mb	120 Mb/s	6 μs
Cineca T3E-300	128	600 Mflops/s	128 Mb	480 Mb/s	4 μs
NERSC T3E-900	512	900 Mflops/s	256 Mb	480 Mb/s	4 μs

Table 7. Total calculation times for test cases E and F on T3D.

number of processors	test case E	test case F
32	1514 s	—
64	911 s	1109 s
128	626 s	742 s
256	—	657 s

Table 8. Total calculation times for test cases F, G and H on T3E-300.

number of processors	test case F	test case G	test case H
32	545 s	1376 s	6250 s
64	370 s	833 s	2985 s
128	297 s	595 s	1666 s

interprocessor communication bandwidth together with the rather high latency make efficient programming of this machine a complex task. Tables 7, 8, 9 and 10 present the timings of these test cases, though as an attempt is made to do the calculations with a variety of number of processors, these test cases by no means approach the limit of the size of calculations which are possible, particularly on the T3E-900, which has a relative large amount of memory on each processor.

The results shown in Table 7 obviously do not demonstrate linear scaling, although they suggest that more than 99% of the CPU work is being done in parallel, and that the parallel efficiency on 32 processors is over 80%, around 65% on 64 processors and falling to just below 50% on 128 processors. Table 8 shows that the results for the same test case (F) on the T3E-300 scale slightly less well, and Table 9 for the T3E-900 demonstrates a further deterioration. This is easilier understood if we consider the considerable increase in the speed of processors from the T3D to the T3E (and seen clearly in the timings, say on 64 processors, for this test case), which although also accompanied by an increase in bandwidth and decrease in latency, are not quite matched by it. On the T3E-300 the results of the larger test case G, show similar scaling to test case F on the T3D, while those on the T3E-900 again show slightly worse scaling.

It is test case H, which has over one billion uncontracted configurations and is probably more characteristic of calculations for which the MPP ICMRCI is designed, that starts to display good speed-ups, particularly on the T3E-300. Unfortunately, it is not possible to run this test case on the T3D, due to the quite small amounts of memory on each processor. Although a great deal of data is distributed in the program, certain quantities, such as configuration lists and the matrices required to calculate the coupling coefficients, must be held in local memory in order that the program can run efficiently.

Table 9. Total calculation times for test cases F, G and H on T3E-900.

number of processors	test case F	test case G	test case H
16	602 s	1420 s	6113 s
32	357 s	824 s	3297 s
64	264 s	542 s	1844 s
128	—	414 s	1225 s

Table 10. Total calculation times for test cases C and D on the SP2.

number of processors	test case C	test case D
1	3950 s	14885 s
2	2624 s	8586 s
4	1583 s	4913 s
8	839 s	2736 s

The results for the SP2 shown in Table 10 are certainly not as good as those on the T3D/T3E. The rather limited number of processors available (8 being the maximum) to run one job, makes the selection of a suitable test case difficult.

CONCLUSION

We have presented a fully parallel version of the MOLPRO ICMRCI program. We have demonstrated that it is possible to obtain good speed ups on large problems, despite the fact there has been no major restructuring of the existing complex code, without sacrificing program portability.

The first stage in the parallelisation involved the substitution of the fairly large amounts of disk i/o, by the reading and writing of data to global arrays, using the GA subroutine libraries. The next stage of the parallelisation, in which the work was partitioned between the processors, has been carried out at the subroutine level. Several different strategies have been used, ranging from a coarse-grain decomposition of the trial vector, to a much finer-grain decomposition over the all external two-electron integrals, or intermediate configurations.

The resulting parallel program cannot be simply categorized as being either a replicated data or a distributed data code. While a large proportion of the data is distributed over the processors via the use of the global arrays, within each separate set of tasks, there are cases where some of the data is distributed, but other cases where it is not.

ACKNOWLEDGEMENTS

This work has been carried out under the auspices of the Chemical Reactions and Energy Exchange Processes HPCI Grand Challenge Consortium, supported by EPSRC (grant number GR/K41656). The financial support of the EEC as part of the TMR network "Potential Energy Surfaces for Spectroscopy and Dynamics", contract No. FMRX-CT96-088 (DG 12 – BIUO) is acknowledged. PJK and AJD are grateful to R.J. Harrison, H.-J. Werner and R.J. Allan for helpful discussions. The practical assistance of Drs. R.J. Allan and G.G. Balint-Kurti is acknowledged. This research was performed in part using the Molecular Science Computing Facility in the Environmental Molecular Sciences Laboratory at the Pacific Northwest National Laboratory (PNNL).

REFERENCES

1.   A.J. van der Steen, *Overview of recent Supercomuters*, National Computing Facilities Foundation, The Hague, The Netherlands (1997)

2.   J.J. Dongarra and H. Meuer and E. Stohmaier, The 1995 TOP500 Report, *Supercomputer*, 12:1 (1996)
      Top500 list at, *http://parallel.rz.uni-mannheim.de/top500.html* .
3.   R.J. Harrison and R. Shepard, *Ann. Rev. Phys. Chem.*, 45:623 (1994)
4.   M.F. Guest and P. Sherwood and J.H. van Lenthe, *Theor. Chim. Acta.*, 84:423, (1993).
5.   M.F. Guest and R.J. Harrison and J.H. van Lenthe and L.C.H. Corler, *Theor. Chim. Acta.*, 71:117 (1987).
6.   M. Schüler and T. Konvar and H. Lischka and R. Shepard and R.J. Harrison, *Theor. Chim. Acta.*, 84:489 (1993).
7.   H. Dachsel and H. Lischka and R. Shepard and J. Nieplocha and R.J. Harrison, *J. Comp. Chem.*, 18:430 (1997).
8.   R. Kobayashi and A.P. Rendell, *Chem. Phys. Lett.*, 265:1 (1997).
9.   W. Meyer, in: *Modern Theoretical Chemistry*, H.F. Schaefer III ed., Plenum Publishing Company, New York, (1977)
10.   H.-J. Werner and E.-A. Reinsch, *J. Chem. Phys.*, 76:3144 (1982)
11.   H.-J. Werner and P.J. Knowles, *J. Chem. Phys.*, 89:5803 (1988).
12.   P.J. Knowles and H.-J. Werner, *Chem. Phys. Lett.*, 145:514 (1988).
13.   H.-J. Werner, in: *Ab Initio Methods in Quantum Chemistry II*, K.P. Lawley ed., John Wiley and Sons (1987).
14.   P.J. Knowles and H.-J. Werner, Internally Contracted Multiconfiguration Reference Configuration Interaction Calculations for Excited States, *Theor. Chim. Acta.*, 84:95-103 (1992)
15.   MOLPRO is a package of *ab initio* programs written by H.-J. Werner and P.J. Knowles, with contributions from R.D. Amos, A. Berning, D.L. Cooper, M.J.O. Deegan, A.J. Dobbyn, F. Eckert, C. Hampel, T. Leininger, R. Lindh, A.W. Lloyd, W. Meyer, M.E. Mura, A. Nicklass, P. Palmieri, K. Peterson, R. Pitzer, P. Pulay, G. Rauhut, M. Schütz, H. Stoll, A.J. Stone, and T. Thorsteinsson.
16.   J. Nieplocha and R.J. Harrison and R.J. Littlefield, in: *Proceedings of Supercomputing 1994*, IEEE Computer Society Press, Washington, DC, (1994).
17.   R.J. Harrison, *Int. J. Quantum Chem.*, 40:847 (1991).
18.   B. Roos and P. Taylor and P.E.M. Siegbahn, *Chem. Phys.*, 48:157 (1980).
19.   H.-J. Werner and P.J. Knowles, *J. Chem. Phys.*, 82:5053 (1985).
20.   P.J. Knowles and H.-J. Werner, *Chem. Phys. Lett.*, 115:259 (1985).
21.   E.R. Davidson, *J. Chem. Phys.*, 17:87 (1975).
22.   *MPI: A message passing interface standard*, Message Passing Interface Forum, University of Tennesee, Knoxville, Tennesee, USA (June 12, 1995)

MOLECULAR PROPERTIES FROM FIRST PRINCIPLES

C.J. Adam, S.J. Clark, G.J. Ackland and J. Crain

Department of Physics and Astronomy, The University of Edinburgh,
Edinburgh EH9 3JZ, U.K.

ABSTRACT

The purpose of this paper is to illustrate the application of first principles methods using pseudopotentials, density functional theory and plane wave basis sets, to calculate physical properties of large-scale organic molecules. It has been recently demonstrated that these techniques, which have been adapted from electronic structure calculations on periodic solids, can predict successfully molecular properties of molecules such as intra-molecular potentials and molecular dipoles. Until now, unfavourable system size scaling has precluded the application of conventional computational methods to large-scale molecules.

INTRODUCTION

The development of accurate first principles methods for the determination of molecular properties has been an enduring challenge in computational chemistry. The fundamental quantum-mechanical equations that govern the behaviour of electrons and atomic nuclei are well known and in principle allow us to calculate the energetics of the material without the use of adjustable or empirical parameters. However as a typical molecule represents a large assembly of interacting electrons and nuclei, the exact ground state wavefunction depends on the coordinates of all the electrons together in an inconceivably complicated way. Theoretical approaches must therefore include approximations and incur a consequent loss of accuracy. The degree of this depends on the extent on which the approximations incoporate all the physical features involved at the quantum mechanical level. Their implementation and scaling factors also determines the relative computational efficiency of the approach and hence the the maximum size of molecular systems that can be feasibly investigated.

Density functional theory (DFT) is now gaining wide acceptance amongst computational chemists. There is mounting evidence that suggests DFT can give remarkably accurate results for molecular structures and electronic properties and is able to describe a variety of molecular bonds. However the significant advantage of DFT is that for increasing system size, the computational effort scales between N^2 and N^3 while correlated Hartree-Fock methods giving results of similar quality scale at least with a

Figure 1. Calculated valence charge distribution for the 38 atom liquid crystal molecule 4-4'
pentyl-cyanobiphenyl (5CB) as reconstructed from the electron wavefunctions corresponding to
occupied bands. The two phenyl groups, hydrocarbon chain and cyano end group are clearly visible.
The charge density is evaluated on a discrete grid with a basis set containing 2.5 million plane waves.
The entire calculation (including structural relaxation) took 3 hours on the Edinburgh 512 node
CRAY-T3D.

fifth power. These features of DFT in conjunction with the completeness properties of
plane wave basis sets, opens exciting opportunities for the study of large and complex
molecular systems, previously considered out of reach for ab initio calculations.

Nearly all physical properties of molecules are either related to total energies or
differences between total energies. Once the total energy as a function of the positions
of the atoms is known, many other quantities such as equilibrium structure, bonding,
conformational dynamics, inter-molecular potentials, molecular multipoles, polarisability, as well as vibrational characteristics can be determined. In general, it is a difficult
task to extract routinely the relevant molecular properties from experiment, although
a combination of diffraction, optical spectroscopy and NMR can provide some quantitative information [1, 2, 3, 4]. Reliable first-principles computer simulations therefore
provide an attractive and versatile alternative route to extracting molecular properties and afford the opportunity to investigate molecules yet to be synthesized. Such
calculations require efficient *ab initio* algorithms, such as those based on Kohn-Sham
DFT, and an understanding of the relationship between the respective molecular properties. For example in a calculation of the electrostatic properties attention must be
also paid to the conformational dynamics as geometrical shape will greatly influence
the electronic distribution and molecular multipoles.

The paper is organised as follows. The basic theory and computational implementation is first described, with particular emphasis on solving for the molecular
electronic structure (an example of which is shown in Figure 1). Then the calculation
of vibrational and intra-molecular molecular properties is then discussed.

METHOD

First Principles Techniques

As already discussed, in order to accurately determine the electronic structure
of a molecular system it is necessary to use a first principles method which is both
efficient and includes all the physical features present. The approximations used in this
work include Kohn-Sham DFT to model electron-electron interaction, pseudopotential
theories to model the electron-ion interactions, supercells to enforce periodic boundary
conditions and iterative minimisation techniques to relax the electronic co-ordinates.

Density functional theory as developed by Kohn and Sham [5] simplifies the many body electron problem by mapping a system of strongly interacting electrons onto a series of single electron equations where each particle is moving in an effective non-local potential. The many body effects of exchange and correlation are included via density approximations of which the generalised gradient [6] is the most recent and successful. However in addition, two further simplifications are employed to make the calculation of the ground state wavefunction more manageable. The core electrons and the strong nuclear potentials are removed and replaced with pseudopotentials which act on a set of pseudowavefunctions. The physical properties of molecules are primarily dependent on the valence electrons and an explicit calculation of the tightly bound core electrons is unnecessary. The pseudopotential is much weaker than the all electron potential and so requires fewer basis functions to describe its properties. This approximation allows the use of a plane wave basis set which is poorly suited to describing the rapidly oscillating electron wavefunctions in the core region.

A plane wave basis set has some important advantages in comparison to the conventional localised orbitals typically used. The convergence properties towards completeness can be easily tested and it can be used for all molecules irrespective of the consituent atoms with equally accurate results. Plane waves also facilitate the use supercells, the periodic boundary conditions which allows the expansion of the plane waves in accordance with Bloch's Theorem. The calculation is performed with the molecule is placed in the centre of a periodically repeating supercell, the size of which must be taken to be sufficiently large so as to ensure the molecules are effectively isolated from each other. The ground state total energy is then found using efficient iterative procedures such as pre-conditioned conjugate gradients.

Codes and Algorithms

The Cambridge-Edinburgh Total Energy Package (CETEP code)[7, 8] has been modified and implemented on the massively parallel Cray T3D at the Edinburgh Parallel Computing Centre. This uses *ab initio* pseudopotentials (in Kleinman -Bylander form) [9] and the generalised gradient approximation [6], to the exchange and correlation energy. More details of this method as applied to molecules and the modifications to CETEP can be found in a previous paper[10]. The *ab initio* pseudopotentials used for the constituent atoms in the molecules under investigation are determined according to the scheme of Lin *et al* [11]. The minimisation of energy is carried out from an initial random assignment of the plane wave co-efficients which are distributed over different processors.

CALCULATION OF MOLECULAR PROPERTIES

Optimisation of Molecular Structures

The Hellmann-Feynman force on each atom in the molecule is calculated and the atoms are moved under the influence of these forces until no force component exceeds some tolerance. If structure optimisation is all that is required a typical value of the residual forces is 0.1 eV/Å. The calculation of vibrational frequencies, however, demands convergence of the forces to much greater precision, typically 0.01 eV/ Å. For these calculations, molecular point group symmetry is not enforced. Usually there will be a wide variety of different conformations that a molecule can sample due to rotations of molecular segments about bond directions. The conjugate gradients minimisation

Table 1. Calculated and experimental geometries for Benzene.

Method	r(CC)(Å)	r(CH)(Å)	Ref.
DFT-PW	1.396	1.089	present work
CCSD	1.393	1.082	[20]
MP2	1.398	1.086	[17]
MP2	1.390	1.079	[18]
MP2	1.390	1.080	[19]
exp	1.393-1.398	1.081-1.090	[18]

routine which is employed to optimise the geometry will not necessarily locate the global minimum of this conformational energy landscape[12].

Benzene Benzene has been studied in considerable detail by a variety of quantum chemical methods and it is therefore an excellent test case for this simulation method. Benzene exhibits regular hexagonal planar symmetry with six carbon atoms joined by σ bonds and six remaining p-orbitals which overlap to form a delocalised π bond over all six carbon atoms.

The initial geometry of the molecule placed in the supercell, before optimisation was performed, was a configuration in which the C-C and C-H bonds were 1.33 Å and 1.00 Å, respectively. The total energy convergence was tested with respect to supercell size to ensure the molecule was isolated. In this case the dimensions 10 Å× 10 Å× 10 Å were adequate, giving a separation of at least 6Å between any pair of atoms in neighbouring cells. The plane wave energy cutoff was set at 700 eV for which energies were converged to better than 0.0001 eV/atom. The structural optimisation proceeded until the residual forces on the ions were below 0.01 eV/Å. This gave a relaxed structure in which the C-C and C-H bondlengths were 1.396Å and 1.089Å, respectively (as compared to experimental bondlengths of 1.393 - 1.398 Å and 1.081-1.090 Å respectively [13]). Note that this is a 0K calculation, therefore temperature effects such as anharmonic bond-lengthening at higher temperatures are not included. A summary of these results and those obtained by other methods are given in Table 1. To within our numerical accuracy, all C-C bonds are of equal length as are all C-H bonds although no external constraint is applied to enforce this symmetry. This relaxed structure was then used to determine the vibrational properties according to the method discussed in the next section.

Vibrational Modes

Once the equilibrium structure of a molecule has been determined, the dynamical properties can be investigated by displacing each atom in the molecule and calculating the resultant forces on all atoms. The magnitude of the displacements must be chosen such that they are small enough to be within the harmonic regime but large enough to give rise to measurable restoring forces. The effect of anharmonicity can be greatly reduced by making positive and negative displacements about the equilibrium positions and averaging the corresponding restoring forces. This procedure can be used to construct a dynamical matrix of dimension $3N \times 3N$ where N is the number of atoms in the molecule. Subsequent diagonalisation of this matrix gives directly all $3N - 6$ normal mode vibrational frequencies and their associated eigenvector displacement patterns [14, 15]. The results of the calculation on benzene for the vibrational properties are summarised in Table 2, along with comparison to experimental data and other calculations.

Table 2. This table shows the experimentally observed frequencies of benzene and those from the DFT plane wave calculations (the units are in cm^{-1}). Also shown, are results from several different commonly used quantum chemistry methods and basis sets. Agreement with experiment to better than 2-3% is obtained in most cases. Some quantum chemistry methods do not calculate all the frequencies. Also calculated are the phonon eigenvectors from which mode symmetries have been assigned as shown. Results from the most commonly used HF-MP2 method are shown with progressive 'improvements' to localised basis set. Note that, for example, the B_{2u} mode at around 1300cm^{-1} actually gets worse with increased sophistication. This indicates that, even for the most 'reliable' localised basis sets, convergence is not being achieved.

Sym.	Obs. [a] (cm^{-1})	DFT [b] PW	HF [c] (631G)	HF [c] (631G*)	HF [a] (6311++)	MP2 [c] (631G)	MP2 [c] (631G(dp))	MP2 [d] (TZ2P+f)	CCSD [e] (cc-VTZ)	DFT [f] (631G*)	DFT [g] (TZ)
A_{1g}	994	1006	1088	1084	1070	1000	1015	1018	1012	1051	1004
A_{1g}	3191	3260	3392	3390	3352	3213	3240	3242	3228	3114	3101
A_{2g}	1367	1351	1539	1508	1491	1412	1367	1374	1391	1318	1314
E_{2g}	608	597	689	666	662	632	610	608	613	606	602
E_{2g}	3174	3208	3357	3359	3323	3179	3215	3217	3204	3096	3065
E_{2g}	1607	1607	1769	1791	1774	1626	1645	1637	1672	1645	1610
E_{2g}	1178	1181	1320	1294	1280	1234	1199	1195	1207	1157	1150
B_{1u}	1010	1008	1138	1097	1090	1046	1009	1039	1025	1017	993
B_{1u}	3174	3199	3346	3348	3311	3173	3204	3218	3189	3092	3065
B_{2u}	1309	1339	1385	1352	1335	1377	1451	1461	1304	1411	1379
B_{2u}	1150	1120	1271	1197	1180	1229	1173	1178	1166	1129	1125
E_{1u}	1038	1042	1151	1142	1128	1073	1063	1074	1071	1043	1039
E_{1u}	1494	1476	1662	1652	1629	1527	1509	1515	1528	1475	1462
E_{1u}	3181	3224	3377	3371	3341	3203	3231	3238	3221	3108	3091
B_{2g}	990	993	-	-	1118	-	-	-	-	-	985
E_{2u}	967	962	-	-	1099	-	-	-	-	-	952
B_{2g}	707	722	-	-	769	-	-	-	-	-	713
E_{2u}	398	399	-	-	448	-	-	-	-	-	399
A_{2u}	674	667	-	-	755	-	-	-	-	-	664
E_{1g}	847	833	-	-	954	-	-	-	-	-	830
Av. % Dev. to B_{2g}		3.07	10.69	8.97	7.86	4.47	3.90	4.37	3.21	4.27	4.25
Av. % Dev. for all modes		2.44	-	-	9.13	-	-	-	-	-	3.31

[a] [13] [b] [15] [c] [16] [d] [19] [e] [20] [f] [21] [g] [22]

253

Intra-molecular degrees of freedom

Molecular flexibility depends on chemical structure and can strongly influence molecular properties such as vibrational modes and multipole moments. The degree of flexibility is sensitive to several intra-molecular degrees of freedom, of which bond stretching, bond bending and internal rotation are the main contributors. First principles calculations provide a way to investigate these degrees of freedom individually or the molecular flexibility as a whole. The results can be used to both improve the accuracy of molecular mechanics methods, which rely on parameterised force fields and give insight into the relationship between molecular properties such as dipole moment and internal-rotation.

Internal-Rotation and Molecular Dipole

Once the relaxed electronic structure for a given molecular conformation has been found, the calculation of the molecular dipole moment can be made. It can be determined by taking the vector difference between the locations of the centroids of the electronic and ionic charge distributions. The valence charge density is computed on a discrete grid and it is obtained directly from the wavefunctions. A fuller discussion of the method can be found in a previous paper[10]. The aim here is to illustrate the strong relationship that can exist between internal-rotation and molecular dipoles using the example of 2-2' difluorobiphenyl.

The molecule 2-2' difluorobiphenyl consists of two phenyl groups with a fluorine atom replacing one of the hydrogens on each group as shown in Fig. 2. The addition of fluorine atoms on the lateral hydrogen sites creates highly dipolar regions situated on carbon fluorine bonds. Consequently the strong electrostatic interactions, arising from accumulation of charge on the fluorine atoms and their increased steric repulsion, can significantly affect the internal dynamics and overall molecular shape. Therefore the torsional potential for rotation about the inter-ring C-C bond will be a sensitive function of torsional angle, since this governs the relative positions of the fluorine atoms. Also the highly electronegative fluorine will attract electrons from the delocalised π-orbitals which may reduce the energy cost of distortion away from the 0° conformation. The conformationally dependent dipole moment magnitude and the torsional potential are shown in Figure 3.

Transferability of Internal-Rotation Potentials

Parameterisation of molecular mechanics force fields requires an understanding of the transferability of fragment potentials to real molecules. If all the natural distances, bond angles and torsional potentials in a force field are chosen to reproduce a particular structure then that structure will be reproduced exactly. However these parameters are not guaranteed to reproduce the structures of related molecules. Therefore implicit in the development of force field parameters is the ability to define the criteria required for transferability in between different molecular systems and to quantify the mechanisms which reduce it. Ab initio calculations are typically the only approaches which can provide information about detailed shapes of the potential energy curves[23].

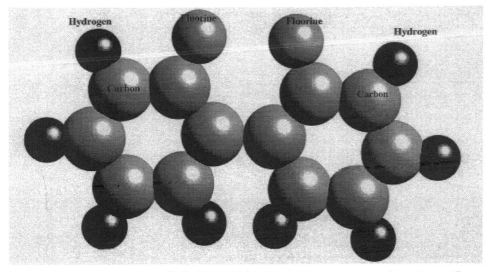

Figure 2. Illustration of 2-2' difluorobiphenyl in its lowest energy conformation at 0^0.

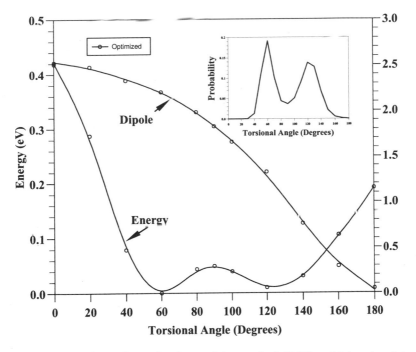

Figure 3. This Figure shows the torsional potential curve for 2-2' difluorobiphenyl as determined using full molecular geometry optimisation for each conformer. Calculations were performed at each 20^0 interval from 0^0 (cis) to 180^0 (trans). The molecular dipole as a function of torsional angle is also shown on the same plot. The curve (inset) shows the relative probabilities of the conformations and the slight preference of the 60^0 to the 120^0 minimum at 300K.

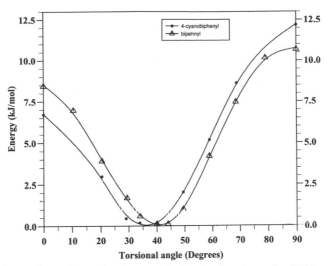

Figure 4. This figure shows the torsional potentials for rotation about the C-C inter-ring bond for 4-cyanobiphenyl and biphenyl. For each torsional angle, the geometry of the remaining degrees of freedom was optimised to obtain accurate total energies. For example the bondlength of the inter-ring C-C bond in biphenyl, changes from 1.472 Å in the coplanar conformation to 1.485 Å in the perpendicular conformation. The bondlength in 4-cyanobiphenyl follows a similar trend and only differs from the biphenyl bondlength by a few thousands of an Å. However, although the variations in structure between biphenyl and 4-cyanobiphenyl were slight, the potential for 4-cyanobiphenyl is markedly different from the torsional potential of biphenyl.

Biphenyl and 4-cyanobiphenyl

Biphenyl and 4-cyanobiphenyl are similar in structure as they both have a high degree of electronic conjugation due to the two phenyl groups. The only difference is 4-cyanobiphenyl has a highly electronegative cyano group in place of one of the hydrogens furthest from the inter-ring bond. A typical molecular mechanics force field would calculate identical potentials for rotation about this bond for both systems and hence assume transferability. However as Figure 4 shows, the torsional potentials for both systems as calculated from first principles are not identical, hence reducing transferability.

In general the shape of the potentials are governed by a trade-off between steric repulsion and increased conjugation, which is maximised when the rings are coplanar. Electronic charge transfer towards the highly electronegative CN group results in subtle bonding and structural effects which influence with this trade-off. These effects can only be quatified by first principles calculations. They can change both the barrier heights and position of the minimum. The minimum in biphenyl is found to occur at 42.0° which is in good agreement with other *ab initio* work [24, 25] and the value of 44.4 ± 1.2° from gas phase electron diffraction experiments [3, 26]. The position of the torsional potential minimum for 4-cyanobiphenyl is at 37.0°, which also agrees well with NMR studies on the condensed phase [27].

Transferability Of Molecular Properties From The Gas To Condensed Phase

One of the most important practical questions concerning such molecular properties is the degree to which they are relevant to environments which differ from those in which they have been determined. It should be emphasised that this issue is relevant for experiment, theory and computer simulation. In general there may be a variety of mechanisms which compromise transferability. For example, the intramolecular dynamics may be influenced by charge transfer between functional groups or electrostatic interactions and steric effects. In particular it has been shown that molecular models lacking polarisation can lead to significant errors concerning important issues such as conformational equilibria[28]. In the condensed phase there is also the prospect of coupling between molecular structure and orientational order which will also influence flexibility. However, as yet, reliable models which can accurately predict the degree of coupling in the condensed phase are still in development [29, 30].

Work progressing in Edinburgh is currently trying to address this problem by developing a charge density model for solute ordering in nematic liquid crystals. The model is based on the decomposition of a mean field orientational potential according to the contributions from individual surface elements. The underlying assumption of the model is that each surface element contributes independently to the orientational potential. The surface elements are used to calculate a surface integral which contains parameters to describe the orientational properties of the system investigated[31]. Essential to the success of such a model is the realistic representation of the molecular surface accessible to the solvent. The electronic structure calculations performed with CETEP provide a highly reliable and accurate molecular surface. The surface is first parameterised by 3D Contouring the charge distribution and then implemented in a calculation of the surface integral[32].

CONCLUSIONS

Until now a lack of detailed knowledge about molecular properties such as the intra-molecular degrees of freedom, has lead to a relatively poor understanding of the relationship between the microscopic properties of individual molecules and the observable macroscopic properties of the bulk material. It is well known, for example, that subtle alterations in molecular flexibility can have profound effects on the stability and properties of liquid crystalline phases[33]. Although a number of empirical rules have been formulated, fundamental understanding of this relationship has been hindered by a lack of information about these molecular properties.

This paper have shown that the DFT method is capable of determining electronic, structural and dynamic properties of large-scale molecules to a level of accuracy previously attainable only for very small molecules. Work is progressing on calculating other molecular properties such as polarisability and also the transferability of these properties between the condensed and gas phases.

ACKNOWLEDGEMENTS

We are grateful to the Edinburgh Parallel Computing Centre for access to the T3D. We are also grateful to Mark Wilson of Durham University, Alberta Ferrarini of the Padua University. CJA would like to thank EPSRC.

REFERENCES

1. J.W. Emsley, G.De Luca, G. Celebre and M. Longeri,*Liq. Cryst.* 20:569 (1996).
2. J.W. Emsley, M.I.C. Furby and G.Deluca, *Liq. Cryst.* 21:877 (1996).
3. A. Almenningen, O. Bastiansen, L. Fernholt, B.N. Cyvin, S.J. Cyvin and S. Smadal, *J. Mol. Struct.* 128:59 (1985).

4. H.C. Hsueh, H. Vass, F.N. Pu, S.J. Clark, C.K. Poon and J. Crain, *Europhys. Lett* 38:107 (1997).
5. W. Kohn and L.J. Sham,*Phys. Rev.* 140:A1133 (1965).
6. J.P. Perdew, J.A. Chevary, S.H. Vosko, K.A. Jackson, M.R. Pederson, D.J. Singh and C. Fiolhais, *Phys. Rev. B* 46:6671 (1992).
7. L.J. Clarke, I. Stich and M.C. Payne, *Comput. Phys. Commun.* 72:14 (1992).
8. J.A. White, and D.M. Bird, *Phys. Rev. B* 50:4954 (1994).
9. L. Kleinman and D.M. Bylander,*Phys. Rev. Lett.* 48:1425 (1982).
10. C.J. Adam, S.J. Clark, G.J. Ackland, and J. Crain *Phys. Rev. E*, 55:5641 (1997).
11. J.S. Lin, A. Qteish, M.C. Payne and V. Heine, *Phys. Rev. B* 47:4174 (1993).
12. S.J. Clark, C.J. Adam, D.J. Cleaver and J. Crain. *Liq. Cryst..*22:477 (1997).
13. L. Goodman, A.G. Ozkabak and S.N. Thaker, *J. Phys. Chem.*95:9044 (1991).
14. J.C. Decius and R.M. Hexter, *Molecular Vibrations in Crystals* McGraw-Hill, Inc., USA (1997).
15. S.J. Clark, C.J. Adam, G.J Ackland, J. White and J. Crain, *Liq. Cryst.* 22:469 (1997).
16. H. Guo and M. Karplus, *J. Chem. Phys.* 89:4235 (1988).
17. L. Goodman, A.G. Ozkabak and S N. Thakur, *J. Phys. Chem.* 95:9044 (1991).
18. P.E. Maslen, N.C. Handy, R.D. Amos and D. Jayatilaka, *J. Chem Phys.* 97:4233 (1992).
19. N.C. Handy, P.E. Maslen, R.D. Amos, J.S. Andrews, C.W. Murray and G.J. Laming, *Chem. Phys. Lett.* 197:506 (1992).
20. L.J. Brenner, J. Senekowtsch and R.E. Wyatt, *Chem. Phys. Lett.* 215:63 (1993).
21. E. Albertozzi and F. Zerbetto, *Chem. Phys.* 164:91 (1992).
22. A. Bérces and T. Ziegler, *J. Chem. Phys.* 99:11417 (1995).
23. C.J. Adam, S.J. Clark, M.R. Wilson and J. Crain, *Mol. Phys.* (1997) in press.
24. M. Rubio, M. Merchan and E. Orti, *Theor. Chim. Acta* 91:17 (1995).
25. S. Tsuzuki and K. Tanabe, *J. Phys. Chem* 95:139 (1991).
26. O. Bastiansen and S. Samdal,*J. Mol. Struct.* 128:115 (1985).
27. J.W. Emsley, T.J. Horne, G. Celebre, G. Deluca, and M. Longeri, *J. Chem. Soc., Faraday Trans.* 88:1679 (1992).
28. W.L. Jorgensen, N.A. McDonald, M. Selmi and P.R. Rablen, *J. Am. Chem. Soc.* 117:11809 (1985).
29. J.W. Emsley, G.R. Luckhurst, and C.P. Stockley, *Proc. Roy. Soc. Lond.* A381:139 (1982).
30. M.R. Wilson, *Liq. Cryst.* 21:437 (1986).
31. A. Ferrarini, G.J. Moro, P.L. Nordio, and G.R. Luckhurst, *Molec. Phys.* 77:1 (1992).
32. C.J. Adam, A. Ferrarini, M.R. Wilson, G.J. Ackland and J. Crain,*Phys. Rev. E* (1998) submitted.
33. S.M. Kelly, *Liquid Crystals* 20:493 (1996).

MASSIVE PARALLELISM: THE HARDWARE FOR COMPUTATIONAL CHEMISTRY?

M.F. Guest [1], P. Sherwood [1] and J.A Nichols [2]

[1] Department for Computation and Information,
CCLRC Daresbury Laboratory, Daresbury,
Warrington WA4 4AD, Cheshire, UK
[2] High Performance Computational Chemistry Group,
Environmental Molecular Sciences Laboratory,
Pacific Northwest National Laboratory,
PO Box 999, Mail Stop K1-90, Richland, WA. 99352, USA

INTRODUCTION

Computational chemistry covers a wide spectrum of activities ranging from quantum mechanical calculations of the electronic structure of molecules, to classical mechanical simulations of the dynamical properties of many-atom systems, to the mapping of both structure-activity relationships and reaction synthesis steps. Although chemical theory and insight play important roles in this work, the prediction of physical observables is almost invariably bounded by the available computer capacity.

The potential of massively parallel computers, with hundreds to thousands of processors (MPPs), to significantly outpace conventional supercomputers in both capacity and price-performance has long been recognised. While increases in raw computing power alone will greatly expand the range of problems that can be treated by theoretical chemistry methods, it is now apparent that a significant investment in new algorithms is needed to fully exploit this potential. Merely porting presently available software to these parallel computers does not provide the efficiency required, with many existing parallel applications showing a deterioration in performance as greater numbers of processors are used. New algorithms must be developed that exhibit parallel scalability (i.e., show a near linear increase in performance with the processor count). Such improvements promise substantial scientific and commercial gains by increasing the number and complexity of chemical systems that can be studied.

Organised in five sections, this paper aims to analyse just how far the potential of MPPs has been realised, and what is needed in terms of software development to provide the appropriate applications. Following a consideration of the computational requirements for molecular computations as a function of the level of theory and required accuracy (second section), we present an analysis of the cost effectiveness of current top-end MPP solutions against both desk-top and mid-range departmental resources.

We identify two distinct categories of application, *Grand Challenge* and *Throughput*. The former is taken to include those chemical systems of sufficient complexity that neither desk top nor mid-range resources provide a computationally viable solution. *Throughput* applications are defined to include chemical systems that are amenable to treatment by more modest resources, but where time to solution demands a top-end capability. We define an MPP performance metric applicable to both application categories that emerges from this consideration of price-performance.

THe third section considers *Grand Challenge* applications, and the need to address parallel scalability from the outset in developing molecular modelling software applications. In assessing the algorithmic impact of this requirement, we overview the development of the NWChem package at the Pacific Northwest National Laboratory as an example of the provision of capabilities designed to facilitate "new science", but in cost effective fashion. The fourth section turns to the treatment of *Throughput* applications, and considers the efforts required to migrate existing applications to MPP platforms so as to achieve the performance metric of the second section. We consider the parallel development of the GAMESS–UK package [28] at the CCLRC Daresbury Laboratory as a typical example of what is achievable from this more modest approach to exploiting MPP platforms. We present performance figures for a number of *Grand Challenge* and *Throughput* calculations on representative molecular systems using a variety of methods routinely employed by the computational chemist, including self-consistent field Density Functional Theory (DFT) and Second-order Moller Plesset Perturbation theory (MP2).

METHODS AND COST-PERFORMANCE ISSUES

The complexity of computational chemistry, and the demands that the discipline imposes on any computational resource, are illustrated in Figure 1 where the relative computing power required for molecular computations at four different levels of theory (and accuracy) are shown: configuration interaction (CI), Hartree-Fock (HF), density functional, and molecular dynamics. Each of these levels of theory formally scale from N^2 to N^6 with respect to computational requirements. While standard screening techniques which in essence ignore interactions for atoms far apart can bring these scalings down significantly, it is clear that the major challenge facing the computational chemist is the development of methods that exhibit superior scaling properties to those shown in Figure 1. We should not, of course, lose sight of this goal in our efforts to maximally exploit current methodology on present and future hardware.

In assessing alternative hardware platforms for computational chemistry, we need to quantify in some fashion the cost effectiveness of a top-end MPP solution, given that such a solution should be able to offer capabilities not possible on either desk top or mid-range resources i.e., a solution that will facilitate "new science", but in cost effective fashion. Table 1 compares the cost of three HPC solutions, associated with the desk-top, with a departmental or mid-range resource (e.g., a 16 processor Origin 2000 from Silicon Graphics, with 4 GByte RAM), and with a top-end MPP solution comprising perhaps 500 or 1000 nodes (e.g., an IBM SP P2SC/160 or Cray T3E/900, with 64-128 GByte RAM). We also consider the associated capabilities, defined in terms of CPU, memory and Input/Output (I/O) rate. In each case we have normalised these attributes relative to that available on the desk-top. We have assumed the most cost effective desk-top solution to be a 200 MHz PentiumPRO PC, and have costed that to include 256 MByte RAM and 9 GByte disk. Allowing for a factor of two in this, or for local variations in costs, does not change the final conclusion; the cost ratios of

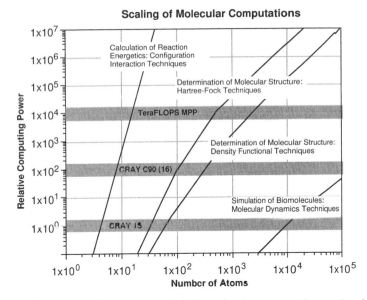

Scaling of Molecular Computations

Figure 1. The relative computing power required for molecular computations at four levels of theory. The formal scaling for configuration interaction, Hartree-Fock, density functional, and molecular dynamics is: N^6, N^4, N^3, and N^2, respectively. Standard screening techniques which in essence ignore interactions for atoms far apart can bring these scalings down significantly.

Table 1 lend little or no support to the repeated claims that MPP solutions are cost effective. This is clear when we compare the actual increase in capability as we move from desk-top to the top-end national resource. We see factors of 500/1000 increase in CPU capability, 250/500 increase in available memory, and 20/100 in achievable I/O bandwidth, each factor far below the associated cost ratios.

What is clear from this rather crude analysis is that scalability, with the exploitation of not only CPU, but also memory and I/O bandwidth in scalable fashion, is crucial to the twin goals of cost-effectiveness and the solving of *Grand Challenge* problems. Without that scalability, there is nothing to justify a top-end resource against, say, a number of SGI Origin 2000's; funding for the latter is often easier to find and provides the principle investigator with satisfaction of local ownership and control. Such mid-range provision would, we suspect, be the preferred route unless the top-end resource really does provide the opportunity for *Grand Challenge* science not possible at the departmental or desk-top level. So what level of scalability justifies a top-end resource? A pragmatic approach suggests that the latter should provide a number of

Table 1. Relative costs and Capabilities of three HPC solutions

Specification	Usage	Units of Cost	Hardware Attributes		
			CPU	Memory	I/O
Top-end MPP (T3E-P2SC/500-1000)	HPC community	3000	500-1000	250-500	20-100
mid-range SMP (Origin 2000/16)	Department	60	20	10-20	20-50
Desktop (200Mhz PentiumPRO)	Single User	1	1	1	1

groups with simultaneous access to at least a two-orders of magnitude improvement over the desk-top, and an order of magnitude improvement over computation available at the departmental level. Such comparisons should allow for the rapid evolution of desk-top and mid-range technology, with some form of technology refresh at the top-end enabling these factors of 100 and 10 to be sustained throughout the lifetime of the machine. Another pointer comes from considering time-to-solution for *Throughput* applications. Perhaps the longest viable desk-top run would be of the order of 1 month; a user is unlikely to dedicate his own machine for longer periods. Given a 100-fold performance improvement, such a run could be accomplished over-night.

Thus in our analysis of the parallel developments conducted to date, we will be looking for scalability and performance at the above levels. Furthermore, we require these levels to be available from some fraction of the top-end resource, rather than the whole machine; this would enable several groups to be simultaneously delivering that level of performance. If a given group has to wait a week to access that level, rather than overnight, the value of the top-end resource will rapidly diminish.

GRAND CHALLENGE APPLICATIONS; THE NWChem PROJECT

The background and role of computational simulations and modelling within the Environmental Molecular Sciences Laboratory (EMSL [1]) has been covered by Dr. David A. Dixon, and we refer the reader to his contribution in these proceedings for more detail. We note here the work of the High Performance Computational Chemistry (HPCC) group within the EMSL in developing a new generation of molecular modelling software to take full advantage of the parallel computing power of MPPs. In addition to the development of new algorithms that exhibit parallel scalability (i.e., show a near linear increase in performance with the number of processors), work has focused on creating the high-level data and control structures needed to make parallel programs easier to write, maintain, and extend. These efforts have culminated in a package called NWChem that includes a broad range of functionality; algorithms for HF self-consistent field (SCF) calculations [2, 3], a four-index transformation [5], several forms of second-order perturbation theory [6, 7], density functional theory(DFT), and molecular dynamics with classical, quantum mechanical, or mixed force-fields, are already in the code and others, such as multiconfiguration SCF (MCSCF) and higher-level correlated methods are under development. A wide range of properties are also available: gradients, Hessians, electrical multipoles, NMR shielding tensors, etc.

In outlining these developments, we focus below on the design philosophy, structure, and tools which make NWChem an effective environment for the development of computational chemistry applications. Although this focus has been on the efficient use of parallel computers in chemistry, almost all of the ideas and tools presented here are generally applicable. We provide, by way of example, an outline of the development and performance of a highly efficient and scalable algorithm for conducting SCF-DFT studies on molecular systems, illustrated through calculations of a zeolite fragment.

Design Philosophy and Software Development Tools

The philosophy and structure of NWChem evolved in response to experiences gained with other computational chemistry packages, and a number of constraints. Initial experiments led to the conclusion that while it was possible to modify existing software to implement parallel algorithms on a "proof of principle" basis, the effort involved in turning them into fully scalable (distributed rather than replicated data) codes was tantamount to writing the codes over from scratch. At the same time, as in most software development efforts, limitations were in effect on both personnel and time available to the development effort. The success of the project relied on having a system which was easy to program, and in particular one in which new parallel algorithms could be readily prototyped.

In order to meet the twin requirements of both *maintainability* (the ability to transition from locally developed parallel programming tools to vendor-supported, standard-based implementations) and *portability* (finding a programming model which can be used effectively across the variety of MPP platforms), ideas were adopted from the object-oriented (OO) style of software development [8]. Examples of the adoption of the basic principles of this methodology (abstraction, encapsulation, modularity, and hierarchy) have been presented elsewhere [4]. While there are OO programming languages, such as C++ [9], which allow the formal aspects of the methodology to be taken straight into the actual code, the project chose a compromise approach in the development of NWChem. Object-oriented ideas were used at the design stage, while the implementation remained a mixture of Fortran and C, languages much more familiar to current computational chemists than C++. Experience suggested that while it may take extra time, a thorough, careful design phase is probably the most important aspect of the OO methodology, and this effort is quickly recouped in easier implementation and fewer instances of modification of code after implementation. A design based on OO principles can be implemented in a non-OO programming language with only a small amount of discipline on the programmer's part.

A number of low-level tools were selected or developed to facilitate the development of NWChem. Although some were designed specifically to meet the needs of computational chemistry applications, they remain generally applicable. We focus attention below on the tools that target the development of distributed data applications and distributed linear algebra, noting that a discussion of the building blocks behind these tools, notably message passing and memory allocation, has appeared elsewhere [4].

Global Arrays. A vital consideration in using MPPs is how data are stored. So-called replicated-data schemes require that a copy of each data item in the program be stored on each processor, so that the size of the problem that can be handled is limited by the memory of a single processor. In distributed-data applications each processor holds only a part of the total data; in such cases, the problem size is limited only by the total memory of the machine, allowing much larger problems to be treated. Efforts at PNNL have focused on distributed data applications; relying on a replicated data implementation of the DFT module, for example, would limit the size of calculation to perhaps 1800 basis functions (see section 4), far short of the project's requirements of treating molecular systems with 10,000 basis functions.

No emerging standards for parallel programming languages (notably just High Performance Fortran (HPF-1)) provide extensive support for multiple instruction multiple data (MIMD) programming [11]. The only truly portable MIMD programming model is message passing, for which a standard interface is now established [10, 12]. It is, however, very difficult to develop applications with fully distributed data structures using the message-passing model [13, 14]. The shared-memory programming model offers increased flexibility and programming ease but is less portable and provides less control over the inter-processor transfer cost. What is needed is support for one-sided asynchronous access to data structures (here limited to one- and two-dimensional arrays) in the spirit of shared memory. With some effort this can be done portably [15]; in return for this investment, a much easier programming environment is achieved, speeding code development and improving extensibility and maintainability. A significant performance enhancement also results from increased asynchrony of execution of processes [16]; with a one-sided communication mechanism, where each process can access what it needs without explicit participation of another process, all processes can operate independently. This approach eliminates unnecessary synchronization and naturally leads to interleaving of computation and communication. Most programs contain multiple algorithms, some of which may naturally be task-parallel (e.g., Fock matrix construction), and others that may be efficiently and compactly expressed as data-parallel operations (e.g., evaluating the trace of a matrix product). Both types of parallelism must be efficiently supported. Consideration of the requirements of the SCF algorithm, the parallel COLUMBUS configuration interaction program [17], second order many-body perturbation theory [7] and parallel coupled-cluster methods [18] led to the design and implementation of the Global Array (GA) toolkit [15] to support one-sided asynchronous access to globally-addressable distributed one- and two-dimensional arrays.

This toolkit provides an efficient and portable "shared-memory" programming interface for distributed-memory computers. Each process in a MIMD parallel program can asynchronously access logical blocks of physically distributed matrices, without need for explicit co-operation by other processes. Unlike other shared-memory environments, the GA model exposes the programmer to the non-uniform memory access (NUMA) timing characteristics of the parallel computers and acknowledges that access to remote data is slower than to local data. From the user perspective, a global array can be used as if it were stored in shared memory, except that explicit library calls are required to access it. The information on the actual data distribution can be obtained and exploited whenever data locality is important. Each process is assumed to have fast access to some "local" portion of each distributed matrix, and slower access to the remaining "remote" portion. Remote data can be accessed through operations like "get", "put" or "accumulate" (floating point sum-reduction) that involve copying the globally accessible data to/from process-private buffer space. The toolkit provides operations for (i) the creation, and destruction of distributed arrays, (ii) the synchronization of all processes, (iii) inquiries about arrays and their distribution, and (iv) primitive operations, such as get, put, accumulate, atomic read and increment, gather and scatter, and direct access to the local portion of an array, A number of BLAS-like data-parallel operations have been developed on top of these primitives.

Additional functionality is provided through a variety of third party libraries made available by using the GA primitives to perform the necessary data rearrangement. These include standard and generalized real symmetric eigensolvers (PeIGS, see below), and linear equation solvers (SCALAPACK) [19]. The $O(N^2)$ cost of data rearrangement is observed to be negligible in comparison to that of $O(N^3)$ linear-algebra operations. These libraries may internally use any form of parallelism appropriate to the host computer system, such as co-operative message passing or shared memory.

The GA interface has been designed in the light of emerging standards. In particular HPF-1 and subsequent revisions will provide the basis for future standards definition of distributed arrays in Fortran. A long term goal must be to migrate to full language support, and to eliminate as much as possible the practice of parallel programming through subroutine libraries. The basic functionality described above (create, fetch, store, accumulate etc.) may be expressed as single statements using Fortran90 array notation and the data-distribution directives of HPF. However, HPF currently precludes the use of such operations on shared data in MIMD parallel code. There is reason to believe that future versions of the HPF standard will rectify this problem, as well as provide for irregular distributions, which has been found to lead to a significant increase in performance in computational chemistry applications.

Distributed Matrices and Linear Algebra. Many electronic structure computations are formulated in terms of dense matrices of size roughly N by N, where N is the number of basis functions. Two distinct classes of operations are performed on these matrices: random access to small blocks, for constructing matrices as a function of many one- and two-electron integrals, and linear algebra operations on the entire matrix, such as eigensolving, Cholesky decomposition, linear system solution, inversion, and matrix-matrix multiplication. Both types of operations must work on distributed matrices if the resulting application is to be truly scalable.

A particular focus of our distributed linear algebra work has been the development of a scalable, fully parallel eigensolver whose numerical properties satisfy the needs of the chemistry applications. This package, called PeIGS, solves dense real symmetric standard (Ax=lx) and generalized (Ax=lBx) eigenproblems. The numerical method used is multisection for eigenvalues [20] and repeated inverse iteration and orthogonalization for eigenvectors [21]. Accuracy and orthogonality are similar to LAPACK's DSPGV and DSPEV [22]. Unlike other parallel inverse iteration eigensolvers, PeIGS guarantees orthogonality of eigenvectors even for arbitrarily large clusters that span processors. Internally, PeIGS uses a conventional message passing programming model and column-distributed matrices. However, it is more commonly accessed through an interface provided by the GA toolkit. The necessary data reorganization is handled by the interface, and is very fast compared to the $O(N^3/P)$ linear algebra times.

PeIGS is both fast and scalable – on a single processor it is competitive with LAPACK, and parallel efficiency remains high even for large processor counts. For example, in one of our SCF applications, the standard eigenproblem Ax=lx was solved for all eigenpairs of a 2053 by 2053 matrix. This computation required only 110 seconds on 150 processors of an Intel Paragon computer, a time-to-solution estimated as 87 times faster than LAPACK for the same problem on 1 processor.

Chemistry Applications on MPPs – the DFT-SCF Module

A wide range of methods have been implemented in NWChem, and more are planned, representing the core functionality required of any general-purpose computational chemistry package. The novel aspects of their implementation are the parallel algorithms employed, along with other ideas for reducing the scaling of the resource requirements of calculations as a function of their size. Complete descriptions of all of the algorithms employed in NWChem are not possible in the space available here, and are being published separately e.g., QCSCF [35], MP2 [7, 36, 37], RISCF [31, 30, 38], RIMP2 [6]. In order to demonstrate how the structure and tools described above are used by a high-level application, we sketch below the operation of the DFT module.

An essential core functionality in any electronic structure program is the direct self consistent field (SCF) module. While the application of direct SCF to large molecules is well suited to MPPs, targeting systems in excess of 1,000 atoms and 10,000 basis functions requires a re-examination of the conventional algorithm and assumed memory capacities. The majority of previous parallel implementations use a replicated-data approach [11] which is limited in scope since the size of these arrays will eventually exhaust the available single processor memory. The parallel direct DFT within NWChem (as in the HF module) distributes these arrays across the aggregate memory using the GA tools. This ensures that the size of systems treated scales with the size of the MPP and is not constrained by single processor memory.

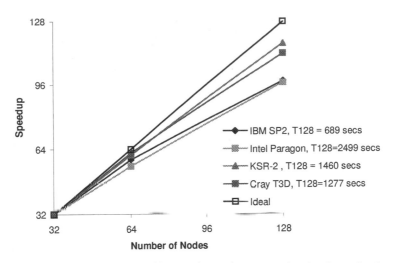

Figure 2. Parallel scaling of the NWChem Fock matrix construction for the zeolite fragment $Si_{28}O_{67}H_{30}$ using density functional theory (in the local density approximation) on a variety of massively parallel systems.

The software developed is a MPP implementation of the Hohenberg-Kohn-Sham formalism [32] of DFT. This method yields results similar to those from correlated *ab initio* methods, at substantially reduced cost. It assumes a charge density and approximations are made for the Hamiltonian (the exchange correlation functional), in contrast with traditional molecular orbital methods that assume an exact Hamiltonian and choose approximations to the wavefunction [33]. The Gaussian basis DFT method in NWChem breaks the Hamiltonian down into the same basic one- and two-electron components as traditional HF methods, with the latter component further reduced to a Coulomb and an exchange-correlation term. The treatment of the former can be accomplished in identical fashion to that used in traditional SCF methods, from a fitted expression similar to that found in RI-SCF [31], or from the commonly used Dunlap fit [34]. DFT is distinguished from other traditional methods, however, by the treatment of the exchange-correlation (XC) term. This term is typically integrated numerically on a grid, or fit to a Gaussian basis and subsequently integrated analytically.

It is instructive to elaborate on the computation/communication requirements of these time-consuming components. The computationally dominant step, the construction of the Fock matrix, is readily parallelized as the integrals can be computed concurrently. A strip-mined approach is used where the integral contributions to small blocks of the Fock matrix are computed locally and accumulated asynchronously into the distributed matrix. By choosing blocking over atoms, the falloff in interaction between distant atoms can be exploited, while simultaneously satisfying local memory constraints. The three time-consuming steps in the construction of the Fock matrix are the fit of the charge density (FitCD), the calculation of the Coulomb potential (VCoul), and the evaluation of the XC potential (VXC). FitCD and VCoul typically consume similar amounts of floating point cycles, primarily evaluating three-center two-electron integrals at a rate of about 300-400 flops each. Few communications are required beyond

Table 2. Time in wall clock seconds for the construction of the primary components of the NWChem DFT Fock matrix: FitCD, VCoul, VXC, diagonalisation, plus the Total iteration time

Machine	Nodes	Component				Total
		FitCD	VCoul	VXC	Diag	
IBM SP2	32	631	683	666	130	2110
	64	324	346	358	120	1148
	128	163	174	244	108	689
Intel	32	2368	2806	2080	332	7586
Paragon	64	1197	1402	1540	216	4355
	128	639	797	990	163	2499
KSR-2	32	1497	1806	1718	319	5340
	48	1050	1198	1151	243	3642
	64	771	933	873	218	2795
Cray T3D	32	1702	2078	535	150	4465
	64	830	1014	339	116	2299
	128	426	520	228	103	1277

a shared counter and global array accumulate; these components are easily parallelized and display high efficiencies. VXC requires far fewer flops (evaluating gaussian functions numerically on a grid, as opposed to analytical integrations) but approximately the same communication efforts. This results in a higher communication/computation ratio then the preceding two components.

We illustrate these effects by considering the performance of the DFT module in calculations of the zeolite fragment $Si_{28}O_{67}H_{30}$ (with 1673 basis functions) on representative MPP systems (the IBM-SP2, Kendall Square Research KSR-2, the Intel Paragon and Cray T3D). Figure 2 shows the impressive speedup achieved using up to 128 processors to build and diagonalise the Fock matrix. Considering the total time to solution (Table 2), we see that the IBM SP2 is approximately twice as fast as the Cray T3D and KSR-2, which in turn are almost twice the speed of the Paragon. The speedup curves present a somewhat different picture, with the T3D and KSR-2 exhibiting significantly higher efficiencies, particularly for the VXC construction, than the IBM SP2 and Paragon. The high efficiencies on the KSR-2 and T3D are a direct reflection of the hardware/software shared memory.

THROUGHPUT APPLICATIONS: THE GAMESS–UK PROJECT

GAMESS–UK [28] represents a typical established electronic structure code, comprising some 500K lines of Fortran that permits a wide range of computational methodology to be applied to molecular systems; other such codes include the GAUSSIAN series [29], CADPAC [23], GAMESS-US [24], MOLPRO [25], Turbomole [26], HONDO [27] etc. Present functionality lies at varying levels of approximation, including SCF, CASSCF, MCSCF, GVB, MP2/3/4, CCSD(T), MRDCI, Direct-CI, Full-CI, and analytic second derivatives. The package is available on vector hardware (e.g., Cray C90 and J90, and Fujitsu VPP/300), on workstations from all the leading vendors, and on parallel hardware from Cray/SGI (Cray T3D and T3E, and SGI Origin series) and from IBM (the SP series). There are over 110 academic and industrial user sites.

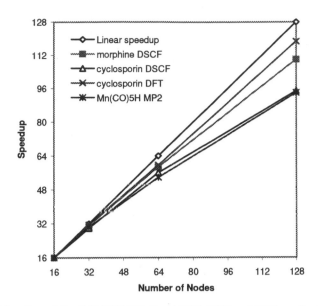

Figure 3. Parallel scaling of the GAMESS-UK direct-SCF, DFT and MP2 gradient modules in calculations on Morphine, Cyclosporin and $Mn(CO)_5H$ on the Cray T3E/900 (see text).

Parallel Implementations of the SCF, DFT and MP2 Modules

Initial parallel SCF implementations of GAMESS–UK included those for the Intel iPSC-2 and iPSC/860 hypercubes, with more recent developments undertaken within the Esprit-funded EUROPORT 2 project, IMMP (Interactive Modelling through Parallelism). EUROPORT's principle aim was to provide exemplars of commercial codes that demonstrated the potential of parallel processing to industry; the primary goal was to establish a broad spectrum of standards-based medium-scale parallel application software rather than targeting the highest levels of scalability and performance.

In contrast to NWChem, both SCF and DFT modules are parallelised in a replicated data fashion, with each node maintaining a copy of all data structures present in the serial version. While this structure limits the treatment of molecular systems beyond a certain size, experience suggests that it is possible on machines with 128 MByte nodes to handle systems of up to 1,800 basis functions. The main source of parallelism in the SCF module is the computation of the one- and two-electron integrals and their summation into the Fock matrix, with the more costly two-electron quantities allocated dynamically using a shared global counter. The result of parallelism implemented at this level is a code scalable to a modest number of processors (around 32), at which point the cost of other components of the SCF procedure starts to become relatively significant. The first of these addressed was the diagonalisation, which is now based on the PeIGS module from NWChem.

Once the capability for GA [15] is added, some distribution of the linear algebra becomes trivial. As an example, the SCF convergence acceleration algorithm (DIIS - direct inversion in the iterative subspace) is distributed using GA storage for all matrices, and parallel matrix multiply and dot-product functions. This not only reduces the time to perform the step, but the use of distributed memory storage (instead of disk) reduces the need for I/O during the SCF process. Timings for a number of Direct-

Table 3. Time in wall clock seconds for a variety of Direct-SCF, DFT and MP2-gradient Calculations using GAMESS–UK

Machine	Nodes	Calculation			
		Morphine Direct-SCF 6-31G**	Cyclosporin Direct-SCF 3-21G	Cyclosporin DFT 6-31G	Mn(CO)$_5$H MP2-gradient TZVP
IBM SP/P2SC-160	16	715	3627	9008	6955
	32	379	1946	4711	3967
	64	212	1121	2552	2372
	128	136	735	1430	1482
Cray T3E/900	16	873	3869	10305	8035
	32	447	2068	5489	4229
	64	238	1102	2899	2390
	128	127	653	1456	1368

SCF and DFT calculations conducted on an IBM SP/P2SC-160 and Cray-T3E/900 are shown in Table 3. The Direct-SCF calculations on morphine used a 6-31G** basis of 410 functions, those on cyclosporin a 3-21G basis of 1000 functions, while the DFT calculations on cyclosporin using the B-LYP functional were conducted in a 6-31G basis. Figure 3 shows the speedups achieved using up to 128 Cray T3E processors; speedups of 94.8, 110.0 and 118.7 are obtained on 128 nodes for the morphine 6-31G** SCF, the cyclosporin 3-21G SCF, and cyclosporin 6-31G DFT calculation, respectively. Considering the total time to solution, we see that the IBM SP2 is typically faster than the Cray T3E for up to 64 processors, although the T3E exhibits significantly higher efficiency than the SP and is some 10% faster with 128 processors.

Substantial modifications were required to enable the MP2 gradient to be computed in parallel [39]. Specifically, the conventional integral transformation step has been omitted, with the SCF step performed in direct fashion and the MO integrals generated by recomputation of the AO integrals, and stored in the global memory of the parallel machine. This storage and subsequent access is managed by the GA tools. The basic principle by which the subsequent steps are parallelised involves each node computing a contribution to the current term from MO integrals resident on that node. For some steps, however, more substantial changes to the algorithms are required. In the construction of the Lagrangian (the RHS of the CPHF equations), for example, MO integrals with three virtual orbital indices are required. Given the size of this class of integrals, they are not stored, the required terms of the Lagrangian being constructed directly from AO integrals. A second departure from the serial algorithm concerns the MP2 two-particle density matrix. This quantity, which is required in the AO basis, is of a similar size to the two-electron integrals and is stored on disk in the conventional algorithm, but generated as required during the two-electron derivative integral generation from intermediates stored in the GAs.

We consider the performance of the MP2-gradient module in an MP2 geometry optimisation of the Mn(CO)$_5$H molecule (with 217 basis functions) on the IBM SP/P2SC-160 and Cray T3E/900. Figure 3 shows a speedup of 94 achieved using 128 T3E processors to perform the complete optimisation, involving 5 energy and 5 gradient calculations. Considering the total time to solution (Table 3), we again see that the IBM SP2 is faster than the Cray T3E for up to 64 processors, although the T3E exhibits significantly higher efficiency than the SP.

CONCLUSIONS

We have described work in progress to develop the molecular modelling software applications, NWChem and GAMESS–UK, for both present and future generations of MPPs. The emphasis throughout the development of NWChem, targeting the *Grand Challenge* class of application, has been on scalability and the distribution, as opposed to the replication, of key data structures. An important part of this effort has been the careful design and implementation of a system of supporting libraries and modules using ideas from OO methodology. The DFT module provides just one example of the development of efficient parallel algorithms and the high level of performance which can be achieved through this approach. A performance analysis of calculations on the zeolite fragment $Si_{28}O_{67}H_{30}$ suggests that our goal of scalability, and hence the ability to treat more complex species in cost-effective and routine fashion, is well in hand. Based on experience gained with the GAMESS-UK code, we have shown that the less demanding treatment of *Throughput* applications may be accomplished by judicious porting to MPP platforms, with our cost-performance metric realised by adoption of the GA tools.

With the next generation of MPP systems likely to provide an improvement by a factor of three in the key hardware attributes e.g., CPU, communication latencies and bandwidth, we are confident that our on-going software development strategy will lead to both the type of calculations described above being performed interactively, a far cry from the present situation on conventional supercomputers, coupled with the ability to handle larger and more complex species. Important problems in biotechnology, pharmaceuticals and materials await just such advances. As stressed at the outset, however, raw computing power alone will not be sufficient to achieve these goals; continuing theoretical and computational innovation will be the real key to realising the full potential of the teraFLOPs computers scheduled for the late 1990's.

REFERENCES

1. Additional information on the EMSL is available via WorldWide Web URL*http://www.pnl.gov:2080*.
2. I.T. Foster, J.Jl Tilson, A.F. Wagner, R. Shepard, D.E. Bernholdt, R.J. Harrison, R.A. Kendall, R.J. Littlefield and A.T. Wong, Toward high-performance computational chemistry: I. Scalable Fock matrix construction algorithms, *J. Comp. Chem.* 17:109-123 (1996).
3. R.J. Harrison, M.F. Guest, R.A. Kendall, D.E. Bernholdt, A.T. Wong, M. Stave, J.L. Anchell, A.C. Hess, R.J. Littlefield, G.I. Fann, J. Nieplocha, G.S. Thomas, D. Elwood, J. Tilson, R.L. Shepard, A.F. Wagner, I.T. Foster, E. Lusk and R. Stevens, Toward high-performance computational chemistry: II. A scalable self-consistent field program, *J. Comp. Chem.* 17:124-132 (1996).
4. M.F. Guest, E. Apra, D.E. Bernholdt, H.A. Fruechtl, R.J. Harrison, R.A. Kendall, R.A. Kutteh, X. Long, J.B. Nicholas, J.A. Nichols, H.L. Taylor, A.T. Wong, G.I. Fann, R.J. Littlefield and J. Nieplocha, *Future Generation Computer Systems* 12:273-289 (1996).
5. A.T. Wong, R.J. Harrison and A.P. Rendell, Parallel Direct Four Index Transformations, *Theor. Chim. Acta* 93:317-31 (1996).
6. D.E. Bernholdt and R.J. Harrison, Large-Scale Correlated Electronic Structure Calculations: the RI-MP2 Method on Parallel Computers, *Chem. Phys. Lett.* 250:477-84 (1996).
7. D.E. Bernholdt and R.J. Harrison, Orbital invariant second-order many-body perturbation theory on parallel computers. An approach for large molecules, *J. Chem. Phys.* 102:9582-89 (1995).
8. G. Booch, *Object-Oriented Analysis and Design*, 2nd edition, Benjamin/Cummings Publishing Co., Inc. (1994).
9. B. Stroustrup, *The C++ Programming Language*, 2nd edition, Addison-Wesley Publishing Company (1991).
10. The MPI Forum, *MPI: A Message-Passing Interface Standard*, University of Tennessee (1994).

11. R.A. Kendall, R.J. Harrison, R.J. Littlefield and M.F. Guest, in: *Reviews in Computational Chemistry*, K.B. Lipkowitz and D.B. Boyd eds., VCH Publishers, Inc., New York (1994).

12. The MPI Forum, A message passing interface, in: *Supercomputing '93*, pp878-83, IEEE Computer Society Press, Los Alamitos, California, Portland, OR (1993).

13. M.E. Colvin, C.L. Janssen, R.A. Whiteside and C.H. Tong, *Theoretica Chimica Acta* 84:301-14 (1993).

14. T.R. Furlani and H.F. King, *J. Comp. Chem.* (1995).

15. J. Nieplocha, R.J. Harrison and R.J. Littlefield, Global arrays; A portable shared memory programming model for distributed memory computers, in: *Supercomputing '94*, IEEE Computer Society Press, Washington, D.C. (1994).

16. M. Arango, D. Berndt, N. Carriero, D. Gelernter and D. Gilmore, *Supercomputer Review* 3(10) (1990).

17. M. Schuler, T. Kovar, H. Lischka, R. Shepard and R.J. Harrison, *Theoretica Chimica Acta* 84:489-509 (1993).

18. A.P. Rendell, M.F. Guest and R.A. Kendall, *J. Comp. Chem.* 14:1429-39 (1993).

19. J. Choi, J.J. Dongarra, L.S. Ostrouchov, A.P. Petitet, R.C. Whaley, J. Demmel, I. Dhillon and K. Stanley, *LAPACK Working Note: Installation Guide for ScaLAPACK, Department of Computer Science, University of Tennessee, Knoxville, TN,* (1995). Available from *anonymous@ftp:ftp.netlib.org* .

20. S.S. Lo, B. Phillipps and A. Sameh, *SIAM J. Sci. Stat. Comput.* 8(2) (1987).

21. G. Fann and R.J. Littlefield, Parallel inverse iteration with reorthogonalization, in: *Sixth SIAM Conference on Parallel Processing for Scientific Computing (SIAM)*, pp409-13 (1993).

22. E. Anderson, Z. Bai, C. Bischof, J. Demmel, J.J. Dongarra, J. Du Croz, A. Greenbaum, S. Hammerling, A. McKenney, S. Ostrouchov and D. Sorensen, *LAPACK User's Guide*, SIAM (1992).

23. R.D. Amos, I.L. Alberts, J.S. Andrews, S.M. Colwell, N.C. Handy, D. Jayatilaka, P.J. Knowles, R. Kobayashi, N. Koga, K.E. Laidig, P.E. Maslen, C.W. Murray, J.E. Rice, J. Sanz, E.D. Simandiras, A.J. Stone and M.-D. Su, *CADPAC, Issue 6*, University of Cambridge, (1995).

24. M.W. Schmidt, et al., *QCPE Bulletin* 7:115 (1987).

25. MOLPRO is a package of *ab initio* programs written by H.-J. Werner and P.J. Knowles, with contributions from R.D. Amos, A. Berning, D.L. Cooper, M.J.O. Deegan, A.J. Dobbyn, F. Eckert, C. Hampel, T. Leininger, R. Lindh, A.W. Lloyd, W. Meyer, M.E. Mura, A. Nicklass, P. Palmieri, K. Peterson, R. Pitzer, P. Pulay, G. Rauhut, M. Schütz, H. Stoll, A.J. Stone and T. Thorsteinsson.

26. R. Ahlrichs, M. Br, M. Hser, H. Horn and C. Klmel, Electronic structure calculations on workstation computers: The program system TURBOMOLE, *Chem. Phys. Letters* 162:165 (1989); R. Ahlrichs and M.v. Arnim, TURBOMOLE, parallel implementation of SCF, density functional, and chemical shift modules, in: *Methods and Techniques in Computational Chemistry: METECC-95*, E. Clementi and G. Corongiu, eds.

27. M. Dupuis, J.D. Watts, H.O. Villar and G.J.B. Hurst, The general atomic and molecular electronic structure system HONDO: version 7.0., *Comput. Phys. Commun.* 52:415 (1989); M. Dupuis, A. Farazdel, S.P. Karma and S.A. Maluendes, HONDO: A general atomic and molecular electronic structure system, in: *MOTECC*, E. Clementi ed., ESCOM, Leiden (1990).

28. GAMESS-UK is a package of *ab initio* programs written by M.F. Guest, J.H. van Lenthe, J. Kendrick, K. Schoffel and P. Sherwood, with contributions from R.D. Amos, R.J. Buenker, M. Dupuis, N.C. Handy, I.H. Hillier, P.J. Knowles, V. Bonacic-Koutecky, W. von Niessen, R.J. Harrison, A.P. Rendell, V.R. Saunders, and A.J. Stone. The package is derived from the original GAMESS code due to M. Dupuis, D. Spangler and J. Wendoloski, *NRCC Software Catalog*, Vol. 1, Program No. QG01 (GAMESS), 1980; M.F. Guest, R.J. Harrison, J.H. van Lenthe and L.C.H. van Corler, *Theo. Chim. Acta* 71:117 (1987); M.F. Guest et al., Computing for Science Ltd., CCLRC Daresbury Laboratory, Daresbury, Warrington WA4 4AD, UK.

29. M. Frisch et al., Gaussian, Inc., 4415 Fifth Avenue, Pittsburgh, PA 15213, USA (1992).

30. M. Feyereisen, G. Fitzgerald, A. Komornicki, *Chem. Phys. Lett.* 208:359-63 (1993).

31. O. Vahtras, J. Almlof and M. Feyereisen, *Chem. Phys. Lett.* 213:514 (1993).

32. P. Hohenberg and W. Kohn, *Phys. Rev. B.* 136:864-71 (1964); W. Kohn and L.J. Sham, *Phys. Rev. A.* 140:1133-38 (1965).

33. E. Wimmer, *Density Functional Methods in Chemistry*, J.K. Labanowski and J.W. Andzelm, eds. pp7-31 Springer-Verlag, Berlin (1991).

34. B.I. Dunlap, J.W.D. Connolly and J.R. Sabin, *J. Chem. Phys.* 71:3396, 4993 (1979).

35. G.B. Bacskay, *Chem. Phys.* 61:385 (1982).

36. P. Pulay, *Chem. Phys. Lett.* 100:151-4 (1983).
37. P. Pulay and S. Saebo, *Theor. Chim. Acta* 69:357-68 (1986);
    S. Saebo and P. Pulay, *Annu. Rev. Phys. Chem.* 44:213–236 (1993).
38. H.A. Fruechtl and R.A. Kendall, A Scalable Implementation of the RI-SCF Algorithm, *Int. J. Quant. Chem.* 64:63-69 (1997).
39. G.D. Fletcher, A.P. Rendell and P. Sherwood, A parallel second-order Moller-Plesset gradient, *Molec. Phys.* 91:431-38 (1997).

ATOMIC PHYSICS

THE MULTIPHOTON AND ELECTRON COLLISIONS CONSORTIUM AND THE HELIUM CODE

K.T. Taylor, J.S. Parker and E.S. Smyth

Department of Applied Mathematics and Theoretical Physics,
Queen's University, Belfast BT7 1NN, U.K.

INTRODUCTION

The Multiphoton and Electron Collisions Consortium is formed from groups of scientists working at problems demanding High Performance Computing resource at the following institutions:

- The Queen's University of Belfast

- Durham University

- Imperial College London

- Oxford University

- University College London

- CLRC Rutherford Appleton and Daresbury Laboratories

On the multiphoton side the Imperial College/Oxford groups have used the Cray T3D in the first calculation[1] that includes sufficient dimensionality to allow for the effect of the laser's magnetic field at relativistic intensities. The objective in the T3D grand-challenge work carried out over the last three years by the Queen's University Belfast group was to build MPP algorithms and a production code, HELIUM, to solve the time-dependent Schrödinger equation for a laser-driven atomic 2-electron system in its full generality. This objective has been achieved and is the principal topic discussed in this paper.

On the electron collision side the participants at QUB and DL have acquired considerable experience since 1994 in: devising parallel algorithms; in exploiting optimized linear algebra libraries such as SCALAPACK and PEIGS; and in re-engineering and optimizing scalar/vector FORTRAN codes for parallel architectures. In particular, attention has focused on the use of distributed memory MIMD architectures, including the Cray T3D, to alleviate a number of computational bottlenecks in R-matrix computations. First the use of the PEIGS library, developed at PNL in the USA, has enabled the full diagonalisation of Hamiltonian matrices of order 17,000 to be treated on the T3D. Second, two propagator approaches have been studied to obtain parallel decompositions which are load-balanced and scale to large numbers of processors. A block-cyclic data distribution and the SCALAPACK library formed the basis of a successful implementation of the Burke-Baluja-Morgan propagator. A very different decomposition using pipelines of processors was used for a Light-Walker propagator. These methods have been demonstrated to be capable of accurately integrating several thousand equations with good load balancing and scaling. In addition, a special hybrid code has been developed from these two approaches to maximize efficiency for calculations when large numbers of scattering energies are required. Finally, a prototype parallelized 2-D propagator program has been developed on the T3D. Here the 2-D configuration space is partitioned into regions and each region mapped onto one or more processing elements. During an energy independent initialisation phase a local Hamiltonian matrix is constructed and diagonalised in each region. In the propagation phase an energy dependent global R-matrix is propagated from one region to another across a pipeline of processing elements. The range of problems investigated has resulted in the development of a number of efficient and scalable object-based F90 programs which use MPI to achieve MIMD functionality within the SPMD programming model. More details are presented in the contribution by Sunderland and others in this volume.

The remainder of this article is arranged as follows. The next two sections give details of the algorithms which have been implemented in the HELIUM code and on quality proving of the code. A section in which we present some results demonstrating the excellent scalability of the HELIUM code on the Cray T3D and T3E supercomputers is followed by one giving a forward look for the Consortium.

THE HELIUM CODE

In this and subsequent sections we discuss the design, implementation and performance of a finite-difference code, HELIUM, for the numerical solution of the time-dependent Schrödinger equation (TDSE) for laser-driven 2-electrons atomic systems.

From the outset the HELIUM program was required to satisfy a number of criteria:

- Accurately model the physical system

- Have a modular design for easy testing and maintenance

- Be portable across many different computer systems, serial and parallel

- Obtain a high performance to reduce run lengths to acceptable levels

Some of these criteria, if not in complete opposition, certainly conflict with others. For example, it is difficult to have a portable program that is optimized for different computer architectures, and the problem is particularly acute on parallel architectures. The decision to parallelize the code arose from estimates of the software's memory requirements and floating point overhead. It was clear that the only publicly available machines that could satisfy these requirements were MPP's.

Let us begin by reviewing the physical problem and the structure of the time-dependent Schrödinger equation. Our primary interest is in the physics of atom-laser interactions particularly in the intense field limit where both electrons are violently stripped from the atom in a small number of field periods. With the assumption that the field is linearly polarized, symmetry about the axis of polarization reduces the TDSE to a 5 dimensional, time-dependent PDE. Nuclear mass is assumed infinite, but no further simplifications are considered justified.

It is generally straightforward to parallelize the matrix-vector product, $H\Psi$, where H is the time-dependent Hamiltonian for the problem and Ψ the time-dependent wavefunction. Parallelizing the propagation of Ψ, namely the transformation of $\Psi(t)$ to $\Psi(t + \delta t)$, is a far more challenging problem. A Krylov-subspace method (H is written in the space spanned by $\Psi, H\Psi, H^2\Psi, ...$) proved to be a very satisfactory solution, with good stability properties, and very little inter-processor communications overhead beyond that required for $H\Psi$.

The requirement for high accuracy stems from a need to obtain reliable predictions of double ionization rates and high harmonic yields. For example, the high frequency features of the spectrum of light scattered by an atom are typically 4 to 8 orders of magnitude less prominent than low frequency features of the spectrum, and not surprisingly, calculations of harmonic spectra can prove sensitive to numerical errors as small as 10^{-8} in the atomic wavefunction Ψ. The need for high accuracy, though not uncommon in atomic physics, is less typical of engineering domains like computational fluid dynamics where many of the techniques used to integrate high order partial differential equations were pioneered. As a rule the requirement for high accuracy translates to a requirement for a high order propagator. A simple example illustrates how the order of the numerical integration can be crucial to efficiency. Suppose we insist on local truncation errors ϵ of order 10^{-12} in the calculation of $\Psi(t + \delta t)$ and compare a 12th order propagator with a 2nd order propagator. By definition, the local truncation error of the 12th order method scales as δt^{12}, which we write $\epsilon_{12} = \alpha_{12}\delta t_a^{12}$. Similarly, $\epsilon_2 = \alpha_2\delta t_b^2$. If we were to insist that $\epsilon_{12} \leq 10^{-12}$ and $\epsilon_2 \leq 10^{-12}$, and suppose for the moment that $\alpha_{12} = \alpha_2 = 1$, then the step size of the 12th order method would be $\delta t_a = 0.10$, and that of the 2nd order method. $\delta t_b = 0.0000010$. As we shall see, the computational overhead of the 12th order method is 6 times that of the 2nd, so that the 12th order method is $10^6/6$ faster than the 2nd. In practice, $\alpha_{12} \neq \alpha_2$, and the analysis is more complicated, but as we shall show, it is straightforward to measure α_{12} and α_2 from the working code, and if we insist $\epsilon_{12} \leq 10^{-12}$ and $\epsilon_2 \leq 10^{-12}$, then we find that the 12th order is $10^4/6$ faster than the 2nd order method. If we are willing to accept larger local truncation errors, then for some value of ϵ the 2 methods are of equal efficiency. In our code, that value of ϵ is about 10^{-4}. Whether or not a local truncation error of 10^{-4} can be tolerated in general is difficult to verify, though in some calculations it is adequate. The high order propagator has the merit of removing from the problem a large class of possible numerical errors at no extra computational expense.

Using HELIUM the 2-electron TDSE is solved on a finite-difference (FD) lattice in the two radial variables, with a partial wave expansion to treat the angular variables[2,3], where $|L, M, l_1, l_2 >$ represents a two-electron angular partial wave for the He wavefunction, with total angular momentum L, total $z-$component of angular momentum M and individual angular momenta l_1 and l_2 for the two electrons. In the present work a linearly polarized laser field is used; hence rotational symmetry about the field polarization vector means M is constant, and equal to zero for the atom initially in its ground state. In this case the angular variables are handled by just three quantum numbers l_1, l_2 and L.

The problem becomes that of solving N 2D time-dependent partial differential equations, where N is the number of partial waves retained in the basis. Thus N is determined by the maximum angular momentum l_{max} that each electron is allowed to have, and the maximum total angular momentum of both electrons, $L_{max} = 2 \times l_{max}$. The value of l_{max} necessary to achieve converged results is of course both frequency and intensity dependent. In the frequency range we have considered (approximately 0.15 a.u. to 0.25 a.u.), $l_{max} = 7$ is acceptable up to a peak laser electric field intensity of approximately 1.6×10^{15} W/cm^2. At higher intensities, up to 3.2×10^{15} W/cm^2, over the same frequency range, a $l_{max} = 9$ is necessary, corresponding to a basis set of 385 partial waves.

For all current production runs, a unitary propagator, based on the Lanczos/Arnoldi method, has replaced the Taylor series propagator used in the original version of the HELIUM program. The Taylor series is retained as an option within the code as it provides a rigorous means of checking the results from the Lanczos/Arnoldi propagator. The Taylor series fails catastrophically if incorrect parameter values (e.g. δt too large) are used, while the unitary propagator will continue but give incorrect results in similar circumstances.

Before we consider the details of the Lanczos/Arnoldi method, it is useful to review the Taylor series propagator. The Taylor series is an explicit method. It is relatively simple to implement and propagates Ψ in time as follows:

$$\Psi(t + \delta t) = c_0 \Psi(t) + c_1 H \Psi(t) + c_2 H^2 \Psi(t) + \ldots + c_n H^n \Psi(t)$$

where c_n are the Taylor series coefficients, and H the Hamiltonian. The elements of the Hamiltonian matrix can be written as:

$$< l_1, l_2, L|H|l_1', l_2', L' > =$$

$$\left(-\frac{d^2}{dr_1^2} + \frac{l_1(l_1+1)}{r_1^2} - \frac{2Z}{r_1} \right) \delta_{l_1,l_1'}, \delta_{l_2,l_2'}, \delta_{L,L'}$$

$$+ \left(-\frac{d^2}{dr_2^2} + \frac{l_2(l_2+1)}{r_2^2} - \frac{2Z}{r_2} \right) \delta_{l_1,l_1'}, \delta_{l_2,l_2'}, \delta_{L,L'}$$

$$+ \sum_l \frac{r_<^l}{r_>^{l+1}} < l_1, l_2, L|P_l(cos(\theta_{12}))|l_1', l_2', L' >$$

$$+ 2iA(t) \left(\frac{d}{dr_1} - \frac{l_1(l_1+1) - l_1'(l_1'+1)}{2r_1} \right) f^1_{l_1,l_2,L,l_1',l_2',L'}$$

$$+ 2iA(t) \left(\frac{d}{dr_2} - \frac{l_2(l_2+1) - l_2'(l_2'+1)}{2r_2} \right) f^2_{l_1,l_2,L,l_1',l_2',L'}$$

Where $f^i_{l_1,l_2,L,l_1',l_2',L'} = < l_1, l_2, L|cos(\theta_i)|l_1', l_2', L' >$ and where, $\frac{d}{dr_1}, \frac{d}{dr_2}, r_1, r_2$ are all FD operators, $A(t)$ is the electric field vector potential. $< l_1, l_2, L|cos(\theta_2)|l_1', l_2', L' >$ is proportional to $\delta_{l_1,l_1'}, \delta_{l_2,l_2'\pm1}, \delta_{L,L'\pm1}$. Angular momentum selection rules denoted by the Kronecker δ's govern interactions between partial waves.

The Hamiltonian consists of three major parts:

- The atomic Hamiltonian which includes the interaction between each electron and the nucleus. This consists of the two pure coulomb attraction terms and 2nd derivative operators, d^2/dr^2 for each electron. A 5 point finite-difference method is used for the 2nd derivative. The Atomic Hamiltonian does not invoke interaction between different partial waves.

- The electron-electron interaction term $\frac{1}{r_{12}}$. This is expanded in the well known series:

278

$$\frac{1}{|r_1 - r_2|} = \sum_l \frac{r_<^l}{r_>^{l+1}} P_l(cos(\theta_{12}))$$

Originally limited to 2 terms, the HELIUM program now allows terms up to any limit desired to be retained.

- The interaction Hamiltonian consisting of first derivative terms d/dr multiplying the time dependent vector potential and accounting for the interaction between each electron and the laser field in the momentum gauge. A 4 point finite-difference method is used for each first derivative. This part causes each partial wave to interact with up to 8 other partial waves.

The Krylov subspace is the space spanned by the vectors $\Psi, H\Psi, ..., H^n\Psi$. Gram-Schmidt, with iterative refinement, is used to orthonormalize the vectors, yielding vectors $Q_0, Q_1, ..., Q_n$, where $Q_0 = \Psi/|\Psi|$. If we define Q to be a matrix formed from the $n + 1$ column vectors $(Q_0, Q_1, ..., Q_n)$, then the Krylov subspace Hamiltonian is $h = Q^+HQ$. As Lanczos showed, h can be calculated simultaneously with Q, at no extra cost, and its eigendecomposition can be used as the first step in an iterative scheme to calculate eigenvalues of H. More recently, it has been appreciated that h can be used as a replacement for H in a wide variety of applications[4-8]. In the present application, the TDSE, $\dot{\Psi} = iH\Psi$, is written in the Krylov subspace, $\dot{\Psi}_k = -ih\Psi_k$, and the series solution for $\Psi_k(t + \delta t)$ can be calculated in negligible time since h, with $n = 12$, is a 13×13 matrix:

$$\Psi_k(t + \delta t) = e^{-ih\delta t}\Psi_k(t)$$

The desired solution then is $\Psi = Q\Psi_k(t+\delta t)$. The method proves attractive for two reasons. First, the computational overhead rises linearly with n. A 12th order method ($n = 12$), for example, is 6 times slower than a 2nd order method, and in our application, and most applications for general reasons, the work is almost entirely in the calculation of $\Psi, H\Psi, ..., H^n\Psi$. Even on a parallel machine, we have found the calculation of h and Q to be less than 10% of the total overhead. Second, the Krylov subspace solution $Q\Psi_k$ is good to n th order in δt. In other words, for $m \le n$, $Qh^m\Psi_k = H^m\Psi$; i.e. the first n time-derivatives of Ψ_k equal the first n time-derivatives of Ψ. As a result, the method may be viewed as a means of finding a Unitary propagator that is correct to order n in δt. The Lanczos/Arnoldi propagator allows a much larger value of δt to be used, typically 3 times that possible with the Taylor series propagator.

The finite-difference lattice is a 2-dimensional (typically 270×270) grid representing the 2 radial variables $|r_1|$ and $|r_2|$. Each of the three operators appearing in the Hamiltonian, d^2/dr^2, d/dz and $1/r_{12}$, is $|r_1|$ and $|r_2|$ dependent and must be cast in FD form. Each of these operators presented special difficulties. The $1/r_{12}$ operator, for example, which is represented as the series $\sum_l r_<^l/r_>^{l+1} P_l(cos\theta_{12})$, has a singularity at $r_1 = r_2$, which falls exactly on the diagonal elements of the lattice. It is far from clear a *priori* that the series representation of $1/r_{12}$ will converge to the correct value as $\delta r \to 0$ and $k \to l_{max}$. Care must be taken to verify that the truncated series approximates $1/r_{12}$ adequately, and to estimate the truncation error. The FD operators d^2/dr^2 and d/dr prove to be large sources of error in the atomic systems we are concerned with, and their design is specially tuned to the particular problem.

Now, we need to consider two of the basic parameters in the code, the FD lattice spacing, δr, and the timestep used in the propagation, δt. Physical considerations constrain the values these take to within certain limits. Upper limits on both δr (the finite-difference grid spacing) and δt (the timestep interval) are set by the need to model high energy excitation induced by the intense fields. To good approximation we can think of these as plane-waves, $\Psi \simeq e^{i k \cdot r - \omega t}$, where the energy is $\hbar \omega$, and k satisfies $\hbar^2 k^2 / 2m = \hbar \omega$. In intense field experiments ionizing electrons with energies well over 100 eV have been detected. We believe that the lattice, at a minimum, should be able to model excitations as high as 200 eV.

So, what is an appropriate value for δr? Let E_{max}, the maximum energy we wish to model, be 200 eV. Since $E_{max} = \hbar \omega_{max}$, in atomic units where $m = \hbar = 1$ and 1 a.u. $= 27.212$ eV, we have that $\omega_{max} = 7.350$ a.u. From the relation for k above, we have that $k_{max} = \sqrt{2 \omega_{max}}$, i.e. $k_{max} = 3.834$. From $\lambda = 2\pi / k$, we have $\lambda_{min} = 1.639$ a.u.. Thus, with $\delta r = 0.3333$ a.u., we have 5 grid points per λ_{min}. This is acceptable but far from ideal. A smaller spacing is desirable, however, because ionizing wavepackets can extend a considerable distance from the nucleus. The number of grid points must be increased if we decrease δr, to maintain an adequately large box size in each radial variable. If the radial box size is too small, wavefunction can be reflected back from the boundary towards the region near the nucleus. This can cause spurious population, affecting ionization rates and other quantities characterizing the system. In course of developing the program we have progressed from a box extending to 50 Bohr in each electron's radial co-ordinate with a FD mesh spacing of 0.5 Bohr, to a 90 Bohr box with a FD mesh spacing of 0.333 Bohr. This represents an increase in the total number of grid points in our largest runs from 3.85 million to 28.07 million. On 256 processors of the Cray T3D, a 14 field period run requires approximately 17000 PE hours.

To maintain convergence, δt must be reduced as δr is decreased. For the Taylor series there is a quadratic relationship. Hence if δr is halved, δt must be reduced by a factor of 4 and the number of grid points in the radial box must be increased by a factor of 4 to get to the same box size in both radial variables. The consequence is an increase in execution time by at least a factor of 16.

We believe that the small extent of the radial box is a larger source of errors than the current grid spacing. Thus any increase in application performance and reduction in memory usage (coming from further development of the HELIUM code), and any increase in CPU power and available memory (coming from newer supercomputers) will go primarily to enlarge the extent of the radial box.

As well as the constraints on δt coming from the numerical methods used, there is an upper limit set by the need to sample physical quantities on a time scale that is short compared to the period T_{min} of oscillation of the highest energy excitation, where $T_{min} = 2\pi / \omega_{max}$ For $\omega_{max} = 7.35$ a.u. we have $T_{min} = 0.8549$. We use $\delta t = 0.015$, hence there are about 60 timesteps per period T_{min}. The calculation of the acceleration, the population within the radial box (used for calculation of ionization rates) and of correlation is computationally expensive and it is usually performed only every 5 program timesteps. Hence there are 12 data values produced per period of oscillation of the highest energy excitation.

The simplest FD method for d^2/dr^2 is a 3-point operator. However, this results in a 30% error in the Helium ground state. As $\delta r \to 0$, these errors should disappear. But constrained to $\delta r = 0.3333$ a.u. the 3-point method is unacceptable. A 5-point method has a smaller error, but it also has the desirable characteristic that it can be tuned to give the right answer, even on the states with the largest errors. The largest error occurs in the s states, i.e. those states in which l is zero, due to the fact that they have a non-zero derivative at the $r = 0$ boundary (called the inner boundary). For each electron a fictitious $\Psi(-|r|)$ is used to tune d^2/dr^2 at the inner boundary for the lowest energy states of the He$^+$ ground state. This turns out to remove most of the error from the ground state of He as well. The error is moved to $1sns$ states, because ns has the wrong derivative at the inner boundary for the d^2/dr^2 operator tuned to $1s$. However, the error in ns is much smaller than the error in the ground state for an un-tuned d^2/dr^2 operator, due in part to the greater radial extent of the ns states, and in part to the ns states being correctly screened by the correctly modelled $1s$ (He$^+$) part.

For $l > 0$ states, $r\Psi \rightarrow r^{l+1}e^{-\alpha r}$ as $r \rightarrow 0$, hence the derivative $= 0$ at $r = 0$, and no tuning of d^2/dr^2 is necessary. The error is small, especially for $1snp$ states, again due to correct screening by the correctly modelled $1s$ (He$^+$) part. Any choice of $r \rightarrow 0$ tuning for d^2/dr^2 must approach the $\delta r = 0$ limit of 1 for s states.

The HELIUM code now uses Fortran 90. The restriction of 1 partial wave per processor has been removed and the number of partial waves per processor can now be any multiple of 2. Communication between partial waves is implemented using PVM, MPI, Cray SHMEM, and shared memory array copies. This has allowed the HELIUM program to be ported to run on other systems, without requiring a very large number of processors. The program could run on any workstation, although few workstations are sold with sufficient memory to handle an acceptable number of partial waves and an sufficiently large radial grid size. Most runs are still performed on the EPCC Cray T3D, although the EPCC T3E is also used. The EPCC J90 and the QUB/TCD SP2 are useful for test purposes, but neither machine has sufficient memory to handle production runs.

Memory requirements are large and reducing memory usage is now a major priority. The number of states in the basis set that must actually be stored can be reduced using symmetry. Within the full basis set, there exists pairs of partial waves which are symmetric, e.g. $|1,1,0>$ and $|1,0,1>$. To save memory and reduce the number of processors required, one partial wave in each symmetric pair can be discarded. When the missing partial wave is required in the calculation, it can be created from the transpose of its pair. Using this truncated basis set allows $l_{max} = 9$ runs (385 partial waves in the complete basis set) to be done on 256 processors with 1 partial wave on each processor. Even so, the memory requirements can be substantial. With $l_{max} = 9$ and a 270 x 270 FD grid, even using the truncated basis, the total memory required approaches the 16 GB limit on a 256 processor Cray T3D partition.

VERIFYING CORRECTNESS OF NUMERICAL METHODS

One of the more difficult problems in the development of large-scale numerical simulations is that of proving the quality of the code. For the case in hand this can be addressed in part by comparing the results from HELIUM with those from independent calculations in special limits. Each term in the FD Hamiltonian has been tested independently in this way. The Hamiltonians of the individual electrons have been tested by removing the laser field and discarding the $1/r_{12}$ term; the equation is then that for He$^+$, and it has been verified that the ground state energy is -2.00 a.u. as $\delta r \rightarrow 0$. The $1/r_{12}$ term has been tested by discarding the laser field, and verifying that the ground state energy approaches the established non-relativistic limit of -2.9037 a.u. as $\delta r \rightarrow 0$. Finally the interaction Hamiltonian has been tested by discarding $1/r_{12}$: the TDSE equation which results is then equivalent to that describing two non-interacting laser-driven He$^+$ ions, and it has been verified that results for multiphoton ionization

Table 1. He ground state energy calculated by the
HELIUM code for various values of δr and N.

N terms	Separation of grid points (Bohr radii)				
	0.5	**0.4**	**0.3333**	**0.12**	**0.06**
2	-3.07429	-2.96848	-2.93050	-2.90034	-2.90103
3	-3.07851	-2.97163	-2.93313	-2.90207	-2.90268
4	-3.07979	-2.97259	-2.93391	-2.90250	-2.90306
5	-3.08036	-2.97302	-2.93427	-2.90266	-2.90320
6	-3.08068	-2.97326	-2.93457	-2.90274	-2.90326
7	-3.08086	-2.97340	-2.93447	-2.90279	-2.90329

in this limit are identical to those from an independently written one-electron code for
He$^+$. The independent code employs a pure basis set method[9]. The dipole moments
and eigenenergies of the basis set are integrated to 12 significant figures, ensuring highly
accurate numerical results, and the close agreement with the FD He$^+$ results is a stringent test of the numerical quality of the finite-differencing. Energies of the eigenstates
of the FD helium Hamiltonian are an additional indicator of quality.

To accurately study the evolution of the helium wavefunction over time as the
atom interacts with the laser pulse, it is crucial to accurately model the helium ground
state wavefunction to provide a starting point for the calculation. The easiest means
of achieving a good ground state is by integrating the diffusion equation. The diffusion
equation consists of the wave equation, without the laser interaction part of the Hamiltonian, and with t replaced by it. On integration, all states apart from the ground
state dissipate rapidly. One disadvantage of the diffusion equation is that it can take
quite a long time for the wavefunction to settle down into the helium ground state.
The Lanczos/Arnoldi method offers the possibility of accelerating this process. A LA-
PACK eigenvalue/eigenvector subroutine, SSTEQR, is used to select the eigenvector of
h whose eigenvalue is closest to the value desired, i.e. the energy of the helium ground
state. This eigenvector is used as a starting point for the next Lanczos/Arnoldi iteration. By this process of iterative refinement, the eigenvector retained approaches the
helium ground state wavefunction. The process is an order of magnitude more rapid
than integrating the diffusion equation. However, the diffusion equation must be briefly
integrated to smooth the ground state and get the ground state energy converged in
the last few decimal places.

The calculation of the helium ground state energy depends upon the grid spacing, δr and the number of terms, N, in the series expansion of the electron-electron
interaction. For a converged, non-relativistic, infinite nuclear mass model, the best
possible values for the He ^1S ground state and the $^1P_0$ ($1s2p$) states are -2.9037 a.u.
and -2.12395 a.u. respectively.

The helium ground state wavefunction predominantly exists very close to the nucleus. As discussed in the previous section, it is very sensitive to the first few points
on the grid. The correction, proportional to $Z \times \delta r$, at the inner boundary in the
5-point FD representation of the d^2/dr^2 operator was used to calculate the He ground
state energy for various values of δr and N. Table 1 shows that the HELIUM program
converges to the correct value as δr goes to zero, and as N increases. The value of most
interest to us is $\delta r = 0.3333$ and $N = 3$, as these are the parameters we currently use
in production runs. The value with correction of -2.93313 a.u. compares to a value of
-2.28583 a.u. without the use of the correction.

It is possible to do even better by adding a higher order correction, proportional
to $Z^4 \times \delta r^4$, tuned for the particular values of δr and N used. With $\delta r = 0.3333$ a.u.

Table 2. Execution time (secs) for $l_{max} = 9$ with 385 partial waves in full basis set and a 150×150 radial FD grid.

Number of processors	Time T3D SHMEM_GET	Time T3D SHMEM_PUT	Time T3E-900 SHMEM_GET	Time T3E-900 SHMEM_PUT
256	440.72	404.06	–	–
128	797.88	704.32	–	–
64	1477.75	1281.83	463.38	473.40
32	–	–	813.56	850.40

Table 3. Execution time (secs) for $l_{max} = 5$ with 91 partial waves in full basis set and a 150×150 radial FD grid.

Number of processors	Time T3D SHMEM_GET	Time T3D SHMEM_PUT	Time T3E-900 SHMEM_GET	Time T3E-900 SHMEM_PUT
64	372.34	330.19	117.37	119.50
32	677.80	588.12	213.13	216.84
16	1106.00	969.625	352.65	357.19
8	–	–	623.10	638.69

and $N = 3$, the FD ground state energy of He is -2.9043 a.u., and the FD $1s2p\ ^1P_0$ state is -2.12384 a.u.

All runs are repeated with $L_{max} = 14$ (204 partial waves) and $L_{max} = 18$ (385 partial waves). Results are found, in the frequency range under consideration, to be insensitive to these variations of L_{max} up to 16×10^{14} W/cm^2.

HELIUM PERFORMANCE

To test the scalability of the HELIUM program, several test runs were performed on various numbers of processors on the EPCC T3D and T3E. The range over which scalability can be tested is limited by memory size, and the fact that the largest number of processors available on the EPCC T3E is 64. 512 processors are available on the EPCC T3D, but all runs to date have required a maximum of 256.

The test runs were performed with a 150×150 radial FD grid, a 14th Order Lanczos/Arnoldi propagator, over 50 program timesteps (a 14 field period run would typically require about 6000 timesteps). Three terms were retained in the electron-electron interaction series. Runs were repeated using both SHMEM_GET and SHMEM_PUT for interprocessor communication on both systems. The results clearly show that SHMEM_PUT is considerably faster than SHMEM_GET on the T3D, while on the T3E SHMEM_GET is faster. All times are in seconds.

Another important test is how well the program scales with increasing grid size. To test this, the $l_{max} = 5$ (91 partial waves in full basis set) test case described above was extended to a 300×300 radial FD grid. With four times as many grid points to model as in the 150×150 grid test case, the HELIUM program is limited by memory to a maximum of just 1 partial wave per PE on the T3D and a maximum of 2 partial waves per PE on the T3E.

On the T3D, runtime increased by a factor of about 4.4, while the runtime on the T3E increased by a factor of between 3.91 and 4.00.

Table 4. Execution time (secs) for $l_{max} = 5$ with 91 partial waves in full basis set and a 300×300 radial FD grid.

Number of processors	Time T3D SHMEM_GET	Time T3D SHMEM_PUT	Time T3E-900 SHMEM_GET	Time T3E-900 SHMEM_PUT
64	1638.76	1424.90	461.29	478.19
32	–	–	833.53	851.10

THE FORWARD LOOK

The Multiphoton and Electron Collisions Consortium is looking forward eagerly to the provision of a minimum 100 GFlop/s sustained, massively parallel machine through HPC97. Additional groups from various institutions have augmented the current profile of the Consortium in the putting forward of topical problems in modern atomic, molecular and optical physics that demand this level of computing power.

One of the most exciting of the new problems that HPC97 will enable the Consortium to tackle is the interaction of femto-second, intense laser pulses with the hydrogen molecule H_2. This builds on the experience of the QUB group in producing a code for the analogous two- electron system of laser-driven helium, as detailed above, but builds also on the expertise of the Durham group[10] in molecules generally but especially that gained through study of the one-electron H_2^+ molecule in femto- second laser pulses. The interest in H_2 comes about through the inter-nuclear vibrational motion which occurs on a time scale commensurate with that of the laser pulse. In contrast the rotational motion is so slow that it can be neglected. The molecular axis will initially be taken aligned with the polarization axis of the electric field of the laser. The vibrational motion thus brings into play one extra degree of freedom over that which occurs in laser-driven helium.

ACKNOWLEDGEMENTS

This work is supported in part by EPSRC but also by the Department of Education Northern Ireland through the provision of a Research Studentship to one of us (ES).

REFERENCES

1.   U.W. Rathe, C.H. Keitel, M. Protopapas and P.L. Knight, *J. Phys. B: At. Mol. Opt. Phys.* 30:L531 (1997).
2.   J. Parker, K.T. Taylor, C.W. Clark and S. Blodgett-Ford, *J. Phys. B: At. Mol. Opt. Phys.* 29:L33 (1996).
3.   S. Blodgett-Ford, J. Parker, C.W. Clark in: *Super-Intense Laser-Atom Physics* vol. 316 of NATO Advanced Study Institute, Series B:Physics, Plenum, New York (1993).
4.   J.A. Scott, *ACM Transactions on Mathematical Software* 21:432 (1995).
5.   H. Tal-Ezer and R. Kosloff, *J. Chem. Phys.* 81:3967 (1984).
6.   T.J. Park and J.C. Light, *J. Chem. Phys* 85:5870 (1986).
7.   H. van der Vorst, *J. Comput. Appl. Math.* 18:249 (1987).
8.   E. Gallopoulos and Y. Saad, *SIAM J. Sci. Stat. Comput.* 13:1236 (1992).
9.   J. Parker and C.W. Clark, *J. Opt. Soc. Am. B* 13: 371 (1996).
10. M. Plummer and J. McCann *J. Phys. B: At. Mol. Opt. Phys.* 30:L401 (1997); L.B. Madsen and M. Plummer, *J. Phys. B: At. Mol. Opt. Phys.* 31:87 (1998)

APPLICATION OF 6DIME: (γ,2e) ON HE

J. Rasch,[1] Colm T. Whelan,[2] S.P. Lucey,[2] and H.R.J. Walters[3]

[1]Institut de Physique, Laboratoire de Physique Moléculaire et des Collisions, Technopôle Metz 2000, Rue Arago, France
[2]Department of Applied Mathematics and Theoretical Physics, University of Cambridge, Silver Street, Cambridge, CB3 9EW, England
[3]Department of Applied Mathematics and Theoretical Physics, The Queen's University of Belfast, Belfast, BT7 1NN, England

INTRODUCTION

In many areas of atomic, molecular and nuclear physics one needs to evaluate highly oscillatory integrals of the form,

$$\int d^3r_f \, f(\mathbf{r}_f) \int d^3r_s \, g(\mathbf{r}_s) \, V(\mathbf{r}_s, \mathbf{r}_f), \tag{1}$$

where $f, g, V \in \mathbb{C}$ and where V has a singularity no worse than $1/|\mathbf{r}_s - \mathbf{r}_f|$. With the advent of the 6 Dimensional Integral Method, (6DIME) it is now possible to numerically evaluate any integral of the form (1) if g is a bounded function[1]. This method has already been applied successfully to (e,2e) processes on H,[2] and H$^-$,[3] targets. In this paper we will demonstrate the applicability of the 6DIME to the calculation of (γ,2e) in which a linearly polarised photon ionises both electrons from a Helium atom. The two exiting electrons are detected in coincidence, thus providing one with a complete kinematical description of the collision.

Recently, there have been a number of theoretical calculations applied to this problem[4, 5, 6, 7, 8]. The former of these approaches has been to model the final 3 body Coulomb state by a 3 Coulomb wavefunction that treats each two body subsystem of the 3 body system exactly though assumes that each subsystem develops independently of the other two. This approach, due to restrictions of the numerical methods employed, could only evaluate relatively simple ground state Helium wavefunctions of the form,

$$\Psi_a(\mathbf{r}_s, \mathbf{r}_f) = n(e^{-ar_s - br_f} + e^{-br_s - ar_f}) \, e^{-c|\mathbf{r}_s - \mathbf{r}_f|}, \tag{2}$$

where $a, b, c \in \mathbb{R}$ are determined variationally and n is a normalisation constant.

By using the numerical approach of the 6DIME one is no longer constrained as to the sophistication of wavefunctions that one can employ, for instance one can use highly correlated wavefunctions in *both* the initial and the final states of a system. Hence our approach is similar[4, 5], except that we have the freedom to use very much

High Performance Computing
Edited by R. J. Allan *et al.*, Kluwer Academic / Plenum Publishers, New York, 1999

more sophisticated initial and final state wavefunctions in our calculations. Atomic units are used throughout.

THEORY

The triple differential cross section, (TDCS) for a double ionisation event on a He atom can be written in the form,

$$\frac{d\sigma}{d\Omega_s d\Omega_f dE_s} = \frac{4\pi^2\alpha}{\omega} |\langle \psi_b(\mathbf{r}_s, \mathbf{r}_f)|H_{per}|\psi_a(\mathbf{r}_s, \mathbf{r}_f)\rangle|^2, \tag{3}$$

where ω is the incoming photon energy and α the fine structure constant. The two electrons ionised from the ground state of He are detected in solid angles $d\Omega_j$, with energy $E_j = k_j^2/2$ and momentum $\mathbf{k}_j$, conjugate to the spatial coordinates $\mathbf{r}_j$; $j = s, f$. The anti-symmetrised initial and final state wavefunctions are represented by Ψ_a and Ψ_b respectively.

Setting the scalar field $\phi = 0$ and adopting the gauge $\nabla \cdot \mathbf{A} = 0$ and neglecting terms of the order $1/c^2$ one can show that , $H_{int} = \exp(i\mathbf{p} \cdot \mathbf{r}_s)\,\hat{\mathbf{e}} \cdot \nabla_s + \exp(i\mathbf{p} \cdot \mathbf{r}_f)\,\hat{\mathbf{e}} \cdot \nabla_f$. Assuming that the influence of the term $\mathbf{p} \cdot \mathbf{r}_j \ll 1$ in calculations is small, equation (3) becomes,

$$H_{per} = \sum_{j=1}^{2} \mathbf{A} \cdot \mathbf{p}_j + \frac{A^2}{c^2}. \tag{4}$$

If one assumes that the prefactor $1/c^2$ makes the matrix elements relative to the second term in equation (4) small, then using standard expressions for $\mathbf{p}$ and $\mathbf{A}$, [21] one can write

$$H_{per} = -i \sum_{j=1}^{2} e^{i\mathbf{p} \cdot \mathbf{r}_j}\hat{\mathbf{e}} \cdot \nabla_j, \tag{5}$$

where $\hat{\mathbf{e}}$ is the photon polarisation vector. At the incident energies studied here, $E_\gamma = 88.975$, 99.975eV, the influence of the term $\mathbf{p} \cdot \mathbf{r}_j \ll 1$ in calculations is small, hence the electric dipole approximation, $\exp(i\mathbf{p} \cdot \mathbf{r}_j) = 1$ is valid. In this manner one arrives at an expression for the TDCS,

$$\frac{d\sigma}{d\Omega_s d\Omega_f dE_s} = \frac{4\pi^2\alpha}{\omega} |\langle \Psi_b(\mathbf{r}_s, \mathbf{r}_f)|\hat{\mathbf{e}} \cdot \nabla_s + \hat{\mathbf{e}} \cdot \nabla_f|\Psi_a(\mathbf{r}_s, \mathbf{r}_f)\rangle|^2, \tag{6}$$

commonly known as the velocity dipole approximation. The dependency of the results on the gauge that one adopts shall be discussed in detail elsewhere[16].

Recently, systematic attempts to find good approximate analytical solutions to the 3 body continuum problem have been carried out[9, 10, 11]. In[10, 11] the highly correlated final state wavefunctions have been designed in a drive to correctly model the asymptotic region in which either two or all of the inter-particle distances tend to infinity. In this work we shall use the 3 Coulomb adiabatic wave function[9], Ψ_b^{ad}, this wavefunction attempts to incorporate properties of the 3 body triangle in regions closer to the target than the asymptotic regime by making the Sommerfeld parameters dependent upon the shape of the 3 body triangle. The anti-symmetrised form of Ψ_b^{ad} with incoming boundary conditions appropriate to the asymptotic final state can be written,

$$\Psi_b^{ad}(\mathbf{r}_s, \mathbf{r}_f) = \frac{\sqrt{2}N}{(2\pi)^3} e^{i\mathbf{k}_s \cdot \mathbf{r}_s + i\mathbf{k}_f \cdot \mathbf{r}_f} \,_1F_1[i\beta_s, 1, -i(k_s r_s' + \mathbf{k}_s \cdot \mathbf{r}_s)] \tag{7}$$

$$_1F_1[i\beta_f, 1, -i(k_f r_f + \mathbf{k}_f \cdot \mathbf{r}_f)] \,_1F_1[i\beta_{sf}, 1, -i(k_{sf} r_{sf} + \mathbf{k}_{sf} \cdot \mathbf{r}_{sf})] \;,$$

where

$$Z_{sf} = (1 - \eta), \quad Z_s = -Z + \frac{(1 - Z_{sf})r_s}{(r_s^2 + r_f^2)r_{sf}}, \quad Z_f(\mathbf{r}_s, \mathbf{r}_f) = Z_s(\mathbf{r}_f, \mathbf{r}_s), \tag{8}$$

and

$$\eta = \left(\frac{3 + \cos^2 4\gamma}{4} \frac{r_{sf}}{r_s + r_f}\right)^2, \tag{9}$$

where $\tan \gamma = r_s/r_f$. The Sommerfeld parameter has the form, $\beta_j = Z_j/k_j$, Z is the nuclear charge, $\mathbf{r}_{sf} = \mathbf{r}_s - \mathbf{r}_f$ is the inter-electron coordinate and $\mathbf{k}_{sf} = (\mathbf{k}_s - \mathbf{k}_f)/2$ its conjugate momentum. In calculations the final state wavefunction Ψ_{ad}^- has been normalised to the asymptotic flux, N. It is important to stress that differential cross sections including wavefunctions of the form Ψ_{ad}^-, in which the Sommerfeld parameters are dependent on the integration variables can only be calculated using a numerical approach such as the 6DIME. In[4] a 3 Coulomb wavefunction, Ψ_{BBK}^- with constant Sommerfeld parameters, $Z_s = Z_f = 1$, $Z_{sf} = 1/2$ was used, where the authors approximated the 3 body problem as a combination of 3 two body interactions in which each two body subsystem acts independently of the other two. Therefore, Ψ_{BBK}^- does not account for any explicit 3 body interactions. At low incident energies it has been shown[9] by studying the radial behaviour of Ψ_{BBK}^- that the application of Ψ_{BBK}^- is no longer valid as the parts of the Hamiltonian that this approximation neglects are no longer small. A consequence of this is that the TDCS reported[4] are far too small for detected electron energies up to $40eV$. The fact that the authors still retain a reasonably good angular description of $(\gamma, 2e)$ events in this region is not surprising since one would still expect the two body interactions to play important roles in the physical processes.

To describe the initial state wavefunction we use a highly correlated, 50 parameter optimised Kinoshita wave function[12] of the form below,

$$\Psi_a^{kin} = e^{-\nu s} \sum_{l,m,n=0}^{N} c_{l,m,n} \, s^l \, p^m \, q^n, \tag{10}$$

where $p = u/s \in (0,1)$ and $q = t/u \in (-1,1)$ and $s = r_s + r_f$, $t = r_s - r_f$ and $u = |\mathbf{r}_s - \mathbf{r}_f|$. In this application the series is truncated at $N = 11$, $c_{l,m,n}$, $\nu \in \mathbb{R}$ and $l, m, n \in \mathbb{N}$. This wavefunction predicts an extremely accurate ground state energy for Helium, 2.90372437240 though in achieving this value[12] no longer enforces the Kato cusp conditions[13] that were originally included in this wavefunction[14]. As pointed out[15] the main contribution to the binding energy comes from a spatial region whose distance from the nucleus is of the order of the characteristic binding radius, it is reasonable therefore to expect Ψ_a^{kin} to be correct in this region. However, spatial regions much closer to the nucleus are also relevant to our calculations, where constraints such as the Kato cusp conditions may be relevant. The importance of including the Kato cusp conditions in the initial state wavefunction has been discussed in[16] in which it is shown that the TDCSs calculated using Ψ_a^{kin} converge to the TDCSs calculated using a wavefunction that includes the Kato cusp conditions in its formulation[17].

RESULTS

In Figure 1 we compare our calculations to experimental data[18]. In this experiment the two ionised electrons are detected in coincidence in a plane perpendicular to the

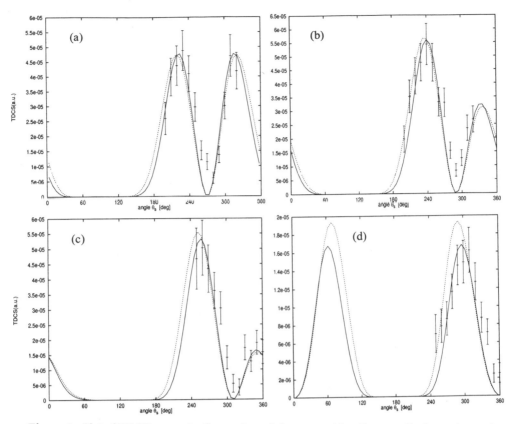

Figure 1. Plot of TDCS versus θ_s. Comparison of theory to arbitrarily normalised experimental data of[18]. Equal energy sharing regime, $E_s = E_f = 5eV$, $\theta_f = $ (a) $90°$, (b) $110°$, (c) $130°$ and (d) $180°$. The solid and dashed lines approximate Ψ_b^- by Ψ_{ad}^- and Ψ_{bbk}^- respectively.

photon direction. The photon is plane polarised and one electron is observed at a fixed angle θ_f with respect to the polarisation vector while the TDCS is measured as a function of the angle of the second electron, θ_s relative to the polarisation direction. We adopt the convention that both θ_s and θ_f are measured anti-clockwise to the polarisation direction. The linear polarisation of the incident photon is not complete[18] the major axis of polarisation is in the direction $\theta_s = 0°$ and the minor axis is at right angles to this and in the ionisation plane. Using[19] the observed TDCS can be written in the form,

$$TDCS_{obs} = \frac{1}{2}[(1 + S_1)TDCS_{major} + (1 - S_1)TDCS_{minor}], \qquad (11)$$

in the case of[18] the Stokes parameter $S_1 = 0.83$.

The experimental data is normalised at each θ_f to the theory using Ψ_b^{ad} in the final state, the results using Ψ_b^{bbk} are scaled upwards by a factor of 17.2 so that the maximum of the TDCS in Figure 1(a) is equivalent to the maximum using Ψ_b^{ad} this scaling is held constant for the other 3 figures. On the whole very good agreement is found using Ψ_b^{ad} between theory and experiment for all Figure 1. The angular distributions produced by the two final state wavefunctions are very similar, however if one looks at Figure 1(a), Ψ_b^{ad} predicts a separation between the two maxima $\approx 10°$ smaller than Ψ_b^{bbk}, a similar effect can be observed in Figures 1(b,c,d). The ratios of the two maxima are also different, the effect of Ψ_b^{ad} being to enhance the relative size of the smaller peak. One can see clearly that the TDCS calculated using Ψ_b^{bbk} demonstrate a differing inter-normalisation trend as one increases θ_f with the absolute size of the Ψ_b^{bbk} TDCS increasing relative to Ψ_b^{ad} as one increases θ_f.

At slightly higher detected electron energies, $E_s = E_f = 10eV$, (Figure 2) the experimental data is fitted to the solid line at the point $\theta_s = 230°, \vartheta_f = 110°$ on Figure 2(b) and this normalisation is then fixed for the other 3 plots, as the experimental data are inter-normalised. The results calculated using Ψ_b^{ad} are in excellent agreement with the experiment for all four cases. The TDCSs calculated using Ψ_b^{bbk} has been multiplied by a factor of 4.35, from Figure 2 it can be seen that Ψ_b^{bbk} predicts a slightly different trend in the relative size of the TDCS for different θ_f. However, this deviating behaviour is within the errorbars of the experiment.

The absolute value using Ψ_b^{ad} in the final state can be compared to absolute experimental data[20] at $E_s = E_f = 10eV$, $40° \leq \theta_f < 65°$ and $0° \leq \phi_f < 20°$, where ϕ_f is angle of the "f" detected electron out of the ionisation plane. In order to compensate for the large angular windows in this data we have calculated our TDCS over several portions of this angular range and averaged over these results. The agreement in absolute value between data and theory is poor, the theoretical TDCS is approximately 3 times larger than experiment.

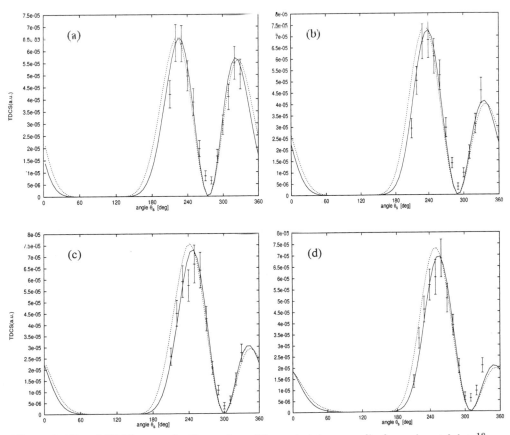

Figure 2. Plot of TDCS versus θ_s. Comparison of theory to inter-normalised experimental data[18]. Equal energy regime, $E_s = E_f = 10eV$, $\theta_f = $ (a) 95°, (b) 110°, (c) 120°, (d) 130°. Curves are as in Figure 1.

CONCLUSION

In conclusion we have shown that by being able to include both highly correlated initial and final state wavefunctions in our calculation it is possible to reproduce the angular behaviour of TDCS very well at small electron energies above threshold. We have demonstrated that compared to the inter-normalised data in Figure 2, the different final state wavefunctions predict different trends with respect to θ_f. Furthermore, by comparison to the inter-normalised experiment at $E_s = E_f = 10eV$ it has been shown that Ψ_b^{ad} models the experimental data very well. Though compared to the absolute experimental data[20] at the same energy the absolute size of our calculations are shown to be in poor agreement with experiment.

In S. P. Lucey[16] we will present a further critique of the use of these ansatz based functions in $(\gamma,2e)$ studies. Our primary interest in this paper was the exploitation of the 6DIME numerical technique and we have found it to robust and flexible. Clearly, it allows us the possibility to use fully numerical wave functions and liberates us from any analytic constraints. It is our intention to exploit the full power of this method in more advanced calculations.

ACKNOWLEDGEMENTS

We gratefully acknowledge support from the Royal Society, the EC (FMBICT972172), NATO(CRG 950 407). We are grateful for the use of the Hitachi parallel High Performance Computer at the Cavendish Laboratory, Cambridge, U.K. and to the authors of[18, 20] for making their experimental data available to us. This work was (partially) supported by the National Science Foundation through a grant for the Institute for Theoretical Atomic and Molecular Physics at Harvard University and Smithsonian Astrophysical Observatory.

REFERENCES

1.   J. Rasch, S.P. Lucey, Colm T. Whelan and H.R.J. Walters, *Comp. Phys. Comm.* 101:197 (1997).
2.   S.P. Lucey, J. Rasch, Colm T. Whelan and H.R.J.Walters in: *Proceedings of the 14th International Conference on Application of Accelerators in Research and Industry* Denton, 215 (1996).
3.   S.P. Lucey et al., in: *Coincidence Studies of Electron and Photon Impact Ionization*, proceedings of the International Conference, Belfast 1996, Colm T. Whelan ed., University of Cambridge (1997).
4.   F. Maulbetsch and J.S. Briggs, J. Phys. B. 26:L647 (1993).
5.   F. Maulbetsch, M. Pont, J.S. Briggs and R. Shakeshaft, J. Phys. B. 28:L341 (1995).
6.   M. Pont and R. Shakeshaft, *Phys. Rev. A* 51:R2676 (1995).
7.   M. Pont, R. Shakeshaft, F. Maulbetsch and J.S. Briggs, *Phys. Rev. A* 53:3671 (1996).
8.   A. Huetz, P. Selles, D. Waymel and J. Mazeau, J. Phys. B. 24:1917 (1991).
9.   J. Berakdar, *Phys. Rev. A* 53:2314 (1996).
10. A.M. Mukhamedzhanov and M. Lieber, *Phys. Rev. A.* 54:3078-85 (1996).
11. P.A. Macri, J.E. Miraglia, C.R. Garibotti, F.D. Colavecchia and G. Gasaneo, *Phys. Rev. A.* 56:3518 (1997).
12. T. Koga and S. Morishita, *Z. Phys. D.* 34:71 (1995).
13. T. Kato, *Trans. Amer. Math. Soc.* 70:212 (1951).
14. T. Kinoshita, *Phys. Rev.* 105:1490 (1957).
15. Z. Teng and R. Shakeshaft, *Phys. Rev. A* 49:3597 (1994).
16. S.P. Lucey, J. Rasch, Colm T. Whelan and H.R.J. Walters, (1998) in preparation.
17. C. Le Sech, J. Phys. B. 30:L47 (1997).

18. T.J. Reddish, J. Wightman and S.J. Cvejanovic, in: *Coincidence Studies of Electron and Photon Impact Ionization*, proceedings of the International Conference, Belfast 1996, Colm T. Whelan ed., University of Cambridge (1997).

19. O. Schwarzkopf and V. Schmidt, J. Phys. B. 28:2847 (1995).

20. T. Vogt, R. Dörner, O. Jagutzki, C.L. Cocke, J. Feagin, M. Jung, E.P. Kanter, H. Khemliche, S. Kravis, V. Mergel, L. Spielberger, J. Ullrich, M. Unverzagt, H. Bräuning, U. Meyer, and H. Schmidt-Böcking, in: *Coincidence Studies of electron and photon impact ionisation*, Colm T. Whelan and H.R.J. Walters eds., Plenum, (1997).

21. B.H. Bransden and C.J. Joachain. *Physics of Atoms and Molecules*, Longman, London and New York (1983).

PARALLELIZATION OF ATOMIC R-MATRIX SCATTERING PROGRAMS

A. Sunderland,[1] P.G. Burke,[2] V.M. Burke,[1] and C.J. Noble[1]

[1]Department for Computation and Information,
CLRC Daresbury Laboratory,
Warrington WA4 4AD, UK.
[2]Department of Applied Mathematics and Theoretical Physics,
Queen's University of Belfast,
Belfast BT7 1NN, Northern Ireland.

INTRODUCTION

R-matrix theory has been the basis of computer programs that have described a wide range of atomic, molecular, optical and surface processes for more than twenty-five years[1]. These processes now include electron scattering by atoms, ions and molecules, atomic and molecular photoionization, free-free transitions, atomic and molecular multiphoton processes, positron scattering by atoms and molecules, electron scattering by molecules physisorbed on surfaces and electron energy loss in transition metal oxides. The R-matrix programs written to describe these processes have been used to calculate data of importance in many applications including those in laser physics, controlled thermonuclear fusion and plasma physics, atmospheric physics and chemistry and astronomy[2]. These programs are now used world-wide and have formed the basis of several international collaborations including most notably the international Opacity Project[3]

In spite of this success it has become clear in the past few years that outstanding problems of crucial importance in many applications cannot be tackled by the present generation of R-matrix programs. The most important problems that need to be addressed are:

1. to extend the theory and programs to enable electron and photon interactions with atoms, ions and molecules to be accurately calculated at intermediate energies which extend from the ionization threshold to several times this threshold. The main difficulty in this energy region is the need to accurately represent the infinity of ionizing channels that then become open. Accurate cross sections at intermediate energies which include both excitation and ionization cross sections are urgently required in the modelling of many laboratory and astrophysical plasmas. Recently a new R-matrix with pseudo-states approach has been developed[4]

which has been used with success in treating a number of light atoms and ions including H, He$^+$, He, Be$^+$, Be, B^{2+} and B. However more complex targets present a formidable computational problem involving the necessity of including hundreds or even thousands of coupled channels to yield accurate results.

2. to extend the theory and programs to enable electron and photon interactions with complex open shell atoms and ions to be accurately calculated. Accurate cross sections involving open 3d-shell targets, for example in astronomy, where observed lines of low ionization stages of the iron-peak elements Fe, Co and Ni are central to the analysis of many astronomical objects. They also arise in laser physics and plasma physics where, for example, lines of Ni-like ions are being considered as the basis for X-ray lasers. Recently, a new R-matrix approach which is more efficient for treating these open shell atoms and ions has been proposed[5] which has yielded encouraging results for electron collisions with Ni$^+$, Ni^{2+}, Ni^{4+} and Co$^+$. However, again such targets lead to many hundreds of coupled channels posing major computational problems.

In order to address these problems and similar ones that arise in other areas of current interest, two new program packages are being developed to run on current and future MPPs. The first package RMATRX II[5] treats the internal region in R-matrix theory where the scattered or ejected electron lies within the charge distribution of the target atom or ion. The second package FARM[6] treats the external region in R-matrix theory where the scattered or ejected electron lies outside the charge distribution of the target atom or ion.

In this paper we describe recent work on the RMATRX II and FARM programs to develop a parallel program package running on the Cray T3D. The implementation is based on the use of new features of Fortran90 and the use of an object-oriented programming style where parallelism is accomplished using MPI.

In the following sections we outline the basic steps in this calculation and describe the parallelization strategy that has been adopted. To provide a concrete example of the type of calculations to be treated we cite the case of electron scattering by the iron-peak element Ni^{3+} which is currently being studied.

PARALLEL DECOMPOSITION

Inner Region R-Matrix Calculation

The inner region calculation requires the construction and diagonalization of the scattering Hamiltonian matrix using a basis of multi-electron functions formed by antisymmetrizing products of target wave functions and single-electron continuum functions. Russell–Saunders coupling is used in RMATRX II, so that a Hamiltonian matrix is constructed for each combination of the total orbital angular momentum, L, total spin angular momentum, S, and the parity π. The target wave functions are expressed in terms of Slater determinants and are obtained separately from the scattering calculation. The continuum functions are described in terms of numerically-defined orbitals.

In the present implementation this internal region calculation has been only partially parallelized. The Hamiltonian matrix elements are evaluated serially using a single node of the Cray T3D. For the models presently under consideration this part of the computation requires relatively little computational time. However it becomes much more time consuming in models which use more accurate target wave functions involving large numbers of configurations. Work is underway, using newly developed algorithms, to fully parallelize this part of the calculation using a data decomposition approach.

The program distributes columns of the Hamiltonian matrix across the nodes of the parallel computer and it is then diagonalized using the parallel diagonalizer PEIGS[10]. This routine returns all eigenvalues and eigenvectors for a dense Hermitian matrix and has been tested for matrices of dimension up to 17000 on the T3D. Each eigenvector is returned on a single computational node. This allows the reduced-width amplitudes which define the boundary R-matrix to be computed without further transfer of data between nodes. These amplitudes may be written out to disk files if required. This stage of the calculation is load-balanced and communication losses are limited to those inherent in the diagonalizer. By associating the reduced-width amplitude with the corresponding R-matrix eigenvalue the R-matrices required in the subsequent calculation may be constructed without sorting the results, thereby avoiding potentially large communication costs.

External Region R-Matrix Calculation

The R-matrix at the boundary determined by the solution of the inner region scattering equation, is used as inner boundary conditions for the integration of the external region equations. These reduce to sets of coupled second-order linear differential equations. The integration of this equation set must be taken out to a sufficiently large radial distance that an asymptotic expansion of the wave function is feasible. This distance may, in principle, be very large because of residual long-range dipole couplings between the scattering channels. The asymptotic boundary conditions determine the scattering matrix and the scattering observables.

For reasons of numerical stability, R-matrix propagator or logarithmic derivative methods are preferred for integrating the external region equations. The radial distance between the inner region boundary and the asymptotic matching point is divided into sectors. Within each sector R-matrices relating the solution at the inner and outer boundaries are constructed by diagonalizing a sector Hamiltonian matrix. The R-matrix from the inner region boundary can then be propagated from one sector to the next to the asymptotic boundary.

Parallelization of the external region calculation is therefore complicated because of the large amounts of data which must be stored, particularly for large equation sets. The variations in the architecture of the present parallel computers (memory and disk resources) make it difficult to choose a single parallel decomposition which will be efficient in all cases. Three distinct approaches have been developed which must be selected depending on the characteristics of the physical problem and of the computer[11].

For problems of around 500 coupled equations it is feasible to use an approach based on the Burke-Baluja-Morgan R-Matrix propagator[12] in which the Hamiltonian matrix is constructed on a shifted Legendre basis. As these matrices have dimensions which are 8-10 times the number of scattering channels the approach has large storage requirements and is efficient only if the number of scattering energies at which the calculation must be repeated is very large. These conditions are amply fulfilled for this range of scattering problems. Typical problem sizes are indicated by Table 1 which illustrates the number of target states and scattering channels which might be included.

The results required, for example, by astrophysicists, take the form of thermally averaged (effective) collision strengths, Υ where

Table 1. Number of target states and coupled channels for different Ni ions including states with the ground configuration together with 4s and 4p states and 4s, 4p and 4d states.

Target Ion	Ground Config.	4s,4p		4s,4p,4d	
		Target States	Maximum Channels	Target States	Maximum Channels
Ni^+	$3d^9$	27	73	54	161
Co^+, Ni^{2+}	$3d^8$	69	194	141	420
Fe^+, Co^{2+}, Ni^{3+}	$3d^7$	100	315	204	678
$Fe^{2+}, Co^{3+}, Ni^{4+}$	$3d^6$	136	403	272	841
$Fe^{3+}, Co^{4+}, Ni^{5+}$	$3d^5$	108	342	212	705
$Fe^{4+}, Co^{5+}, Ni^{6+}$	$3d^4$	80	228	152	454
$Fe^{5+}, Co^{6+}, Ni^{7+}$	$3d^3$	34	97	61	185

$$\Upsilon(T) = \int_0^{E_{max}} \Omega(E) e^{-\frac{E}{KT}} d\left(\frac{E}{KT}\right), \tag{1}$$

T denotes the temperature and $\Omega(E)$ is the collision strength defining the probability for a transition to a particular final state. The evaluation of this integral requires the integrand to be calculated at a very large number of energy points, of the order of several thousand, to accommodate the rapidly varying collision strengths due to the overlapping Rydberg series of resonances converging on each channel threshold. As an illustration Figure 1 shows the collision strength as a function of energy calculated by Teng et al.[7] for the $^5D^e$ - $^3P^e$ transition in electron scattering by Ni^{4+}.

The external region calculation proceeds in two distinct stages. The first uses data decomposition as each of the sector Hamiltonian matrices in turn is distributed across all the nodes and diagonalized by the PEIGS code. The sector reduced width

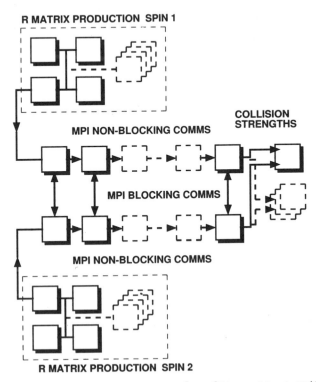

Figure 1. Collision strengths for the $^5D^e$ - $^3P^e$ transition in Ni^{4+}

amplitudes may then be stored on disk files before beginning the second phase of the calculation.

The second stage of the external region calculation produces the effective collision strength for a predefined set of temperatures for a single partial wave, L, S and π combination. In this stage nodes are allocated for specific purposes. A group of nodes is required to read the internal region data from disk and produce the R-matrix corresponding to the internal region boundary radius for each scattering energy, E. A chain of nodes is set up, one node for each propagation sector, forming a pipeline along which the R-matrix is passed as it is propagated from one sector to the next. The R-matrix is finally passed to one of a group of nodes each dedicated to calculating collision strengths. The calculation proceeds for several thousand scattering energies, the stream of R-matrices passing through the chain and fanning out to a final node. The rate at which R-matrices emerge from the end of the chain determines the number of nodes in the final group. The overall scheme is illustrated in Figure 2.

A number of algorithmic improvements have been incorporated into the program. Significant computational savings are obtained by taking advantage of the decoupling of channels associated with target states with differing spin quantum numbers. This requires the pairing of nodes within a pipeline. Other savings result from the reuse of data corresponding to different $SL\pi$ partial waves. The approach scales and latency can largely be hidden by using non-blocking communication between the nodes of the pipeline.

This parallelization may provide the basis for large scale parallel calculations of electron-atom collision phenomena. The codes have been fully implemented and tested on the Cray T3D.

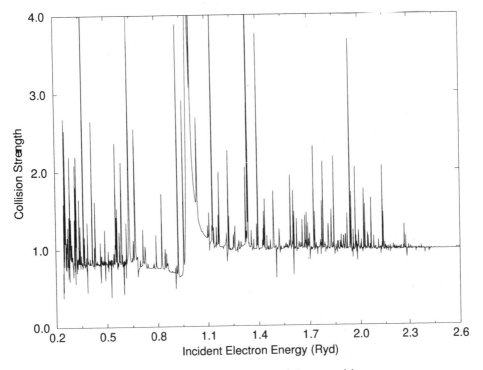

Figure 2. External Region Parallel Decomposition.

Table 2. Number of coupled channels for each $LS\pi$ combination for Ni^{3+} including states with the ground configuration together with 4s, 4p and 4d states.

L	Total Parity $(-1)^L$ Total Spin S				Total Parity $(-1)^{L+1}$ Total Spin S			
	1	3	5	7	1	3	5	7
0	67	98	37	6	58	97	42	3
1	185	282	111	14	176	281	116	11
2	285	435	170	20	276	434	175	17
3	359	547	211	23	350	546	216	20
4	408	618	234	24	399	617	239	21
5	435	656	245	24	426	655	250	21
6	417	672	249	24	438	671	254	21
7	451	677	250	24	442	676	255	21

CALCULATIONS

The calculation described above was for a single partial wave. In reality the effective collision strengths must be summed over all partial waves or until such a high L that the sum has converged or can be simply "topped up" by extrapolation. For values of L above about 11 the internal region calculation can be simplified as the exchange between the target electrons and the scattered electron becomes negligible and the spin of the target states becomes a good quantum number. In Table 2 we illustrate how the size of the problem changes during the calculation. For given L the number of channels differs greatly with total spin, S. As L increases, the number of channels increases for each S until it reaches a maximum. Thus the calculation needs to be driven through the partial waves and the results collated. In addition the machine has to be optimally configured for each stage of the calculation and for each partial wave.

The problem is to develop a robust model of the computation which predicts the optimum number of nodes which should be devoted to each aspect of the calculation based on the size of the calculation and the number of processor nodes in the available pool. This is particularly subtle for the second phase of the external region calculation described above. The model is expressed in terms of simple functional forms each predicting the time to complete one phase of the calculation. The model needs to be incorporated within the program to enable the calculation to be continually reconfigured according to the widely differing cases to be treated within a given calculation. The type of timing data required to empirically fit the constants within the model is illustrated in Table 3.

Here, the problem size is defined in terms of the number of scattering channels (the numbers in parentheses indicate the number of channels for each contributing target spin). The column headed *Initial R-matrix* shows the time to compute a single R-Matrix from the internal region results. The *Final R-Matrix* column indicates the rate at which R-matrices are produced at the final node of the pipeline and the column headed *Collision Strength* gives the timing to produce a collision strength from the final R-matrix. The ratio of these times determines the number of processor nodes which should be devoted to each task in order to produce load-balancing and avoid stalling of the pipeline.

To control the sequence of the complete calculation we have chosen to use the PERL scripting language[13]. This language is designed specifically for gluing together computationally intensive components and provides a range of facilities which transcend

Table 3. R-Matrix Propagation Times.

Number of Channels	Initial R-Matrix (sec)	Final R-matrix (sec)	Collision Strength (sec)
120 (102/18)	0.13	1.34	11.47
130 (108/22)	0.16	1.48	13.43
194 (108/86)	0.32	1.81	26.76

those available in simple UNIX shells. The property, shared with shells, of being (normally) interpreted rather than compiled greatly speeds the development process and facilitates flexibility in treating the overall computational path. It is believed that much of the tedious checking and bookkeeping previously performed manually in these calculations will be gradually taken over as the scripts become more elaborate. This may also provide a short-cut to the functionality described in the work on the GRACE interface[14].

Although we have found it necessary to completely reorganize the structure of the programs used and have taken the opportunity to improve many of the underlying algorithms the basic approach of R-Matrix theory appears entirely suitable for use on distributed-memory parallel computers. Good efficiency and scaling properties can be achieved and there is every expectation that this work will result in a new range of large-scale scattering calculations which have not been feasible on serial computers for example in electron scattering by ions of iron-peak elements and in scattering at intermediate energies.

ACKNOWLEDGEMENTS

The work is part of the programme of the Multiphoton and Electron Collisions Consortium.

REFERENCES

1. P.G. Burke, The Calculation of Atomic Properties, in *Computational Physics*, J.E. Crow and M.G. Haines, eds., published in the IOP and Physical Society Conference Digest Series, London (1970).
2. P.G. Burke and K.A. Berrington, *Atomic and Molecular Processes: an R-matrix Approach*, IOP Publishing, Bristol (1993).
3. M J Seaton, Atomic data for opacity calculations, *J. Phys. B: At. Mol. Opt. Phys.* 20:6363 (1987).
4. K. Bartschat, E.T. Hudson, M.P. Scott, P.G. Burke and V.M. Burke, Electron atom scattering at low and intermediate energies using a pseudo-state/R-matrix basis, *J. Phys. B: At. Mol. Opt. Phys.* 29:115 (1996).
5. P.G. Burke, V.M. Burke and K.M. Dunseath, Electron-impact excitation of complex atoms and ions, *J. Phys. B: At. Mol. Opt. Phys.* 27:5341 (1994).
6. V.M. Burke and C.J. Noble, FARM - a flexible asymptotic R-matrix package, *Comput. Phys. Commun.* 85:471 (1995).
7. H. Teng, M.S.T. Watts, V.M. Burke and P.G. Burke, Electron impact excitation of complex atoms and ions IV: forbidden transitions in Ni^{4+}, *J. Phys. B: At. Mol. Opt. Phys.* 31:1355 (1998).
8. V.M. Burke and C.J. Noble, Recent developments in R-matrix programs for electron-atom scattering, *Comput. Phys. Commun.* 84:19 (1994).
9. *MPI: A Message-Passing Interface Standard*, Message Passing Interface Forum communication (1995).
10. D. Elwood, G. Fann and R. Littlefield, *Parallel Eigensolver System*, User Manual available from *anonymous@ftp://pnl.gov* PNNL (1994).

11. A.G. Sunderland, J.W. Heggarty, C.J. Noble and N.S. Scott, Parallelization of R-matrix propagation methods on distributed memory computers, *Comput. Phys. Commun.* (1998) submitted.

12. K.L. Baluja, P.G. Burke and L.A. Morgan, R-matrix propagation program for solving coupled second-order differential equations, *Comput. Phys. Commun.* 27:299 (1982).

13. L. Wall, T. Christiansen and R.L. Schwartz, *Programming Perl*, O'Reilly & Associates, Bonn (1996).

14. N.S. Scott, J. Johnston, V.M. Burke, C.J. Noble and D.W. Busby, GRACE: the problem specification state. II, *Comput. Phys. Commun.* 84:317 (1994).

PARTIAL WAVE INTEGRALS

J. Rasch[1] and Colm T. Whelan[2]

[1]Institut de Physique, Laboratoire de Physique Moléculaire et des Collisions, Technopôle Metz 2000, Rue Arago, France
[2]Department of Applied Mathematics and Theoretical Physics, University of Cambridge, Silver Street, Cambridge, CB3 9EW, England

INTRODUCTION

Perhaps the most common problem faced in collision theory is the evaluation of cross sections in the scattering from a symmetric radial potential $V(r)$, and the most common approach is to to expand the potentials in multipoles and the wave function in partial waves. One is then faced with the problem of evaluating integrals over highly oscillatory integrands when these integrals are required to high accuracy, since such is the character of the expansion that situations where cancellations can occur between near equal terms are common. These problems are particularly acute for integrals involving large values of the angular momentum.

In this paper we will consider these integrals, as they are met in both relativistic and non-relativistic problems and consider the analytic and numerical methods that have been employed to evaluate them. The characteristic integral that one needs to evaluate in the non-relativistic case is[1, 2, 3, 4, 5]

$$(1) \quad I(l_1, l_2, \lambda, l_0, l_h) := \int_0^\infty \int_0^\infty u_{l_1}^a(k_1 r_1) u_{l_2}^b(k_2 r_2) \frac{r_<^\lambda}{r_>^{\lambda+1}} u_{l_0}^0(k_0 r_1) w_{l_b}(r_2) r_1^2 r_2^2 \, dr_1 \, dr_2$$

where one assumes that w_{l_b} is an exponentially decaying function and each of the functions $u_{l_0}^0$, $u_{l_1}^a$ and $u_{l_2}^b$ can be expressed asymptotically in the form

$$(2) \qquad u(kr) = \cos(\delta_l) f_l(kr) + \sin(\delta_l) g_l(kr) \qquad r \geq R$$

where

$$(3) \qquad f_l(kr) = \begin{cases} j_l(kr) & \text{spherical Bessel function} \\ F_l(k,r) & \text{regular spherical Coulomb function} \end{cases}$$

$$(4) \qquad g_l(kr) = \begin{cases} n_l(kr) & \text{spherical Neumann function} \\ G_l(k,r) & \text{irregular spherical Coulomb function} \end{cases}$$

and δ_l denotes the lth scattering phase shift.

High Performance Computing
Edited by R. J. Allan *et al.*, Kluwer Academic / Plenum Publishers, New York, 1999

For large values of l the functions u are approximately zero in the region of finite w_{l_b} and one may make the Bethe approximation[2, 6]

$$(5) \qquad \frac{r_<^\lambda}{r_>^{\lambda+1}} = \frac{r_2^\lambda}{r_1^{\lambda+1}}$$

After making this approximation the major computational difficulties become concentrated in the integral

$$(6) \qquad J_{0,\infty}(k_1, l_1, k_0, l_0, \lambda) := \int_0^\infty r_1^{-\lambda+1} u_{l_1}^a(k_1 r_1) u_{l_0}^0(k_0 r_1) \, dr_1$$

The above analysis can be generalised to relativistic problems and in particular has been applied to the study of (e,2e) processes on the K and L shells of heavy metal targets[']. In this case the key integral (1) generalises to

$$(7) \mathcal{J} = \int_0^\infty dr_1 r_1^2 \int_0^\infty dr_2 r_2^2 \left\{ j_n(\zeta r_<) \left[-y_\lambda(\zeta r_>) + i j_\lambda(\zeta r_>) \right] \right\} \phi_1(r_1)\phi_0(r_1)\phi_2(r_2)\phi_b(r_2)$$

where all the ϕ's represent a large or small component solution of the Dirac equation, as defined in Ref. 7, ϕ_b is a bound state, the other three correspond to continuum wave functions. A further complication arises in the relativistic case since one needs to use the full QED propagator[7, 8] in place of the Coulomb potential. The terms in the curly brackets involving $j_n(\zeta r_<)$, $y_\lambda(\zeta r_>)$, $j_\lambda(\zeta r_>)$, where $\zeta := \frac{E_0 - E_1}{c}$ come from the multipole expansion of the propagator. They add a further two oscillatory functions to the integral, i. e. in equation (7) we have now got to deal with an integral over 5 highly oscillatory functions and one bound state. One may use eventually the same argument as before for large values of the angular momentum[7] to write

$$(8) \qquad r_> \approx r_1, \qquad r_< \approx r_2$$

i. e. make the Bethe approximation, and thus reduce the problem to the solution of an integral including a bound function ϕ_b and integrals of the form

$$(9) \qquad \mathcal{J}_0 = \int_0^\infty dr_1 \phi_1(r_1)\phi_0(r_1)b_\lambda(\zeta r_1)$$

where $b_\lambda(\zeta r_1)$ are spherical Bessel functions of the first of second kind. Even for the lower angular momentum values integrals of the type (7) and (9) occur but then the lower bound of the integral is replaced by a value $R \geq 0$, i. e.

$$(10) \qquad J_R = \int_R^\infty dr_1 u_{l_1}^a(k_1 r_1) u_{l_0}^0(k_0 r_1)$$

$$(11) \qquad \mathcal{J}_R = \int_R^\infty dr_1 \phi_1(r_1)\phi_0(r_1)b_\lambda(\zeta r_1)$$

Now if R is sufficiently large then one may replace the u's and ϕ's by their asymptotic forms.

ANALYTIC RESULTS FOR FINITE RANGE POTENTIALS

Let us now replace the u's ϕ's and b's by spherical Bessel functions of the first kind then we need to evaluate

$$(12) \qquad J_R = \int_R^\infty r^{-\lambda+1} j_l(kr) j_{l'}(k'r) dr$$

$$(13) \qquad \mathcal{J}_R = \int_R^\infty r^2 j_l(kr) j_{l'}(k'r) j_\lambda(\zeta r) dr$$

where $R \geq 0$. It will be convenient to assume $k \neq k'$. Let us first consider the special case $R = 0$. Whelan[6] has shown that

$$(14) \qquad J_0 = \frac{\overline{\pi}}{kk'} \sum_{\mathcal{L}} \mathcal{Q}_{\lambda \mathcal{L}}(k, l, k', l') Q_{\mathcal{L}}(\chi) \frac{1}{2^\lambda \Gamma(\lambda + \frac{1}{2})}$$

where the sum of $\mathcal{L}$ is finite and the coefficients $\mathcal{Q}_{\lambda \mathcal{L}}(k, l, k', l')$ are defined in Seaton[9, 10], $Q_{\mathcal{L}}(\chi)$ is a Legendre function of the second kind, $\chi := \frac{k^2 + k'^2}{2kk'} > 1$. The great advantage of the result (14) is that it is exceedingly easy to accurately compute the value of J_0. A numerical technique for evaluating the Q's based on a backward recurrence relation has been devised[6] and successfully implemented[10].

Whelan[11] also gave an analytic solution for the more general integral J_R for arbitrary non-negative values of R and an analytic result[12] was obtained for the integral

$$(15) \qquad \mathcal{J}_0 = \int_0^\infty r^2 j_l(kr) j_{l'}(k'r) j_\lambda(\zeta r) dr$$

These results are important in themselves, the integrals occur frequently in atomic collision theory[1], in structure calculations[13] and nuclear physics problems[14] and crucially they also provide us with analytic results which can be used to benchmark the numerical methods for the general case.

CONTOUR INTEGRATION

Often one is interested in integrals of the form (10) and (11) where each function in the integrand allows for an asymptotic expansion of the from (2). In this case the analytic form of the integrand can be exploited by rotating the integration contour into the complex plane where the integrand usually exhibits exponentially increasing or decreasing behaviour. This method has been used by various authors[15, 14, 16]. However, as was pointed out by Rasch[17], for high accuracy a merely asymptotic representation of the functions is usually not sufficient for applications of current interest[18]. In the following we shall highlight the main points for securing high accuracy and refer the reader to the literature for more details[17].

We shall restrict ourselves to the treatment of the integral J_R, equation (10), since the treatment of $\mathcal{J}_R$, equation (11), is a strait forward generalisation of the simpler problem. The functions f_l and g_l in (2) can be expressed asymptotically in terms of Hankel functions

$$(16) \qquad \chi_l = \frac{1}{2} t \left(\mathcal{H}_l^+ + s \mathcal{H}_l^- \right)$$

where

$$(17) \quad t = \begin{cases} 1 & \text{for } \chi_l = j_l(kr), F_l(k,r) \\ \frac{1}{i} & \text{for } \chi_l = n_l(kr), G_l(k,r) \end{cases} \qquad s := \begin{cases} 1 & \text{for } \chi_l = j_l(kr), F_l(k,r) \\ -1 & \text{for } \chi_l = n_l(kr), G_l(k,r) \end{cases}$$

Because of the similarities between Bessel and Coulomb functions we will treat both on an equal footing and therefore let $\mathcal{H}$ denote a Hankel function for the two cases:

$$(18) \qquad \mathcal{H}_l^\pm := \begin{cases} H_l^\pm e^{\pm i \sigma_l} & \text{for } \chi_l = F_l(k,r), G_l(k,r) \\ h_l^\pm & \text{for } \chi_l = j_l(kr), n_l(kr) \end{cases}$$

The exact asymptotic form for the case of a Coulomb function is (see e.g. 19):

$$(19) \qquad H_l^\pm = \pm \frac{1}{i} e^{\pm ikr \mp il \frac{\pi}{2} \mp i\eta \ln(2kr)} e^{\pm i \sigma_l} v(l + 1 \pm i\eta, -l \pm i\eta, \pm 2ikr)$$

303

where

$$(20) \qquad v(\alpha, \beta, z) := \sum_{n=0}^{\infty} \frac{\Gamma(n+\alpha)\Gamma(n+\beta)}{\Gamma(\alpha)\Gamma(\beta)} \frac{1}{n! z^n}$$

Inserting equation (16) and (2) into J_R in (10) gives rise to 16 integrals of which, however, only the two integrals $\int_R^{\infty} \mathcal{H}_{l_f}^+ \mathcal{H}_{l_i}^+ r^{-\lambda+1} \, dr$ and $\int_R^{\infty} \mathcal{H}_{l_f}^+ \mathcal{H}_{l_i}^- r^{-\lambda+1} \, dr$ need to be calculated. All others can be obtained by complex conjugation and using the correct phase factors.

It was shown[17] that for many cases using (20) allowed one to achieve at least 8 figure accuracy in `real*8`. Earlier treatments using such an expansion achieved only about 3 digits[15]. This method of rotating the contour works well for a variety of k and l values and it has even been successfully applied to relativistic electron impact ionisation (see 7) where k values as large as ≈ 250 in a. u. arise. However as pointed out[17] there can be a problem with using an expansion of the form (19), when one has to evaluate integrals where $k_0 \approx k_1$ this method can fail. In this case the monopole $\int_0^{\infty} j_l(k_1 r) j_l(k_0 r) r^{+1} \, dr$ and also dipole integrals die off only very slowly as can be clearly seen by looking at the analytic result (14). The value for $\chi = \frac{k_0^2 + k_1^2}{2 k_0 k_1}$ is in this case very close to 1 and therefore close to the singularity of the Legendre polynomial $Q_{\mathcal{L}}$. A large number of these integrals with high l-states need therefore to be calculated. As shown[17, 18] between 400 and 1000 l-states need to be taken into account. For such large l-states the asymptotic expansion of v (20) will eventually fail to converge.

A method to overcome this problem is to use a subtraction scheme of the form

$$(21) \qquad \int_R^{\infty} f_{l_1} f_{l_0} r^{-\lambda+1} \, dr = \int_0^{\infty} f_{l_1} f_{l_0} r^{-\lambda+1} \, dr - \int_0^R f_{l_1} f_{l_0} r^{-\lambda+1} \, dr$$

In the case of Bessel functions the first integral on the left hand side can be calculated up to machine precision for any value of $k_0, l_0, k_1, l_1, \lambda$ by using equation (14). In the case of Coulomb functions it is possible to express the most important monopole and dipole terms by terms of hypergeometric $_2F_1$ functions which can be evaluated to high accuracy[17]. The second integral however can be calculated using numerical integration. Rasch and Whelan used the standard Clenshaw-Curtis and Gauss-Legendre integration schemes. In the case of Legendre integration the integration interval was split into n evenly spaced subintervals. Each subinterval was evaluated using an mth order Gauss-Legendre quadrature[20]. This method avoids the problem of calculating a high order quadrature rule which is difficult to achieve numerically but also exhibits numerical instabilities for highly oscillatory integrals. Using this method[17] Rasch and Whelan achieved at least 8 figures, and typically 10, accuracy by comparison with the analytical results of Whelan[6, 11].

CONCLUSIONS

Methods for evaluating the highly oscillatory integrals characteristic of collision problems have been analysed. We have shown that in a number of cases general analytic results can be found and in all other cases it is possible to accurately and efficiently generate these integrals numerically.

ACKNOWLEDGMENTS

We gratefully acknowledge support from the Royal Society, the EC (FMBICT972172), NATO(CRG 950 407). We are grateful for the use of the Hitachi parallel High Performance Computer at the Computer Laboratory, Cambridge, U.K. This work was (partially) supported by the National Science Foundation through a grant for the Institute for Theoretical Atomic and Molecular Physics at Harvard University and Smithsonian Astrophysical Observatory.

REFERENCES

1. Colm T. Whelan, M.R.C. McDowell, and P.W. Edmunds, Electron impact excitation of atomic hydrogen, *J. Phys. B: At. Mol. Phys.* 20:1587 (1987).
2. A. Burgess, D.G. Hummer, and J.A. Tully, electron impact excitation of positive ionsm *Phil. Trans. Roy. Soc. A* 266:225 (1970).
3. K. Alder, A. Bohr, T. Iluus, B. Mottelson, and A. Winther, Study of nuclear structure by electromagnetic excitation with accelerated ions, *Rev. Mod. Phys.* 28:432 (1956P.
4. D.H. Madison, R.V. Calhoun, and W.N.A. Shelton, Triple-differential cross section for electron-impact ionization of helium, *Phys. Rev. A* 16:552-62 (1977).
5. J. Rasch and Colm T. Whelan, On the numerical evaluation of a class of integrals occurring in scattering problems, *Comput. Phys. Commun.* 101:31 (1997).
6. Colm T. Whelan, On the Bethe approximation to the reactance matrix, *J. Phys. B: At. Mol. Phys.* 19:2343 (1986).
7. S. Keller, Colm T. Whelan, H. Ast, H.R.J. Walters, and R.M. Dreizler, Relativistic distorted-wave Born calculations for (e,2e) processes on inner shells of heavy atoms, *Phys. Rev. A* 50:3865 (1994).
8. H. Ast, S. Keller, R.M. Dreizler, Colm T. Whelan, L.U. Ancarani, and H.R.J. Walters, On the position of the binary peak in relativistic (e,2e) collisions, *J. Phys. B: At. Mol. Phys.* 29:L585 (1996).
9. M.J. Seaton, The evaluation of partial wave integrals in the Born approximation, *Proc. Phys. Soc.* 77:184 (1961).
10. A. Burgess and Colm T. Whelan, BETRT - a procedure to evaluate the cross section for electron-Hydrogen collisions in the Bethe approximation to the reactance matrix, *Comp. Phys. Commun.* 47:295 (1987).
11. Colm T. Whelan, On the use of the Bethe approximation in the treatment of long-range multipole interactions in electron–neutral-atom collisions, *J. Phys. B: At. Mol. Phys.* 20:6641 (1987).
12. Colm T. Whelan, On the evaluation of integrals over three spherical Bessel functions, *J. Phys. B: At. Mol. Phys.* 26:L823 (1993).
13. I.P. Grant and H.M. Quinez, A class of Bessel function integrals with application in particle physics, *J. Phys. A* 26:7547 (1993).
14. K.T.R. Davies, M.R. Strayer, and G.D. White, Complex-plane methods for evaluating highly oscillatory integrals in nuclear physics: I, *J. Phys. G* 14:961 (1988).
15. N.C. Sil, M.A. Crees, and M.J. Seaton, Integrals involving products of Coulomb functions and inverse powers of the radial coordinate, *J. Phys. B: At. Mol. Phys.* 17:1 (1984).
16. K.T.R. Davies, Complex-plane methods for evaluating highly oscillatory integrals in nuclear physics: II, *J. Phys. G* 14:979 (1988).
17. J. Rasch and Colm T. Whelan, On the numerical evaluation of a class of integrals occurring in scattering problems, *Comput. Phys. Commun.* 101:31 (1997).
18. J. Rasch, M. Zitnik, L. Avaldi, Colm T. Whelan, G. Stefani, R. Camilloni, R.J. Allan, and H.R.J. Walters, Theoretical and experimental investigation of the triple differential cross sections for electron impact ionization of Kr(4p) and Xe(5p) at 1keV impact energy, *Phys. Rev. A* 56:4644 (1997).
19. C.J. Joachain, *Quantum Collision Theory*, North-Holland (1987).
20. W.H. Press, S.A. Teukolsky, W.T. Vetterling, and B.P. Flannery, *Numerical Recipes in Fortran*, Cambridge University Press, Cambridge (1992).

MOLECULAR ROTATION-VIBRATION CALCULATIONS USING MASSIVELY PARALLEL COMPUTERS

Hamse Y. Mussa,[1] Jonathan Tennyson,[1] C. J. Noble,[2] and R.J. Allan[2]

[1]Department of Physics and Astronomy,
 University College London, London WC1E 6BT, UK
[2]CLRC Daresbury Laboratory, Daresbury, Warrington WA4 4AD, UK

INTRODUCTION

Calculations of rotation-vibration spectra of small molecules by direct solution of the nuclear motion Schrödinger equation are beginning to make a major impact in the area of experimental spectroscopy[1]. However, although the techniques used to solve these problems are now well developed, there are applications, even for three atom systems on a single potential energy surface, which present a formidable computational challenge. Such systems may possess $10^5 - 10^6$ bound states.

One class of methods which have proved very successful at treating large nuclear motion problems are based on a particular finite element representation of the problem known as the Discrete Variable Representation[2]. Here we describe how the DVR3D package of Tennyson and co-workers[3], has been ported to massively parallel computers in manner appropriate for tackling large problems. Wu and Hayes [4] have also addressed the problem of performing rotation-vibration calculations for three atom systems on a Cray T3D machine. Our method is designed to tackle larger problems than theirs and is therefore represents a more radical change from algorithms used on sequential machines. A preliminary version of the algorithm discussed here has been published[5].

PDVR3DR : AN OVERVIEW

The DVR3D program suite[3] contains a number of modules. Program DVR3DRJ solves the rotationless $J = 0$ vibrational problems using either atom-diatom Jacobi coordinates or Radau coordinates. It also solves the effective 'vibrational' problems required for a two-step treatment of the rotational excitation problem[6]. Programs ROTLEV3 and ROTLEV3B are driven by DVR3DRJ and solve the fully coupled rotation-vibration problem respectively for Jacobi and a symmetrized grid in Radau coordinates. A final module, DIPOLE3, computes transition dipoles but will not concern us here.

In this article parallel versions of the programs DVR3DRJ and ROTLEV3B are presented. However parallelization requires a number of changes in the solution algorithm which were best achieved by splitting program DVR3DRJ. The parallel version thus consists of PDVR3DR and PROTLEV3R which solve the nuclear motion problem for symmetrized Radau coordinates with a bisector embedding[7], and PDVR3DJ and PROTLEV3J which are appropriate for Jacobi coordinates. Parallelizing the Radau co-ordinate program is the more difficult task and the only one we discuss here.

When determining parallelization procedures it is necessary to consider both the construction of the scientific problem being addressed and the architecture of the computer being employed. The codes themselves are not molecule or problem specific but it is helpful to use specific examples as these define typical limits of a calculation. Here we consider water at its dissociation limits as the prototype system to be treated in symmetrized Radau coordinates. The DVR3D programs have been adapted relatively easily to run very efficiently on parallel shared memory architectures. However these machines have few processors and the benefits in throughput are limited.

The intended architecture is a parallel message passing computer. PDVR3DR and PROTLEV3R are designed for parallel and distributed memory computers which use a Single Program Multiple Data (SPMD) loading strategy. Particular architectures used are the 26-processor IBM SP2 machine at Daresbury Laboratory, the Edinburgh Parallel Computer Centre (EPCC) Cray Machines – the T3D which has 512 processors, and the T3E which has 256 application processors –, and finally the 128 processor Cray-T3E in Bologna. For algorithmic reasons, described below, PDVR3DR program is usually run on 64 or 128 processors of the T3D, 16 to 64 of the EPCC Cray-T3E, 8 to 16 of the Daresbury IBM SP2 and 16 to 128 of the Bologna Cray-T3E while PROTLEV3R can use any number of processors.

Originally the programs were written using PVM[8] but we switched to MPI[9] when this became available. The DVR3D program suite was written in FORTRAN77, but modifications use Fortran 90 constructs.

The T3D and the SP2 were used for development purposes. Since the intention of the work is to build and diagonalize large Hamiltonian matrices, our work has been done mainly on the T3D and T3E's. The results presented here are for the Cray-T3D, similar results were obtained on the IBM SP2 when the size of the problem matched the available memory.

VIBRATIONAL CALCULATIONS

For an AB_2 molecule, Radau coordinates[10] consist of two symmetry related stretching coordinates and an included angle. For molecules, such as water, where the central atom is significantly heavier than the others ($m_A >> m_B$), these coordinates are similar to those given by the two AB bond distances, r_1 and r_2, and the $A\hat{B}A$ bond angle, θ. Coordinates (r_1, r_2, θ) are represented using DVR grids (α, β, γ).

The DVR3D programs use sequential diagonalization and truncation[2, 3] to create a compact final Hamiltonian matrix. For such schemes, the order in which this diagonalization and truncation is performed can be crucial in determining computational requirements[11]. Because the Radau coordinate symmetrization procedure used by DVR3DRJ is based on symmetrizing DVR grid points and not basis functions[12], the stretches need to be treated equivalently and hence together, and the logical choice is to treat the angular θ coordinate last.

The basic method of distributing the DVR calculation over N processors was to place $\frac{n_\gamma}{N}$ (n_γ = number of angular grid points) on each processor. Since $\frac{n_\gamma}{N} = 1$ is faster than $\frac{n_\gamma}{N} > 1$, this coordinate was usually represented by either $n_\gamma = 64$ or 128 'active' grid points, but calculations where $n_\gamma = 48, 80, 96$ have also been performed.

In the standard energy-selected diagonalization and truncation procedure, the number of functions used to construct the final Hamiltonian varies with γ. This algorithm leads to an poorly load-balanced calculation, so the selection criterion was modified. In the modified version, an equal number of eigenstates of the two-dimensional problem, n^{2D}, were retained for each angular 'active' grid point. Inactive grid points were simply dropped from the calculation. One could include all γ grid points in a calculation, but for water geometries near linear OHH configurations are very high energy, and indeed the Tennyson and Sutcliffe[7] strategy for implementing the bisector embedding involves dropping at least the point nearest linearity. In practice our water calculations drop about 10% of the points, but tests we have performed on ozone involved dropping more than half. For ozone points are dropped from both ends of the angular grid.

Construction and diagonalization of intermediate, 2D Hamiltonians is performed simultaneously on each processor using a standard diagonalizer. We use NAG routine F02ABF [13] for this purpose. In this approach, labelled SDTA for Sequential Diagonalization Truncation Approach, the size of the final Hamiltonian, $\mathbf{H}^{\text{SDTA}}$, is given by

$$N^{3D} = n^{2D} \times n_\gamma \tag{1}$$

and

$$\mathbf{H}^{\text{SDTA}} = \mathbf{E}^{2D} + \mathbf{C}^{(2D)} \mathbf{L} \mathbf{C}^{(2D)\text{T}} \tag{2}$$

Where $\mathbf{E^{2D}}$ are the eigenvalues of the 2D Hamiltonian in (α, β) at each angular grid point γ and $\mathbf{C^{(2D)}}$ are the corresponding eigenvectors. $\mathbf{L}$ represents the angular kinetic energy terms[3, 12].

Construction of the final 3D Hamiltonian matrix requires all the vectors, $C_j^{2D}(\alpha\beta : \gamma), j = 1, n^{2D}$, obtained for the n_γ active angular grid points. A strip of the Hamiltonian is constructed on each processor which guarantees that one set of vectors is always correctly located. For blocks off-diagonal in γ, a second set of vectors has to be imported from another processor if they are not locally available.

Time-wise the most critical step is diagonalization of the final 3D Hamiltonian. A survey of available parallel diagonalizers has been conducted by Allan[14] and discussed by us[5].

Standard real symmetric diagonalisers, of which PeIGS[15] proved the best, limit the size of calculation which could be attempted due particularly to workspace considerations[5]. We therefore explored the possibility of using iterative diagonalisation procedure. Such an approach has been advocated by a number of other workers[4, 16], although, in contrast to many other studies, we require eigenvectors as well as eigenvalues. The parallel iterative eigensolver PARPACK[17] is designed to obtain a few eigenvalues of a large sparse matrix. To diagonalize the Hamiltonian matrix, $\mathbf{H}$, iteratively, as is usual, one needs to perform the matrix-vector product

$$\mathbf{y} = \mathbf{Hx} \tag{3}$$

However, there are three options as far as this operation is concerned:
a) $\mathbf{H}$ is obtained without any intermediate diagonalisation and truncation. In this approach the Hamiltonian is mainly block diagonal. Operation (3) can therefore be performed without forming $\mathbf{H}$ by using only the non-zero elements since the zero elements do not contribute. This operation is now given by $\mathbf{y} = (\mathbf{elements} > \mathbf{0})\mathbf{x}$ and takes advantage of the natural sparseness of the DVR based Hamiltonian. Such an approach has been used very successfully on sequential machines[16]. It involves treating very sparse matrices of dimension 100,000 or larger. On parallel machines vector $\mathbf{x}$ is distributed over the processors and broadcast to the other processors. BLAS routines

are used to give **y**. This procedure is very efficient and requires little interprocessor communication.

b) **H** is obtained by the SDTA procedure. It is possible to use the iterative diagonaliser to diagonalize the same matrix as we tested the real symmetric diagonalisers on[5]. **x** is distributed as in (a) and **y** is obtained by using BLAS2 routine sgemv.

c) **x** is distributed as in (a) and **H** is obtained by the SDTA procedure, not formed explicitly. Thus

$$\mathbf{y} = \mathbf{H}^{\mathbf{SDTA}}\mathbf{x} = (\mathbf{E}^{\mathbf{2D}} + \mathbf{C}^{(\mathbf{2D})}\mathbf{L}\mathbf{C}^{(\mathbf{2D})\mathbf{T}})\mathbf{x} \tag{4}$$

$$\mathbf{y} = \mathbf{E}^{\mathbf{2D}}\mathbf{x} + (\mathbf{C}^{(\mathbf{2D})}\mathbf{L}\mathbf{C}^{(\mathbf{2D})\mathbf{T}})\mathbf{x} \tag{5}$$

One virtue of method (a) is that in its simplest form it is very easy to program. However the symmetrisation of the radial grids points employed by DVR3D presents a significant complication as it destroys much of the structural simplicity of the un contracted Hamiltonian matrix. This simplicity is important for the efficiency of this algorithm[16]. For this reason we implemented this method without using the radial symmetry. In this case the size of the matrix is given as the product of the number of grid points $n_\alpha n_\beta n_\gamma$. Distributing the matrix as before over the angular grid, γ, there is a local memory balance between the number of eigenvectors required and the size of the matrix stored on each processor. The maximum size for which we performed calculations was for 100 eigenstates of a matrix of dimension 102400. These proved considerably more expensive to obtain than method (b). One reason for the difference is that the speed of PARPACK convergence is dependent on the eigenvalue distribution of the Hamiltonian matrix[4]. For us the speed of convergence is related to the $\mathbf{H}^{\mathbf{SDTA}}$ Hamiltonian eigenvalues distribution via

$$r = \frac{|\lambda_{suw} - \lambda_{lw}|}{|\lambda_{max} - \lambda_{suw}|} \tag{6}$$

Where suw is the smallest unwanted eigenvalue, lw is the largest eigenvalue required, and max is the largest eigenvalue of the Hamiltonian matrix. Larger r gives faster the convergence. The SDTA affects neither λ_{suw} nor λ_{lw}, but it reduces the λ_{max}. So the SDTA increases r and one can say then that the SDTA procedure not only reduces the size of the Hamiltonian, but is also improves the conditioning of the matrix and by doing so it speeds up convergence.

An advantage of (c) over (a) is faster convergence, and its advantage over (b) is less memory requirement. However, it has a disadvantage over (b) in that the number of matrix-vector operations needed is much bigger. Since the calculation is a trade-off between memory requirement and CPU-time, and our main concern when using PARPACK as the eigensolver is the CPU-time rather than memory, only (a) and (b) have been implemented. Implementing (c) would be straightforward and can be done if needed.

Using the PARPACK to diagonalise the contracted Hamiltonian, method (b), has one immediate advantage over use of PeIGS. PARPACK has little workspace requirement meaning that, in practice, matrices of twice the size could be diagonalized.

ROTATIONAL EXCITATION

Eigenvectors from the vibrational step of the calculation are required as a basis for computing rotationally excited states. It is necessary to transform the eigenvectors obtained from the final STDA diagonalization back to vectors which give the amplitude of the wavefunction on the original, raw grid points. As only h ($< N^{3D}$) eigenvectors need to be transformed, it is necessary to redistribute the eigenvectors so that those required are spread equally between the processors.

The transformation can be written

$$d^{ik}_{\alpha\beta\gamma} = \sum_{j=1}^{n^{2D}} C^{2D}_j(\alpha\beta:\gamma)C^{3D}_i(j,\gamma), \qquad i = 1,2,3...,h, \tag{7}$$

The 2D vectors, $\underline{C}^{2D}$ are mapped on to processors using γ as they are generated during the Hamiltonian construction. The 3D vectors, $\underline{C}^{3D}$, are mapped by distributing vectors between processors. The multiplication is performed using MPI broadcast-reduce routines. If one wants to analyse the wavefunctions of the vibration, the results are saved on disk. Since I/O is not parallel, one processor is used to perform the I/O to avoid bottle-necks. When considering states of rotational excitation J, it is necessary to repeat all the above steps for each k, the projection of J on the body-fixed z-axis, required. In general, if both even and odd Wang parity rotational states are to be calculated using the same first step vectors, then $J + 2$ separate calculations are performed, one for $k = 0,1,\ldots J$ plus an extra $k = 1$ calculation as this must be treated as a special case[7.]

Saving the transformed eigenvectors in eq. (7) for each k [5] causes disk space problems for even low J calculations. For example, we have access to 2GB only of disk space and to accurately calculate $J = 2$ up to dissociation requires 3.5 GB for the transformed eigenvectors. Therefore, PDVR3DR has been modified to keep transformed eigenvectors in core and then construct the off-diagonal blocks in k, which are contributed by the Coriolis operators, $\hat{H}_{kk'}$. In addition to overcoming the disk space problem, there is another benefit to this – computationally, there is no limit to how many J's one can calculate provided there is enough CPU-time. However, in real memory terms this costs 30 to 50 MBs extra for a typical calculation.

For symmetrized Radau coordinates, symmetry considerations dictate that the Cartesian axes of the system should be placed so x bisects the angle θ and the z axis is then in the plane of the molecule and perpendicular to x. This is the so-called bisector embedding[7]. In this embedding, the Coriolis coupling links blocks of the matrix labeled by k with those for with $k \pm 1$ and $k \pm 2$[7].

The block construction step consists of a series of transformations of the form

$$B^{kk'}_{ii'} = < k,i|\hat{H}_{kk'}|k',i' > = \sum_{\alpha\beta\gamma\gamma'} d^{ik}_{\alpha\beta\gamma} H_{kk'}(\alpha\beta\gamma\gamma')d^{i'k'}_{\alpha\beta\gamma'} \tag{8}$$

using the vectors created in eq. (7). The Coriolis operators are diagonal in the radial DVR's, $(\alpha\beta)$, but not in the angular DVR, γ. Full details are given elsewhere[3, 7]. The transformation uses h ($< N^{3D}$) eigenvectors from each k calculation[18]. In the present procedure it was assumed that h is the same for all k.

Because a four-dimensional transformation, eq. (8), is used to construct each matrix element, this step can be quite time consuming. It is therefore necessary to consider carefully how one might parallelize it. Goldfield and Gray[19] mapped the triatomic rearrangement reaction problem using k (or Ω) blocks. However in our case there is insufficient memory to treat more than one J at a time and only for large J are there enough k blocks to make distributing over the processors a possibility. Alternatively one can distribute over the stretching coordinates $(\alpha\beta)$ or the angle γ or the eigenvectors h. Parallelizing over $(\alpha\beta)$ is difficult because of the symmetrization. Conversely parallelizing over γ is superficially attractive as this is done in the first step and the number of γ's has already been chosen to map conveniently onto the number of processors. However there are problems with distributing on γ. In this mode, each processor

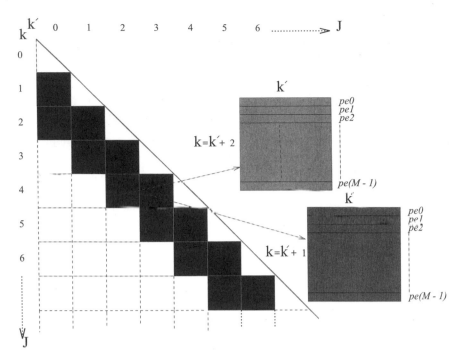

Figure 1. Structure of the fully coupled Hamiltonian diagonaliseded by PROTLEV3R. Only non-zero elements, those on the diagonal and shaded blocks, for the lower triangle are computed. The blocks are spread over the processors as indicated in the enlargements.

will build contributions to every element of the block, which will have to be assembled on a single processor once constructed. More seriously, a typical set of h vectors $\underline{d}^{ik}$ takes about 0.5 GB of memory, so these vectors must be distributed. This method leads to an excessive amount of inter-processor communication. We therefore distribute the building of the blocks over the M processors by placing a $\frac{h}{M}$ bra and $\frac{h}{M}$ ket vectors on each processor. Each processor thus builds rows (columns) of each block, see eq. (8). This is done using the vectors in local memory for the bra and for portion of the ket and vectors from other processors for the other portion of the ket. These following steps show how this is done:

Step 1: For $k = 0$; create $\mathbf{d}^{k=0}$ using eq. (7).

Step 2: For $k = 1$; create $\mathbf{d}^{k=1}$ using eq. (7).

form $\mathbf{B}^{k=0,k'=1}$ using eq. (8), $k' = k + 1$.

Step 3: For $k = 2$ to J; create $\mathbf{d}^k$ using eq. (7).

form $\mathbf{B}^{k,k'}$ using eq. (8), $k' = k + 2$.

replace $\mathbf{d}^{k-2}$ by $\mathbf{d}^{k-1}$.

form $\mathbf{B}^{k,k'}$ using eq. (8), $k' = k + 1$.

This algorithm ensures that not more than three sets of vectors, $\mathbf{d}^k$, are retained in memory at one time.

The diagonal elements, which correspond to the eigenvalues generated in each k calculation, and the built blocks, $\mathbf{B}^{k,k'}$, are sent to one processor which then writes them to disk. Both the data transfer between processors and writing on the disk are done in big chunks to avoid any an unnecessary overheads.

In the final step of the solution procedure, PROTLEV3B, off-diagonal and diagonal elements are read back from the disk. The fully coupled Hamiltonian is then formed by distributing the blocks and the diagonal elements over the processors. The resulting structure, see Fig. 1, is appropriate when the matrix is to be diagonalized iteratively(suitable for both (a) and (b)). This algorithm for treating rotationally excited states is an improvement on our previous procedure[5] because it considerably reduces I/O and disk storage both of which are a serious problem with the old procedure.

PERFORMANCE

As shown by Mussa *et al.*[5], PDVR3DR scales very well when building the Hamiltonians as the number of processors is increased. However, it scales less well when diagonalizing the Hamiltonian on 128 processors or more. This is due to that PeIGS(the diagonaliser used for the scaling test)which is based on Householder transformation. There are several reasons for this:

- Most time is spent in the Householder reduction of the Hamiltonian to tridiagonal form.

- Repeated inverse iteration and orthogonalisation is used to give orthogonal eigenvectors. This can result in both poor load balancing in its algorithm and in creating large overheads.

- Finally, a matrix of 3200 is used for the scaling so the message passing latency must also be an other factor as the number of processors increases. This factor must be less important as the size of the Hamiltonian increases because of the trade-off between the latency and the bandwidth.

CONCLUSION

We have developed parallel programs for treating the vibration-rotation motion of three-atom systems using either Jacobi or Radau coordinates. These programs are based on the published DVR3D program suite[3] which is designed for computers with traditional architectures. Significant algorithmic changes were required, in particular to reduce I/O and disk usage in the original programs and to produce a load balanced algorithm. The new suite shows good scalability and can be used for more challenging calculations. However diagonalization remains a bottleneck in these calculations. Tests of presently available software favour the use of iterative diagonalisation procedures, although further significant improvements in real symmetric matrix diagonalizers could alter this view.

Our parallel DVR suite has enabled us to calculate all the bound vibrational states of water in one wall clock hour using the EPCC Cray T3D. Using the modifications discussed above we have been able to calculate bound rotation-vibration states of water all the way to dissociation for $J > 0$. Previously such calculations took times measured in days or even in weeks. Results for both rotationless and rotationally excited states are being prepared for publication[20].

ACKNOWLEDGEMENTS

This work was performed as part of the ChemReact High Performance Computing Initiative (HPCI) Consortium. We thank members of HPCI Centre at Daresbury Laboratory for their help.

REFERENCES

1. O.L. Polyansky, N.F. Zobov, S. Viti, J. Tennyson, P.F. Bernath and L. Wallace, *Science*, 277:346 (1997).
2. Z. Bacic and J. C. Light, *Ann. Rev. Phys. Chem.* 40:469 (1989).
3. J. Tennyson, J.R. Henderson and N.G. Fulton, *Computer Phys. Comms.* 86:175 (1995).
4. X.T. Wu and E.F. Hayes, *J. Chem. Phys.* 107:2705 (1997).
5. H.Y. Mussa, J. Tennyson, C.J. Noble, R.J. Allan, *Computer Phys. Comms.* 108:29 (1998).
6. J. Tennyson and B.T. Sutcliffe, *Mol. Phys.* 58:1067 (1986).
7. J. Tennyson and B.T. Sutcliffe, *Intern. J. Quantum Chem.* 42:941 (1992).
8. *Parallel Virtual Machine - The PVM3 Users' Guide and Reference Manual* is available from *anonymous@ftp://netlib2.cs.utk.edu/pvm3/ug.ps*
9. *Messuye Passing Interface*, The MPI Standard is available from *anonymous@ftp://netlib2.cs.utk.edu* . Note: it is a very large document!
10. D.It. Johnson and W P Reinhardt, *J. Chem. Phys.* 85:4538 (1986).
11. J.R. Henderson, C.R. Le Sueur, S.G. Pavett and J. Tennyson, *Computer Phys. Comms.* 74:199 (1993).
12. J. Tennyson and J.R. Henderson, *J. Chem. Phys.* 91:3815 (1989).
13. *NAG Fortran Library Manual, Mark 17*, Vol. 4 (1996).
14. R.J. Allan and I.J. Bush, *Parallel Application Software on High Performance Computers: Parallel Diagonalisation Routines.* Edition 3 (Daresbury Laboratory HPCI Centre, 22/8/96) available on WWW URL *http://www.dci.clrc.ac.uk/Publications*
15. G. Fann, D. Elwood and R.J. Littlefield, PeIGS *Parallel Eigensolver System, User Manual*, available via ftp from *anonymous@ftp://pnl.gov* .
16. M.J. Bramley and T. Carrington Jr, ftp from J. Chem. Phys. 99:8519 (1993).
17. K. Maschhoff and D. Sorensen *A portable implementation of ARPACK for distributed memory parallel architectures* Preliminary proceedings, Copper Mountain Conference on Iterative Methods (1996)
18. J. Tennyson, ftp from *J. Chem. Phys.* 98:9658 (1993).
19. E.M. Goldberg and S.K. Gray, ftp from Computer Phys. Comms. 98:1 (1996).
20. H.Y. Mussa and J. Tennyson, to be published.

ENVIRONMENTAL MODELLING

MODELLING CLIMATE VARIABILITY ON HPC PLATFORMS

Lois Steenman-Clark and Alan O'Neill

Centre for Global Atmospheric Modelling
Department of Meteorology
University of Reading
Whiteknights
Reading RG6 6BB

INTRODUCTION

UGAMP, the UK Universities' Global Atmospheric Modelling Programme, is a Natural Environment Research Council (NERC) funded community research programme which brings together university atmospheric groups with the aim of addressing high priority issues in climate research. Participating institutions in UGAMP come from seven universities: Cambridge, East Anglia, Edinburgh, Leicester, Imperial College London, Oxford and Reading as well as the Rutherford Appleton Laboratory (RAL). UGAMP has a close working relationship with weather centres including the European Centre for Medium Range Weather Forecasts (ECMWF) and the UK Meteorological Office (UKMO) as well as oceanographic centres such as the Southampton Oceanographic Centre (SOC). The UGAMP consortium includes some hundred research scientists and students, who use a wide range of numerical models as research tools for controlled experimentation.

The challenge to UGAMP is to develop models which are the foundation of the effort to understand climate variability. One of the most striking characteristics of the climate is its variability on a broad range of time scales. Superimposed on the periodic phenomena, such as the daily and seasonal variations, are many different types of irregular variations. To model phenomena which have time scales of months or years requires model experiments to be run for decades to ensure that a significant signal can be obtained. Such model experiments place huge demands on the national HPC resources. One example of climate variability on a time scale of years is the El Niño, an event in the Pacific Ocean in which the sea temperatures rise sharply on the eastern side and which have a strong influence on the weather patterns throughout the world. The 1997 El Niño, which is the strongest yet recorded, has been headline news because of its world wide economic and social effects. Figure 1, which shows the growth of El Niño events since 1950, illustrates the variability of these sudden but quite natural changes in the ocean currents and the movement of air. It is not certain what sets off this

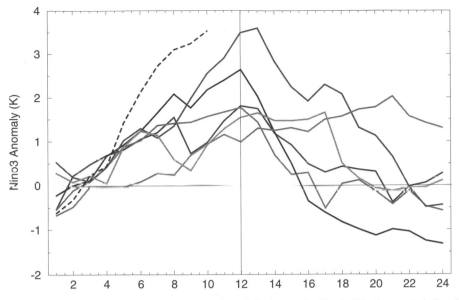

Figure 1. The development of major El Niño events since 1950 characterised by the Niño3 anomaly index plotted over the months of their duration. The 1997 El Niño event is shown as a dotted line.

change. Modelling such phenomena inevitably involves both atmosphere and ocean so these problems cannot be solved by modelling one of these systems in isolation.

UGAMP's research is driven by the need for a better understanding of the atmosphere and its links with other systems such as the ocean. It is this understanding that is necessary to produce future models with greater predictive power. To this end UGAMP uses a wide range of numerical models of varying complexity to study the complex processes that occur in the atmosphere.

MODELS AND TOOLS

Simple models are used to study particular features of the climate system. The results from these models are often easier to interpret, so they are much used to test specific hypotheses. These models, which were developed at the University of Reading, have been successfully used to tackle fundamental questions such as how the Asian Monsoon affects the European climate. The time of year at which the Mediterranean countries are driest, June to August, coincides precisely with the time when South Asia is wettest during the monsoon. The mechanism that links these two phenomena has now been identified showing that the strength of the monsoon can have an impact on the climate of Europe and possibly explaining why high lake levels and lush vegetation in North Africa over the last 140,000 years tended to occur when the Asian monsoon was weak for long periods of time. These simple models can now run at low resolutions on PC's and at very high resolutions on the Edinburgh Parallel Computing Centre (EPCC) Cray T3D. The work of parallelising UGAMP simple global atmospheric model was carried out under the Engineering and Physical Sciences Research Council (EPSRC) HPC initiative.

General circulation models (GCMs) are top-of-the-range models that attempt to simulate the global atmosphere circulation in as much detail as practicable. They give a much more complete representation of the atmosphere and consequently require more HPC resources. They are, at times, themselves difficult to understand and so UGAMP has invested a

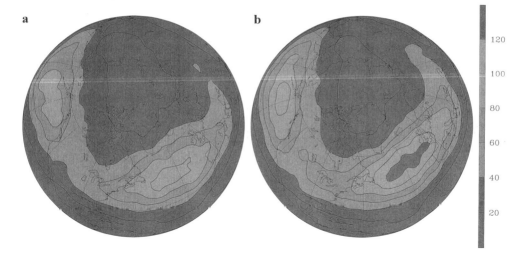

Figure 2. From an analysis of the UKMO Unified Model (UM) experiments with a) present day and b) doubled carbon dioxide levels, the North Atlantic storm track, as shown by the high pass eddy kinetic energy at 250 hPa, shows that the most noticable change over the globe is an increase in the storms reaching the UK.

considerable effort in developing software for analysing the output from these models. UGAMP uses top-of-the-range forecast models from both the UKMO and ECMWF adapted for long range climate simulations. With these models UGAMP has shown, for example, how increasing atmospheric carbon dioxide can strengthen the winter North Atlantic storm track, shown in Figure 2 and how changes in the sea surface temperatures due to El Niño can influence the course of the Asian Monsoon.

UGAMP is aiming to increase understanding of key phenomena and how their representation in forecast models can be improved.

An important component of the UGAMP science plan is improving the simulation of atmospheric chemistry. Chemical transport models, which attempt to represent the complex set of chemical reactions that occur in the atmosphere, can trace the movement of chemical species using the winds and temperatures derived from CGMs or from observations or forecasts. Such models can be used either to provide forecasts for planning observational campaigns or for simulating past chemistry scenarios, such as ozone loss in Northern Hemisphere winters, to evaluate and develop the chemistry schemes in the model. These chemical transport models have now been parallelised to enable high resolution runs on MPPs. Chemical transport models are now beginning to be coupled to atmospheric models to look at, for example, the importance of heterogeneous processes, which occur on surfaces such as clouds or aerosols, on global tropospheric chemistry.

Coupled ocean - atmosphere models are an essential tool to advance the understanding of natural climate variability. These models, which are currently under development in UGAMP, are being used to look at, for example, the natural variability of storm tracks. The North Atlantic storm track, which is an average of weather systems such as depressions, varies from month to month and from year to year and there are indications of variability on longer decadal timescales. There is a strong indication that decadal variability over the North Atlantic involves interactions between the ocean and the atmosphere.The mechanisms involved are poorly understood and this understanding needs to be improved to ensure that such variability represented in climate forecast models.

As well as developing models, UGAMP has invested a considerable effort in developing software tools for analysing the output from these models. Novel diagnostics of model output or observed data can bring insight to the understanding the complex processes. UGAMP has also developed tools which can automate the tracking of atmospheric features such as individual weather systems. These tools can help in the characterisation of phenomena such as the North Atlantic storm track and can help in the analysis of the natural variability of the atmosphere. Contour advection, which is a technique for following the very fine structures which develop in chemical concentrations, was developed in UGAMP and independently at the Massachusetts Institute of Technology, USA and is now used widely to simulate chemical transport and ozone destruction.

The use of these models and tools is aided and enhanced by a range of seemingly mundane tools which pack and unpack data, store and restore data from archive, change grids, change data formats etc., without which running models on today's HPC platforms would be tedious and time consuming.

CURRENT HPC RESOURCES

All the HPC facilities to which UGAMP has access to are shown in Table 1.

UGAMP has access to all the national academic HPC platforms as well as local departmental and University research facilities. NERC has also supplemented the national resources to enable particularly demanding environmental codes to be run. Through

Table 1. HPC facilities used by UGAMP

HPC Platform	Service Centre	Access
Cray J932 32 processors	RAL	All research councils
Cray T3D 512 PEs	EPCC	All research councils
Cray T3E-900 256 PEs	EPCC	PPARC/EPSRC/ NERC
Fujitsu VPP300 3 PEs	RAL	NERC
Cray T3E-900 850 PEs	UKMO	Collaborative agreement
Fujitsu VPP700 116 PEs	ECMWF	Collaborative agreement
NEC SX4 2 processors	Danish Met. Office	EC contract
Hitachi SR2201 224 PEs	University of Cambridge	University research facility
SGI Origin 2000 12 processors	University of Reading	University research facility

collaborative agreements UGAMP also has access to forecast facilities at the meteorological agencies. From Table 1 it can be seen that UGAMP has access to a wide range of different architectures from a number of manufactures. Each of these services presents different challenges in providing UGAMP scientists with the software, data and support at a level which can sustain their scientific programs.

HPC ISSUES

UGAMP embraces a broad range of computational modelling and involves a large number of researchers, which means that the consortium has to deal with a number of difficult HPC issues. The three issues that are currently of concern are; firstly, access to sufficient HPC resources: secondly, porting codes and ensuring that the HPC resources are used efficiently and finally, handling data.

Modelling climate variability requires experiments to be run that are very resource intensive both in CPU and data requirements. Ensembles of experiments, each slightly different, are run to test the sensitivity to some aspect of the model or to the initial data supplied to the model. To perform these ensemble experiments, where each run may be for several model years, demands a high throughput capability, large amounts of CPU time and the ability to handle large volumes of data. To remain on the leading edge of research in this area and to participate in national and international research programs, these experiments need to be performed in a time critical frame which is not possible with current national HPC resources. An illustration of the size of the HPC requirement can be seen from the seasonal forecasting project at ECMWF, where ensembles of twelve to fifteen coupled ocean-atmosphere models are regularly run for six model months, or longer . Details of this seasonal forecasting project being used to study the current El Niño event at ECMWF can be obtained from their web site (*http://www.ecmwf.int*). The HPC resources required for such a project are far beyond those available to the UK academic community.

Even increasing vertical resolution of atmospheric models to study the stratosphere can overwhelm the current national HPC resources. For example, to explore the impact of the stratosphere on the tropospheric climate requires an accurate representation of the stratosphere in atmospheric GCMs. Multi-year simulations using the UKMO Unified Model (UM) are being used in this study, requiring a significant increase in the vertical resolution of the model. Increasing the vertical resolution of the model from the standard 19 levels to 58 levels has a significant impact on the HPC resources required to run these experiments.

To ensure that UGAMP can exploit all the HPC resources available and make use of these resources effectively requires a considerable support effort. UGAMP has benefited from the pioneering efforts of the meteorolgcial agencies such as the UK Meteorological Office and ECMWF to parallelise their model codes. However for UGAMP, even moving to new architectures such as the EPCC Cray T3D and now the Cray T3E requires manpower which would detract from the research effort were it not for support programs such as the research councils' HPC initiative. This initiative provided UGAMP with extra support effort to allow the atmospheric and off-line chemical models to exploit MPP platforms such as the EPCC Cray T3D and T3E, efficiently and effectively. Work carried out under the HPC initiative included the investigation of the scalability on the Cray T3D of the ECMWF atmospheric model, the Integrated Forecast System (IFS), at different horizontal and vertical resolutions, as shown in Figure 3. Work on the chemical transport models is described in a poster given by Cate Bridgeman, University of Cambridge at this meeting.

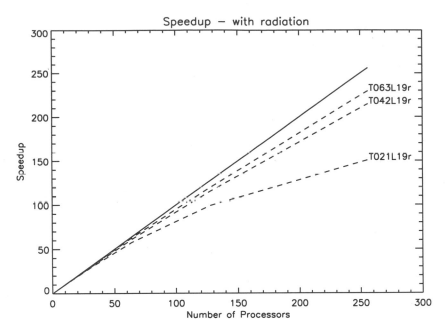

Figure 3. Speedup of the IFS model at different spatial resolutions on the EPCC Cray T3D.

The data volumes generated by climate models can be very large and are a problem to deal with especially when the models generating this data have been optimised to run quickly and efficiently on HPC platforms. UGAMP has, in general, stored model experiment data for several reasons. Firstly, because of the current funding model for national HPC resources it is easier to store data than to re-run experiments as and when the data is required. Secondly, the compilers and operating systems, to which model runs are sensitive, are upgraded when convenient to the majority of users so there cannot be a guarantee to reproduce results when an experiment is re-run. As a large research group there are many students who can be trained initially by analysing experimental model data which can be considerable simpler than setting up the model experiments. One of UGAMP's research strengths is the development of novel diagnostics techniques which can be applied retrospectively to existing data to give new insight to old experiments.

The objective of the ECMWF Re-Analysis (ERA) project was to produce a new, validated 15 year data set of assimilated data for the period 1979 to 1993. This data set is now accessible to all NERC funded researchers at RAL via the British Atmospheric Data Centre (BADC) and it provides an important observed data set for UGAMP. The data set is being analysed for a wide variety of projects from the study of climate variability in the tropics to tracking features in the North Atlantic to characterise the storm track. Fast access to this very large data set is key to many UGAMP researchers.

The data sizes generated by experiments depend on the experimental setup and vary considerably with the vertical resolution, the number of fields output as well as the frequency of output. Typical data sizes generated by current atmospheric and oceanographic models at different resolutions are shown in Tables 2 and 3.

Table 2. Typical atmospheric experiment data sizes for a 10 model year run with data output four times per model day.

Spatial Resolution	Data Sizes (Gbytes)
climate	36.5
seasonal	74.8
climatology	98.1
forecast	324.1

Table 3. Typical ocean experiment data sizes for a 10 model year run with data output once per model day.

Spatial Resolution	Data Sizes (Gbytes)
$4° x 4°$ (global)	5.8
$1° x 1°$ (Atlantic)	21.5
$1° x 1°$ (global)	133.2
$1/4° x 1/4°$ (Atlantic)	347.1

Handling such data sets can cause significant bottlenecks in the experiment process. The experiment process, schemtically shown in Figure 4, begins by reading data to initialise a model, which may consist of more than one component. As the model runs information is shared either via disk or, if memory allows, through memory resident file systems.

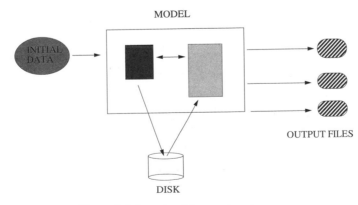

Figure 4. Schematic of the experiment process

Data is output at regular intervals throughout the model run and is then stored to create the time history of the experiment. Different data handling bottlenecks can occur for different models. For example, reading the initial data can be difficult for the chemical off-line model as this data set consists of fields from the time history of the GCM run which needs to be read in regularly for the advection of the chemical species. Coupled models, which communicate between model components via a coupler, exchange information via disk. This can hold up the model runs if the disk input/output is not sufficient for the speed of the models. Archiving and restoring model output can be a major bottleneck in the experiment process if the archive process similarly does not match the speed of the model run.

Data handling bottlenecks are specific to the HPC platform. For example, in moving from the EPCC Cray T3D to the Cray T3E a factor 4 speed up of the model run is expected but the archiving procedure has not been updated and so the full effect of this speed up is not felt in the experiment process. These data handling bottlenecks are often time consuming and expensive to solve but above all they do not get the recognition they deserve.

CONCLUSIONS

UGAMP has benefited from the research councils' HPC initiative in that it provided the level of support which meant we could explore all the issues related to the running of climate models on HPC platforms. It is to be hoped that there will be future initiatives to ensure that new academic HPC platforms are efficiently and effectively exploited.

The support of the national HPC centres, the meteorological agencies and the HPC manufacturers is invaluable to UGAMP. Future collaborations, it is hoped, will ensure that information concerning the running of climate models on HPC platforms will be shared.

Finally, new academic HPC services need to be well balanced, with disk and mass storage systems balanced to match the performance of the HPC platforms. Given such an HPC service, UGAMP would be in a strong position to utilise its full potential to further the models used to study climate variability.

THE UK OCEAN CIRCULATION AND ADVANCED MODELLING PROJECT (OCCAM)

Beverly A. de Cuevas, David J. Webb, Andrew C. Coward,
Catherine S. Richmond and Elizabeth Rourke

Southampton Oceanography Centre
Empress Dock
Southampton, SO14 3ZH

ABSTRACT

The key advance in ocean modelling in the last ten years has been the ability to resolve the fine structure of the major oceanic features such as the Gulf Stream. This was first done for basin-sized regions of ocean, as in the UK Fine Resolution Antarctic Model (FRAM) project, using vector super-computers. With the OCCAM project using the Cray T3D at Edinburgh, we have extended this to the global ocean.

The ability to represent small scale features adds greatly increased realism to the model description of the world's oceans. The model results now compare well with observations at sea and from satellites. This is not true for low resolution models. For example in the Pacific Ocean, OCCAM shows excellent agreement with the known structure of the deep western boundary current and in the timing and structure of tropical instability waves near the equator. As with FRAM, the model estimates of sea level variance in the eddy rich regions of the ocean are in good agreement with satellite observations. However the eddy energy is too small in the quiet regions of the ocean. This is being investigated in a model run with high frequency wind forcing.

It is important in climate studies to have accurate estimates of the heat and freshwater transports of the ocean. Predictions from low resolution models, where the Gulf Stream is broad and sluggish, contain serious errors. These models are not able to resolve features such as Agulhas eddies, which bring warm, salty water from the Indian Ocean into the Atlantic and are ultimately responsible for the Atlantic thermohaline circulation and the mild climate of Europe. With the results from OCCAM, we are quantifying the effect of different processes in the overall transport of heat and fresh water among the oceans of the world.

As a further result of the increased confidence we have in the current models, we are using them to increase our understanding of the physical processes taking place in the ocean. The OCCAM model is being used to study the physics behind fluctuations in the strength of the Antarctic Circumpolar Current and the Indonesian throughflow. As part of a large

High Performance Computing
Edited by R. J. Allan *et al.*, Kluwer Academic / Plenum Publishers, New York, 1999

modelling and observational program, it is being used to investigate the mixing of Mediterranean Water in the Atlantic and its effect on the Azores Current. Finally, the OCCAM model is also being used to assimilate measurements from satellite radar altimeters. These constrain the surface currents in the model and result in a further improvement in the model's representation of the present state of the ocean.

INTRODUCTION

The natural scale of the major ocean features, the currents and mesoscale eddies, is small compared to the size of the oceans. Whereas ocean basins are 5000 km or more across, major currents such as the Gulf Stream are 50 km or less in width. Similarly the major ocean eddies, which correspond to the high and low pressure regions of the atmosphere, are only 100 km to 200 km across.

In order to represent the small scale features correctly, ocean models must have a horizontal resolution of 30 km or less. Until recently, it was possible to run high resolution models of only limited regions of the world's ocean. In the UK, the Natural Environment Research Council (NERC) funded Fine Resolution Antarctic Model (FRAM) was the first high resolution (1/4° latitude by 1/2° longitude) model of the Southern Ocean (The FRAM group, 1991). This model produced new insights regarding the dynamics of the Antarctic Circumpolar Current but was recognised to have limitations, both in the physics used and associated with the open boundary at 24°S. The procurement of the Cray T3D has enabled us to develop and run the first fully global high resolution ocean model in the UK.

The UK Ocean Circulation and Climate Advanced Modelling Project (OCCAM) is a NERC Community Research Project hosted by the James Rennell Division for Ocean Circulation, Southampton Oceanography Centre, in collaboration with researchers at the Universities of East Anglia, Exeter, Southampton and Keele. The primary aims of the project are to develop better schemes to represent the key physical processes involved in climate studies and to use the model to increase our understanding of the world's oceans. This paper will give a brief description of the model and focus on the analysis which is currently being undertaken.

THE MODEL

The Bryan-Cox-Semtner code is one of the most widely used general ocean circulation models. First developed in the late 1960s (Bryan, 1969), it has undergone periodic modifications either to include better ocean physics or to take advantage of the latest developments in computer architecture (Semtner, 1974; Cox 1984). The code is most widely available as the GFDL Modular Ocean Model (MOM) code (Pacanowski et al., 1990). For OCCAM, this version was re-written for use on generic message passing systems (Webb, 1996; Webb et al., 1997). An explicit free surface, a development of the Killworth et al. (1991) scheme, was added to replace the original stream function code and an improved method was implemented for the vertical advection of momentum (Webb, 1995).

The OCCAM model (Gwilliam et al., 1997) has a horizontal resolution of 1/4° x 1/4° with 36 levels in the vertical, ranging from 20 metres near the surface to 255 metres at depth. The model bathymetry is based on the DBDB5 depth dataset (U.S. Naval Oceanographic Office, 1983) with important sills and straits modified according to more recent data (Thompson, 1995). In order to avoid Fourier filtering at high latitudes or an artificial island at the North Pole, a two-grid scheme was developed (Coward et al., 1994). A latitude-longitude grid is used for one model, covering all the world's oceans except the Arctic and the North Atlantic, which are handled by a second, rotated latitude-longitude grid with pseudo-poles on the equator in the Pacific and Indian Oceans. The models match at the equator in the Atlantic, where the longitude lines of the first grid become the pseudo-latitude lines of the second. The model run was started with the Bering Strait closed, the only communication between the two model grids being through the Atlantic. Later a channel model was included to represent the flow through the Bering Strait.

The model was initialised with potential temperature and salinity fields derived from the Levitus (1982) annual mean dataset, and zero velocity and sea level everywhere. It was run in robust diagnostic mode for the first four years with a relaxation time scale of 30 days for the surface level and 360 days at all other depths. Surface forcing was provided by interpolation between monthly mean wind stress from a 1986-1988 ECMWF climatology (Siefridt and Barnicr, 1983), which was initialised from zero to full strength over 1/6 year. Surface buoyancy flux was provided by relaxation to Levitus monthly mean potential temperature and salinity (Levitus et al., 1994; Levitus and Boyer, 1994) with a time scale of 30 days.

Improvements made to the model physics include the Philander and Pacanowski (1981) vertical mixing scheme, important for an accurate representation of processes near the equator, a freshwater flux formulation to simulate the effect of precipitation and evaporation on sea level, and a higher order advection scheme (Webb et al., 1998). The model has been run for 14 years and data for the period years 8.0-12.0 is being used for most analysis purposes. The results from an additional run with six-hourly wind forcing using the ECMWF re-analysis data for the period 1992-1993 and from a run at higher resolution also form part of the analysis dataset.

MODEL RESULTS

Sea Level Variability

Sea level variability from the model run forced with monthly climatological winds has been compared with that from the Topex/Poseidon satellite measurements. Agreement is good in regions of high eddy activity, such as the Agulhas retroflection region (Figure 1a,b), the Gulf Stream and the Kuroshio, where sea level variability is high. However, it appears that the eddy energy in the model is not being transmitted away from these source areas and in the 'eddy-quiet' regions of the world ocean, the model shows significantly less variability than satellite observations. The run with six-hourly wind forcing (Figure 1c) shows a noticeable increase in variability at the low end of the scale, where it more closely resembles the satellite derived variability.

OCCAM has been developed as part of the UK contribution to the international World Ocean Circulation Experiment (WOCE), designed to improve the ocean models used for climate research. The behaviour of the model is being checked against the observations which are a major part of the experiment. It is also being used to interpret the representativeness of the results from individual cruises.

Pacific Ocean Volume Fluxes

Saunders and Coward (1996) have compared the volume fluxes across 12 WOCE sections in the Pacific Ocean with observations. In the South Pacific the volume flux reflects the exchange of water between the Pacific and Indian Oceans through the Indonesian throughflow. The annual mean throughflow in the model is 12 Sv with a seasonal cycle of between 5 and 8 Sv ($1Sv = 10^6$ m^3s^{-1}). Because of the large seasonal variability related to the Indian monsoon, there is a large scatter in measurements. Estimates published recently by Fieux et al., (1994, 1996) give a transport of 18.6 ± 7 Sv during August 1989 and -2.6 ± 7 Sv in February 1992 and the OCCAM results fall within this range. Estimates of

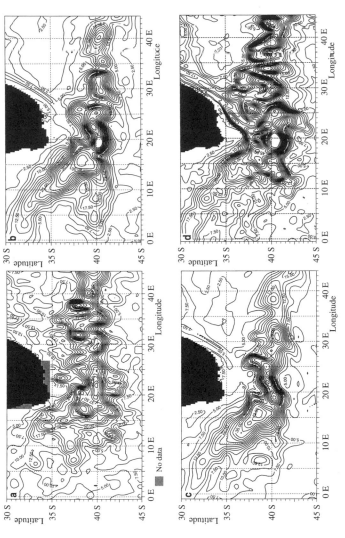

Figure 1. Sea level variability (cm) in the Agulhas retroflection region south of South Africa from: (a) Topex/Poseidon satellite altimeter data for 1993, (b) OCCAM 1/4° model forced with ECMWF wind stress climatology, (c) OCCAM 1/4° model forced with six-hourly ECMWF re-analysis wind stress and (d) OCCAM 1/8° model forced with ECMWF wind stress climatology.

Table 1. Abyssal Transport (Sv) through the Samoa Passage and adjacent regions (10°S).

Location	Robbie Ridge	Samoa Passage	Manihiki Plateau
Longitude	175°W	170°W	160°W
Roemmich et al.	1.1	7.8	2.8
OCCAM (>4km)	1.3	5.9	1.8

precipitation minus evaporation obtained from the divergence of the volume flux between sections have been compared with the tabulated values of Baumgartner and Reichel (1975). The model gives an excess of evaporation over precipitation for the whole South Pacific and the mid-latitude North Pacific, in agreement with measurements, although the magnitude over the whole Pacific is less than observed.

Saunders and Coward also compared the model with measurements of the cold Antarctic Bottom Water which enters the Pacific around the Campbell Plateau, south-west of New Zealand and flows north. Roemmich et al. (1996) measured the deep flow at 10°S across the Samoa Passage (170°W) and found additional flows on Robbie Ridge (175°W) and at 160°W, east of the Manihiki Plateau. The results are listed in Table 1. The agreement is surprisingly good, as the Roemmich results are from a single survey and the OCCAM results are a 3-year mean.

Equatorial Pacific

The equatorial Pacific is an important region in the global climate system. The sea surface temperature (SST) in the east is markedly cooler than in the west by about 5°C in the annual average and more than 10°C at specific times. This temperature contrast drives, and is itself maintained by, a large-scale atmospheric circulation with warm air rising in the west and cool air descending in the east. Disruption of this circulation, the El Niño-Southern Oscillation, affects the biological productivity of the eastern Pacific coastal waters and has a larger, possibly global, scale effect on weather. Although this coupled ocean-atmosphere system is quite well understood in broad terms, there are small scale processes still to be investigated.

Tropical instability waves, large eddies centred at about 4° north and south of the equator, have a significant impact on local meridional heat and momentum fluxes. Westward propagating signals, whose periods of about 30 days and wavelengths of about 1000 km, confirm to be tropical instability waves, can be identified in the OCCAM near-surface temperature field. At 4°N these waves weaken and slow down in spring, in agreement with observations (Wilson and Leetma, 1988).

Satellite data and model output show increased SST variability west of the Galápagos Islands on seasonal time scales. In the model, cold water west of the islands spreads out westwards in northern autumn to enhance the Pacific cold tongue along the equator. Similar behaviour has been observed in hydrographic data and can be inferred from individual satellite images. The model is being used in conjunction with the observations to investigate this phenomenon and the possible enhancement of the tropical instability waves by the presence of the islands.

Meridional Heat Transport

Most of the energy reaching the earth from the sun is received in the equatorial and sub-tropical regions but longwave radiation from the earth into space occurs relatively uniformly at all latitudes. A large poleward transport of heat is necessary in order to maintain a steady state balance. Over the continents this occurs exclusively in the atmosphere, but over the oceans the transport is partitioned between the atmosphere and the ocean. The relative importance of the oceans and atmosphere in this energy transfer is one of the major areas of study by meteorologists and oceanographers because of its importance in climate research.

Traditionally meridional oceanic heat fluxes have been calculated by indirect methods using atmospheric observations and analyses. Some direct measurements have been made at specific latitudes and particular seasons. Ocean models provide a useful tool for estimating these fluxes globally.

Figure 2 shows the mean meridional heat transport and its variance for the OCCAM year 8.0 - 12.0 dataset. The global mean is away from the equator as expected but it exhibits a large seasonal variability, particularly in the Pacific and Indian Oceans. In the Atlantic, the heat transport is northwards at all latitudes. This anomalous result is in agreement with all direct calculations of the heat transport in the South Atlantic.

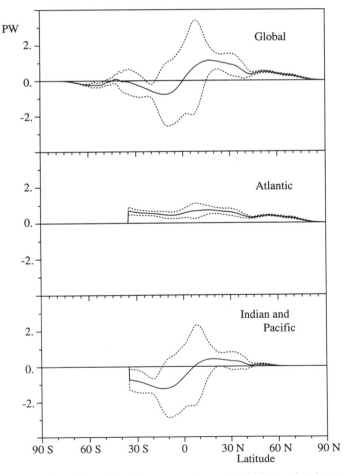

Figure 2. OCCAM 1/4° model meridional heat transport: mean (solid line) and variance (dotted lines).

The total transport may be split into the thermohaline circulation (overturning or vertical circulation) and the gyre (or horizontal) circulation (Figure 3). In the Atlantic Ocean the thermohaline circulation dominates everywhere, except north of 50°N where warm water flows north on the eastern side of the basin (west of Scotland) and cold water flows south on the western side (the Labrador coast). Globally the picture is more complicated. The thermohaline circulation generally takes heat away from the equator, except in the Southern Ocean. Here the model has a large equatorward transport, which is not adequately compensated by the poleward gyre circulation. The transport due to diffusion, which is a representation of sub-grid scale eddy processes, is also plotted. It is negligible everywhere except in the region of the Antarctic Circumpolar Current.

An alternative method is to split the total transport into that due to the mean temperature and velocity fields and that due to the transients or fluctuations from the mean (Figure 4). Away from the equator, the transients will primarily be due to the eddy field. Globally the model has insufficient eddy activity at 45°N and 45°S. In the Atlantic, the transients play a small role, most of the heat transport being due to the mean fields. Again the lack of eddies is evident at 45°N.

A comparison between the meridional heat transport in the latitude range covered by both the FRAM and OCCAM models shows that although both the mean and transient components have similar patterns, the maxima in both occur about 3° further south in FRAM, The magnitude of the transient flux is similar in the two models, but the maximum northwards flux due to the mean flow in OCCAM is of order 0.2 Pw higher than in FRAM, giving a total northwards total transport between 40°S and 45°S.

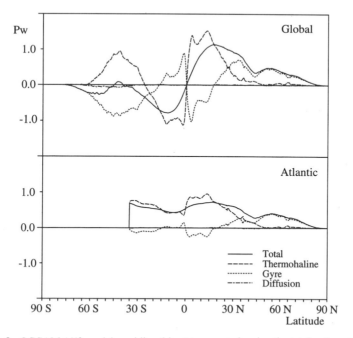

Figure 3. OCCAM 1/4° model meridional heat transport showing the total and contributions due to the thermohaline circulation, gyre circulation and diffusion.

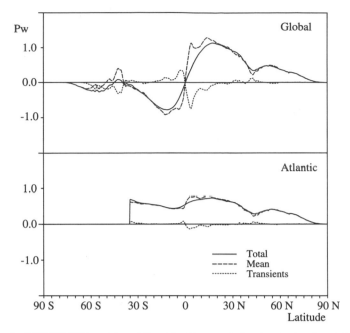

Figure 4. OCCAM 1/4° model meridional heat transport showing total and contributions due to the mean circulation and fluctuations from the mean.

A major difference between the two models is the surface wind forcing, which for FRAM was the Hellerman and Rosenstein (1983) monthly climatology. The ECMWF climatology has a stronger eastwards component in the latitude range 45°S to 60°S, particularly in the Indian Ocean but also to a lesser degree in the Pacific and Atlantic Oceans. This will produce a stronger northwards Ekman transport in the surface layers of the ocean and contribute at least partly to the differences in heat transport between the models. This is being investigated further in the run using six-hourly winds.

The Mediterranean Outflow

The OCCAM 1/4° model includes the Mediterranean Sea. However even at this resolution, important processes near the Straits of Gibraltar which determine the properties of the water spreading into the North Atlantic are not fully resolved. This high salinity water is an important component of the global thermohaline circulation that maintains our current climate. The computing power is not yet available to run climate models with ocean components at the resolution of or finer than the OCCAM model. Therefore an important use of the model is to test out how such key processes can be parameterised in models which are less able to represent them.

The OCCAM model maintains a realistic volume flux of high salinity water out of the Mediterranean (Figure 5) and this forms a tongue of high salinity water, which spreads out westwards and northwards into the North Atlantic. Given that the mixing and entrainment processes are not well represented, it is likely that this water is "over-mixed" and simulations over longer time periods would result in a water mass less dense than that observed in reality. One idea being investigated is to couple a "stream-tube" model to the OCCAM global model.

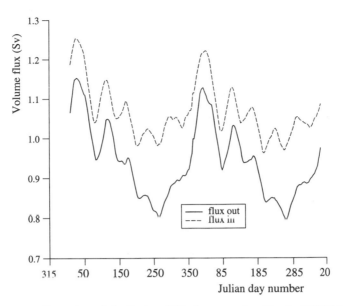

Figure 5. The volume fluxes through the Straits of Gibraltar as predicted by the OCCAM global ocean model using monthly climatological surface forcing.

A version of the Price and Baringer (1994) stream-tube model has been coded in a form suitable for use with an ocean general circulation model (OGCM). This model is designed to use topographic information at finer resolutions than the OGCM grid and to model a plume descending through ambient fields set by the OGCM. Entrainment is permitted but detrainment (and hence feedback to the OGCM) occurs only through the end of the tube. The test cases reported by Price and Baringer have been reproduced and tests with ambient fields set by OCCAM are currently taking place.

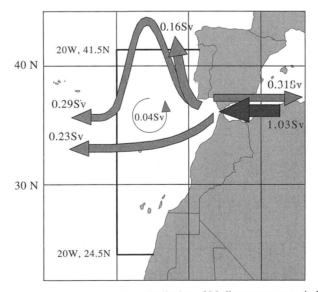

Figure 6. Schematic summary of the distribution of Mediterranean water in OCCAM. Note that of the 0.45Sv originally travelling north, 0.29Sv returns to exit through the western face.

Comparisons of the predicted fate of the Mediterranean water with observations is also of interest to several on-going EU projects. The results of a preliminary study of the fluxes across some completed observational sections is shown in Figure 6.

1/8° Resolution Model

In order to study the effect of resolution on the representation of key features such as the Gulf Stream, Kuroshio and the Agulhas eddies, and the background eddy field of the ocean, a model at a resolution of 1/8°, requiring all 512 processors of the Cray T3D at Edinburgh, has been developed and run for about 500 days. Initial results are encouraging and show improvement in small scale features, even though it is still too early in the run for Rossby waves to have travelled across the ocean basins. The variability in sea surface height calculated from this run (Figure 1d) is higher in the eddy rich parts of the global ocean and appears to be spreading further from the source regions than is found with the coarser resolution model.

CONCLUSION

The enhancement of high performance computing facilities in the UK gained by the procurement of the Cray T3D has enabled the NERC research community to develop, with its 1/4° model, the first fully global high resolution model, including the whole of the Arctic and marginal seas such as the Mediterranean. This model has been successfully run for 14 simulated years. The results are highly realistic and compare well with observations. The model dataset provides a rich tool to advance our understanding of oceanic processes and the role of the ocean in the climate system.

The upgrade of the T3D to 512 processors then allowed us to develop the first 1/8° global model. Although only a short run has been carried out so far, this has produced an even more realistic representation of the global ocean. Together with modern observational instruments, such as satellite altimeters and autonomous floats, models like OCCAM are making significant advances in our understanding of the ocean and its role in the global environment.

ACKNOWLEDGEMENTS

The UK Ocean Circulation and Advanced Modelling Project (OCCAM) is a Community Research Project supported by the Natural Environment Research Council. Computations were performed on the Cray T3D at the Edinburgh Parallel Computer Centre. Optimisation of the code was funded by Cray Research (UK) Ltd.

REFERENCES

Bryan, K., 1969, A numerical method for the study of the circulation of the world ocean, *J. Comput. Phys.* 4: 347-376.

Cox, M.D., 1984, A primitive-equation 3-dimensional model of the ocean, *GFDL Ocean Group Tech. Rep.* 1, Geophysical Fluid Dynamics Laboratory/NOAA, Princeton University, Princeton.

Coward, A.C., Killworth, P.D., and Blundell, J.R., 1994, Tests of a two-grid world ocean model, *J. Geophys. Res.* 99:22725-22735.

Fieux, M., Andrié, C., Delecluse, P., Ilahude, A.G., Kartavseff, A., Manlisi, F., Molcard, R., and Swallow, J.C., 1994, Measurements within the Pacific-Indian Oceans throughflow region, *Deep Sea Res.* 41: 1091-1130.

Fieux, M., Ilahude, A.G., and Molcard, R., 1996, Geostrophic transport of the Pacific-Indian Oceans throughflow, *J. Geophys. Res.* 101: 12421-12432.

Gwilliam, C.S., Coward, A.C., de Cuevas, B.A., Webb, D.J., Rourke, E., Thompson, S.R., and Döös, K., 1997, The OCCAM global ocean model, in: *Proceedings of the Second UNAM-Cray Supercomputing Conference: Numerical Simulations in the Environmental and Earth Sciences,* Mexico City, 1995, Garcia-Garcia, F., Cisneros, G., Fernández-Egularte, A. and Álvarez, R. eds., Cambridge University Press, Cambridge.

Hellerman, S., and Rosenstein, M., 1983, Normal monthly wind stress over the world ocean with error estimate, *J. Phys. Oceanogr.* 13:1093.

Killworth, P.D., Stainforth, D., Webb, D.J., and Paterson, S.M., 1991, The development of a free-surface Bryan-Cox-Semtner ocean model, *J. Phys. Oceanogr.* 21:1333-1348.

Levitus, S., Climatological atlas of the world ocean, 1982, *NOAA Prof. Paper* 13, Geophysical Fluid Dynamics Laboratory, Princeton University, Princeton.

Levitus, S., Burgett, R., and Boyer, T., 1994, *World Ocean Atlas, Volume 3 Salinity, NESDIS 3,* NODC, Washington.

Levitus, S., and Boyer, T., 1994, *World Ocean Atlas, Volume 4 Temperature, NESDIS 4,* NODC, Washington.

Pacanowski, R.C., and Philander, S.G.H., 1981, Parameterization of vertical mixing in numerical models of tropical oceans, *J. Phys. Oceanogr.* 11: 1443-1451.

Pacanowski, R.C., Dixon, K., and Rosati, A., 1990, The GFDL Modular Ocean Model users guide, version 1.0 *GFDL Ocean Tech. Rep.* 2, Geophysical Fluid Dynamics Laboratory/NOAA, Princeton University, Princeton.

Price, J. F., and Baringer, M., 1994, Outflows and deepwater production by marginal seas, *Progress in Oceanogr.* 33:161 200.

Saunders P.M., and Coward, A.C., 1996, An investigation and validation of a global ocean model (OCCAM), in: *International WOCE Newsletter* 24, WOCE International Project Office, Southampton.

Semtner, A.J., 1974, A general circulation model for the world ocean, *Dept. of Meteorology, University of California Tech. Rep.* 9, Los Angeles.

Siefridt, L., and Barnier, B., 1993, *Banque de Connées AVISO Vent/flux: Climatologie des Analyses de Surface du CEPMMT, Rapp.* 91 1430 025, Toulouse.

The FRAM Group (Webb, D.J. et al), 1991, An eddy-resolving model of the Southern Ocean, *EOS, Trans. Am. Geophys. Union* 72(15): 169-174.

Thompson, S.R., 1995, Sills of the global ocean: a compilation, *Ocean Modelling,* 109: 7-9, (unpublished manuscript).

U.S. Naval Oceanographic Office, and the U.S. Naval Ocean Research and Development Activity, *DBDB5 (Digital Bathymetric Data Base-5 minute grid),* 1983, U.S.N.O.O., Bay St. Louis.

Webb, D.J., 1995, The vertical advection of momentum in Bryan-Cox-Semtner ocean general circulation models, *J. Phys. Oceanogr.* 25: 3186-3195.

Webb, D.J., 1996, An ocean model code for array processor computers, *Computers and Geoscience* 22: 569-578.

Webb, D.J., Coward, A.C., de Cuevas, B.A., and Gwilliam, C.S., 1997, A multiprocessor ocean general circulation model using message passing, *J. Atm. Oceanic Technology* 14: 175-183.

Webb, D.J., de Cuevas, B.A., and Richmond, C.S., 1998, Improved advection schemes for ocean models, *J. Atm. Oceanic Technology* (in press).

Wilson, D. and Leetma, A., 1988, Acoustic doppler current profiling in the equatorial Pacific in 1984, *J. Geophys. Res.* 93: 13947-13966.

THE SOUTHAMPTON - EAST ANGLIA (SEA) MODEL:
A GENERAL PURPOSE PARALLEL OCEAN CIRCULATION MODEL

Matthew I. Beare

School of Mathematics
University of East Anglia
Norwich NR4 7TJ
England

ABSTRACT

If ocean modellers are to continue doing useful research particularly in the area of global and climate modelling, it is essential that they effectively utilise today's computing technology. Increasingly, this means parallel computers. To ease this transition, the Southampton - East Anglia model has been developed and is suitable for running in a wide range of configurations, on a wide range of platforms; from scalar workstations to clusters of workstations and massively parallel processor systems, rather than vector processors. Using high-level message passing routines the technical intricacies of parallelism are, as much as possible, hidden from the users, thus allowing them to concentrate on the ocean science, but at the same time enabling them to maximise the utilisation of whatever compute resources they have available to them.

BACKGROUND

Many ocean models run today are three-dimensional finite difference models, based on the primitive equations of motion, as described by Bryan (1969). Models of particular importance include the Cox (1984) model, an ocean code optimised for efficient running on vector supercomputers, and more recently, the Modular Ocean Model (MOM), a flexible research tool that uses C-language pre-processor directives to allow modularised physical options to be included or excluded as required (Pacanowski *et al.*, 1991 and Pacanowski, 1995). Both these codes partition the model domain into *slabs* (latitudinal cross-sections), performing all calculations pertaining to each slab, before proceeding onto the next. The computation required to complete all of the slabs represents a timestep. An array processor version of MOM, with a reduced set of options, has been adapted by Webb (1996) and subsequently named MOMA (Modular Ocean Model - Array processor version). Although not strictly a parallel code, MOMA is arranged into *columns*, such that the model arrays can be vectorised in the vertical and decomposed in the horizontal. In contrast to Bryan's rigid-lid approximation, which requires a stream function formulation for the barotropic mode, MOMA adopts a free surface formulation (Killworth *et al.*, 1991). This is better suited to parallelism, since the finite difference representation only requires nearest neighbour information and does not require line integrals to be calculated around islands. It is MOMA that has been used as the base model for the OCCAM (Ocean Circulation and Climate Advanced Modelling) project; a truly global (including the Arctic) ocean model, which has been parallelised and optimised specifically for running on the Cray T3D (Gwilliam *et al.*, 1995). As part of a collaborative project between the Southampton Oceanography Centre and the University of East Anglia, MOMA has also been developed into a general purpose, portable, parallel ocean circulation model. It is this model that has become known as the Southampton - East Anglia (SEA) model.

As part of an on-going project the aim of SEA is to provide an ocean model code that is capable of running on a variety of platforms, both parallel and sequential, with reasonable efficiency. The technical intricacies of parallelism are, as much as possible, hidden from the user, presenting researchers with a familiar code with which to work. Hence allowing them to progress seemlessly from traditional scalar computers, to clusters of workstations and even massively parallel processors (MPPs). As well as portability and efficiency, particular emphasis has been placed on adaptability and user-friendliness. Optional physical schemes, parameterisations and architectural optimisations are coded as modules and can be included into the model via C-language pre-processor directives. A configuration utility will automatically parallelise the model, depending on the number of available processors, and a post-processing program allows model output to be archived, in a format suitable for use with the OCCAM graphical visualisation programs.

PARALLEL IMPLEMENTATION

The model implements parallelism using a two (or one) dimensional geometric data decomposition with respect to the ocean surface. The resulting *sub-domains* are each allocated to a processor leading to a single program multiple data (SPMD) parallel model. The finite difference representations of the ocean physics, used by the model, require first and second order derivatives, spanning three grid points. Data therefore needs to be shared at the boundaries of the sub-domains, to guarantee coherency, requiring communication with up to eight adjacent processors. For this purpose an outer halo is defined so that the boundary data from neighbouring processors can be stored and used in a convenient fashion, as illustrated in figure 1.

Using a suite of high-level communication routines, the technical issues required to pass the boundary data are hidden from the user, so the only visible aspect of parallelism are the calls to these routines. To ease understanding, an association between inter-process communication and the application of cyclic boundary conditions is encouraged, since the aim of both schemes is to copy data from one place to another, to ensure data coherency. The application of cyclic boundary conditions is required when the model domain is global and data arrays must be cyclically connected. As with message passing, additional rows are allocated to allow data to be copied from one end of the array to the other, and it is a concept that ocean modellers are well accustomed to. Therefore, by making this association within the code, the parallel ocean model remains familiar to those used to working with sequential models. In keeping with other physical module options of the model, parallelism is offered as a compiler option, allowing the model to be easily run in either sequential or parallel modes. This is illustrated by the following simplified code overview of the *step*

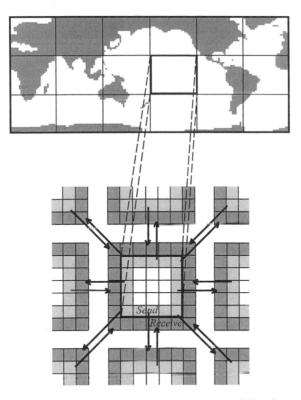

Figure 1. When a two dimensional geometric data decomposition is applied to the ocean surface, each sub-domain (enclosed by, and including, the *send* halo) is unique, whilst the *receive* halo is used to share boundary data with up to eight neighbouring sub-domains.

subroutine, which is responsible for calling all the various physical routines required to evolve the dynamic variables one timestep. This is the only place where parallelism is readily visible to the user, with the underlying physical subroutines remaining unchanged.

```
      subroutine step
            initialise timestepping variables
#ifdef parallel
            call routines to predict (boundary) ocean variables
            send boundary data to adjacent processors
            call routines to predict (inner) ocean variables
            receive boundary data from adjacent processors
#else
            call routines to predict (all) ocean variables
            apply cyclic boundary conditions
#endif
      end subroutine
```

To reduce the time that processors spend waiting for messages to arrive, the ocean variables within the boundary (send) halos are calculated prior to those within the inner domain. Hardware permitting, this allows inter-processor communications to overlap (and be hidden by) the computation of the majority of interior ocean variables.

To aid portability, the high-level communication routines offer a choice of two message passing environments: the popular *de facto* standard, PVM (Parallel Virtual Machine, Sunderam, 1990) and the emerging formal standard, MPI (Message Passing Interface, MPI Forum, 1994). It is very likely that one, or both, of these environments will be provided (or can be easily installed) on whatever computers are available, including distributed-memory parallel computers, most Unix workstations and PCs (running Linux or Windows).

Given that many institutions already possess networked workstations, and that PVM and MPI are public domain packages, the cost of setting up a local parallel platform, suitable for running the SEA model, is expected to be just a few man hours.

MODEL PERFORMANCE

To demonstrate the potential performance of SEA, two pseudo global models, spanning 70°S to 70°N, are defined. Each model has 32 vertical depth levels, one with a two degree horizontal resolution (domain size of 180×71×32) and the other a one degree horizontal resolution (domain size of 360×141×32). The models are run (with no model options selected) for one day using a 45 minute timestep, on a cluster of sixteen DEC Alpha 3000/300 workstations, connected via FDDI (Fibre Distributed Data Interface). The elapsed time, when run on one to sixteen workstations, is recorded and plotted in figure 2.

The parallel efficiency on P processors, $PE(P)$, relative to P_0 processors, where $ET(P)$ is the elapsed time on P processors, may then be determined from the equation:

$$PE(P) = \frac{ET(P_0)}{ET(P)} \times \frac{P_0}{P} \times 100.$$

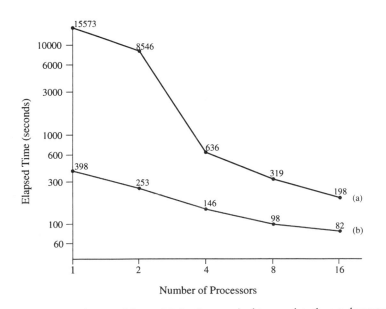

Number of Processors

Figure 2. For an initial one day run of the model, the time required to complete the run decreases as the number of workstations used is increased, for both (a) one degree and (b) two degree horizontal resolutions.

340

Ideally we would expect to achieve *linear* speed-up, a parallel efficiency of 100%. Assuming $P_0 = 1$, we find that $PE(16) = 492\%$ (*super-linear*) for the one degree run, whilst for the two degree run $PE(16) = 30\%$ (*sub-linear*). The super-linear speed-up, exhibited by the one degree model, is due to the model being *memory-bound*, leading to a significant overhead when data is paged in and out of virtual memory. When running in parallel on four or more workstations, however, the memory requirements per processor are reduced, such that memory paging is eliminated. Considering that lack of memory is often a prominent limitation when running large ocean models, it can therefore, in some cases, be very advantageous to run the model on a small number of workstations.

To further analyse the parallel efficiency of the model we can ignore the effects of eliminating memory paging by choosing $P_0 = 4$. Now, for the one degree model $PE(16) = 80\%$. This could still be conceived as being a reasonable performance, but inefficiencies are apparent. The two degree model emphasises these inefficiencies more readily, as can be seen in table 1. With less work per processor, the overheads associated with parallelism become increasingly dominant as more processors are used. If the workstations are to be used efficiently, then these overheads need to be addressed.

The parallel efficiency of the SEA model is improved through the adoption of various optimisation techniques, as discussed by Sawdey *et al.* (1995) and Beare and Stevens (1997). In keeping with the model design these optimisations have been introduced as optional modules that can be chosen at compile time.

Optimising Communications

A significant bottleneck of the model is the calculation of the (depth-averaged) barotropic velocities and free surface height. Due to the relatively fast speed of external gravity waves, the calculation of these two-dimensional variables requires a much smaller timestep (approximately 50-100 times smaller) than the three-dimensional (depth-dependent) baroclinic and tracer variables, and is therefore timestepped separately. Despite the relatively small computation required to complete a barotropic timestep, communication is necessary at the end of each of these timesteps and the overheads of this can be substantial, increasing as more processors are used.

The technique adopted to reduce this bottleneck is one that trades the time-expensive communication for a small amount of computation. An additional outer halo is defined for the free surface arrays so that two halos worth of data can be transferred between neighbouring sub-domains. With this extra information, each sub-process is able to complete two barotropic timesteps before needing to communicate, thus halving the frequency of communication. Figure 3 illustrates the gain in performance when using this optimisation on our cluster of workstations.

This optimisation of the free surface communications tends to be better suited to platforms with slower interconnects, such as workstation clusters, and has shown to be not

Table 1. Taking $P_0=1$, the parallel efficiency of the 1 degree model is seen to display a super-linear speed-up, due to the elimination of memory paging, whilst the 2 degree model is dominated by overheads due to parallelism and displays a sub-linear speedup.

	PE(1)	*PE(2)*	*PE(4)*	*PE(8)*	*PE(16)*
1 degree model	100%	91%	612%	610%	492%
2 degree model	100%	79%	68%	51%	30%

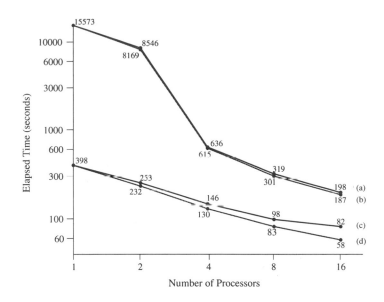

Figure 3. Compared to the initial one day model runs, for the (a) one and (c) two degree models, the free surface communications optimisation gives rise to an improved parallel efficiency as shown by (b) and (d) respectively.

as beneficial when running on MPP systems with very fast communications. Increasing the number of halos transferred, thus reducing communication frequencies further, has also been investigated, but the benefits soon diminish and the code becomes complicated and messy to implement as a general scheme. Since interconnects are likely to improve in future, it was decided to offer only the two-halo optimisation scheme as an option. Furthermore, this two-halo scheme fits in neatly with other optional physical modules, some of which also require two halos of data to be communicated, such as the QUICK tracer advection scheme (Farrow and Stevens, 1995), the numerical method of which spans five grid points instead of three. The communication routines of the SEA model are therefore designed to transparently handle single or double halo transfers, upon request.

Load Balancing

A second bottleneck emerges from having an imbalanced work load. Despite the regular domain decomposition used, each sub-domain does not necessarily contain the same amount of work due to presence of *land* points, which have no associated work. For example, for our one degree global model, there are a total of 1624320 grid points, of which only 1106370 (68%) are *sea* points, and these are unevenly distributed. For sixteen workstations, say, 69148 sea points per processor would be an ideal distribution, but a regular domain decomposition leads to a minimum of 29600 and a maximum of 97954. The performance of the model is therefore slowed by the time taken to step forward 97954 grid points. Two possible solutions are catered for in the SEA model, depending on the number of processors being used.

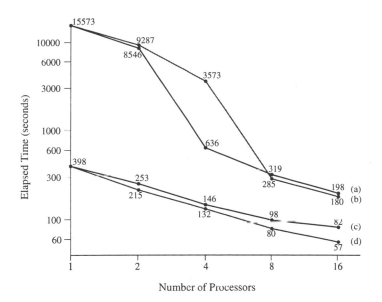

Figure 4. Using variable sized sub-domains the processor load is better balanced and can lead to an improved performance for the (b) one degree and (d) two degree models, when compared to the initial runs, (a) and (c) respectively, using a regular domain decomposition.

For a small number of processors (less than 64, say) an option to vary the sizes of each sub-domain (for one-dimensional decompositions only) is offered. If selected, the model will automatically adjust the extents of each sub-domain in an attempt to balance the load more evenly between processors. Now, for sixteen workstations, the maximum number of sea points allocated to any processor is just 70656. The performance benefits that this gives are seen in figure 4, which shows the speed-up achieved when using this load balancing technique on our cluster of Alpha workstations. Caution is, however, required when using this option, if memory is limited. As observed when using four workstations, for the one degree model, the technique may re-introduce memory paging, due to the increased size of some sub-domains, and this can be more detrimental than an imbalanced load.

For a larger number of processors (greater than 64, say) it is less efficient, and in some cases not possible, to use a one-dimensional decomposition. However, when using two-dimensional decompositions some processors may be allocated sub-domains that contain no sea points at all. For instance, if using 256 processors, dividing the one degree model into 256 sub-domains using a 16×16 decomposition, results in 41 sub-domains containing just land points. The maximum number of sea points in any one sub-domain is 6624, compared to an ideal of only 4320 sea points. The SEA model seeks to eliminate idle processors by over-partitioning the domain and ignoring those sub-domains containing no sea points. This also means that the remaining sub-domains are smaller, reducing the maximum work load per processor. Using this technique, a 20×15 decomposition results in 300 sub-domains, of which 256 contain *some* work and the maximum number of sea points is reduced to 5760. Figure 5 shows how, on the Cray T3D, both load balancing techniques can be used to improve the performance of the model, using the one-dimensional load balancing for 1-32 processors and the two-dimensional load balancing for 64-256 processors. The advantage

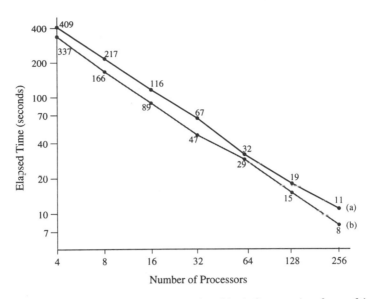

Figure 5. The model is easily ported to the Cray T3D and on this platform a series of runs of the one degree model illustrates how the performance of the model, with (a) no optimisations selected, is improved when using (b) the load balancing options (the one-dimensional technique on 4-32 nodes and the two-dimensional technique on 64-256 nodes).

of the two-dimensional technique is that as the number of available processors increases, beyond 256, the more effective it will be.

DISCUSSION

The SEA model can quickly adapt to a wide variety of parallel platforms. By providing parallelism as a compiler option and hiding the technical intricacies from the user, it is anticipated that the model can act as a useful tool and allow ocean modellers to easily progress from scalar to parallel computers. Using a combination of one or more of the optional optimisations, the model can be quickly tailored to run efficiently on whatever resources are available.

It has been demonstrated that the SEA model, on a small cluster of workstations, can provide a cost effective resource for researchers working on projects without access to state-of-the-art MPPs. To highlight the possibilities of such a platform, when using all sixteen of our workstations (which are five years old now), a one year run of the two degree model is completed in less than six hours. The one degree model takes just 18 hours to complete a similar run, thus allowing useful science to be undertaken on local resources.

With access to MPP systems the model can be easily ported. On a 256-node Cray T3D, the one degree model completes a one year run in under an hour. Using a finer, half degree resolution, and a 22.5 minute timestep, just over six hours is needed to model a year.

Although not discussed in this paper, a major bottleneck facing the SEA model now, is that of i/o. Other models, such as the OCCAM model, have also faced similar problems (Coward, 1998). Currently, each SEA process is responsible for reading in and writing out data pertaining to its own sub-domain. This is achieved using a number of sub-files, each suffixed with a process number. A suite of utility programs are therefore provided to perform pre and post processing functions on the datasets, for analysis purposes or porting to other configurations or architectures. In OCCAM, a different approach is chosen. A separate i/o process runs concurrently and acts as a channel through which data is collected and distributed. However, Coward discusses how, for very fine-resolution models, the i/o process is unable to cope with the size of the full 3D restart datasets and, for these, a similar approach to SEA is adopted, whilst maintaining the i/o process for writing out smaller 2D snapshot datasets and reading in 2D climatological surface forcing data. The primary problem in the SEA model is the reading in of the 2D surface forcing fields. When using many processors, if frequent updates to the fields are required (i.e. daily or hourly time scales), then pre-processing these files into sub-files is both messy and impracticable. This issue will be complicated further if the SEA model is to be coupled with an atmospheric model. An i/o process, similar to that used in OCCAM may therefore be the solution, but implementing it in a manner that allows the model to be transparently run in sequential mode or parallel mode, without over-complicating the code, may be non-trivial.

Further information about the SEA model and the source code can be obtained from the World Wide Web (Beare, 1996).

ACKNOWLEDGEMENTS

This work would not have been possible without the help of Dave Stevens, at the University of East Anglia. I would also like to thank the members of the OCCAM core team at Southampton Oceanography Centre for their support. M.I. Beare has been partially funded by NERC research grants GST/02/1012 and GST/02/1455.

REFERENCES

Beare M.I., 1996, *SEA: Installation and User Guide*, (see *http://www.mth.uea.ac.uk/ocean/SEA*).

Beare M.I. and Stevens D.P., 1997, Optimisation of a parallel ocean general circulation model, *Annales Geophysicae* 15:1369-1377.

Bryan K., 1969, A numerical method for the study of the circulation of the world ocean, *Journal of Computational Physics* 4:347-376.

Coward A.C., 1998, Experiences with a very high resolution global ocean model on a 512 processor Cray T3D, *Physics and Chemistry of the Earth*, in press.

Cox M.D., 1984, *A primitive equation, three-dimensional model of the ocean*, GFDL Ocean Technical Report No.1, Geophysical Fluid Dynamics Laboratory/NOAA, Princeton Univ., NJ.

Farrow D.E. and Stevens D.P., 1995, A new tracer advection scheme for Bryan and Cox type ocean global circulation models, *Journal of Physical Oceanography* 25:1731-1741.

Gwilliam C.S., Coward A.C., de Cuevas B.A., Webb D.J., Rourke E., Thompson, S.R. and Doos, K., 1995, The OCCAM global ocean model, *Proceedings of the second UNAM-Cray Supercomputing Conference: Numerical Simulations in the Environmental and Earth Sciences*, Mexico City.

Killworth P.D., Stainforth D., Webb D.J. and Paterson S.M., 1991, The development of a free-surface Bryan-Cox-Semtner ocean model, *Journal of Physical Oceanography* 21:1333-1348.

MPI Forum, 1994, MPI: a message passing interface standard, *International Journal of Supercomputer Applications and High Performance Computing*, Special Issue on MPI, 8 (3/4).

Pacanowski R.C., Dixon K. and Rosati A., 1991, *The GFDL modular ocean model users guide: verion 1.0*, GFDL Ocean Technical Report No.2, Geophysical Fluid Dynamics Laboratory/NOAA, Princeton Univ., NJ.

Pacanowski, R.C., 1995, *MOM2 documentation, user's guide and reference manual*, GFDL Ocean Technical Report No.3, Geophysical Fluid Dynamics Laboratory/NOAA, Princeton Univ., NJ.

Sawdey A., 1995, O'Keefe M., Bleck R. and Numrich R.W., The design, implementation and performance of a parallel ocean circulation model, *Proceedings of the Sixth ECMWF Workshop on the use of Parallel Processing in Meteorology*, G.Hoffman and N.Kreitz, eds., World Scientific, 523-548.

Sunderam V.S., 1990, PVM: A framework for distributed parallel computing, *Concurrency: Practice and Experience* 2:315-339.

Webb D.J., 1996, An ocean model code for array processor computers, *Computers and Geosciences* 22:569-578.

HIGH RESOLUTION MODELLING OF AIRFLOW OVER THE ISLE OF ARRAN

Alan Gadian[1], Ian Stromberg[1] and Robert Wood[2]

[1]Department of Physics, UMIST, Manchester, M60 1QD

[2]now at UK Meteorological Office, Bracknell, Berks

ABSTRACT

A numerical simulation of the airflow over the Island of Arran, is described and the results, particularly the vertical velocity fields, are qualitatively compared with airborne observations made using a light aircraft. The sophisticated Clark meso-scale model is used in a nested construction, with an outer domain size of 64km*80km and with a spatial resolution of only $\Delta y=100m$ in the inner domain. This study demonstrates that meso-scale meteorological models, for the first time can realistically simulate meteorological flow on such scales, using parallel supercomputers. These preliminary results also indicate that the model is beginning to resolve the observed atmospheric turbulence structure on the scale of 20m, and that it is now feasible to prescribe these flows.

INTRODUCTION

The Isle of Arran, approximately 30 km long by 15 km wide, is located off Scotland's west coast at 5.3°W and 55.5°N. Towards the North of the island there is a peak, Goat Fell, of 870m. During April 1996, a field campaign, led by Professor Mobbs, Leeds University made surface observations, and the UMIST Cessna collected aircraft data. This paper discusses the results of a complex meso-scale model, and makes a preliminary comparison with some of the observations.

Over recent decades there have been many analytical, observational, and with increasing computing power, also numerical studies of airflow over isolated hills. Baines (1995) provides a useful summary of the topographic effects in stratified flows, but Jackson and Hunt (1975), Mason and Sykes (1979), Carruthers and Hunt (1990) and Broad (1995) are a few of many specific papers in this area. Most of these studies are two-dimensional, due to the complexity of the simulation problem and a lack of computer power. A major limitation of the two-dimensional approach is, however, that horizontal divergent flows are not possible. With advances in computer hardware in the recent years, three dimensional simulations of the air flow are not only possible for idealised flows, but also for complex terrain. Smolarkiewicz et al (1988), for example, simulates the airflow over the much larger island of Hawaii, which supports the limited observational data about the general flow pattern, and the development of shedding twin-vortexes leeward of the island.

In numerical forecasting, the role of stationary gravity waves in atmospheric flows is of much interest. Orographic gravity wave drag parameterisation schemes are considered inadequate and selective absorption at critical layers are important. (e.g. Shutts, 1997). Likewise, gravity wave momentum fluxes are sub-grid scale for general circulation models and need improved parameterisations.

Wave breaking can also be an important atmospheric phenomena. In 1992, a DC-8 lost 19 feet of wing plus an engine whilst passing over the Rockies in an area of wave breaking. Directional wind shear can act as a critical layer where wave energy is absorbed. Severe turbulence occurs in these regions and this is an important phenomena which is not clearly understood.

This study will demonstrate that meso-scale models can accurately simulate these flows, and thus provide useful tools to increase understanding of these processes. A brief description of the model is first provided, followed by a description of the model set-up and the computational structure. Some results are then presented followed by a summary and conclusions.

THE CLARK MODEL

The Clark model (Clark 1979) has previously been used to study problems of air flow over complex terrain, for example by Smolarkiewicz et al. (1988), Smolarkiewicz and Clark (1985), Clark and Farley (1984) and Rasmussen (1989). In brief, it is a 3D finite-difference, non-hydrostatic model written in the anelastic form (Lipps and Hemler, 1982). It is based on the Navier Stokes equations of motion, including the effects of the Coriolis force, the pressure gradient force, buoyancy forces, and shear stress. The continuity equation is included, and also equations governing heat and moisture conservation (see Bruintjes et al. 1994 for a detailed description). The Clark model uses a terrain-following co-ordinate transformation based on the work of Gal-Chen and Somerville (1975), in particular being able to converge towards stable solutions for steep topography. An approximation to free-slip boundary conditions for the velocity components and zero-flux-type conditions on all thermodynamic variables are employed at the upper boundaries. Reflection of waves at the upper lid of the domain are treated with a Rayleigh damping and Newtonian cooling sponge layer above, in our experiment, 11km.

The model further includes optionally (i) treatment of the microphysics, where the warm rain parameterisation is based on the Kessler scheme (1969), and the ice microphysics scheme on the work of Koenig and Murray (1976), (ii) initialisation of the Ekman layer processes, (iii) terrain depending surface heating (sensible and latent heat fluxes) (Clark and Gall 1982, Thielen and Gadian, 1996). The surface and latent heating fluxes and the ice microphysical parameterisations are not implemented to reduce computing requirements.

THE MODEL SET-UP

The Clark model is typical of several meso-scale models now available. However it is capable of resolving steep orography, with its terrain following co-ordinate system. The

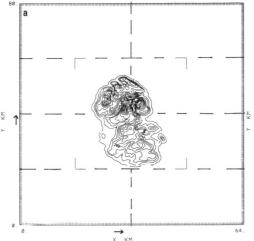

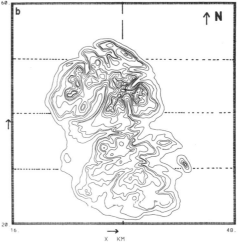

Figure 1(a). A topographical map of the Isle of Arran, showing the outer domain (1) with the island in a central position. The domain is split into sub-areas 1 to 8. The height contours are displayed at 70m. intervals, the maximum height of the topography is 755.2m, and the grid-point horizontal spatial resolution of 400m. The inner domain is delineated by the central square box.

Figure 1(b). A topographical map of the Isle of Arran showing the inner domain (2). The domain is split into sub-areas 9-16. The height contours are displayed at 80m. intervals, the maximum height of the topography is 794.3m, and the grid-point horizontal spatial resolution of 100m. The heights at individual positions agree between domains (1) and (2).

nesting structure enables bi-directional transfer of parameter variables, and is thus ideally suitable for airflow studies over complex terrain. In this study, there are two domains, an inner high resolution domain, which encompasses the island, and an outer region over the sea. The variable vertical grid spacing enables the concentration of grid points in the lower atmosphere, where increased resolution is essential.

50m resolution topographical data is used to produce topographical height fields for the two domains. Each individual point was averaged with the eight surrounding data points to produce a smooth field. This effectively reduces the peak height by 10%, but generally the model heights are within 2% of the measured values. Figures 1(a) and 1(b) show the computational and orographical set-up.

The inner nested domain, 32*40km in the x and y directions has:

- 8 sub-models, each 16*10km in the x and y directions,
- a horizontal resolution of $\Delta y = 100$m and 162*102 points in the x and y directions in each sub-model,
- a time step of $\Delta t = 2$s,
- a 24 point variable vertical grid of approximately 40m near the ground, to about 200m at the upper boundary at 2.6km,
- a fully interactive nesting structure which feeds the parameter values back between the domains.

The outer domain, 64*80km in the x and y directions has:

- 8 sub-models, each 16*10km in the x and y directions,
- a horizontal resolution of $\Delta y = 400$m and 82*52 points in the x and y directions in each sub-model,
- a time step of $\Delta t = 4$s,

- a 40 point variable vertical grid of approximately 100m near the ground, to about 2km above 8km,
- a lid at 23km, but with a Rayleigh damping and a Newtonian sponge cooling of the temperature perturbations above 11km.

The simulation is initialised with single sounding, Figure 2, which is largely derived from the aircraft observations and from a radio sonde ascent taken at 9:00am in the morning. The data initialisation will be shown to be the major limitation in the inter-comparison with the aircraft observations.

COMPUTATIONAL ASPECTS

The logistics of the computations separate the calculations into generator and analysis phases.

Data Generation

Each simulation is made for 74 minutes of "real time", corresponding to approximately 10^6 single processor cpu seconds, and about 110 Mega Words of memory on the RAL 32 processor Cray J90 computer.

Memory management is a crucial part of the program and an important operational advantage as regards memory usage. Word packing is implemented, to varying degrees, under user control, for all grid-point variables. All computations involve having only the adjacent x-z slices in memory. This is necessary to be able to attain the resolution with the present resources, and is one of the advantages of the model.

The model is initially spun up for "70 minutes" until the flow is fully developed. This is attained with 7 submitted jobs. Each produces a history file of about 220 megabytes which can be used for analysis and as initiation data for subsequent simulations. Finally the program was run for "4 minutes", with dumps at 1 minute intervals to look at the flow in detail.

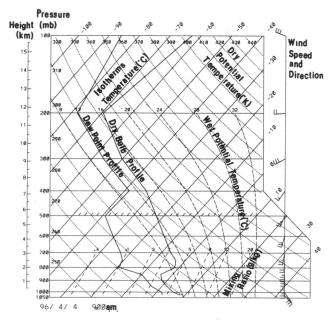

Figure 2. A thermodynamic diagram of the data used to initiate the model. The data is taken from aircraft observations and from the sonde launched at 9:00am on 4th April 1996, by Stephen Mobbs.

The program has been set up to autotask, utilising up to 8 cpus, if available. This model has 16 sub-models, the inner domain sub-models interacting with the 8 outer domain sub-models arguably making 8 cpus optimal. Even so, much effort was required to achieve this level of autotasking. Including compilation (about 1 hour), the single job cpu time requirement is $0.13*10^6$ seconds, (an average concurrent cpu usage of 6.8). Simultaneously, simpler calculations are also being carried out, in conjunction with Dr. G. Shutts, UK Meteorological Office, using only 1 domain, with 8 sub-models, but with an "ideal smooth hill", to contrast theoretical and model produced results. An average concurrent cpu usage of 7.8 is achieved for these runs.

Data Analysis

The analysis code is run separately and is a single tasking cpu process. Contour analysis and production of wind vectors requires scalar operations. A graphics package (NCAR graphics), suitable for meteorological fields, is used to display the stored data in the history files. Analysis of each timestep requires approximately $5*10^3$ cpu seconds.

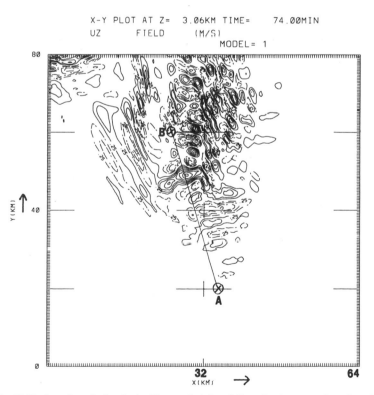

Figure 3. An X-Y plot of vertical velocity U_z at a height of 3km for the outer domain. The contour intervals are at 0.25ms^{-1} intervals, with negative contours being displayed as dashed lines. The maximum and minimum velocities for this domain's U_z field are 3.7ms^{-1} and -3.9ms^{-1}. The partition of the domain into 8 sub-models is displayed. The line shows the flight track from point A to B. The horizontal resolution, Δy, of the outer domain is 400m.

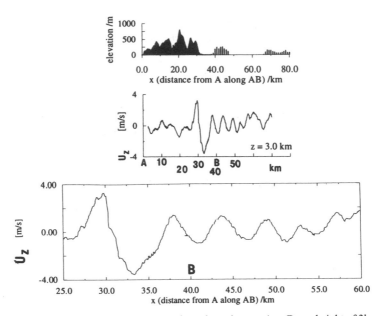

Figure 4. Aircraft measured vertical velocity, U_z, along the trajectory A to B at a height of 3km. Point A is over the southern coast of the island. The flight track continues beyond B, where it crosses the land of the Kintyre Peninsula.

RESULTS

This initial observational programme was conducted over a period of a few weeks, but only a small selection of the results on the 4th April 1996 are selected for comparison. On this day there was a southerly wind, which is the only wind direction which has a smooth fetch over the sea.

Model Results and Observations at 3 km

Here the airflow is well above the boundary layer, (see Figure 2) and relatively smooth. Figure 3 displays an XY plot of the vertical velocity, with the aircraft track displayed as A to B. The model spatial resolution is 400m and the full outer domain data is displayed.

The aircraft continued beyond B, in a straight line, to collect data for another 20km, but this has not been included, due to the fact that after point B, the track passes over the land of the Kintyre peninsula which is not included in the computer simulation. Details of the observational technique are described in Wood (1997 and 1996). Generally the measured vertical velocities (Figure 4) are accurate to less than 0.2ms^{-1}. However, when the aircraft experiences rapid changes of velocity, the errors are larger, the actual magnitude still the subject of analysis.

A detailed comparison is not presented here, as such would be inappropriate. The flight track is not aligned with the model's stationary gravity waves, indicating that the initial input wind data field is erroneous. Much of the wind initialisation data above 1km is taken from the radiosonde which was launched on the west coast and would therefore have been influenced by Arran and the Kintyre Pensinsula, less than 20 km away. Thus the lack of directional alignment could be expected. However, many of the observations do support the validity of the model. The model's largest amplitude wave occurs within 200m of observed position, and the subsequent two wave peaks are accurate within 300m. The maximum

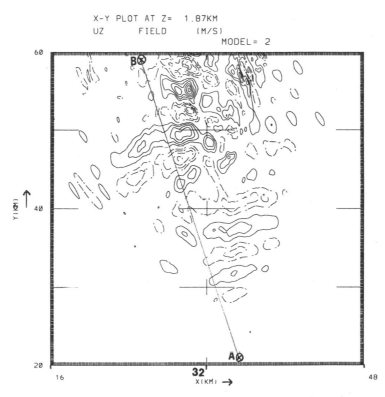

Figure 5. An X-Y plot of vertical velocity U_Z at a height of 1.87km for the inner domain.   The contour intervals are at 0.5ms^{-1} intervals, with negative contours being displayed as dashed lines. The maximum and minimum velocities for this domain's U_Z field are 9.1ms^{-1} and -8.3ms^{-1}.   The partition of the domain into 8 sub-models is displayed. The line shows the flight track from point A to B.   The horizontal resolution, Δy, of the inner domain is 100m.

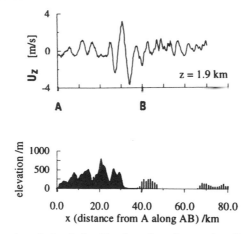

Figure 6. Aircraft measured vertical velocity, U_Z, along the trajectory A to B at a height of 1.9km, as in Figure 4.   Point A is over the southern coast of the island. The flight track continues beyond B, where it crosses the land of the Kintyre Peninsula.

velocity peak at 30km along the track is measured as 3.6ms^{-1} in contrast to the model's 2ms^{-1}. This probably represents a combination of the poor spatial resolution of 400m at this height, the misalignment of the initial wind field and observational problems. The two subsequent peaks at 38km and 43km along the track are in clear agreement with the model's values of 1.5ms^{-1} and 1.6ms^{-1}. There are many important meteorological features which are apparent from the diagrams, and these will be discussed in more detail in subsequent publications.

Model Results and Observations at 1.9 km

Figure 5 presents the model's computed vertical velocity field for the inner domain, with a resolution of $\Delta y=100$m. The observed results again indicate that the initialisation data held for the wind structure was likely to be in error, since the observed track is to the west of the main wave train. Similarly the main peak velocity is 3.6ms^{-1} (Figure 6), compared with the simulated 2.5ms^{-1}. However, again, the location of this
feature is in agreement. The two crests upwind of the tallest peak, Goat Fell, are accurately positioned at approximately 12km and 18km along the aircraft track, and the peak velocities of 1.3ms^{-1} are matched by model values within 10%. The agreement in the observed wavelength structure again implies that the model's Physics is appropriate and that there is a problem with the initialisation wind field's structure. A comparison of the velocity fields between 70 and 74 minutes (not shown), indicates that the gravity wave pattern, as at 3.0km

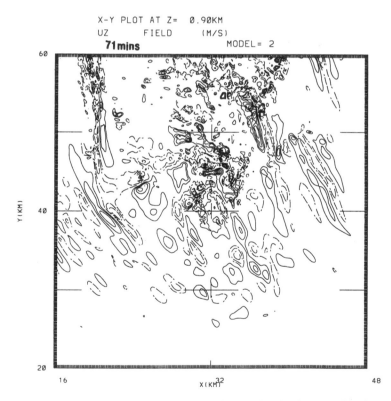

Figure 7. An X-Y plot of vertical velocity U_z at a height of 0.9km for the inner domain, at 71 minutes. The contour intervals are at 0.5ms^{-1} intervals, with negative contours being displayed as dashed lines. The maximum and minimum velocities for this domain's U_z field are 7.6ms^{-1} and -8.1ms^{-1}. The partition of the domain into 8 sub-models is displayed. The horizontal resolution, Δy, of the inner domain is 100m. The flow is turbulent and unsteady.

354

remains stationary related to the topography. There is no evidence of turbulent eddies being advected downstream but the small velocity perturbations (< 1ms^{-1} in magnitude and 1km in diameter) at the sides of the main wave train, do alter shape and magnitude although still fixed in space.

Model Results at 0.9 km

At the height of the highest topography, it is expected that the amount of turbulent activity would be the greatest (Clark et al, 1997) and is reflected in the lack of observational aircraft data, for safety reasons. The flow here is within the turbulent boundary, at about 870mb (see Figure 2). The vertical velocity field, Figure 7, exhibits rapid changes in

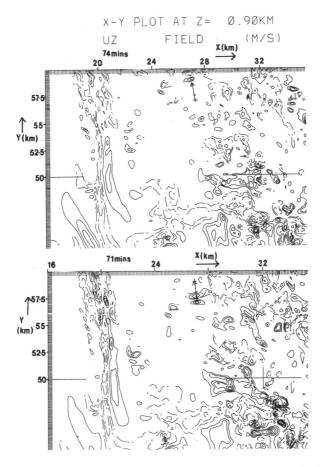

Figure 8. Enlarged X-Y plots of vertical velocity U$_Z$ at a height of 0.9km for the inner domain at times 71minutes (as Fig. 7) and 74minutes (max. and min. velocities 9.1ms^{-1} and -8.3ms^{-1}). The contour intervals are at 0.5ms^{-1} intervals, with negative contours being displayed as dashed lines. Turbulent eddies are resolved, even with the spatial resolution of Δy=100m, and can clearly be observed moving downstream, with typical velocities of 11ms^{-1}

structure, so that over the period of 3 minutes, only the main wave crests maintain their character with a vacillating shape. Figure 8 is an enlarged view of the north west corner of the inner domain, contrasting two velocity fields at 71 and 74 minutes. The main peak at approximately 49km north and 29km east of the lower left hand corner, alters its shape and dimensions, as seen on this figure.

Turbulent eddies can be seen to be advected downstream, at approximately within $\pm30\%$ of the speed of the mean flow. At 58km north and 27km east of the lower left hand corner a small eddy of $\pm2ms^{-1}$ is highlighted and is moving with a phase speed of $11ms^{-1}$. Many of the eddies appear to decay on time scales of the order of 10 minutes. The flow is turbulent and unsteady. At present observation data to confirm or deny such velocity fields are not available.

Meteorologically, there are many features modelled which need further study, and it is planned to produce further publications on this work. However it is apparent that for a more complete simulation the inner resolution should be improved from 100m to 50 m

SUMMARY

Computationally

- A sophisticated, state of the art, meso-scale meteorological model, has been applied to the airflow over the Isle of Arran.
- Detailed realistic flow patterns, have been produced which demonstrate the applicability of the Clark model in particular, and such meso-scale models in general, even with the limited current resolution available.
- With an order of magnitude increase in the memory (to 1-10 gigawords) and cpu power, it should be feasible for the first time to produce realistic flows.
- The ability of the numerical modellors to "describe" flows now exceeds the observational ability of the experimentalists to verify that the flows are realistic on scales produced in the models. A good experimental data set is really required to test the models.

Meteorologically

- The limited results of the vertical velocity fields show that it is now possible to produce realistic meteorology down to the scales of tens of metres. The complex modelled wave structure was consistent with the observations and the turbulent buoyant eddies were observed in the boundary layer.
- Consequently it is now feasible to examine the role of mountain drag, gravity wave dispersion and wavebreaking effects. Paramaterisation of these phenomena is crucially needed to improve the understanding the physics, and hence inclusion in meteorological forecast models.

ACKNOWLEDGEMENTS

The authors are very grateful to the assistance of Bill Hall and Terry Clark in the use of the Clark meso-scale model, and for much assistance by Jutta Thielen and Glenn Shutts. The authors also thank the UK Rutherford Appleton Laboratories Supercomputer Centre, for assistance with the use of the Cray J90, and for the award of resources by NERC.

REFERENCES

Baines P., (1995) *Topographic Effects in Stratified Flows*, Cambridge University Press

Broad A.S., (1995) Linear theory of momentum fluxes in 3-D flows with turning of mean wind with height, *Quart. J. Roy. Met Soc.* 121:1891-1902.

Bruintjes R.T., Clark T.L. and Hall W.D., (1994) Interactions between topographic airflow and cloud/precipitation development during the passage of a winter storm in Arizona, *J. Atmos. Sci.*, 51:48-67.

Carruthers D.J. and Hunt J.C.R., (1990) Atmospheric Processes over Complex Terrain, in: *Am. Met. Soc. Monogr.*, 23, No. 45, ch.5, W. Blumen ed., American Met. Soc.

Clark T.L., (1979) Numerical simulations with a three-dimensional cloud model: lateral boundary condition experiments and multi-cellular severe storm calculations, *J. Atmos. Sci.*, 34:2191-2215

Clark T.L. and Farley D., (1984) Severe downslope windstorm calculations in two and three spatial dimensions using anelastic interactive grid nesting: a possible mechanism for gustiness, *J. Atmos. Sci.*, 41:329-350

Clark T.L. and Gall R. (1982) Three-dimensional numerical simulations of airflow over mountainous terrain: a comparison with observations, *Mon. Weath. Rev.*, 110.766-791

Clark T.L, Keller T., Coen J., Neilley P., Hsu H. and Hall W.D., (1997) Terrain induced turbulence over Lantau island, *J. Atmos Sci.,* 54:1795-1814

Gadian A.M. and Green J.S.A., (1983) A theoretical study of small amplitude waves in the Martian lower atmosphere and a comparison with those on Earth, *Ann. Geophys.*, 239-243

Jackson P.S. and Hunt J.C.R., (1975) Turbulent wind flow over a low hill. *Quart. J. Roy. Met. Soc.*, 101:929-55

Kessler E., (1969) On the Distribution and Continuity of Water Substance in: *Atmospheric Circulations, Meteor. Monogr.* 32, American Met. Soc.

Koenig L.R. and Murray F.W., (1976) Ice-bearing cumulus cloud evolution. Numerical simulation and general comparison against observation, *J. Appl. Met.*, 15:747-762

Lipps F. and Helmer R., (1982) A scale analysis of deep moist convection and some related numerical calculations, *J. Atmos. Sci*, 39:2192-210.

Mason P.J. and Sykes R.I., (1979) Measurements and predictions of flow and turbulence over an isolated hill of moderate slope, *Quart. J. Roy. Met. Soc.*, 111:627-40

Rasmussen R.M, Smolarkiewicz P.K. and Warner J., (1989) On the dynamics of Hawaiian cloud bands: comparison of model results with observations and island climatology, *J. Atmos. Sci.*, 46:1589-608

Shutts G. J., (1997) Stationary gravity wave structure in flows with directional wind shear, *Quart. Jou. Roy. Met Soc., 123,* in press

Smolarkiewicz P.K., (1984) A fully multidimensional positive definite advection transport algorithm with small implicit diffusion, *J. Comp. Phys.*, 54:325-362

Smolarkiewicz P.K. and Clark T.L., (1985) Numerical simulation of the evolution of a three-dimensional field of cumulus clouds, Part I: Model description, comparison with observations and sensitivity studies, *J. Atmos. Sci.*, 42:502-522

Smolarkiewicz P.K., Rasmussen R.M. and Clark T.L., (1988) On the dynamics of Hawaiian cloud bands: Island forcing, *J. Atmos. Sci.*, 45:1872-905

Thielen J. and Gadian A.M., (1996) Influence of different wind directions in relation to topography on the outbreak of convection in Northern England, *Ann. Geophys.*, 1168-81

Wood R., Stromberg I.M., Jonas P.R. and Mill C.S., (1997) Analysis of an air motion system on a light aircraft for boundary layer research, *Jou. Atmos. and Ocea. Res.,* 14:960-968

Wood R., (1996) *Airborne observations of the boundary layer*, Ph. D. Thesis, University of Manchester Institute of Science and Technology, U.K.

DEVELOPMENT OF PORTABLE SHELF SEA MODELS FOR MASSIVELY PARALLEL MACHINES

Roger Proctor, Peter Lockey, and Ian D. James

Proudman Oceanographic Laboratory,
Bidston Observatory, Birkenhead,
Merseyside L43 7RA, U.K.

INTRODUCTION

A major challenge in the effective management and protection of coastal marine environments is the ability to both understand and predict the various processes which characterise the behaviour of these regions. Realistic numerical models of shelf sea areas provide powerful tools which can play a central role in these efforts. A project currently underway at Proudman Oceanographic Laboratory is the development of a three dimensional coupled hydrodynamic-ecological model which will span the full North West European continental shelf, including the entire coastline of Britain.

The coupled model is based on current work with a hydrodynamic model of the southern North Sea region, which has been developed on a sufficiently fine grid to capture many features of the barotropic and baroclinic dynamics observed in this region. Simulation results and a detailed description of the model are reported in Proctor and James (1996). The ecosystem modelling component of the coupled model will make use of an existing and widely used package, the European Regional Seas Ecosystem Model (ERSEM) described in Baretta et al. (1995)

Proposed simulation runs of this coupled model are currently only tractable on top-end computing platforms such as the Cray-T3D. This report give an overview of the project, including a description of the model, scientific results already obtained, and particular experiences on the Cray-T3D.

MODEL OVERVIEW

The model solves the 3D shallow water form of the Navier Stokes equations in spherical polar (latitude-longitude) coordinates. Solution is via a finite difference discretisation, on an Arakawa-B grid, using an explicit Forward Time Centred Space scheme in the horizontal, and an implicit method in the vertical. A depth following sigma coordinate system is used in the vertical, giving the same number of grid points regardless of depth, and allowing for easier inclusion of bottom boundary conditions.

The equations are split into depth-mean and depth-fluctuating components, allowing a longer time step for the 3D depth-fluctuating component (typically 12 times the depth-mean timestep), whilst the timestep for the 2D depth-mean component is fixed by the need to satisfy the Courant condition. Full details of the numerical schemes used are given in Proctor and James (1996).

A requirement for modelling the shallow seas found in continental shelf regions is the ability to represent sharp gradients in variables, such as those associated with fronts, thermoclines, or edges of patches of tracer. Resolving these detailed features is particularly important for the ecosystem model coupling. Here a sophisticated front-preserving advection scheme, the Piecewise Parabolic Method (PPM) described in Colella and Woodward (1984) is used. This scheme was originally developed for modelling shocks in gas dynamical simulations, and has been shown in James (1996) to have good structure preserving properties in the context of shelf sea modelling.

On the open sea boundaries, prescribed values of surface elevation and 3D density are applied in a four point wide relaxation zone, where computed values are made to progressively tend towards the imposed values. At the sea surface, prescribed meteorological data values (wind speed components, temperature, relative humidity, cloud cover, air pressure) are used to calculate surface stress and, through coupling to modelled sea surface temperature, heat flux. These input datasets can be obtained from other model runs, or from observed data.

A number of model runs have been carried out on the Southern North Sea region, where a large amount of physical, chemical, biological, and sediment data was collected during the North Sea Project (Huthnance, 1990), giving a unique opportunity for model validation and development. The area of interest extends from 50°40′N to 55°40′N and

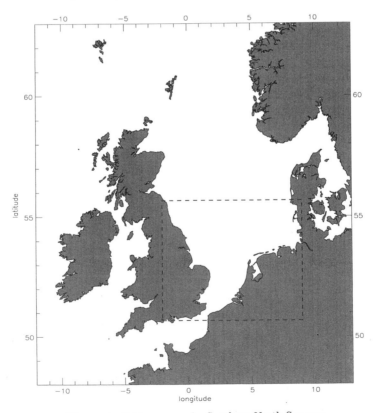

Figure 1. Grid coverage for Southern North Sea area.

2°W to 9°30′E, and the grid coverage is shown by the dashed line in figure 1. The grid contains 226 rows by 346 columns, with 10 sigma levels. The gridsize is $\frac{1}{45}$° latitude by $\frac{1}{30}$° longitude ($\sim 2.4km$), which is the upper limit to resolve the internal Rossby radius.

PARALLEL MODEL

In the parallel version of the model, the 3D grid is partitioned across the available processors in the two horizontal dimensions only, with full vertical columns stored on each processor. This is a commonly used partitioning strategy in ocean models, since horizontal grid dimensions are often much larger (by at least an order of magnitude) than those in the vertical. In addition, the model uses an implicit numerical scheme in the vertical dimension, which is inherently less parallel than the explicit schemes used in the horizontal.

We use a simple partitioning scheme whereby the grid is split into more partitions than processors, then landlocked partitions are discarded. This significantly simplifies programming, whilst giving acceptable (but not optimal) load balance. Figure 2 shows a typical partitioning scheme for 128 processors.

Updating the value on a given grid point via a finite differencing operation, requires knowledge of the values on neighbouring grid points. In the parallel model this implies that finite differencing operations close to the edge of a grid partition require values from other partitions, which are held on other processors. This is generally dealt with by extending the edges of each partition with a 'halo region' whose sole duty is to hold a read-only copy of values on neighbouring processors. Care needs to be taken to ensure that whenever a finite difference operation is carried out, the values in the halo region have been brought up to date with the most recent values on neighbouring processors, and updating by message passing communication with neighbouring processors if necessary.

An important aim in developing the parallel model was to ensure portability over current and future parallel platforms. This has been achieved by making use of the MPI (Message Passing Interface) standard (MPI Forum, 1995) throughout, and avoiding vendor specific language extensions or library calls. Another aim was to keep the

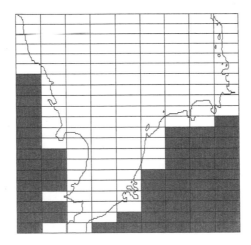

Figure 2. Typical partitioning scheme for 128 processors, for the Southern North Sea region. Each non-landlocked (unshaded) partition is assigned to a single processor.

Table 1. Total times for a short test run of the model on a 24 hour simulation, using different output modes. The time in brackets is the amount of the total time spent dumping output data to disk. For comparison, the time on a single J90 processor is also given. (All times in seconds).

	Processor Count			
	1	32	64	128
T3D: *Single node* output mode	——	721.3 (94.4)	450.8 (152.1)	346.9 (189.6)
T3D: *One node per slab* output mode	——	668.4 (39.9)	340.4 (42.9)	214.9 (56.2)
T3D: *All nodes* output mode	——	668.3 (38.9)	339.4 (38.8)	219.4 (52.9)
J90	7139	——	——	

parallel infrastructure separate from the modelling details as far as possible, and this has been achieved by encapsulating all message passing code inside a small number of routines which carry out higher level operations, such as halo updates or grid redistributions. This also has the advantage of making any parallel infrastructure changes onset of large scale seasonal stratification in the north part of the region, the sharp front separating the stratified waters to the north and the well mixed waters to the south, and internal waves induced by tidal flow over the Dogger Bank. Figure 3 shows the variation in surface salinity in the area of the Rhine delta, taken after 1 month of simulation. The behaviour of the plume of relatively freshwater from the Rhine is successfully reproduced, with the model able to resolve the detailed structure of the eddies formed at the fronts between fresher and ambient seawater, as the plume moves northwards along the coast of the Netherlands.

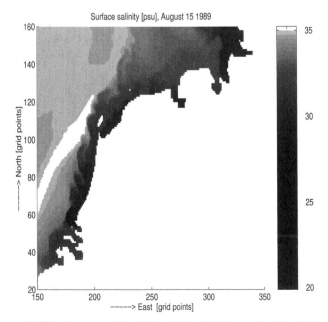

Figure 3. Surface salinity in the area of the Rhine delta after 1 month of simulation (31st March 1989). Clearly visible is the plume of relatively fresh water from the Rhine, with eddies forming as it moves northwards along the Netherlands coast.

FURTHER DEVELOPMENT

Current work is focusing on extending the model to cover the full North West European continental shelf, and development of the coupling mechanism between the hydrodynamic and the ecosystem model. Figure 4 gives a high-level view of the coupling and data transfer between the two submodels. Inclusion of the ecosystem model transparent to the rest of the code, allowing for example the use of alternative communication subsystems such as MPI-2 single sided communications (MPI Forum, 1997), without affecting the rest of the model.

In practice, file I/O, and particularly the dumping of results arrays to disk, has been found to impose a large performance overhead on the parallel model. Each model run produces large amounts of output data, which is dumped to disk for later analysis. Typically 550 Mbytes of data is output per simulated day, with simulations carried out in one month blocks. In order to try and minimise the impact of this operation, the parallel model provides a runtime choice of a number of different modes for file output, allowing tuning for different types of parallel machine. The different output modes introduce varying amounts of parallelism into the output operation. These are summarised below:

- **Single Node Output**. The full grid is collected together on a single node, and then written out to a single file. This is the most convenient scheme, in that output at any given time point ends up in a single file, whose layout is independent of the number of processors in use.

- **One Node per horizontal 'Slab'**. The 2D horizontal slab of grid points at each vertical level is collected on a separate processor, and output to a separate file. This introduces a degree of parallelism into the output operation (equal to the number of vertical levels in the 3D grid), at the expense of having the results data spread over a number of files. However the layout of the files is independent of the number of processors and partitioning scheme in use.

- **All Nodes Output**. Every node outputs its own 3D partition to a separate file. This completely removes the communications required to collect parts of the grid together onto output nodes, at the expense of having results data spread over a large number of files, whose layout is dependent on the number of processors and the partition scheme.

PERFORMANCE RESULTS

Timings for a 24 hour model run on different numbers of nodes, using different output modes, are shown in table 1. This is on a 512 processor T3D, with 2 I/O gateway nodes, and a 256 Mbyte solid state disk cache. In general, the parallel model gives acceptable performance improvements over previous serial runs on vector processor machines, providing a good basis for further model development, particularly increased grid coverage and resolution. However, a noticeable trend in the table is the large overhead involved in performing model output. For model runs on 128 processors, the output represents a full 25% of the execution time for the best case output mode, using one node per horizontal slab of grid points. More detailed analysis shows that most of this output time is spent in the actual fortran write operations, which suggests that I/O is under configured on this particular platform, for this type of I/O intensive application.

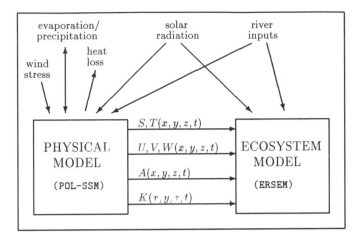

Figure 4. Coupling between the Physical model (POL Shelf Sea Model) and the Ecosystem model (ERSEM). At every timestep, the physical model provides ERSEM with salinity (S), temperature (T), velocity (U,V,W), and diffusivity (A,K) information.

SIMULATION RESULTS

Comparisons of simulation results with observed data from the North Sea show that the model successfully resolves a number of key physical features, including the increases the number of modelled variables by an order of magnitude, requiring a carefully developed data transfer strategy, in order to resolve the critical biological processes, whilst maintaining computational load balancing and efficiency. An initial aim is to carry out a long term (8 year) simulation of ecosystem dynamics over the full shelf region.

REFERENCES

Baretta, J.W., Ebenhoh, W., and Ruardij, P., 1995, The European regional seas ecosystem model. A complex marine ecosystem model, *Neth. J. Sea Res.*, 33 (3/4):233–246,

Colella, P., and Woodward, P.R., 1984, The Piecewise Parabolic Method (PPM) for gas-dynamical simulations, *J. Comput. Phys.* 54:174–201.

Huthnance, J.M., 1990, Progress on North Sea project, *NERC News* 12:25–29.

James, I.D., 1996, Advection schemes for shelf sea models, *J. Mar. Syst.* 8:237–254.

Message Passing Interface Forum, 1995, MPI: A Message Passing Interface Standard, Version 1.1, University of Tennessee.

Message Passing Interface Forum, 1997, MPI-2: Extensions to the Message Passing Interface, University of Tennessee.

Proctor R., and James, I.D., 1996, A Fine-resolution 3D model of the Southern North Sea, *J. Mar. Syst.* 8:285–295.

SATELLITE ALTIMETER DATA ASSIMILATION IN THE

OCCAM GLOBAL OCEAN MODEL

Alan D. Fox,[1] Keith Haines,[1] Beverly A. de Cuevas,[2]
and Andrew C. Coward[2]

[1]Department of Meteorology
University of Edinburgh
Edinburgh, EH9 3JZ
[2]Southampton Oceanography Centre
Empress Dock
Southampton SO14 3ZH

ABSTRACT

Assimilation of satellite altimeter data into ocean models offers the best prospect for testing the ability of ocean models to correctly represent near surface circulation and to extrapolate observations of that circulation in space and time. This will ultimately lead to real time prediction of ocean circulation and of sea surface temperature which will contribute to improving weather and climate forecasts.

The relative scarcity of observations of the ocean, even when satellite data are considered, places special importance on the choice of numerical model in any oceanographic data assimilation system. The 1/4 degree OCCAM global ocean model used here represents the state of the art. This model has been combined with a new approach to assimilating altimeter data developed at Edinburgh, in which emphasis is placed on preserving sub-surface water mass properties.

The basic altimeter assimilation scheme has been tested successfully in an idealised, 'twin' experiment. Further experiments are now underway assimilating TOPEX/POSEIDON altimeter data into the OCCAM model. Early results of a model run assimilating altimeter sea surface height data from 1993 are presented.

INTRODUCTION

The oceans play a vital role in the global climate, as emphasised by the current El Niño event which is affecting climate as far afield as North America and southern Africa. The ability to predict such climate variations requires a combination of ocean observations and numerical models of the global ocean circulation. Data assimilation is the term applied to this

High Performance Computing
Edited by R. J. Allan *et al.*, Kluwer Academic / Plenum Publishers, New York, 1999

process of combining observations and models to obtain a more accurate estimation of ocean circulation than possible with either model or data alone.

The work presented here forms part of the AGORA (Assessment of the Global Ocean ciRculation with data Assimilation systems for climate studies) project. This is funded by the EU Environment Program and its aim is the inter-comparison of global ocean data assimilation systems.

The method described is for the assimilation of satellite altimeter data into a high resolution global ocean model. The model used is the Natural Environmental Research Council's OCCAM Global Ocean Model (Gwilliam et al., 1997), developed at the Southampton Oceanography Centre and run on the Cray T3D at Edinburgh. A simple assimilation scheme for projecting altimeter surface height data in the vertical by lifting or lowering of water columns is used. This has the advantages of conserving water properties on isopycnals, while being simple to implement and consuming little computer time. Some results of experiments assimilating mapped TOPEX/POSEIDON altimeter data are presented.

THE OCCAM GLOBAL OCEAN MODEL

The Ocean Circulation and Climate Advanced Modelling (OCCAM) Project was started in 1992 to develop and run a high resolution global ocean model on the UK Research Councils' first MMP super-computer. It is a three-dimensional global general circulation model based on the GFDL MOM code with the addition of a free surface. The resolution is 1/4° east-west and north-south with 36 levels in the vertical, varying in thickness from 20m at the surface to 255m at depth.

OCCAM was integrated for 14 model years, starting from a climatological state and applying monthly mean conditions at the surface. For the purposes of this project, it has been run with 6 hourly wind forcing for the years 1992-1993 and a repeat 1993 run assimilating TOPEX/POSEIDON altimeter maps has been performed.

TOPEX/POSEIDON ALTIMETRY

Observations of the ocean from space are limited to the sea surface. Observations of sea surface height are of particular importance since near-surface currents are primarily geostrophic - flows are parallel to contours of sea surface height. Altimeter data are most useful when combined with knowledge of the sub-surface water structure. Measured changes in surface currents can then be related to changes at greater depth. Sea surface height measurements are also essential in monitoring possible sea level changes associated with climate change.

The TOPEX/POSEIDON mission is a co-operative project between NASA and the French Space Agency, CNES. The satellite uses advanced radar altimetry to make very precise and accurate observations of sea level. Since its launch on 10th August 1992, it has been providing high quality sea surface measurements to the scientific community. Because of a lack of knowledge of the ocean geoid (the signal due to local variations in the gravitational field), the use of altimeter data by the oceanographic community is limited to anomalies from the mean.

ASSIMILATION METHODS

Altimeter data provides a view of the sea surface height and several methods have been developed to project this information on to the deeper ocean layers. These methods fall into three groups: statistical methods using correlations between sea surface height and ocean fields, nudging techniques, whereby the model sea surface height is forced gradually towards observations and dynamical methods, which rely on simplified, dynamical relationships.

Cooper and Haines (1996) proposed a dynamically based method whereby altimeter sea surface height data are projected on to deeper layers by vertical displacement of the water column by an amount δh. δh is determined from the hydrostatic balance

$$g\rho\delta\eta + g(\rho_b - \rho_s)\delta h = \delta p_b = 0.$$

where g is the gravitational acceleration, ρ is density, η is sea level and p_s and p_b are the pressure at the top and bottom of the water column. Thus a rise in surface height ($\delta h > 0$) results in a downward displacement of the water column ($\delta h < 0$), corresponding, for example, to an anticyclonic baroclinic eddy. Such a technique clearly preserves water properties (temperature, salinity etc.) on isopycnal surfaces.

RESULTS

The Cooper and Haines technique has been applied to the assimilation of mapped TOPEX/POSEIDON 1993 sea surface height anomalies into the OCCAM model. The skill of the model and assimilation scheme at predicting sea surface height over 20 days (Figure 1) is assessed by comparison with predictions based on persistence (i.e. the sea surface height in 20 days will be the same as it is to-day). For periods between 20 and 60 days the model with assimilation beats persistence globally and in all regions except the Antarctic Circumpolar Current. The performance is best at low latitudes, with reduced skill at high latitudes where stratification is less and the flow is more barotropic.

Assimilation has also significantly increased the mesoscale activity in the OCCAM model. This is illustrated in Figure 2 which shows sea surface height anomalies in the tropical Pacific from the TOPEX/POSEIDON satellite, as predicted by the model without assimilation and then with assimilation, 9 days after the last assimilation. The anomaly field produced by the satellite has a range of order ± 20 cm, more than twice that of the model. The model with assimilation now has a range similar to the satellite, showing that the large-scale height signal has been improved, and the highest anomalies occur in the same regions as in the satellite field, evidence that the assimilation is capturing the mesoscale variability correctly.

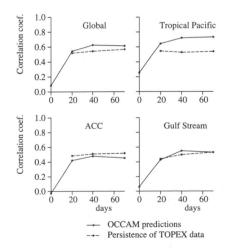

Figure 1. Skill of the model and assimilation scheme of predicting sea surface height over 20 days.

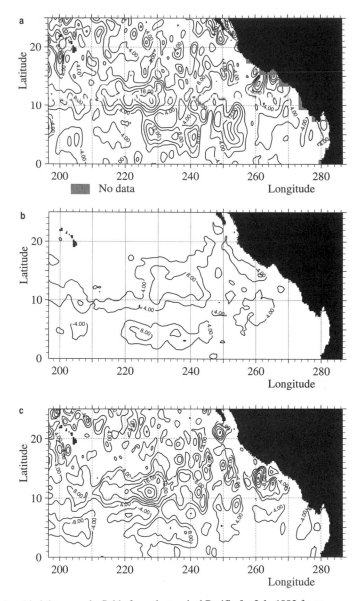

Figure 2. Sea level height anomaly fields from the tropical Pacific for July 1993 from:
(a) TOPEX/POSEIDON measurement, (b) OCCAM model, and (c) OCCAM model with sea surface height assimilation. (The Hawaiian Islands can be seen in the top left corner and the coast of America on the right.)

In the strong western boundary current region of the Gulf Stream, assimilation has corrected the tendency of the model to produce a large northwards loop, evident in the temperature and sea surface height fields where the current leaves the coast and extends into the ocean basin. A similar improvement is found in the representation of the Kuroshio.

CONCLUSION

An assimilation scheme which is simple to implement and computationally inexpensive has been tested in the OCCAM global model. Early results are encouraging with improvements in prediction of the evolution of sea surface height, increased mesoscale activity and the representation of key oceanic features. The next stage will be a detailed comparison between the model results and independent ocean datasets.

ACKNOWLEDGEMENTS

The data assimilation part of this work is funded under the EU Environment Program AGORA. OCCAM is a Community Research Project supported by the Natural Environment Research Council. All model computations were run on the Cray T3D at the Edinburgh Parallel Computer Centre.

REFERENCES

Cooper, M., and Haines, K., 1996, Altimetric assimilation with water property conservation, *J. Geophys. Res.* 101:1059-1077.

Gwilliam, C.S., Coward, A.C., de Cuevas, B.A., Webb, D.J., Rourke, E., Thompson, S.R., and Döös, K., 1997, The OCCAM global ocean model, in: *Proceedings of the Second UNAM-Cray Supercomputing Conference: Numerical Simulations in the Environmental and Earth Sciences, Mexico City, 1995*, García-García, F., Cisneros, G., Fernández-Eguiarte, A. and Álvarez, R. eds., Cambridge University Press, Cambridge.

PARALLELISATION AND PERFORMANCE OF A STRATOSPHERIC CHEMICAL TRANSPORT MODEL

Cate Bridgeman

Department of Chemistry,
University of Cambridge,
Lensfield Road, Cambridge CB2 1EW

ABSTRACT

SLIMCAT is a three-dimensional off-line chemical transport model used to investigate stratospheric processes. The model has been parallelised using MPI and tested on a Cray T3D and a Hitachi SR2201. The parallelisation strategy is discussed and the resulting performance statistics on the two machines are compared.

DETAILS OF THE MODEL

SLIMCAT[1] is a three-dimensional off-line chemical transport model which is widely used within the UK Universities' Global Atmospheric Modelling Programme (UGAMP) community to model stratospheric processes. It consists of two major components: an advection scheme and a chemistry scheme. The advection scheme takes winds and temperatures from meteorological analyses or from other global atmospheric models and uses these to transport chemical species around a global latitude-longitude grid. An isentropic vertical coordinate is used and vertical winds can be generated, if required, by calculating the diabatic heating rate. Two advection schemes are available for use with parallel SLIMCAT. The Prather scheme with conservation of second order moments[2] is a finite volume scheme, based on the exchange of mass between grid boxes, and provides accurate, non-diffusive advection of the chemical species. Also available is a semi-Lagrangian transport scheme taken from the ECMWF's IFS model[3]. This is a three time level scheme which uses linear interpolation to locate the mid-point of the trajectory and quasi-cubic interpolation for the departure point. The main motivation for using a semi-Lagrangian scheme is that it is not limited by the CFL stability criterion which applies to the Eulerian Prather scheme and so can be run at high resolution without the need for a very short time step[4]. The second advantage that the semi-Lagrangian scheme has over the Prather scheme is that it is far less memory intensive, requiring only a single field for each tracer compared with 11 fields per tracer for the Prather scheme.

SLIMCAT can be used simply to transport passive tracers or in conjunction with a comprehensive stratospheric chemistry package[5]. The latter contains a detailed description of stratospheric chemistry with 35 chemical species and over 100 reactions including heterogeneous reactions on polar stratospheric clouds, and liquid sulphate aerosols. Species are grouped together in families for the transport step and this reduces the number of advected species to 25 thus saving CPU and memory. The chemistry is calculated locally for each grid box and so this part of the program, which is the most expensive in terms of CPU, has a high degree of parallelism.

The computation is carried out on a grid of size $LAT \times LON \times LEV$. The program has been parallelised in the "Single Program Multiple Data" (SPMD) style where each processor is allocated a section of this grid (its domain) so that all processors are performing the same computations on their own set of grid points. The domain is decomposed in the horizontal, with each processor dealing with a specific set of latitude-longitude points and all levels. For the transport step a processor needs to know about neighbouring parts of the grid owned by other processors in order to calculate the quantities of tracers blown into its domain from outside. This is done by creating a halo of grid points around the domain which the processor knows about but does not perform any computations on. In order to keep the data associated with these halo points up to date, the processors need to communicate with their neighbours before each advection time step. The size of halo required depends on the length of the time step and the resolution of the grid since it must contain all points from which tracers may enter the domain during the time step. The time taken to perform the communication depends both on the size of the message to be communicated and on a startup time during which the processors set up a communication channel. Thus communication time increases as the time step increases but more importantly, as the number of processors increases. The latter occurs for two reasons. Firstly more processors means more halos to be filled and secondly communication may be required not just between adjacent processors but also between more remote neighbours. This can result in large overheads and a consequent decrease in the efficiency of the program. Communication is achieved in parallel SLIMCAT via the message passing interface (MPI).

EFFECTS OF CHANGING THE GRID RESOLUTION AND ADVECTION SCHEME

The motivation for the parallel implementation of SLIMCAT is to allow experiments to be run at resolutions impossible on conventional architectures. The move to a high resolution grid will improve the representation of global transport, particularly in the regions around the polar vortices where gradients in species concentration are high and thin filaments of vortex air can break away and be transported to mid latitudes

Figure 1 shows an O_3 tracer after 10 days advection at low and high resolution with a time step of 1800 seconds. At T42 the semi-Lagrangian scheme performs poorly, failing to produce even large scale structure effectively. At T213, however, differences between the two advection schemes are slight. The semi-Lagrangian scheme is 10-20% quicker than the Prather scheme using the same time step. It does not perform well at low resolutions, but at high resolutions, where memory considerations become important, it is a sensible choice.

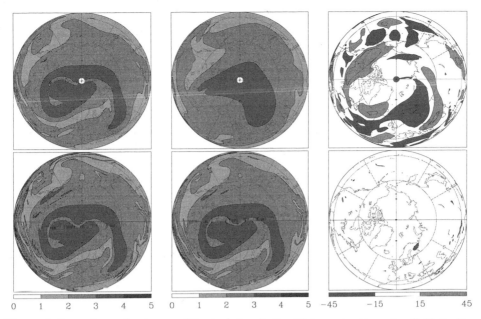

Figure 1. Ozone tracer (parts per million by volume) after 10 days advection using (left) Prather 2nd order moments advection scheme, and (middle) Semi-Lagrangian advection scheme. Percentage differences between the two schemes are shown on the right. Top row is at low (T42 = $2.8° \times 2.8°$) and bottom row at high (T213 = $0.6° \times 0.6°$) resolution.

PERFORMANCE OF THE PARALLEL SLIMCAT MODEL

The difficulties inherent in quantitatively assessing the performance of parallel programs are discussed at length in "Designing and Building Parallel Programs" by Ian Foster[6]. The execution time T of a parallel program (defined as the time elapsed between the first processor beginning and the last processor completing the task) depends on a number of variables. The principal ones are the size of the problem, N (grid resolution and number of layers in this case) and the number of processors P. Execution time is often not the most effective way of assessing performance. Since T varies with problem size, even on a single processor, it is difficult to compare performance at different N. Speedup S removes this size dependence and is defined as $S = T_1/T_P$, where T_P is the execution time on P processors. Efficiency E is a related quantity, $E = S/P$.

Figure 2 shows speedup S, and efficiency E as a function of the number of processors P and the choice of advection scheme. Times are based on a 5 day run at resolution T42 with time step 1800 seconds using the Cray T3D. On the left are the times for the full code and on the right those for transport and chemistry separately. Speedup increases up to 128 processors after which communication costs outweigh the advantage of each processor dealing with fewer grid points and speedup is seen to fall for 256 processors.

Figure 3 shows the performance of the model as the size of the problem changes. The calculations were performed for a 5 day run using a time step of 1800 seconds, SLT advection scheme on the Cray T3D. Figure 3a shows a linear increase in execution time T as the number of levels increases (resolution T42 and 32 processors). This linearity is as expected since changing the number of levels results in a linear increase in the

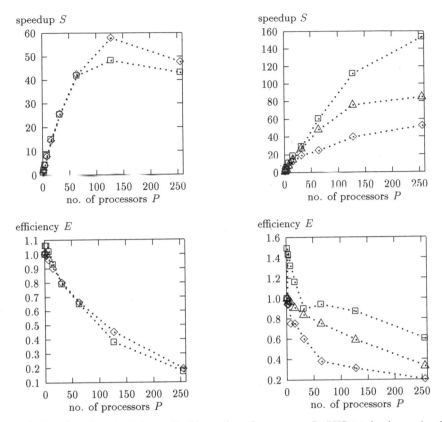

Figure 2. Speedup S, and efficiency E with number of processors P. LHS total values using SLT $\Diamond$ and Prather $\square$ advection schemes. RHS values for advection (SLT $\Diamond$, Prather $\square$) and chemistry $\triangle$ parts of code are shown separately.

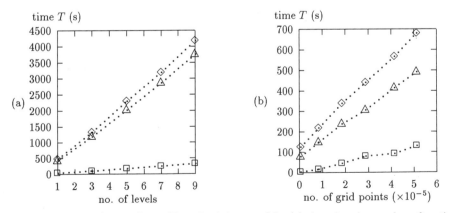

Figure 3. Execution time as the problem size is increased by (a) changing the number of vertical levels and (b) changing the resolution of the grid.

Table 1. CPU time in seconds for 5 days advection with full chemistry on 1 level at T42 and time step 1800s. Efficiencies given in brackets.

NPE	total		transport		chemistry	
	T3D	SR2201	T3D	SR2201	T3D	SR2201
1	12172.2 (1.00)	3479.5 (1.00)	634.6 (1.00)	274.6 (1.00)	11236.2 (1.00)	3166.3 (1.00)
4	3070.6 (0.99)	869.0 (1.00)	169.5 (0.94)	89.3 (0.77)	2864.3 (0.98)	750.0 (1.04)
8	1581.6 (0.96)	452.6 (0.96)	105.8 (0.75)	47.9 (0.72)	1450.5 (0.97)	389.0 (1.02)
64	287.9 (0.66)	108.4 (0.50)	26.2 (0.38)	30.0 (0.14)	233.5 (0.75)	68.9 (0.72)

problem size and in the amount of communication necessary. Figure 3b shows the increase in execution time T as the resolution of the grid increases (for 128 processors). The linearity of these graphs is more surprising and encouraging since a change the number of grid points requires a change in shape of the domain of each processor and hence a non-linear change in the number of processors owning halo points. This shows that time spent in communication between processors is a minor part of the total execution time.

COMPARISON BETWEEN MACHINES

SLIMCAT was parallelised using the MPI standard which is designed to be highly portable. The message-passing library was written so that the domain can be divided between any number of processors. Parallel SLIMCAT was initially developed on a Cray T3D situated at the Edinburgh Parallel Computing Centre (EPPC) and has now been successfully ported to a Hitachi SR2201 at the Cambridge University High Performance Computing Facility (HPCF). Substantial rewriting of the message passing routines was necessary for the program to run on the Hitachi SR2201. Message passing originally relied on buffered communication, i.e. the sending processor writing the message into a buffer from which the receiving processor would read it. This method worked well on the Cray T3D where sufficient buffer resources are provided but failed on the Hitachi SR2201 where buffer space, which is shared between all users of the machine, was frequently unavailable. The library was rewritten so that communication no longer relies on buffering, and tested thoroughly both on the Hitachi SR2201 and the Cray T3D. The program should now port without further changes to other parallel machines. The Cray T3D allows the use of up to 512 nodes while the Hitachi allows the use of a maximum of 64.

The times for transport (SLT) and chemistry and total time for the two machines are compared in table 1. The associated efficiencies are given in brackets. The efficiencies of the chemistry scheme are similar on the two machines but efficiency drops off rapidly for the transport section of the program, as expected since most of the message passing occurs here. When the program is run on 64 processors, the transport part performs poorly on the Hitachi SR2201, with an efficiency of only 0.14 compared with 0.38 on the Cray T3D. Communication overheads are clearly dominating in this case. This problem disappears if the model is run at higher resolution so that each processor is allocated more grid points and the ratio of communication to useful computation is reduced.

CONCLUSIONS

SLIMCAT has been adapted to run on parallel architectures and thoroughly tested on a Cray T3D and a Hitachi SR2201 where it is available for use by the UGAMP community. A semi-Lagrangian transport scheme has been included which enables the model to be run at high resolution with a longer time step than was possible using the Prather advection scheme and is also less memory intensive. This work will allow experiments to be run at resolutions which were prohibitively expensive both in terms of CPU and memory on conventional architectures.

Acknowledgements

The work was funded by NERC HPCI grant GST/02/1000. The initial parallelisation of the Prather advection and MIDRAD radiation schemes was undertaken by B. Edgington, Dept. of Meteorology, University of Reading, UK.

REFERENCES

1. M.P. Chipperfield, M.L. Santee, J. Froidevaux, G.L. Manney, W.G. Read, J.W. Waters, A.E. Roche, and J.M. Russel, Analysis of UARS data in the southern polar vortex in September 1992 using a chemical-transport model, *J. Geophys. Res.* 101:18861 (1996).
2. M.J. Prather, Numerical advection by conservation of second-order moments, *J. Geophys. Res.* 91:6671 (1986).
3. H. Ritchie, C. Temperton, A. Simmons, M. Hortal, T. Davies, and D. Dent, Implementation of the semi-lagrangian method in a high-resolution version of the ECMWF forecast model, *Mon. Weather Rev.* 123:489 (1985).
4. F.X. Giraldo and B. Neta, A comparison of a family of eulerian and semi-lagrangian finite element methods for the advection-diffusion equation, in: *Computer Modelling of Seas and Coastal Regions III*, J.R. Acinas and C.A. Brebbia, eds., Southampton (1997).
5. M.P. Chipperfield, *The TOMCAT offline chemical transport model part I. Stratospheric chemistry code*, (UGAMP internal report 44a), Centre for Atmospheric Sciences, University of Cambridge (1996).
6. I. Foster. *Designing and Building Parallel Programs*, Addison-Wesley, USA (1995).

COMPUTATIONAL ENGINEERING

SUPERCOMPUTING AND APPLICATIONS IN GERMAN RESEARCH AND INDUSTRY

Alfred Geiger, Roland Rühle

High-Performance Computing-Center Stuttgart (HLRS)
Allmandring 30, D-70550 Stuttgart
e-mail: geiger@hlrs.de

INTRODUCTION

High Performance Computing in German research is undergoing a major transition from tens of middle-class centers to four generally accessible high-end centers. Two of these centers are already established: The Research-Center in Juelich (HLRZ) and the High-Performance Computing-Center in Stuttgart (HLRS). Whereas Juelich has its traditions in nuclear research and therefore focusses on applications from high-energy physics, Stuttgart is traditionally oriented towards engineering. It was therefore a natural step to establish the center on the base of a joint-venture between government and industry, thus enabling synergetic effects in acquisition, operation and use of supercomputers.

The main application-fields of the center in Stuttgart are CFD, reacting flows, structural mechanics and electromagnetics. In most of these applications it could be proven, that large-scale engineering-simulations can scale very well to hundreds of nodes on scalar as well as on vector-based parallel systems.

In addition to these generally accessible centers, specialised centers in several governmental and industrial sectors such as climate research and aerospace are available to their specific user-communities.

HPCN-REQUIREMENTS IN ENGINEERING AND RELATED INDUSTRIES

Although desktop-computers get more and more powerful, the need for HPCN in industries like automotive and aerospace is growing rapidly. The tendency of the last years which had the goal to do the same simulations, which were originally done on supercomputers, much cheaper on workstations and clusters, was broken by the availability of better simulation-methods requiring orders of magnitude higher performance.

Programs like the ASCI-initiative in the US, despite gaining visibility through hardware, have their main objective in improving physical modeling, numerics and algorithms in many sectors and therefore have a big impact also on engineering and industry.

Using high-end HPC-equipment today is a key discipline for competitiveness in quality of products and time to market. However creating capabilities in HPCN always means creating capacities, often beyond the requirements of the needs of a single company or institution. An orientation at capability-requirements and economic use of the created capacity is in most cases only possible in a cooperation of different institutions.

HWW: A NEW CONCEPT FOR HPCN IN ENGINEERING

As a consequence of the above analysis, three intensive users of HPCN (Stuttgart University, debis Systemhaus and Porsche AG) joined their forces in operating supercomputers under the roof of a newly founded company, hww (Hoechstleistungsrechner fuer Wissenschaft und Wirtschaft Betriebs GmbH - High-Performance Computing-Center for Academia and Industry Ltd.). The University of Karlsruhe joined in recently.

Due to the geographical distribution of the partners and even more of the users, it was clear from the beginning, that the whole infrastructure had to be set up as a multisite infrastructure:

- No one of the sites had the possibility to install all the big systems with their high requirements in electricity, climate and space.
- Machines and administrators are at different places.
- The industrial and academic users are globally distributed. Therefore acceptable speed of networks, comfort of use and security had to be established.
- The participants of projects are themselves distributed, therefore there is a clear need for cooperative working-methods

The hww actually operates MPPs from Cray and IBM and PVPs from Cray and NEC. Whereas the MPPs are mainly used by customers from universities and scientific laboratories, industrial users are dominant on the parallel vector systems. To keep these production-systems free from small test-jobs, HLRS additionally operates two systems for code-development and testing, an intel Paragon and a Hitachi SR2201.

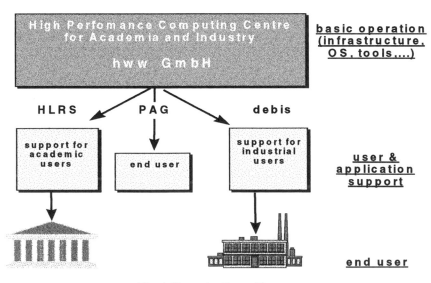

Fig.1 Organisation of hww

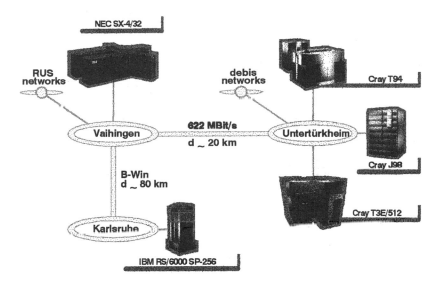

Fig.2 Distributed Hardware-Configuration of hww

APPLICATIONS

The following section gives an overview about the main application-fields of the hww production-systems and a brief insight into the targets and problems of specific applications. A list and abstracts of all non-classified projects running on the hww systems can be found on the WWW (http://www.hlrs.de/projects).

Power-Plant Engineering

The simulation of flow and combustion in coal-fired power-plants has the goal to optimise the performance and to minimise the emmission of pollutants. As coal-fired power-plants are typically in permanent operation, the failure of parts (e.g. burners) has a high probability due to the complexity and physical size of the whole system but should not affect the overall functionality.

The Institute for Power-Plant Engineering (IVD) of Stuttgart University is a leading institution in the simulation of such power-plants on parallel computers. Their code AIOLOS is in heavy use for industrial as well as research projects. As the typical grid-size for such simulations is in the order of millions of grid-cells and due to the complexity of the simulation in such multi-phase flows, this code typically has to run on tens of nodes on PVPs or on hundreds of nodes on MPPs. The achieved sustained performance has to be in the range of 20 – 80 GFLOP/s to get acceptable turnaround-times for design-engineers.

Internal Combustion-Engines

Combustion and Flow simulations for internal combustion-engines are, besides crashworthiness- and acoustics-simulations and acoustics the main type of simulation requiring supercomputers in the automotive industry today. However, other than the simulations related to structural-mechanics, this kind of simulation does in most cases not take place in the development-departments of car-manufacturers, but in the research departments. There are several reasons for this:

- The typical turnaround to simulate a full configuration of an engine today is about 15 weeks, including the pre- and postprocessing efforts. This is unacceptable in the design-process. Even a turnaround of 1 week is at the limits of what could be accepted.
- The simulation-process is dominated by the computational step as well as by the pre-processing. Not only that the geometries and thus the grids are very complicated, also the movement of parts of the grid (opening valves, movement of the piston, sprays) are critical issues. Even semi-automatic tools for the preprocessing-step are still under development.
- The used codes are not on the same level of robustness as codes in structural-mechanics. So they need the scientist, rather than the development engineer, to use them.

It must be a major goal of HPC-centers in engineering to push things forward in this application-field to create an important business for the future.

CFD

The message from the problems we see in the simulation of internal combustion engines is, that to bring CFD calculations in engineering a step forward, the problem of grid-generation and pre-processing in general has to be solved. The reason why simulations in structural mechanics are more succesfully used in the development-process is, that they are based on a nearly fully automatic grid-generation process due to mainly triangular meshes. In CFD such kind of discretisation rarely leads to good results. Structured or blockstructured meshes are not flexible enough and general unstructured meshes on the other side cannot be generated automatically.

What engineers would like to have, are tools that allow calculations to start from the CAD-geometry. Automatic grid-generation alone does not solve this problem, as it does not take into account the properties of the solution itself. Only adaptive methods that are able to use grid-cells of arbitrary shape and connectivity are able to completely handle this issue. Fig.3 shows the converged grid and the results of such a calculation based on the code ceq from Stuttgart University. Such calculations are possible today using the Euler equations. The implementation of

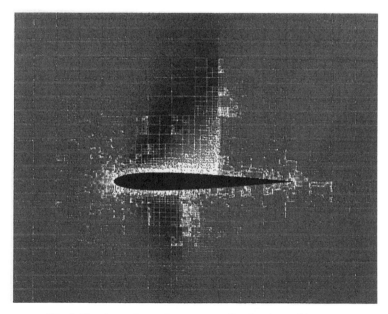

Fig.3 Final mesh and pressure-distribution with ceq

adaptive codes for the Navier-Stokes equations including turbulence-modeling will be a major step for a broader industrial use of CFD and thus is an important research-topic.

Structural Mechanics

Structural mechanics in the classical sense is mainly done on workstations and servers today. Crashworthiness-simulations however are the most important industrial application of supercomputers in engineering. The demand is growing rapidly due to the more complex safety-features and restraint-systems of modern cars and also the growing variety of accident-scenarios that have to be simulated. A critical issue in crash-simulations is, that the complexity of the problem is only partially due to the geometrical size of the computational grid and thus the potential for parallelisation by domain-decomposition is limited. The complexity comes more from the instationary nature of this kind of simulations, especially from the increasing number of contacts during the simulation time. Time however has an inherently sequential nature. Therefore crashworthiness-simulations perform best on a few (2-4) very powerful vecor-processors. Fig. 4 shows, that the performance of MPPs rarely reaches the performance of one actual vector processor, even for an infinite number of nodes. Although the size of the example used for Fig.4 is an order of magnitude below typical simulations that are done today, it describes the situation quite realistically, because also the number of contacts grows with the problem-size.

Distributed Applications and Metacomputing

For industrial users an optimal embedding of HPC into the development-process is a requirement. Most companies have development-groups all over the world, making tools for cooperative working in HPC environments unavoidable, if time to market for the simulated products is a critical issue. In the framework of several EC-funded projects tools were developped to support industrial and academic users of hww in cooperative working and visualisation (http://www.hlrs.de/structure/organisation/vis/covise) as well as in interactive online-simulations. In such scenarios visualisation and simulation are coupled in a distributed environment by a software called COVER and engineers can interact with the simulation (e.g. changing geometry-parameters).

There are problems, that cannot be simulated on even the largest supercomputers in the world. Metacomputing-scenarios, though not very efficient, are the only way in such cases.

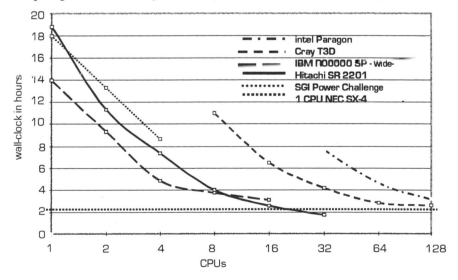

Fig.4 Crash-Benchmark, LS-DYNA3D, 30km/h, 50ms, 28200 elements

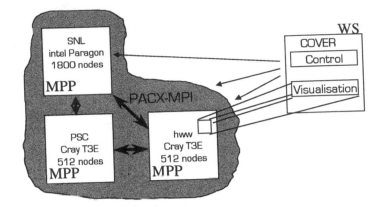

Fig.5 Metacomputing-Scenario for G-WAAT

The G7-countries are supporting several projects in the framework of their GII (Global Information Infrastructure) initiative. The only project targeted at this type of metacomputing-scenarios was G-WAAT (Global Wide Area Application Testbed) in which the Cray T3E of hww was coupled to a similar machine at Pittsburgh Supercomputing Center and an intel system at Sandia National Laboratories. The application scenarios came from molecular-dynamics in granular media (new world-record with 1.8 million particles) and from hypersonic-flow simulations (re-entry).

Whereas the simulations were controlled by the COVER software, the communication was done with a multi-protocol implementation of MPI developped in this G7-project called PACX-MPI (http://www.hlrs.de/people/resch/PROJECTS/PACX.html).

The use of equipment shared by different companies and institutions and cooperative working-methods clearly are in conflict with the security-requirements of each single partner. Therefore sophisticated security-concepts have to be installed. Besides techniques like firewalls and encryption for network-security and Multi-Level-Security on the systems, a strict path of separating different institutions on the systems has to be followed. For distributed and cooperative services it is even more important to establish a clear authentification of all participating instances (machines and users) on an individual basis.

CONCLUSIONS

Industry as well as research-institutions can benefit from a close cooperation in the operating and use of HPC-equipment. Organisational, security and privacy issues can be solved without a high level of functional restrictions for the users. The transition of novel scenarios for problem-solving in engineering from universities to industry is faster and the feedback from industry to universities is more direct.

INVESTIGATION OF SEQUENCING EFFECTS ON THE SIMULATION OF FLUID-STRUCTURE INTERACTION

K.J. Badcock, G S L. Goura and B.E. Richards

Computational Fluid Dynamics Group
Department of Aerospace Engineering,
University of Glasgow, Glasgow G12 8QQ

INTRODUCTION

Unstable interaction between a fluid flow and an aircraft structure can herald disaster, both for the aeroplane and its manufacturer. The simulation of fluid-structure interaction is therefore crucial for risk reduction during the design process. Traditional methods have used linear aerodynamic theory to formulate the stability analysis as an eigenvalue problem. These have the virtue that they are inexpensive, allowing a systematic study of the flight envelope to be carried out. However, the critical instabilities occur outside the validity of linear aerodynamics. For example, in transonic flow the dynamic pressure at which flutter occurs is much lower than at subsonic or supersonic speed [6]. Unfortunately, this is also the speed range in which passenger aircraft cruise and military aircraft manoeuvre. Also important for military aircraft, is the instability induced by vortices [4].

A proper study of these problems is possible using the simultaneous time integration of the compressible flow equations with a structural model. This type of approach has received increasing attention in recent years [3 15 14 16]. Compressible flow simulation is fundamentally nonlinear and is almost always treated in an Eulerian framework. The most common numerical methods used are of the finite volume variety. By comparison, computational structural mechanics (CSD) is better developed than computational fluid dynamics (CFD) because the structural deformations which are typically of interest are linear. A Lagrangian formulation is almost always used with finite element methods being predominant. The structural solution can typically be reduced to a small number of modes which contain most of the motion energy.

The fact that the state of the art numerical methods for CFD and CSD are of different types means that solving the fluid-structure interaction with one code is an unattractive proposition which would entail reworking both the fluid and structural simulation methods. A more popular alternative is to interface well established CFD and CSD methods through some form of coupling procedure. The flow solver passes the required force information to the structure which returns the structure motion to the flow solver. This coupled problem is typically solved using a structural solver and then a flow solver in sequence. The sequencing, which is used for numerical convenience, introduces an additional source of error into the calculation. This error is potentially serious since the energy exchange between the fluid and the structure is altered. The consequence is a reduction in the time step which can be used to accurately calculate a response compared with the flow or structural solvers used separately, hence making an already expensive calculation more costly.

The main cost in a fluid-structure interaction simulation is normally incurred during the flow solution. It is therefore very important to optimise the flow solver. A basic design criterion for a numerical method for unsteady flows is that the time step used should be chosen only on the basis of requirements for accurately following the evolution of the flow without considerations from other numerical factors like stability. The framework introduced by Jameson [1] allows this to be achieved by reformulating the problem of updating the flow as the solution of a modified steady state problem, which can be calculated using the most successful methods for steady state flow analysis. Examples of the application of this method for turbulent [10] and inviscid [12] flows around moving aerofoils have been published.

Within this formulation an iteration is used to update the flow solution. This raises the possibility of building the structural solution into this iteration with the converged flow and structural solutions progressing forward in time together. This paper investigates this approach when applied to the response of an aerofoil moving in pitch and plunge in transonic flow. This problem is attractive for assessing the method because it avoids the difficulties introduced by reforming meshes and passing forces between the non-matching fluid and structure meshes to allow the influence of solution sequencing to be assessed. The mesh for this problem can be rotated and translated rigidly with the aerofoil and the structural solution requires the integrated lift and moment from the flow solution rather than local pressure values. The method is formulated in the next section and results for the test problem are then presented. The main question posed is whether the coupled solution behaves as well (in terms of time accuracy) as the two components of the simulation when used separately.

This work is part of an overall programme to tackle nonlinear aircraft aeroelastic problems. Other components include turbulence model solutions [10], moving grids [12], parallel implementation [9] [8], three dimensional extensions and coupling with the NASTRAN structural solver.

NUMERICAL FORMULATION

The flow is modelled by the planar Euler equations denoted here in the form

$$\frac{\partial \mathbf{w}}{\partial t} + \frac{\partial \mathbf{f}}{\partial x} + \frac{\partial \mathbf{g}}{\partial y} = 0 \tag{1}$$

where $\mathbf{w} = (\rho, \rho u, \rho v, e)^T$ is the vector of conserved variables whose components consist of the density, Cartesian components of velocity and energy respectively. The system is closed by the ideal gas law.

The solution of equation (1) for unsteady flows around moving aerofoils was considered in [12]. The semi-discrete form can be written as

$$\frac{d\mathbf{w}_{i,j}}{dt} = \mathbf{R}_{i,j} \tag{2}$$

where $\mathbf{w}_{i,j}$ denotes the conserved values in a cell in the multiblock mesh and $\mathbf{R}_{i,j}$ the discretisation of the convective terms in equation (1). The discretisation in this paper involves the use of Osher's approximate Riemann solver and MUSCL interpolation along with the Von Albada limiter, a well known combination of methods which provide good resolution of shock waves. A time varying curvilinear transformation is used for the solution. Mesh velocities are calculated by a finite difference between mesh locations at the latest and previous time steps.

The semi discrete form (2) is solved in a time varying domain $\Gamma = \Gamma(t)$. The boundaries of this domain $\partial\Gamma$ consist of the aerofoil surface and the truncated far field. A characteristic treatment is used to implement boundary conditions at the far field. On the aerofoil surface no flow through the aerofoil surface is allowed. The region is divided into a structured mesh which is rotated and translated with the aerofoil in a rigid manner throughout the motion.

The model used to compute the response of the aerofoil, and hence the evolution of Γ, assumes that the aerofoil responds to the flow by moving in pitch and plunge, with a linear restoring force being exerted by the rest of the wing. The equations describing this are [7]

$$\frac{d\mathbf{q}}{dt} = \mathbf{F}(\mathbf{q}, \mathbf{w}) \tag{3}$$

where $\mathbf{q} = (\alpha, h, d\alpha/dt, dh/dt)^T$, α is the aerofoil incidence and h is the vertical displacement non dimensionalised by the semi-chord, measured positive downwards. The vector on the right hand side is $\mathbf{F}(\mathbf{q}, \mathbf{w}) = (q_3, q_4, F_3, F_4)^T$ where, denoting $\tilde{\mathbf{q}} = (q_1, q_2)^T$, $\tilde{\mathbf{F}} = (F_3, F_4)^T$ is given by

$$\tilde{\mathbf{F}} = \mathbf{F}_a(\mathbf{w}) - M^{-1}K\tilde{\mathbf{q}}$$

where

$$M = \begin{bmatrix} 1 & x_\alpha \\ x_\alpha & r_\alpha^2 \end{bmatrix}$$

$$K = \begin{bmatrix} \omega_R^2 & 0 \\ 0 & r_\alpha^2 \end{bmatrix}, \mathbf{F}_a = \begin{bmatrix} -C_L/\beta \\ 2C_M/\beta \end{bmatrix}$$

and $\beta = 4/(\pi\mu\omega_\alpha^R)$. The notation and values used here are

- C_L and C_M are the lift and moment coefficients obtained from the flow solution

- M and K are the mass and stiffness matrices respectively

- $x_\alpha = -0.2$ is the offset between the centre of gravity and the point about which the pitching motion takes place (called the elastic axis), measured negative for the centre of gravity aft of the elastic axis

- $r_\alpha^2 = 0.29$ is the radius of gyration, representing the effect of the moment of inertia about the elastic axis

- $\omega_R^2 = 0.11789$ is the square of the ratio of the natural frequencies of plunging ω_h to pitching ω_α

- $\mu = 10$ is the ratio of the aerofoil to fluid mass

- $\overline{U} = 4b/(U_\infty\omega_\alpha)$ is called the reduced velocity of the problem where U_∞ is the freestream fluid velocity and b is the aerofoil chord length. Increasing values of the reduced velocity indicate an increasingly flexible structure.

Note that the non-dimensionalisation of time for the structural model is with respect to $U_\infty/2b$. The results will be presented in terms of a time non dimensionalised with respect to the chord rather than the semi chord, following the formulation used for the flow solver.

Using the solution of equation (3), the geometry for the flow problem can now be denoted $\Gamma = \Gamma(\alpha, h)$, and hence depends on the structural solution. In return, the structural solution depends on the flow solution through the the lift and moment coefficients. Following the pseudo-time approach of Jameson for the flow solution and using a Runge Kutta solution for the structural solution, the updated flow and structural solutions at time n+1 are calculated from the nonlinear system of algebraic equations

$$\mathbf{R}_{i,j}^{*} = \frac{3\mathbf{w}_{i,j}^{n+1} - 4\mathbf{w}_{i,j}^{n} + \mathbf{w}_{i,j}^{n-1}}{2\Delta t} + \mathbf{R}_{i,j}(\mathbf{w}_{i,j}^{n+1}) = 0 \tag{4}$$

for $\Gamma = \Gamma(\alpha^{n+1}, h^{n+1})$ and

$$\mathbf{q}^{n+1} = \mathbf{G}(\mathbf{q}^{n}, C_{L}^{n}, C_{M}^{n}, C_{L}^{n+1}, C_{M}^{n+1}) \tag{5}$$

where $\mathbf{G}$ indicates the Runge-Kutta solution. If an uncoupled solution is used then lift and moment values at time levels n-1 and n are used to extrapolate for the values at n+1. The updated structural solution is then used to update the flow solution. However, the mismatch between the lift and moment values associated with the flow solution and the extrapolated values used to update the flow solution introduces a source of error into the calculation which is potentially serious since it is associated with the transfer of energy between the fluid and structure which is the crucial feature of the problem. We refer to this method as being sequenced in real time.

This phasing error was removed in [13] by using the same Runge Kutta method to update the flow instead of equation 4. However, using an explicit method to update the flow values incurs a stability restriction on the size of the time step. Using equation (4) is preferable from this point of view since the time step can be chosen on the basis of time accuracy alone. Equation (4) is solved by introducing an iteration $\mathbf{w}_{i,j}^{n+1,m}$ through pseudo time which converges to the updated flow solution. The method used to solve the pseudo time problem involves implicit time stepping and a Krylov type linear solver [2] [12] [10]. Multigrid is an attractive alternative for solving the pseudo steady state problem. An iteration for the structural solution can be introduced so that the latest approximation to the updated lift and moment values is used to calculate a better approximation to the updated pitch and plunge, i.e.

$$\mathbf{q}^{n+1,m+1} = \mathbf{G}(\mathbf{q}^{n}, C_{L}^{n}, C_{M}^{n}, C_{L}^{n+1,m+1}, C_{M}^{n+1,m+1}).$$

The $m + 1$th flow iterate is calculated for the geometry $\Gamma = \Gamma(\alpha^{n+1,m}, h^{n+1,m})$. The mesh velocities required for the transformation are calculated from the mesh locations at time n and pseudo time iterate n+1,m. At convergence the structural solution has been updated using the the correct moment and lift values. The solution is sequenced in pseudo time, with the solution being coupled in real time.

The next section uses a test problem involving limit cycle oscillations due to shock motions to evaluate the influence of using coupled or sequenced solutions in real time on the computed responses and the efficiency of the method.

RESULTS

Forced Pitching

To provide a reference for the coupled solution we first look at the performance of the flow solver alone when applied to calculate the flow over a NACA64A006 following

388

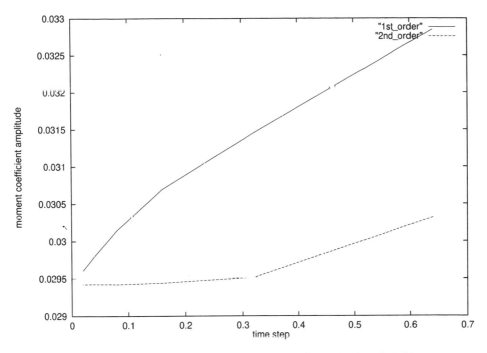

Figure 1. Convergence of maximum moment coefficient for forced pitching.

a pitching motion at similar conditions to the free response examined later (sinusoidal pitching with an amplitude of 3.3 degrees and a reduced frequency of 0.31). The convergence of the maximum moment coefficient during the cycle is shown in Figure 1. Results are shown for the second order flow solver using equation (4) and for a first order method given by

$$\mathbf{R}_{i,j}^* = \frac{\mathbf{w}_{i,j}^{n+1} - \mathbf{w}_{i,j}^n}{\Delta t} + \mathbf{R}_{i,j}(\mathbf{w}_{i,j}^{n+1}) = 0. \qquad (6)$$

The superior convergence of the second order method is clear and a good solution is obtained using a time step of 0.32 (20 steps per cycle) for which the maximum lift is within 0.5% of the fully converged value. The first order solution does not reach this accuracy with 320 steps per cycle and exhibits linear convergence.

Free Response

The main aim of this paper is to examine the effect of fluid and structure sequencing on accuracy and to test whether using solution sequencing in pseudo-time rather than real time yields any accuracy benefits. To examine these issues we use a test problem involving the pitch-plunge response of the NACA64A006 in a freestream at $M_\infty = 0.85$ [7] [11]. The bifurcation behaviour is shown in Figure 2 and shows the response changing from a stable zero position to divergence followed by a finite amplitude limit cycle. The particular example we use here is at $\overline{U} = 1.9$ which features flutter-divergence interaction. The nonlinear nature of the behaviour presents the numerical scheme with a significant test.

A time step refinement was carried out to test time accuracy using the solution which is sequenced in real time, the results of which are shown in Figures 3 and 4. The

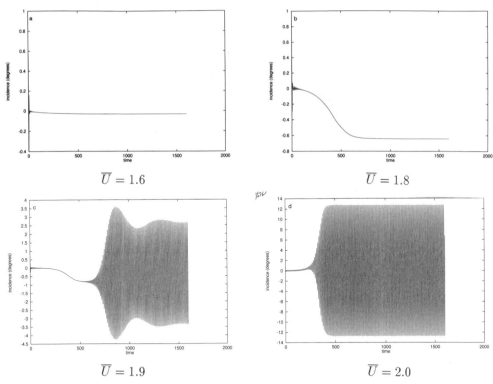

$$\overline{U} = 1.6 \qquad \overline{U} = 1.8$$

$$\overline{U} = 1.9 \qquad \overline{U} = 2.0$$

Figure 2. Behaviour for varying $\overline{U}$.

first point to notice is that the solution appears to be converging to a solution with negative divergence when the traces with $\Delta t = 0.64, 0.32, 0.16$ and 0.08 are examined. However, when the time step is reduced to $\Delta t = 0.04, 0.02$ and 0.01 there appears to be a step change to positive divergence. Initial conditions used here are $\alpha_0 = 1.0$ degrees and $h_0 = 0$. Examples of the influence of numerical parameters and method on the ability to represent nonlinear dynamics have previously been published [5]. It was shown that numerical methods can fail to accurately represent the basins of attraction of a nonlinear system. It appears in the present case that the exact solution for the initial conditions used is the one with positive divergence and that for $\Delta t > 0.04$ the numerical solution is attracted to the spurious solution from the initial conditions used (i.e. negative divergence).

The traces obtained using sequencing in real and pseudo time are compared in Figure 5 for the limit cycle phase of the motion. The comparison shows that sequencing in real time does not change the solution significantly and for $\Delta t \le 0.08$ the solutions appear identical when plotted on the scale used. This value is larger than the time step required to follow the correct solution to the differential system ($\Delta t = 0.02$). Hence, the closer coupling yields no advantage for the current case. The convergence with time step to the extreme values of incidence is shown in Figure 6. The step change in behaviour can be seen clearly, despite the fact that the solution appears to be close to convergence before the jump. The curves indicate that the method which is sequenced in pseudo time is second order accurate and that the sequencing in pseudo time reduces the accuracy to between 1 and 2. The sequenced solution in real time obtained using $\Delta t = 0.08$ is comparable in terms of quality with the solution obtained using $\Delta t = 0.32$ for sequencing in pseudo time, as shown in Table 1. Hence, the solution which is

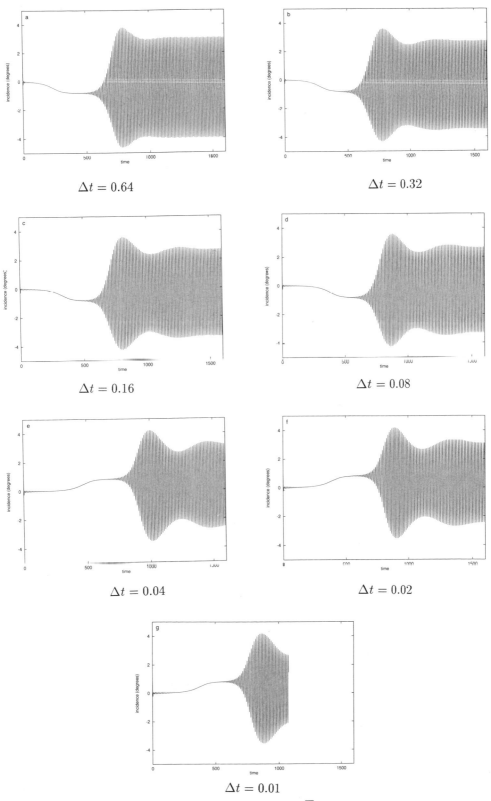

Figure 3. Time Convergence for $\overline{U} = 1.9$.

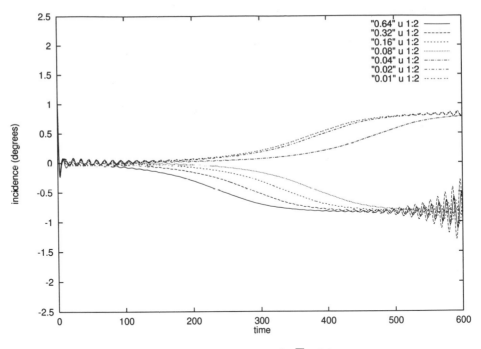

Figure 4. Time Convergence for $\overline{U} = 1.9$.

sequenced in pseudo time is more than twice as efficient in this sense (i.e. time to obtain a given quality of solution.)

From the table it is evident that introducing the structural solver into the pseudo time iterations does not increase the number of steps required to attain a converged flow solution. However, the method which is sequenced in pseudo time requires the solution of the structural equations and the movement of the mesh at each pseudo time step rather than just once per real time step. This increases the cost of each pseudo step by about 40 % in the current case.

CONCLUSIONS AND FUTURE WORK

The results presented in this paper have shown that the simulation of nonlinear dynamics in the time domain requires a careful study of the influence of numerical parameters. The results suggest that sequencing the fluid and structural calculations

Table 1. Summary of time step refinement.

Sequencing	Time Step	α_{min}	α_{max}	iterations/real time step	CPU/unit time
real time	0.64	-4.00	3.05	11.2	5.59
	0.32	-3.48	2.74	10.1	9.98
	0.16	-3.38	2.68	9.2	17.56
	0.08	-3.34	2.65	8.3	29.63
	0.04	-2.56	3.18	7.2	49.94
	0.02	-2.45	3.11	–	–
pseudo time	0.64	-3.05	1.91	11.3	8.06
	0.32	-3.32	2.50	10.1	13.81
	0.16	-3.35	2.63	9.3	24.50
	0.08	-3.33	2.64	8.3	42.63

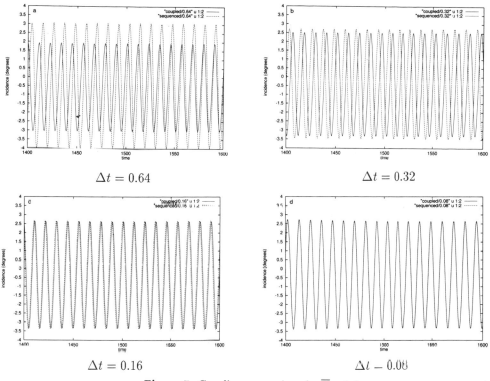

Figure 5. Coupling comparison for $\overline{U} = 1.9$.

in pseudo time gives accuracy benefits for general cases and especially at large time steps. However, the time step required to follow the correct solution for the test problem considered was too small to see any benefit when the method was compared with method sequenced in real time. It is likely that this is due to the very sensitive nonlinear dynamics involved in this case. The implication of this result for the simulation of nonlinear phenomena is clear. Great care needs to be taken in the verification of numerical results, with systematic and detailed parameter studies essential. The computational cost of such studies is likely to be high requiring powerful computers in the process.

Future work for the test case examined involves a detailed study of the influence of initial conditions for different time steps on the final solution behaviour obtained. It

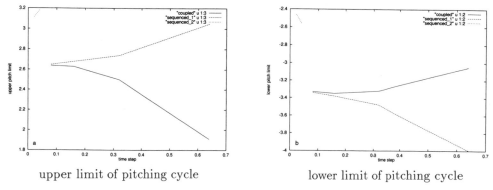

Figure 6. Convergence of limit cycle bounds.

is also of interest to examine exactly why the solution behaviour is so sensitive to the time step that the step change in behaviour is observed. An understanding of this issue would appear useful in contributing to the development of time stepping approaches which can be used with confidence for the simulation of nonlinear phenomena. The determination of the length of time over which small time steps are required to achieve the correct final behaviour is an interesting first question to be answered.

ACKNOWLEDGEMENTS

This work was partly funded by British Aerospace and Defence Evaluation Research Agency. The authors gratefully acknowledge the contribution of the other members of the Glasgow CFD group to this work.

REFERENCES

1. A. Jameson, *Time Dependent Calculations using Multigrid, with Applications to Unsteady Flows past Aerfoils and Wings*, Technical Report, AIAA 91-1596, (1991).
2. F. Cantariti, L. Dubuc, B. Gribben, M. Woodgate, K.J. Badcock and B.E. Richards, Approximate Jacobians for the solution of the Euler and Navier-Stokes equations, *AIAA J.* submitted, Aerospace Engineering Report, 5, Glasgow University, Glasgow, UK, (1997).
3. G.P. Guruswamy, Unsteady aerodynamic and aeroelastic calculations for wings using Euler equations, *AIAA Journal* 28:461-469 (1990).
4. G.P. Guruswamy, Navier-Stokes computations on swept-tapered wings, including flexibility, *Journal of Aircraft* 29:588-597 (1992).
5. H.C. Yee, J.R. Torczynski, S.A. Morton, M.R. Visbal and P.K. Sweby, On spurious behaviour of CFD simulations, in: *AIAA Paper A97-32438*. AIAA, (1997).
6. J.W. Edwards, Computational aeroelasticity challenges and resources, in: *NASA CP 3022*, pp631-637. NASA (1989).
7. K.A. Kousen and O.O. Bendiksen, Nonlinear aspects of the transonic aeroelastic stability problem, in: *29th Structures, Structural Dynamics and Materials Conference*. AIAA (1988).
8. K.J. Badcock, A. Littlejohn, A. Baldwin, R. Watt and M. Robb, A friendly introduction on the web to fluid dynamics, *Active Learning*, 5:26 (1996).
9. K.J. Badcock and B.E. Richards, Implicit Navier-Stokes codes in parallel for aerospace applications, in: *Parallel CFD 95*, p371. California Institute of Technology (1995).
10. K.J. Badcock, F. Cantariti, I. Hawkins, B. Gribben, L. Dubuc and B.E. Richards, *Simulation of unsteady turbulent flows using the pseudo time method*, Aerospace Engineering Report, 21, Glasgow University, Glasgow, UK (1997).
11. K.J. Badcock, G. Sim and B.E. Richards, Aeroelastic studies using transonic flow CFD modelling, in: *CEAS International Forum on Aeroelasticity and Structural Dynamics*. Royal Aeronautical Society (1995).
12. L. Dubuc, F. Cantariti, M. Woodgate, B. Gribben, K.J. Badcock and B.E. Richards, Solution of the Euler unsteady equations using deforming grids, *AIAA J.* submitted, Aerospace Engineering Report 4, Glasgow University, Glasgow, UK (1997).
13. O.O. Bendiksen, A new approach to computational aeroelasticity, in: *32nd Structures, Structural Dynamics and Materials Conference*. AIAA (1991).
14. R.D. Rausch, J.T. Batina, and H.T.Y. Yang, Euler flutter analysis of airfoils using unstructured dynamic meshes, *Journal of Aircraft* 27:436-443 (1990).
15. B.A. Robinson, J.T. Batina, and H.T.Y. Yang, Aeroelastic analysis of wings using the Euler equations with a deforming mesh, *AIAA Journal* 28:781-788 (1991).
16. D.M. Schuster, J. Vadyak, and E. Atta, Static aeroelastic analysis of fighter aircraft using a three-dimensional Navier-Stokes algorithm, *Journal of Aircraft* 27:820-825 (1990).

DIRECT NUMERICAL SIMULATION OF TURBULENT FLAMES

Karl W. Jenkins, [1] W. Kendal Bushe, [2] Laurent L. Leboucher, [3] R. Stewart Cant, [4]

[1] CFD Lab, Engineering Dept, Cambridge University
[2] CTR, Stanford, California
[3] Greenwich University
[4] CFD Lab, Engineering Dept, Cambridge University

ABSTRACT

The present paper is a report on progress in the simulation of turbulent flames using the Cray T3D and T3E at the Edinburgh parallel computing centre, using codes developed in Cambridge. Two combustion DNS codes are described, ANGUS and SENGA, which solve incompressible and fully compressible reacting flows respectively. The technical background to combustion DNS is presented, and the resource requirements explained in terms of the physics and chemistry of the problem. Results for flame turbulence interaction studies are presented and discussed in terms of their relevance to modelling. Recent work on the fully compressible problem is highlighted and future directions are outlined.

INTRODUCTION

Combustion remains central to the production of energy for transport, heating and power generation throughout the world, and is likely to retain its importance well into the next millenium. Nevertheless, concerns over environmentally harmful emissions have led to tough new legislation that places strong constraints on the design of combustion systems. Greater understanding is needed so that fundamental processes can be controlled to ensure clean and efficient combustion.

A difficulty is that practical combustion systems almost always involve turbulent flow, and the coupling between the turbulence, chemistry and heat release is very strong. The interactions are generally three dimensional and time dependent, and are not easily accessible to experimental investigation, even using the latest laser based techniques. Thus computational techniques are especially important in combustion, and the present work aims to simulate turbulent flame behaviour using Direct Numerical Simulation (DNS).

High Performance Computing
Edited by R. J. Allan *et al.*, Kluwer Academic / Plenum Publishers, New York, 1999

COMBUSTION DNS

DNS is a modern technique in the field of computational combustion. It is a method of solving directly the full governing equations, without the need for physical modelling or approximation. The governing equations in this problem are the full Navier-Stokes equations, plus chemistry

In almost all practical combustion systems the physics and chemistry of combustion take place in the gas phase, and in the turbulent flow field which is three dimensional by nature. There is a complex interaction between the chemistry, turbulent flow field and the thermal expansion of the fluid due to heat release, and the combustion process is highly non-uniform in space and time. Therefore the reacting flow contains numerous scales of motion from the largest, limited by the size of the problem, to the smallest which is influenced by viscosity.

DNS of turbulent flames is very expensive in terms of computational resource requirements such as CPU and memory, since all scales of the turbulence and chemistry must be fully resolved. Therefore the use of parallel supercomputers is essential in order to simulate physically relevant problem sizes and hence generate useful information in the form of data and statistics that will be of value to modellers and developers of industrial codes.

However in practice we cannot resolve full scale turbulence and chemistry together. Even with state of the art supercomputers the demand is too great. Therefore a compromise is required, and we choose to simplify the chemistry since we are mainly concerned with the flame turbulence interaction.

The range of turbulent length and time scales encountered in engineering applications often exceed three orders of magnitude. To simulate this problem using DNS requires scaling to Kolmogorov level. The maximum problem size is then determined by the computing resources which at present cannot handle such a range of scales and therefore restricts the attainable Reynolds number. Meanwhile designers need reliable numerical models to solve engineering problems, as legislation for pollutant emissions becomes ever tighter, it is no longer sufficient to design combustion equipment based on traditional empirical methods. Present major combustion models are Reynolds-averaged (RANS), pdf transport and LES. These all require statistical data, validation and guidance. DNS can provide the extra information, especially since the information must be three dimensional and time dependent.

In DNS the terms incompressible and compressible flow are defined in terms of density and pressure. DNS code ANGUS solves an incompressible flow with a low Mach number assumption, where density is a function of temperature and not pressure. The Mach number is taken to be zero everywhere, and all acoustic activity is excluded. Conversely SENGA solves a fully compressible reacting in which density is a function of both temperature and pressure. Therefore acoustic activity is resolved whether or not the Mach number approaches or exceeds unity, enabling the investigation of combustion in high speed flows and wave interaction effects. Most combustion involves a significant density change due to heat release which makes the fully compressible formulation most attractive.

GOVERNING EQUATIONS

The motion and reaction of gases contributing to combustion are governed by conservation equations which describe the transport of properties of the flow. These are partial differential equations expressing the conservation of overall mass, momentum, energy and the mass fractions of individual species. It is assumed that the Fickian law of mass diffusion is applicable and that the Lewis number is unity, also Soret and Dufour effects and thermal radiation are negligible. This results in the following set of equations in Cartesian tensor notation:

$$\frac{\partial \rho}{\partial t} + \frac{\partial \rho u_k}{\partial x_k} = 0$$

$$\frac{\partial \rho u_i}{\partial t} + \frac{\partial \rho u_k u_i}{\partial x_k} = -\frac{\partial P}{\partial x_i} + \frac{\partial \tau_{ki}}{\partial x_k}$$

$$\frac{\partial \rho h}{\partial t} + \frac{\partial \rho u_k h}{\partial x_k} = -\frac{\partial q_k}{\partial x_k} + \frac{\partial P}{\partial t} + u_k \frac{\partial P}{\partial x_k} - \tau_{ki} \frac{\partial u_k}{\partial x_i}$$

$$\frac{\partial \rho Y_\alpha}{\partial t} + \frac{\partial \rho u_k Y_\alpha}{\partial x_k} = \omega_\alpha + \frac{\partial}{\partial x_k}(\rho D_\alpha \frac{\partial Y_\alpha}{\partial x_k})$$

where ρ is the density, u_k is the velocity, P is the pressure, h is the enthalpy, Y_α is the mass fraction of species α and D_α is the diffusivity of species α.
These equations must be supplemented by the thermal and caloric equations of state

$$P = \rho RT \sum_{\alpha=1}^{N} \frac{Y_\alpha}{W_\alpha}$$

$$h = C_p T + \sum_{\alpha=1}^{N} h_\alpha Y_\alpha$$

where N is the number of chemical species, W_α is the molecular weight of species α, T is the temperature, R is the universal gas constant, C_p is the specific heat of the mixture and h_α is the enthalpy of formation of species α.
The viscous forces caused by the fluid motion are described by

$$\tau_{ij} = \mu(\frac{\partial u_i}{\partial x_j} + \frac{\partial u_j}{\partial x_i} - \frac{2}{3}\frac{\partial u_k}{\partial x_k}\delta_{ij})$$

where μ is the viscosity of the fluid and δ_{ij} is the Kronecker delta function.
The energy transport of the flow from heat transfer by conduction and diffusive enthalpy transport is given by

$$q_i = -\lambda \frac{\partial T}{\partial x_i} + \sum_{\alpha=1}^{N} \rho V_{\alpha i} h_\alpha Y_\alpha$$

where λ is the thermal conductivity and $V_{\alpha i}$ is the mass diffusion velocity of species α.
Finally, to close the problem, the source term for the reaction rate of species α is

$$\omega_\alpha = W_\alpha \sum_{j=1}^{N}(\nu''_{\alpha,j} - \nu'_{\alpha,j})B_j T^{n_j}$$

$$\times exp[-\frac{E_j}{R^0 T}] \prod_{\beta=1}^{N}[\frac{\rho Y_\beta}{W_\beta}]^{\nu'_{\beta,j}}$$

where ν'' and ν' are the stoichiometric coefficients for product and reactant species respectively in a reaction j, E is the corresponding activation energy and n_j accounts for non-exponential temperature dependence of the reaction rate.

DNS CODE ANGUS

ANGUS is a three dimensional, incompressible computer code which directly solves the conservation equations for momentum, energy, mass and a reaction progress variable. The turbulent flow field is completely resolved, with the length scale of the grid chosen to be a fraction of the Kolmogorov length, and the time step chosen to be a fraction of a Kolmogorov time.

ANGUS uses a second-order central differencing scheme to evaluate spatial derivatives and an explicit second-order Adams-Bashforth time advancement scheme. However, a Poisson equation has to be solved for pressure in order to enforce continuity under a low Mach number assumption. This can be very expensive computationally on a large grid. Therefore special techniques are required, the fastest being Fourier methods exploiting the Fast Fourier Transform (FFT) algorithm. Applying the Fourier method restricts the grid size to 2^n, where n =1,2,3 ... N. Alternatives for use on arbitary sized grids include preconditioned conjugate gradient and multigrid methods, but the performance penalties are considerable relative to the Fourier Techniques.

The only modelling assumption applied is that the flame is considered premixed with a simple, single step reaction mechanism controlled by Arrhenius kinetics. Therefore the chemical reaction becomes

$$R \rightarrow P$$

where R represents the reactants and P the products. Assuming the species diffusivities are equal, a single reaction progress variable can be used to represent the chemical state of the system (Bray Moss Libby 1985). It is convenient to write the governing equations in non-dimensional form with respect to reference values of velocity u_0, distance l_0, density ρ_0, pressure P_0, temperature difference $T_{ad} - T_0$, specific heat C_{p0}, viscosity μ_0, thermal conductivity λ_0, and diffusion coefficient D_0. The non-dimensional reaction progress variable c has values between zero and one, where c=0 for unburned gas and c=1 for a fully burned gas. Therefore the full equations are

$$\frac{\partial \rho}{\partial t} + \frac{\partial \rho u_k}{\partial x_k} = 0$$

$$\frac{\partial \rho u_i}{\partial t} + \frac{\partial \rho u_k u_i}{\partial x_k} = -\frac{\partial P}{\partial x_i} + \frac{1}{Re}\frac{\partial \tau_{ki}}{\partial x_k}$$

$$\frac{\partial \rho T}{\partial t} + \frac{\partial \rho u_k T}{\partial x_k} = \frac{1}{RePr}\frac{\partial}{\partial x_k}(\lambda \frac{\partial T}{\partial x_k}) - \sum_{\alpha=1}^{N} \frac{h_\alpha}{C_{p0}}w_\alpha$$

$$\frac{\partial \rho Y_\alpha}{\partial t} + \frac{\partial \rho u_k Y_\alpha}{\partial x_k} = w_\alpha + \frac{1}{ReSc}\frac{\partial}{\partial x_k}(\rho D_\alpha \frac{\partial Y_\alpha}{\partial x_k})$$

where (to avoid extra notation) all variables are now understood to be the non-dimensional equivalents of those in the governing equations. The Reynolds, Prandtl and Schmidt numbers are given by

$$R_e = \frac{\rho_0 u_0 l_0}{\mu_0}; P_r = \frac{\mu_0 C_{p0}}{\lambda_0}; S_c = \frac{\mu_0}{\rho_0 D_{\alpha 0}}$$

Finally the non-dimensional reaction rate is given by

$$\omega = B^*\rho(1 - c)exp(-\frac{\beta(1 - T^*)}{1 - \alpha(1 - T^*)})$$

where B^* is the pre-exponential factor, heat release parameter $\alpha = (T_{ad} - T_o)/T_{ad}$, reduced temperature $T^* = (T - T_0)/(T_{ad} - T_0)$ and the Zel'dovitch number $\beta = E(T_{ad} - T_0)/R^0 T_{ad}^2$. Therefore, the whole thermochemical field is controlled by the progress variable c, ranging between zero and one.

ANGUS is in full production mode on the Cray T3D and T3E at Edinburgh parallel computing centre. The code employs 64 processors to solve problems over a 128^3 grid. Runs taking 1500 time steps take approximately 10 hours to complete. Useful results have been obtained using ANGUS, some of which are highlighted later, and further runs are still in progress. We have obtained useful results for code validation, flame curvature, strain rate, stretch rate, alignment between flame and flow field and sub-grid statistics for LES modelling.

DNS CODE SENGA

SENGA is a three dimensional combustion DNS code which solves a fully compressible reacting flow. SENGA was designed to improve efficiency on massively parallel machines such as the T3D and T3E. These machines favour numerical schemes which rely on pointwise rather than global solution methods, since the communication overhead between processors is greatly reduced. Thus, a fully compressible formulation avoiding the need to solve a Poisson equation for pressure as in ANGUS should prove highly efficient. An additional benefit is that all acoustic phenomena are resolved, providing the extra physics which ANGUS cannot provide.

To tackle more realistic problems it is necessary to be able to specify more complex boundary conditions, including inflow and outflow cases. These conditions are implemented using finite difference schemes, whereas ANGUS is restricted to periodic boundary conditions only because of the Fourier pressure solver. Therefore SENGA employs state of the art finite difference numerical schemes to which mimic the accuracy of spectral methods used traditionally for DNS.

Again the governing equations are non-dimensionalised with respect to reference values to give

$$\frac{\partial \rho}{\partial t} + \frac{\partial \rho u_k}{\partial x_k} = 0$$

$$\frac{\partial \rho u_i}{\partial t} + \frac{\partial \rho u_k u_i}{\partial x_k} = -\frac{\partial P}{\partial x_i} - \frac{1}{Re}\frac{\partial \tau_{ki}}{\partial x_k}$$

$$\frac{\partial \rho E}{\partial tk} + \frac{\partial \rho u_k E}{\partial x_k} = -(\gamma - 1)M^2\frac{\partial P u_k}{\partial x_k}$$

$$-(\frac{1}{Re}(\gamma - 1)M^2\frac{\partial \tau_{ki} u_i}{\partial x_k}$$

$$+\frac{\tau^*}{RePr}\frac{\partial}{\partial x_k}(\lambda\frac{\partial T}{\partial x_k})$$

$$-\frac{\tau^*}{ReSc}\frac{\partial}{\partial x_k}(\rho D\frac{\partial c}{\partial x_k})$$

$$\frac{\partial \rho c}{\partial t} + \frac{\partial \rho u_k c}{\partial x_k} = \omega + \frac{1}{ReSc}\frac{\partial}{\partial x_k}(\rho D\frac{\partial c}{\partial x_k})$$

Table 1. ANGUS: Parameters after 1500
time steps

Run Number	1	2	3	4
$[u'/u_l]_{t=0}$	1.0	2.0	5.0	10.0
Fraction Burned	0.283	0.283	0.287	0.289

where M is the Mach number, γ is the ratio of specific heats and τ^* is the heat release parameter, defined as $\alpha = \tau^*/(1 + \tau^*)$ where α is the heat release parameter from the reaction rate term. Additional terms required to complete this problem are the non-dimensional thermal and caloric equations of state, given by.

$$P = \frac{\rho(1 + \tau^* T)}{\gamma M^2}$$

$$E = \frac{(1 + \tau^* T)}{\gamma M^2} + \frac{1}{2} u_k u_k (\gamma - 1) M^2 + \tau^* (1 - c)$$

These equations are solved using high accuracy spatial and temporal differencing schemes. For the three coordinate directions a sixth order compact finite differencing scheme Lele (1992) is used to calculate the spatial derivative terms. This scheme uses a stencil of 5 grid points, 2 points either side of the node at which the differential terms are calculated. This causes a problem at the boundaries, since 2 points would be required outside the domain to calculate the derivative terms at the boundary. Therefore a third order scheme is employed at the boundary and a fourth order Pade scheme at the second point. The temporal derivatives are calculated using a third order explicit Runge-Kutta time stepping scheme. This scheme involves three sub-steps for each main time advancement. A tenth order explicit scheme (for spatial derivatives) is also available and is under test. Finally the boundary conditions are treated using the Navier-Stokes characteristic boundary conditions (NSCBC) formalism of Poinsot and Lele, (1992).

A serial SENGA code has been tested on DEC ALPHA and Silicon Graphics workstations for 32^3 and 64^3 grid sizes in 3D. The parallel version is now under test on a four processor Silicon Graphics machine at the CFD Lab in Cambridge. The code will soon be ported to the T3D, T3E and the Hitachi supercomputer at Cambridge.

SIMULATIONS

Simulations for ANGUS and SENGA were initiated by inseting a laminar flame solution into a fully developed turbulent field.

An initial turbulent field satisfying the conditions of continuity, homogeneity and isotropy was generated and then used as the initial conditions for all simulations.

Figures 1a-e show a back to back flame propagation run from ANGUS. The grid size is 128^3 with a pair of back to back planar flames at the centre. The plots show the flame surfaces at a progress variable value of 0.5. Clearly the flame-turbulence interaction is evident with stretching and wrinkling of the surface as the turbulence decays.

Figures 2 - 5 show the velocity fields for 4 ANGUS runs listed in table 1 at $t/\tau = 0.15$, which equates to 1500 time steps. Only the lower left hand quarter of the plane is shown visible. Five isopleths of reaction progress variable are shown at 0.1, 0.3, 0.5, 0.7 and 0.9, increasing from left to right of the figures.

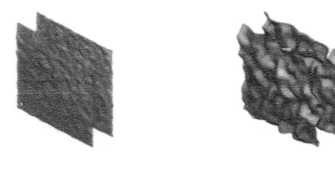

(a) $t/\tau = 0.0051$ (b) $t/\tau = 0.025$

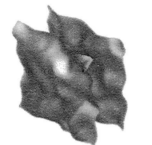

(c) $t/\tau = 0.065$ (d) $t/\tau = 0.105$

(e) $t/\tau = 0.145$

Figure 1. Planar back to back Flame Progagation from SENGA, 128^3 grid

It is clear that the flame is more strongly affected by the higher values of u'/u_l. At the lowest value, the flame is still almost planar, whereas at the highest value, the flame has been warped by the flow. Bushe and Cant (1996) show various results from the same 4 runs at different times.

Figure 6 shows a 3D laminar propagation of (from left to right) density, energy, progress variable, u-velocity, v-velocity and w-velocity, from SENGA on a 32^3 grid. The plots show x-y planes at the centre of the domain. Clearly the progress variable ranges from 0 to 1 for unburned and fully burned cases respectively.

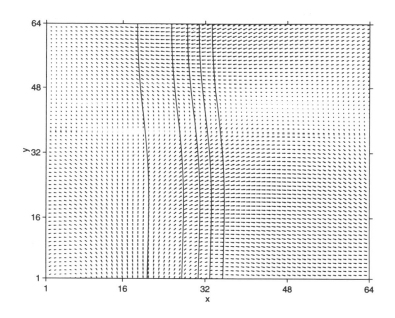

Figure 2. Velocity field for run i at $t/\tau = 0.15$

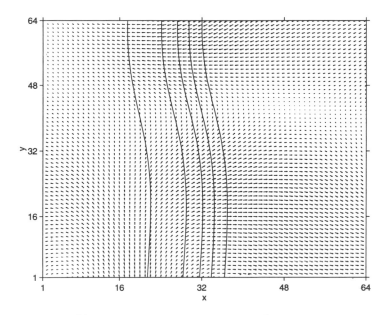

Figure 3. Velocity field for run 2 at $t/\tau = 0.15$

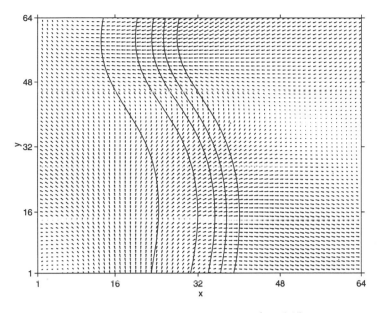

Figure 4. Velocity field for run 3 at $t/\tau = 0.15$

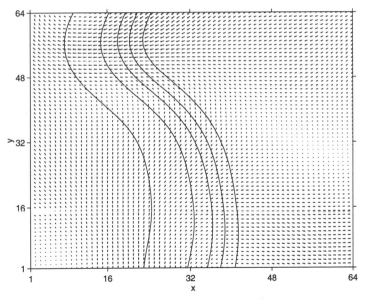

Figure 5. Velocity field for run 4 at $t/\tau = 0.15$

Finally Figure 7 shows a flameball propagating in a turbulent field. This flameball was simulated by setting the progress variable in the form of a 3D Gaussian distribution over a 10^3 grid. This initial 3D ball was then placed in the centre of a 32^3 turbulent field. The plot shows contours of progress variable c , from 0 on the outer core to 1 at the inner, taken at the centre of the domain. The evidence again of wrinkling and stretching due to flame turbulence interactions is clear.

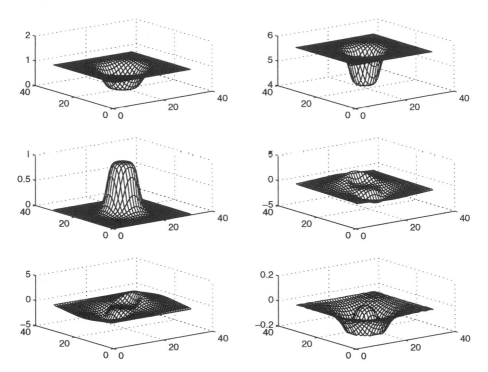

Figure 6. 3D laminar flame propagation from SENGA for ρ, E, c, u, v and w at the centre plane.

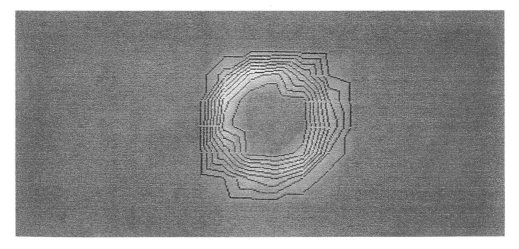

Figure 7. Turbulent flameball from SENGA, progress variable contours from 0 (outer)- 1 (inner)

FUTURE WORK

The next phase in the development of the DNS codes is to port the parallel version of SENGA onto the T3D and T3E. This will enable the investigation of more realistic problems using larger grids with higher Reynolds numbers.

Chemistry is the one area that is simplified in this work. Therefore better chemistry up to a four step mechanism will be investigated.

SENGA utilises inflow and outflow boundary conditions which can also be used to investigate more realistic problems in the form of reactants entering and products leaving the area of interest.

Finally the work carried out will lead to better combustion modelling by providing valuable DNS data.

REFERENCES

Bray, K.N.C., Libby, P.A., Moss, J.B., 1985, "Unified Modelling Approach for Premixed Turbulent Combustion Part 1: General Formulation", *Combust. Flame* 61, 87-102.

Bushe, W.K., Cant, R.S., 1996, "Results of Direct Numerical Simulations of Premixed Turbulent Combustion", *CUED Report CUED/A-THERMO/TR63*, Cambridge University Engineering Department.

Lele, S., 1992, "Compact Finite Difference Schemes with Spectral like Resolution" *J. Comp. Phys.* 103, 16-42.

Poinsot, T.J., Lele, S., 1992, "Boundary Conditions for Direct Simulations of compressible Viscous Flows", *J. Comp. Phys.* 101. 124-129.

UNDERSTANDING TURBULENCE IN FLUIDS USING DIRECT SIMULATION DATA

M. Alam, E. Avital, T.J. Craft[2], S.P. Fiddes[3], H.P. Horton, R.J.A. Howard, D.P. Jones[3], K.H. Luo, N.D. Sandham, A.M. Savill[4], T.G. Thomas, P.R. Voke[5], J.J.R. Williams

Department of Engineering, Queen Mary & Westfield College, Mile End Road, London E1 4NS, U.K.
[2] Department of Mechanical Engineering, UMIST, Manchester, M60 1QD, U.K.
[3] Aerospace Engineering, University of Bristol, University Walk, Bristol, BS8 1TR, U.K.
[4] Department of Engineering, University of Cambridge, Cambridge, CB3 1PZ, U.K.
[5] Department of Mechanical Engineering, University of Surrey, Gilford, GU2 5XH, U.K.

Abstract. Since 1994 the DNS-Turbulence Consortium has been using the Cray T3D and T3E computers at Edinburgh Parallel Computing centre to conduct space-and time-resolved calculations of turbulent flow. A variety of cases have been studied in detail including the first resolved calculation of turbulent boundary layer reattachment following laminar separation, the response of turbulence to three-dimensional distortion and the response of turbulence to separation. The data from these simulations have been used to test and improve turbulence models. In this paper the progress that has been made in the last three years is reviewed.

1. Introduction

Direct numerical simulation (DNS) of the three-dimensional time-dependent Navier Stokes equations provides data for the study of turbulence in fluids, including many quantities that can not be accurately measured experimentally. The largest space scales in a turbulent flow are usually fixed by the geometry, while the smallest scales are determined by viscosity. The difference between the largest and the smallest length scales in turbulence increases as the Reynolds number increases, so simulations with limited memory are limited by Reynolds number. With the advent of reliable high performance massively parallel computers the range of Reynolds numbers and configurations that can be studied is increasing.

A research Consortium was formed in the summer of 1994 to carry out DNS on the new Cray T3D massively parallel computer and use the data from simulations to advance fundamental understanding of turbulence and improve turbulence models. Efficient parallel codes were developed for simulation of compressible and incompressible flows. Numerical methods have included fully spectral methods[1,2] for flows in simple geometry, finite volume schemes[3] for flows in complex geometries and high-order finite difference schemes for compressible flows. Codes are written in Fortran 77 and C. Parallelisation has been accomplished using PVM and MPI. The majority of codes use a global transpose method to cope with the distributed memory architecture. The Consortium has also experimented with HPF. Typically the codes used are well suited to parallel machines and are mainly limited by the per-node performance of the processors. Two codes have been run on the T3E with speed-ups of 3 and 5 recorded in comparison with the T3D. In the following sections progress for specific applications is summarised.

2. Laminar separation bubbles

When a laminar boundary layer encounters a strong enough adverse pressure gradient it separates to form a laminar free shear layer. The separated shear layer undergoes transition to turbulent flow and the turbulent shear layer may reattach to form a closed laminar separation bubble. Laminar separation bubbles are generally undesirable because the locally separated flow degrades aerofoil performance even when the bubble is only a few percent of the chord in length. Furthermore, small changes in freestream condition or angle of attack can cause a short bubble to 'burst' so it covers a large percentage of the chord, which may result in catastrophic stall of the wing. Therefore predictions of laminar separation bubbles are very important for the aircraft design and development process. Despite years of research into laminar separation bubbles the precise mechanism of bursting and the role of critical processes such as the transition and the reattachment on bursting are not clearly understood.

In the current investigation DNS has been employed as a tool to investigate laminar separation bubbles. Two dimensional (2D) DNS is obviously cheaper than 3D DNS but the current authors have shown[4] that 3D DNS must be used to validate bubble models. A fully-resolved 3D simulation with the ability to accumulate statistical data has been run using 128 PE's of the Cray T3D. The adverse pressure gradient required

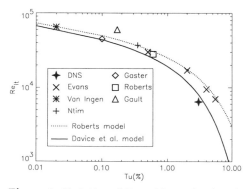

Figure 1. Variation of Reynolds number based on transition length with local turbulence level. Experimental results taken from W.B. Roberts (AIAA 79-0285) and R.L. Davice *et al.* (AIAA 85-1685).

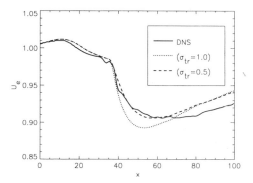

Figure 2. Model comparison of boundary layer edge velocity with DNS data. σ_{tr} is the equilibrium parameter which is fixed up to transition point and allowed to vary according to Johnson & Coakley turbulence model further on.

for flow separation was generated by suction of fluid through the upper boundary of the simulation. Transition in the free shear layer was induced by introduction of a pair of oblique waves upstream of separation. The turbulent free shear layer quickly reattaches and the boundary layer relaxes towards an equilibrium state. The mean bubble is of realistic Reynolds number and its structure is comparable to 'short' leading-edge aerofoil bubbles (see figure 1). In order to assist modelling of the flow a large database of numerous flow quantities have been accumulated. Turbulent quantities such as the budgets of Reynolds stresses and the turbulence kinetic energy along the entire length of the bubble together with a considerable amount of the relaxing boundary layer have been analysed. The findings[5] suggest that the boundary layer relaxation downstream of reattachment is a significant part of a laminar separation bubble and should be accounted for in model calculations.

The data generated under the current investigation is being used by the fifth author to test models. An inverse boundary layer is used, at present incorporating the Johnson and Coakley[6] turbulence model for the post transition region. Referring to figure 2, the model predicts the main features such as the separation and reattachment for a given transition point. Some disagreement remains downstream of reattachment and this could be due to low Reynolds effect or the fact that the model fails to capture boundary layer relaxation.

3. Turbulence in a square duct

The problem considered is the incompressible turbulent flow in a straight, square sectioned, duct. The bulk velocity U_b and duct half–height h are the non-dimensional quantities $(Re_b = 2hU_b/\nu)$.

In the spectral solution algorithm[2], the velocity is written as the sum of basis functions, each of which satisfy the continuity equation and boundary conditions. The basis functions are split into two components, $+$ and $-$, so that they form a complete set in the solution space. The expansion of the velocity vector u becomes

$$u = u^+ + u^- \quad \text{with} \quad u^\pm = \sum_{k=-N_x/2}^{N_x/2-1} \sum_{j=0}^{N_y} \sum_{l=0}^{N_z} \alpha_{kjl}^\pm \phi_{jl}^\pm(y,z,k)e^{ik_x x} \tag{1}$$

After substituting expansion (1) in the Navier-Stokes equation, a set of ordinary differential equations for the coefficients $\alpha_{kjl}^\pm$ are obtained by dot multiplying the Navier-Stokes equations with the test functions and integrating over the entire domain. The

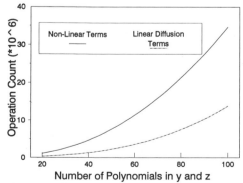

Figure 3. Operation count, per Fourier mode, comparing the evaluation of the non-linear (F_k) and linear terms.

Figure 4. Contours of $\overline{v'v'}$ correlation for bottom left quadrant of simulation S2

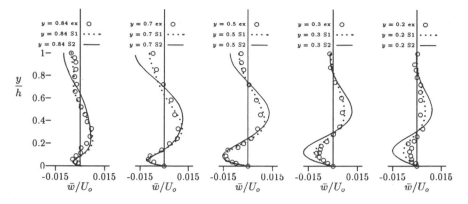

Figure 5. Comparison of simulated horizontal velocity profiles with experiment[7]

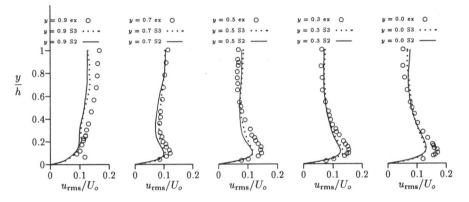

Figure 6. Comparison of simulated streamwise RMS velocity profiles with experiment[7]

orthogonality property of the Fourier series in x means that for each Fourier wave–number k_x, the resulting equation set can be written in matrix form as

$$A_k \frac{d\boldsymbol{\alpha}_k}{dt} = \frac{2}{Re_b} B_k \boldsymbol{\alpha}_k + P_k + F_k \qquad (2)$$

where A_k and B_k are $(2(N_y+1)(N_z+1))^2$ matrices, $\boldsymbol{\alpha}_k$ is the vector with elements α_{kjl}, F_k is the vector representing the non–linear contribution from convection and P_k is the vector representing the mean pressure gradient. During the calculation, the constant pressure gradient is adjusted to ensure constant bulk velocity. The seventh author has developed a fast solution procedure for equation (2) based on ADI and eigenvalue decompositions coupled with the Householder modification formulae. The comparison of operation count for non-linear term evaluation and matrix inversion can be seen in figure 3. Recent work has shown that ADI is not necessary to obtain an efficient solution procedure for the diffusion terms and this will be a basis for future research.

As test cases, two simulations of turbulent square duct flow at a Reynolds num-ber of Re_b=4410 were undertaken. The simulations (referenced S1/S2) used 192/256 Fourier modes in the streamwise direction and 60/92 quasi–orthogonal modes in each of the cross–stream directions respectively. This Reynolds number was chosen as it approximately matched previous DNS[7] and experiment[8]. All the terms required to close the moment equations up to 3rd order have been collected. An example plot of the $\overline{v'v'}$ Reynolds stresses is given in figure 4. The horizontal crossflow velocities and streamwise RMS values are compared with experiment in figures 5 and 6.

4. Databases for turbulence modelling

Although current state-of-the-art DNS can only be applied to low Reynolds number turbulent flows it can still provide extremely useful information for validating turbulence models. This is because DNS is able to provide highly accurate details of all the terms required in statistical modelling. In the past, data obtained from experiments was used to validate turbulence models. This had several drawbacks. Statistics of many of the terms to be modelled are difficult to obtain experimentally and often contain large error margins. This is especially the case for the near wall region of a turbulent flow. DNS however can provide exact results right down to the wall with extremely small error margins.

This Consortium has been involved in generating databases of complex turbulent flows including several based around a channel flow configuration. At this stage the turbulent channel flows investigated are: (i) plane channel at two Reynolds numbers, the larger carried out using 256 processors on the Cray T3D, (ii) a three dimensionally deformed channel in which the channel walls are set into spanwise motion and (iii) a turbulent separation and reattachment caused by introducing body forces into the channel domain. These databases are currently being used directly for turbulence model development. Howard and Sandham[9] identified significant problems in the modelling of the three dimensional flow. The DNS showed breakup of the coherent turbulent streak structures accompanied by drops in the turbulence intensity of the flow. Figure 7 shows the way the maximum turbulence intensity varies with time after the walls have been moved and also the behaviour of current turbulent models under the same conditions. The models tend to show increases in turbulence intensity due to the skewed motion rather than the decrease observed in the DNS. Further analysis of more developed models for complex channel flows was carried out by Howard and Sandham[10]. This work provided some improvement on the modelling for the three dimensional flow and some added understanding of the modelling of the separated flow. One of the aims of the Consortium is to extend the databases available for turbulence modelling development. As outlined above and will be discussed below, several of the flows calculated using DNS codes of this consortium are already contributing to turbulence modelling initiatives. The benefits of this work will become increasingly important as new DNS flows are carried out.

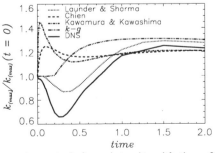

Figure 7. Variation in the maximum turbulence intensity with time after the walls of a turbulent channel flow are skewed in the spanwise direction.

5. Modelling bypass transition to turbulence

Previous analysis of large-eddy simulation (LES) of bypass transition carried out at the University of Surrey on a sharp leading-edge test plate under the influence of variable free-stream turbulence, to deduce mechanisms and corresponding essential modelling strategies[11], has been extended to the case of finite leading-edge separation bubble transition in the absence of external turbulence[12]. Key features of these simulations and in particular the changes in turbulence balances through the bubble and downstream, are qualitatively the same as those obtained by Alam and Sandham[5]. Analysis of these has confirmed earlier speculation[11] that similar mechanisms are involved in each case. The turbulence in the separated shear layer effectively replaces the free-stream turbulence to ensure that a kind of by-pass transition occurs in the internal boundary layer, and therefore the same model treatments should be optimal. Although the separated shear layer imposes correlated u and v fluctuations (and hence a secondary route for generating local u and then v via pressure-strain redistribution), the primary path again appears to be the influence of the dominant imposed v fluctuations, which combine with the local mean shear to generate local uv. Since Reynolds stress treatments can already handle leading-edge impingement and subsequent surface curvature effects, the type of non-local Reynolds stress model treatment proposed in Savill[11] should be equally effective for predicting separation bubble transition, provided the additional length scale is taken to be the correlation scale within the separated shear layer rather than the free-stream length scale used for the sharp leading-edge flow. The main features identified from balances in Savill[11] are confirmed by experimental study of heated transitional flow by Wang and Zhou[13], although they reached somewhat different conclusions regarding the local turbulence generation details.

6. Skewed turbulent flow past a cube

In this application of LES to an engineering problem we are concerned primarily with predicting the unsteady loads acting on a building due to its wind environment. One might suppose that the largest pressures acting on parts of a building might be due to extreme stagnation pressures, but in fact they are the low pressures (or suction) associated with the vorticity shed by separation from the sharp edges of a typical structure. The magnitude of the worst negative pressure coefficient C_p is much larger than the worst positive one, and the probability density function (PDF) of the C_p variation in the separated flow is non-Gaussian and skewed towards negative values. These unsteady low pressures are responsible for much of the wind damage to roof cladding and window glass. To make a realistic prediction of these loads one has to simulate the unsteady flow field surrounding the building and use a computational grid capable of resolving the small scale but very intense vortices shed at the sharp corners; the combination of LES and large scale parallel computers such as the T3D/E is ideally suited for this purpose.

The building is represented by a cubic obstacle placed in a turbulent boundary layer and aligned at 45° to the mean flow. This alignment is particularly interesting because the roof pressure is dominated by a pair of conical vortices attached to the roof at the upwind corner. In a wind loading problem it is important that the turbulent boundary layer match the expected wind environment, thus we match the vertical mean velocity profile characterised by the Jenson number $Je \equiv h/z_o$ where h denotes the scale height of the building and z_o the roughness scale of the upwind terrain, the turbulence intensity $I_u \equiv u_{rms}/u_o$ where u_{rms} denotes the streamwise fluctuation intensity and u_o denotes

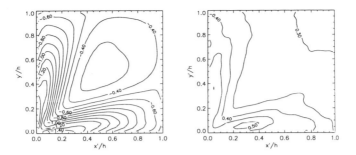

Figure 8. Mean (a) and fluctuating (b) rms pressure coefficient Cp on the roof. The mean flow approaches from the lower left corner.

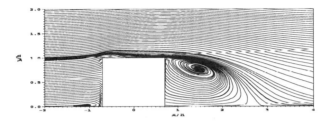

Figure 9. Streamlines in the (x, z) plane (centreline $y = 0$)

the reference velocity at height h, and the ratio L_x^u/h characterising the integral length L_x^u of the approaching turbulence. The flow is insensitive to the Reynolds number $R \equiv u_o h/\nu$, where ν denotes the kinematic viscosity, provided $R \geq 3000$. We set $Re = 10000$, $Je = 60$, $I_u = 0.15$, and $L_x^u/h \sim 1.1$ to simulate a 30-40m high building placed in a typical urban wind environment, and generate this (time dependent) inflow data from a precursor boundary layer simulation. The flowfield around the cube was simulated using our complex geometry code CgLES using a domain decomposition into 239 blocks each using a local $(32, 32, 32)$ grid, giving an effective grid of $(362, 226, 96)$ in the streamwise, spanwise, and vertical directions, i.e. about 3×10^7 degrees of freedom, and run on 64 processors. The code performance is $\sim 1.0 \times 10^{-5}$ processor seconds per degree of freedom per time step on the T3E and is independent of the number of processors used. Further details of numerical methods can be found in Thomas and Williams[3].

Figure 8 shows the mean and fluctuating pressure coefficients on the roof of the cube; this distribution clearly marks out the regions of worst case loading although the PDFs (not shown) provide additional information about extreme values and the load spectra are also available. Figure 9 shows the mean streamlines on the centreline, the downstream re-attachment point, the location of the arch vortex in the near wake, and the upstream horseshoe vortex. Although these structures are clearly marked out by the mean flow, in reality they exist only intermittently. The recirculation bubble is seen to entrain fluid from upstream and therefore, unlike its two dimensional counterpart, cannot form a finite closed surface. Figure 10 shows the surface streak pattern on the floor around the cube indicating the position of the upstream horseshoe vortex, the rear re-attachment nodal point, a pair of saddlenodes, and a pair of converging foci representing the feet of the arch vortex. This pattern agrees with experimental streak patterns. Figure 11 shows a spanwise slice of the mean velocity field in the near wake

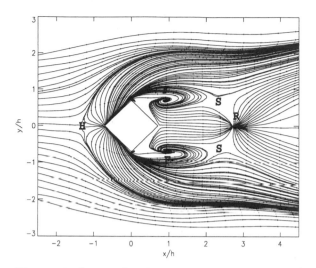

Figure 10. Computed surface streak patterns on the bed ($z = 0$).

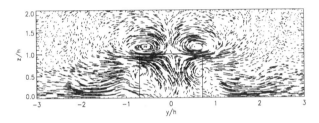

Figure 11. Mean velocity (v, w) at section $x/h = 1$ within the near wake.

showing a four vortex pattern: the upper pair are the tails of the roof conical vortices, and the lower counter-rotating pair represent the tilted legs of the arch vortex.

7. Compressible mixing layers and aeroacoustics

Fundamental understanding of compressible turbulence has always lagged behind its technological applications. The slow progress in the area is largely caused by a lack of techniques to measure accurately the basic quantities in compressible turbulence, such as pressure and density fluctuations, and their correlation with the velocity fields. The DNS of compressible mixing layers conducted by Luo and Sandham[14] showed the first example of transition to fully turbulent flow at high Mach numbers (convective Mach number $M_c = 0.8$). On the Cray T3D a compressible mixing layer at $M_c = 1.1$ was simulated[16] on a computational box of $20.0 \times 45.0 \times 11.5$. The simulation started with 32 processors and ended with 128 processes on T3D, reaching a final grid of $144 \times 384 \times 128$. This well resolved simulation revealed, among other things, the existence of eddy shocklets at $M_c = 1.1$ (Fig. 12), the lowest Mach number at which shocklets have been found in three-dimensional turbulence.

Although the allocation of time on the T3D was initially insufficient to contemplate a parametric study of the effect of Mach number we were able to collaborate with A.W. Vreman of the University of Twente and pool databases to give data at a series of Mach numbers 0.2, 0.6, 0.8, 1.1 and 1.2. Analysis of this was carried out to determine reasons for the reduction in growth rate of the mixing layer as the Mach number is increased. Surprisingly, this was found to be due to reduced pressure fluctuations via

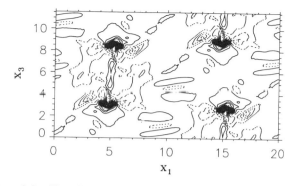

Figure 12. Plan view of the dilatation field in the central plane ($x_2 \approx 0$) at $t = 60.5$, showing the formation of shocklets in the mixing layer at $M_c = 1.1$. ($d_{min} = -2.25$, $d_{max} = 0.44$; 35 contour lines, solid — negative, dotted — positive)

the pressure strain term, rather than through the explicit compressibility terms such as pressure-dilatation or dilatation-dissipation. A new model was constructed to account for the observed phenomena and published[15]. Recent work from the US confirms the conclusions of that paper. Subsequent work[16] focused on modelling the pressure-related terms in the transport equations of the Reynolds stresses and the scalar fluxes, which reached good agreement with the DNS data.

Avital *et al.*[17] showed that the above temporal simulation could be used to generate acoustic information that is identical to that from a spatial simulation of supersonic flow. This is important, as even with a powerful computer like T3D a full spatial simulation of the acoustic near-field is not feasible. Temporal DNS data was processed[18,19] with the aid of the acoustic analogies developed by Lighthill and others. The work has so far enabled a direct comparison between the acoustic field simulated by DNS and that predicted from acoustic analogies. The experimentally observed Mach waves have also been identified in the DNS data. It is anticipated that supercomputing will play a crucial role in the study of compressible turbulence and computational aeroacoustics in the coming years.

8. Conclusions

The Consortium has developed efficient parallel codes and applied these to simulate complex turbulent flows. Highlights of the work can be summarised as follows.
(a) Fully resolved simulations of laminar separation bubbles have been carried out for the first time. In this flow the transition in a separated shear layer is simulated together with the reattaching and relaxing turbulent boundary layer.
(b) Fully spectral simulation of turbulent duct flow has been carried out, generating turbulence data for a flow that exhibits the effects of secondary flow.
(c) Turbulence data has been produced for applications to improvement of turbulence modelling. Datasets range from reduced statistical data of few hundred Kbytes to complete flowfield data from 256^3 calculations that require a Gbyte of storage.
(d) Vortex shedding behind a bluff body in a boundary layer has been computed. Here a wall-mounted cube is exposed corner-on to an atmospheric boundary layer. The re-

sulting flow is characterised by conical roof vortices and complex vortex shedding.
(e) Simulations have been made of compressible mixing layers, leading to fundamental
advances in the understanding of compressibility effects on turbulence and the prediction of aerodynamically generated sound.

REFERENCES

1.   N. D. Sandham and R. J. A. Howard, Direct simulation of turbulence using massively parallel
     computers, *Parallel Com. Fluid Dynamics '97*, A. Ecer *et al.*, Elsevier (1997).
2.   D. Jones, DNS of square duct turbulent flow using a Petrov-Galerkin spectral method. *Report
     97-03*, Department of Aerospace Engineering, University of Bristol (1997).
3.   T. G. Thomas & J. J. R. Williams, Development of a parallel code to simulate skewed flow over
     a bluff body, *J. Wind Eng. Ind. Aerodyn* 67&68 155-167 (1997).
4.   M. Alam and N. D. Sandham, Simulation Of laminar separation bubble instabilities, *Direct and
     Large Eddy Simulation II*, J. P. Chollet *et al.*, Kluwer (1997).
5.   M. Alam and N. D. Sandham, Numerical study of separation bubbles with turbulent reattach-
     ment followed by a boundary layer relaxation, *Parallel Comp. Fluid Dynamics '97*, A. Ecer *et
     al.*, Elsevier (1997).
6.   D. A. Johnson, and T. J. Coakely, Improvements to a nonequlibrium algebraic turbulence model,
     AIAA Journal, Vol.**28**, No. **11**, pp. 2000-2003 (1990).
7.   S. Gavrilakis, Numerical simulation of low-Reynolds-number turbulent flow through a straight
     square duct, *J. Fluid Mech.*, 244:101 (1992).
8.   R. Cheeswright, G. McGrath, & D. G. Petty, LDA measurements of turbulent flow in a duct
     of square cross section at low Reynolds number, *ER1011: Q.M.W., Uni. of London*, (1990).
9.   R. J. A. Howard, and N. D. Sandham, Simulation and modelling of the skew response of turbulent
     channel flow to spanwise flow deformation, *Direct and Large Eddy Simulation II* , J. P. Chollet
     et al., Kluwer (1997).
10. R. J. A. Howard, and N. D. Sandham, Extended algebraic stress models for prediction of three
     dimensional and separating turbulent flow, *Turbulent shear flows 11*, Grenoble, France, 8-10th
     Sept (1997).
11. A. M. Savill. Exploiting DNS data for improved turbulence modelling. *Proc. DRAL Meeting
     on High Performance Computational Engineering in the UK* (1996).
12. P. R. Voke, Z. Yang and A. M. Savill. Large Eddy Simulation and Modelling of Transition
     Following a Leading-Edge Separation Bubble. *In Engineering Turbulence Modelling and Exper-
     iments 3* (Ed. W. Rodi and G. Bergeles) Elsevier pp.601-610 (1996)
13. T. Wang and D. Zhou. Spectral analysis of a boundary-layer transition on a heated flat plate.
     Int. J. Heat and Fluid Flow 17 pp12-21 (1996).
14. K. H. Luo and N. D. Sandham, On the formation of small scales in a compressible mixing layer,
     Fluid Mechanics and its Applications: Direct and Large-Eddy Simulation I, **26**, 335-346, P.R.
     Voke, L. Kleiser and J.-P. Chollet (Eds.), Kluwer Academic Publishers (1994).
15. A. W. Vreman, N. D. Sandham and K. H. Luo, Compressible mixing layer growth rate and
     turbulence characteristics, *J. Fluid Mech.*, **320**, 235-258 (1996).
16. K. H. Luo, Pressure and dilatation effects in high-speed turbulence, *Direct and Large-Eddy
     Simulation II*, **5**, 167-178, J.-P. Chollet, P. R. Voke and L. Kleiser (Eds.), Kluwer Academic
     Publishers (1997).
17. E. J. Avital, N. D. Sandham and K. H. Luo, Mach wave radiation by mixing layers, analysis of
     the sound and source fields, *38th Israel Aerospace conference*, Tel-Aviv and Haifa (1998).
18. E. J. Avital, N. D. Sandham and K. H. Luo, Sound generation using data from direct numerical
     simulations of mixing layers, *AIAA paper 96-1778*, pp. 1-10, 2nd AIAA/CEAS Aeroacoustic
     Conference, May 6-8, Pennsylvania, USA (1996).
19. E. J. Avital, N. D. Sandham and K. H. Luo, Mach wave radiation by mixing layers, Parts I&II,
     Theoret. Comput. Fluid Dyn. (1998, accepted).

PARALLEL PROCESSING AND DIRECT SIMULATION OF TRANSIENT PREMIXED LAMINAR FLAMES WITH DETAILED CHEMICAL KINETICS

R. P. Lindstedt and V. Sakthitharan

Mechanical Engineering Department
Imperial College of Science, Technology and Medicine
Exhibition Road, London SW7 2BX

NOMENCLATURE

Co	Courant number ($Co = u\Delta t/\Delta R$)	
C_p	Specific heat capacity at constant pressure	$J\,kg^{-1}\,K^{-1}$
D_k	Diffusivity of species k	$m^2\,s{-}1$
h	Enthalpy	$J\,kg^{-1}$
h_k	Enthalpy of species k	$J\,kg^{-1}$
$\mathcal{J}_h, \mathcal{J}_h$	Diffusive fluxes of enthalpy	$J\,m^{-2}\,s^{-1}$
$\mathcal{J}_k$	Diffusive flux of species k	$kg\,m^{-2}\,s^{-1}$
Ka	Karlovitz number	—
ΔM	Mass of a cell	kg
M_k	Molecular weight of species k	$kg\,kmol^{-1}$
n	Mole number	$kmol\,kg^{-1}$
N_{reac}	Number of reactions	—
N_{sp}	Number of species	—
P	Pressure	Pa
p	Number of processors	—
R	Eulerian radius	m
R_k	Net rate of formation of species k	$kmol\,m^{-3}\,s^{-1}$
$\mathcal{R}$	Universal gas constant ($= 8.314$)	$kJ\,kmol^{-1}\,K^{-1}$
S	Speedup introduced by parallelisation	—
T	Temperature	K
t	Time	s
u	Velocity	$m\,s^{-1}$

High Performance Computing
Edited by R. J. Allan *et al.*, Kluwer Academic / Plenum Publishers, New York, 1999

Y_k	Mass fraction for species k	—
α	Parameter determining the geometry: 0 – planar, 1 – cylindrical and 2 – spherical	—
η	Efficiency of parallelisation	—
Φ	Equivalence ratio	—
ϕ	Molar concentration	kmol m^{-3}
ψ	Any variable	$[\psi]$
λ	Thermal conductivity	$\text{J m}^{-1}\,\text{K}^{-1}\,\text{s}^{-1}$
μ	Viscosity	N s m^{-2}
ρ	Density	kg m^{-3}
Ξ_{jk}	Stoichiometric coefficient for species k in reaction j	—
ξ_{jk}	Concentration dependence for species k in reaction j	—

INTRODUCTION

Fundamental reactive-diffusive processes form the crucial focal point in both laminar and turbulent combustion and computational investigations of such processes are necessary to provide further understanding of the thermo-chemical structure of flames (e.g. Dixon-Lewis 1990). The solution of the equations of motion coupled with detailed chemical kinetics is becoming recognised as a primary tool in these studies. Recent computational investigations of one-dimensional steady-state flames have extended such investigations to include complex aviation related fuels (e.g. Lindstedt & Maurice 1995; Lindstedt & Maurice 1996) and data obtained from experimental and theoretical (e.g. DNS) investigations has shown that transient effects are of particular importance in determining practical flame characteristics. The computational demands associated with the application of detailed chemical kinetics under transient conditions are considerable and partly account for the present lack of progress for large hydrocarbon systems. The increasing demand for reduced pollutant emissions necessitate the inclusion of large chemical kinetic mechanisms into time-dependent calculation procedures, which in turn will result in computational demands significantly beyond the capabilities of serial machines. The present paper shows how the use of parallel processing coupled with sophisticated numerical techniques enables transient computations at the present time.

GOVERNING EQUATIONS

The present study consists of direct simulations of laminar premixed flames in planar, cylindrical or spherical one-dimensional domains. A Lagrangian frame of reference is used and the corresponding governing equations are given below:

Mass conservation equation

$$\rho = \frac{1}{R^\alpha}\frac{\partial M}{\partial R} \tag{1}$$

Momentum equation

$$\rho\frac{\partial u}{\partial t} = -\rho\frac{\partial}{\partial M}\{R^\alpha \mathcal{J}_u\} - \rho R^\alpha \frac{\partial P}{\partial M} \tag{2}$$

$$\mathcal{J}_u = -\frac{4}{3}\mu\rho R^\alpha \frac{\partial u}{\partial M} \tag{3}$$

Species equations (for a species k where $k = 1 \longrightarrow N_{sp}$)

$$\rho \frac{\partial Y_k}{\partial t} = -\rho \frac{\partial}{\partial M} \{R^\alpha \mathcal{J}_k\} + R_k M_k \tag{4}$$

$$\mathcal{J}_k = -\rho^2 D_k R^\alpha \left(\frac{\partial Y_k}{\partial M} - Y_k \frac{1}{n} \frac{\partial n}{\partial M} \right) - \rho V_c Y_k - \rho \mathcal{W}_k Y_k \tag{5}$$

Thermal diffusion (or the *Soret* effect) have also been included. The thermal diffusion coefficients are evaluated for light species as

$$\mathcal{W}_k = -\rho R^\alpha \frac{D_k \Theta_k}{X_k} \frac{1}{T} \frac{\partial T}{\partial M} \tag{6}$$

where X_k is the mole fraction of species k, D_k is the mixture diffusion coefficient and Θ_k is a thermal diffusion ratio as defined by Chapman & Cowling (1970).

$$R_k = \sum_{j=1}^{N_{reac}} \Xi_{jk} \left[k_j^f \prod_{l=1}^{N_{sp}} \phi_l - k_j^r \prod_{l=1}^{N_{sp}} \phi_l \right] \tag{7}$$

It should be noted that at any point the sum of mass fractions for all species should equal unity while the sum of the flux $\mathcal{J}_k$ for all species should be zero. A correction velocity V_c is defined to ensure the mass balance by satisfying these two requirements (e.g. Miller *et al.* 1982). Thus, the correction velocity V_c is determined from

$$V_c = \frac{-\sum_{k=1}^{N_{sp}} \rho^2 D_k R^\alpha \left(\frac{\partial Y_k}{\partial M} - Y_k \frac{1}{n} \frac{\partial n}{\partial M} \right) - \rho \mathcal{W}_k Y_k}{\sum_{k=1}^{N_{sp}} \rho Y_k} \tag{8}$$

Enthalpy equation

$$\rho \frac{\partial h}{\partial t} = -\rho \frac{\partial}{\partial M} \{R^\alpha \mathcal{J}_h\} + \rho \frac{\partial}{\partial M} \left\{ \sum_{k=1}^{N_{sp}} R^\alpha h_k \left[-\mathcal{J}_k + \mathcal{J}_n \right] \right\} - \frac{\partial P}{\partial t} \tag{9}$$

$$\mathcal{J}_h = -\rho R^\alpha \frac{\lambda}{C_p} \frac{\partial h}{\partial M} \tag{10}$$

$$\mathcal{J}_n = -\rho R^\alpha \frac{\lambda}{C_p} \frac{\partial Y_k}{\partial M} \tag{11}$$

and the equation of state completes the set of governing equations

$$P = \rho \mathcal{R} T \sum_{k=1}^{N_{sp}} \frac{Y_k}{M_k} \tag{12}$$

TRANSPORT PROPERTIES

The transport coefficients for viscosities and binary diffusivities are evaluated using the theory of Chapman and Enskog (see Reid & Sherwood 1960). Thermal conductivities are computed using the approximation suggested by Mason & Monchick (1962). This approximation includes rotational collision efficiency while ignoring the vibrational contributions. Rotational collision numbers and Lennard-Jones potential parameters are taken from the compilations of Svehla (1962). Mixture properties are evaluated using the semi-empirical Wilkes formula (see Reid & Sherwood 1960). Thermodynamic data are computed using JANAF polynomials (e.g. Kee *et al.* 1987).

NUMERICAL SOLUTION PROCEDURE

In order to solve the conservation equations the computational domain is discretised using a finite volume approach featuring a staggered grid formulation. The differencing scheme used is an implicit formulation involving second-order accuracy for the temporal and spatial derivatives. Local grid refinement is introduced and offers two main advantages: (a) The variables can be better resolved in the regions where their spatial gradients are larger than the user-specified critical values; (b) Since the grid is refined only in a specified region, the code performs faster than the one with a uniform grid offering the same resolution. The adaptive gridding technique utilised allows splitting and merging of cells according to a set of criteria based on the chemical structure of the flame.

The set of discretised conservation equations are finally cast into the following form:

$$B_i \psi_{i+1}^{t+1} + A_i \psi_i^{t+1} + C_i \psi_{i-1}^{t+1} = D_i \tag{13}$$

where ψ is any variable to be solved at time step $t+1$ and A, B, C and D are the coefficient matrices at a node i to be solved for the variable ψ. The conservation equations are, however, non-linear. The source terms $M_k R_k$ of the species conservation equations and the second term on the right hand side of the equation for enthalpy (Eq. 9) are the two main sources of non-linearity in the present set of equations.

In order to obtain solutions, a Newtonian linearisation of these terms is utilised. The source term of the species equation can be linearised as follows:

$$R_k = \sum_{j=1}^{N_{reac}} \Xi_{jk} \left[k_j^f \prod_{l=1}^{N_{sp}} \phi_l^{\xi_{jk}} - k_j^r \prod_{l=1}^{N_{sp}} \phi_l^{\xi_{jk}} \right] \tag{14}$$

$$R_k^{\nu+1} = R_k^\nu + \sum_{l=1}^{N_{sp}} \frac{\partial R_k}{\partial \phi_l} \frac{\partial \phi_l}{\partial Y_l} \left\{ Y_l^{\nu+1} - Y_l^\nu \right\} + \frac{\partial R_k}{\partial T} \left\{ T^{\nu+1} - T^\nu \right\} \tag{15}$$

The last term in the above expression is introduced to include the temperature dependence of the species equations. The derivative of R_k with respect to T is evaluated numerically.

The resulting equations for the conservation of the species and enthalpy conform to a structure of a block tri-diagonal matrix. To fully account for the temperature and density dependence of the transport equations of species and enthalpy, the equations of all the other variables such as temperature T and density ρ are also included into the block matrix structure. The density is evaluated from the equation of state and the temperature from an enthalpy balance equation. The non-linear dependences of density and temperature are treated in a similar way.

This block structure is solved using a direct method explained by Hindmarsh (1977). In this method the equations are iterated with respect to the non-linear contributions (such as the rate expressions). The Eulerian radius R is evaluated separately outside the block structure from the conservation of mass.

For incompressible flows the pressure is assumed to be uniform throughout the computational domain. In the present Lagrangian co-ordinate system, the computational domain is allowed to expand or shrink according to the changes in temperature without permitting any change in pressure. Hence, the momentum equation becomes redundant and instead the change in velocity due to the change in density (ρ) and in the size of the cell (Eulerian radius R) is obtained from

$$u = \frac{\partial R}{\partial t} \tag{16}$$

The forward integration time step Δt is controlled using a criterion based on a Courant number Co given by:

$$\Delta t = \text{Co MIN} \left\{ \frac{\Delta R_i}{u_i} \right\} \qquad (17)$$

In the present study the maximum possible time step at each cell is evaluated based upon a global Courant number and the smallest of all time steps is taken to be the value for the next time step Δt. Second-order convergence is achieved for Courant numbers ≤ 0.5.

CODE VERIFICATION AND VALIDATION OF RESULTS

In the present study, code evaluation was performed in two stages: (i) Comparison with an existing Eulerian code and (ii) Comparison of steady state laminar burning velocities with experimental data. The former performs the code verification while the latter validates the results obtained using the present code. For these two cases hydrogen–oxygen–nitrogen mixtures, methane–air and propane–air mixtures were used with a standard detailed chemical kinetic scheme used by Leung & Lindstedt (1995) and Lindstedt & Skevis (1996). The C_1–C_3 subset of this reaction scheme comprises 48 species and 300 reactions. For computations involving hydrogen–oxygen–nitrogen mixtures the corresponding H_2–O_2 sub-mechanism consisting of 9 species and 22 reactions was used.

Comparison with an Eulerian code

Two test cases have been computed using the present code and are compared with the results obtained using an existing Eulerian code (e.g. Lindstedt et al. 1995 and Leung & Lindstedt 1995). In both test cases flame propagation through a 5 mm long tube filled with premixed gas mixtures was studied in a planar geometry. The ignition was initiated at the right-hand boundary and the flame propagated from right to left.

Test Case 1 A lean hydrogen–air flame (with an equivalence ratio $\Phi = 0.8$) was computed at one atmosphere pressure (101.325 kPa) and at a temperature of 298 K. For this test case the flame attained its steady state in about 0.6 ms after starting the computations with arbitrary linear profiles for the species. To reach such a steady state the computations took approximately 1000 time steps with the time step control imposed by the Courant number criterion explained earlier. The minimum time step was set at 1.0E-07 s and this was found to correspond to a Courant number (Co) of 0.35 at most time steps. The computations took approximately fifty minutes on a DEC Alpha 5000/200 workstation.

Figures 1–3 show the comparison of the results obtained using the present code with those from the Eulerian code. The profiles from the present computations are shown at 1 ms after the beginning of the computations. For comparison purposes the profiles from the present computations have been given an offset with respect to the distance such that the profiles match with each other. It should, however, be noted that the same offset has been used for all the profiles from the present computations. Figure 1 shows the temperature profiles and it can be seen that the two profiles coincide well. This is important since any variation in temperature gradient would contribute to changes in the laminar burning velocity. The laminar burning velocity from the present computations is $1.64 \, \text{m s}^{-1}$ in comparison to $1.67 \, \text{m s}^{-1}$ from the earlier calculations.

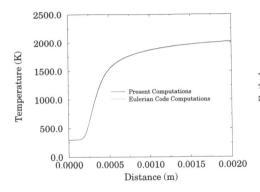

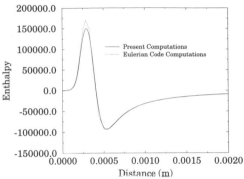

Figure 1. Comparison of temperature profiles for Test Case 1. A fixed offset has been added to the distance for the present computations.

Figure 2. Comparison of enthalpy profiles for Test Case 1. A fixed offset has been added to the distance for the present computations.

Figure 2 shows the comparison of enthalpy profiles and it is clear that the present code with its ability to generate fine grid cells around the high gradient region produces well-resolved profiles of the enthalpy. This accounts for the minor discrepancies observed between the two profiles, noting that the grid cells were coarser in the Eulerian calculations. A comparison of the species profiles is given in Figure 3 which shows excellent agreement between the profiles.

Test Case 2 In this test case a stoichiometric methane–air mixture was computed at one atmosphere pressure (101.325 kPa) and at a temperature of 298 K. The flame attained its steady state in about 5 ms after starting the computations with arbitrary linear profiles for the species. A maximum Courant number (Co) of 0.5 was used to control the time step. The computations took approximately thirty hours on a DEC Alpha 5000/200 workstation.

Figure 4 shows the comparison of the results obtained using the present code with those from the Eulerian code. The profiles from the present computations are shown at 6 ms after the beginning of the computations. For comparison purposes the profiles from the present computations have again been given an offset with respect to the distance such that the profiles match with each other. As with the previous test case the same offset has been used for all the profiles from the present computations. The laminar burning velocity from the present computations is 0.36 m s^{-1} which is identical to the calculations using the Eulerian code. A comparison of the species profiles given in Figure 4 shows excellent agreement between the profiles. Any minor discrepancy observed between these profiles are due to the coarser grid in the Eulerian solution.

Comparison of Laminar Burning Velocities

One-dimensional laminar burning velocities have been computed for hydrogen–air mixtures at a wide range of equivalence ratios (0.2–5.6). For this case a computational domain of 5 mm was used and the ignition was specified at the left-hand boundary. In accordance with the definition of the laminar burning velocity, all these computations were performed in a planar geometry. The Courant number (Co) was limited to 0.20.

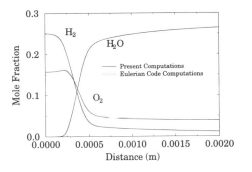

Figure 3. Comparison of species profiles for Test Case 1. A fixed offset has been added to the distance for the present computations.

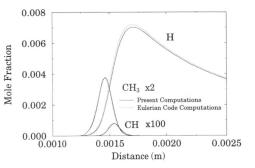

Figure 4. Comparison of species profiles for Test Case 2. A fixed offset has been added to the distance for the present computations.

A comparison of the computed burning velocities with experimental values is shown in Figure 5. The experimental values have been obtained from Egolfopoulos & Law (1990) and Warnatz (1992). The experimental values of the burning velocity have in general been obtained from burner flames and counter-flow flames and extrapolated to "zero" strain rate conditions. The agreement between the computed and experimental laminar burning velocities is excellent.

PARALLELISATION OF THE CODE

Despite the importance of time-dependent effects in pollutant formation and their relevance in turbulent combustion systems, few computational studies of transient laminar flames featuring detailed chemical kinetics scheme have been made. Significant computational demands of such studies are obvious when detailed chemical kinetics schemes required for a number of fuels are considered as given in Table 1. The C_1–C_3 scheme, for example, consists of 48 species and 300 reactions in comparison to the H_2/O_2 scheme which comprises 9 species and 22 reactions. The increases in computational requirement become more acute when flames featuring practical fuels such as kerosene are computed. For practical applicability it is often necessary that soot and pollutant mechanisms are also included in the kinetics scheme. A typical C_{10} scheme would contain about 200 species and over 1000 reactions.

The computational time increases approximately with the square of the number species (N_{sp}^2) and with the number of reactions. Clearly, without the use of parallel processing, even simple transient computations are beyond the capabilities of the present computational resources.

In the present study, since a direct method is used to solve the block matrix structure, the ideal method of parallelising the present serial code is to use domain decomposition, which is easily a universal source of scalable parallelism. The regularity of data structures of the present code makes domain decomposition an ideal choice, although, the irregularity of the domain resulting from grid refinement poses a challenge

Table 1. Practical requirements for different detailed chemical kinetics scheme.

Fuels/Application	Chemical Kinetics Scheme
Propellants	H_2/O_2
Liquefied natural gases (LNG)	C_1–C_3
Liquefied petroleum gases (LPG)	C_4
Endothermic fuels	$> C_7$
Octane	C_8
Kerosene	$> C_{10}$

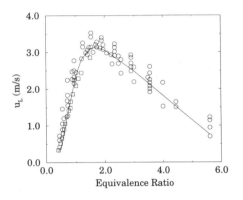

Figure 5. Laminar burning velocities of atmospheric hydrogen–air flames plotted against equivalence ratio. The experimental data have been gathered from Egolfopoulos & Law (1990) and Warnatz (1992).

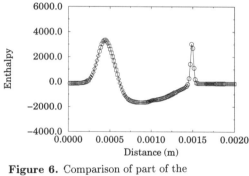

Figure 6. Comparison of part of the enthalpy profiles obtained using 128 processors with that obtained using a single processor. Change of symbol indicates change of processor.

in that without efficient dynamic load balancing the efficiency of decomposition reduces rapidly.

Any practical parallel system requires communication between processors. The time to transfer data between processors is usually the most significant source of parallel processing overhead. Moreover, a processor might idle due to unbalanced work load or because of a synchronisation point at which other processors have not yet arrived. Whatever the cause, the total idle time of all the processors contributes to the parallel processing overhead. This overhead reduces the effectiveness of parallel processing and must therefore be kept at a minimum.

Performance of a parallelised code is measured here with speedup S and efficiency η. Speedup is a measure which captures the relative advantage of the parallel code. It is defined as the ratio of the time taken to solve a problem on a single processor to the time required to solve the same problem on a parallel computer with p identical processors.

$$S = \frac{T_s}{T_p} \tag{18}$$

where T_s is the time taken on a single processor and T_p is the time taken on the parallel computer. Efficiency, on the other hand, shows the fraction of time for which a processor is usefully employed. It is defined as the ratio of speedup to the number of processors used and is often expressed in percentage.

$$\eta = \frac{S}{p} \times 100\% \tag{19}$$

where p is the number of processors used. In a perfectly parallel system, speedup is equal to the number of processors and hence the efficiency is 100 %. In practice, however, speedup is less than the number of processors and efficiency is between zero and 100 %, depending on the degree of effectiveness of processor utilisation.

Speedup and efficiency of the present parallelised code are shown in Table 2. When a single iteration within the block solver algorithm is tested the speedup and efficiency show excellent scalability, which is a measure of parallel algorithm's ability to achieve performance proportional to the number of processors. However, due to increased communications between the processors, for a full transient computation the efficiency drops as the number of processors increases. When local grid refinement is implemented the efficiency drops further since after a few grid refinement calls there is a greater difference in number of cells allocated to each processor. Therefore, the load of each processor must balanced dynamically. Dynamic load balancing is computationally expensive, as there are further communications between the processors which contribute to the parallel processing overhead. In the present implementation local grid refinement is permitted at a controlled frequency which largely depends on the particular application. Following a grid refinement, the cells are completely reallocated to each processor. With this implementation, it is clear that the speedup and efficiency follow very closely with trend of the code without local grid refinement.

Table 2. Performance of the parallel code.

Processors	Block-solver		No Refinement		Refinement	
	Efficiency (%)	Speed-up	Efficiency (%)	Speed-up	Efficiency (%)	Speed-up
2	99.7	1.99	99.3	1.99	98.5	1.97
4	99.7	3.99	97.4	3.89	95.3	3.81
8	99.2	7.94	97.3	7.78	93.4	7.47
16	98.3	15.72	90.9	14.54	87.6	14.01
32	96.4	30.84	84.3	26.98	80.4	25.73
64	93.6	59.89	67.7	43.35	65.8	42.11
128	87.5	111.99	43.3	55.37	40.2	51.46

Parallel Code Validation

It is important that the results obtained using the parallelised code are identical to those obtained with the serial code. As an example, the enthalpy profile obtained using 128 processors is compared with that obtained on a single processor in Figure 6. These computations have been made for CH_4–air mixture at an equivalence ratio of 0.8 in a planar geometry. The comparisons are made after about 10 μs following ignition. It is clear that the parallel algorithm works perfectly without any numerical difficulties.

SPHERICALLY EXPANDING FLAMES

Computations have been performed to study spherically expanding and contracting flames for hydrogen–oxygen–nitrogen systems. For spherically expanding flames the computed values are compared with the experimental data of Kwon *et al.* (1992). The flame is initiated at the centre of a sphere and the resulting flame propagation is investigated. The ignition is initiated by setting arbitrary profiles for the reactants and products with all other species set to zero. In accordance with the experimental data, the computations were performed at initial conditions of 303.975 kPa and 298 K. Equivalence ratios vary from 1.0 to 3.27 at $O_2/(N_2+O_2) = 0.125$. The experimental studies (Kwon *et al.* 1992) utilised schlieren photographic techniques to locate the flame front. Accordingly, the flame radius was extracted from the computations based on the same isocontour of ~ 305 K as used in experiments to locate the position of the flame.

Comparisons of computational results with the experimental data are shown in Figures 7–10 for a number of equivalence ratios. In Figures 7–8 laminar burning velocity variation with the flame radius is shown. The results show the qualitatively different behaviour as the mixture varies from stoichiometric composition to fuel-rich condition. The high rate of strain close to the centre of the sphere results in an increase of the burning velocity for the stoichiometric flame, while the opposite trend is observed for rich flames. In general the computational results agree well with the experimental results, both qualitatively and quantitatively. It is also found that for hydrogen flames it is imperative to include *Soret* (thermal diffusion) effects as the burning velocities are otherwise under-predicted by up to 25%.

Laminar burning velocity variation with respect to the rate of strain is shown in Figures 9–10. The qualitative agreement of the computational results is strikingly good, although the burning velocities are under-predicted by about 15–25 %. It is of interest to note that the variation of laminar burning velocity with rate of strain is not linear.

CLOSING REMARKS

The present paper has outlined a numerical procedure and computer code for the transient computation of laminar flames. The conservation equations for momentum, enthalpy and species are solved in a one-dimensional Lagrangian co-ordinate system. The source terms in the species transport equations have been obtained using detailed

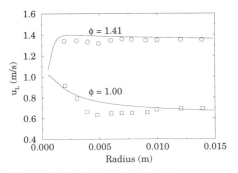

Figure 7. Variation of laminar burning velocity with flame position from the centre of the sphere for hydrogen flames.

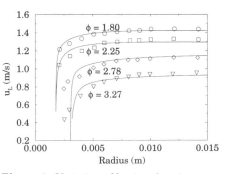

Figure 8. Variation of laminar burning velocity with flame position from the centre of the sphere for hydrogen flames.

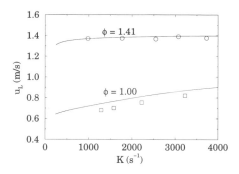

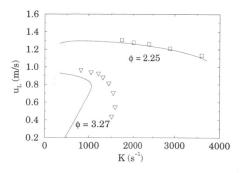

Figure 9. Variation of laminar burning velocity with rate of strain for hydrogen flames.

Figure 10. Variation of laminar burning velocity with rate of strain for hydrogen flames.

reaction schemes. All the conservation equations are solved using an iterative procedure through the use of a direct block tri-diagonal matrix solver. Local grid refinement offers better resolved profiles and faster computations. With the introduction of Newton linearisation for the non-linear terms appearing in the conservation equations due to inter-dependences of species concentration, temperature, enthalpy and density, the present code maintains a second order convergence even as the Courant number approaches 0.5. Furthermore, the solution of discretised equations has a second-order accuracy in both time and space. The parallelised version of the code shows good efficiency and excellent scalability and hence provides the means to perform computationally expensive studies involving higher hydrocarbons of practical interest. The potential of the present technique has been underlined through the computations of spherical flames.

ACKNOWLEDGEMENTS

The present work is part of the (HPCI) Computational Combustion Consortium co-ordinated by K.N.C. Bray (University of Cambridge, U.K.) and D. Emerson (Daresbury Laboratory, U.K.) via GR/K 41601 and is directly supported by EPSRC contracts GR/K 43575 and GR/K 96366. Authors also wish express their gratitude for the excellent support offered by K. Kleese (Daresbury Laboratory, U.K.) in porting the parallel code from the Cray T3D to T3E.

REFERENCES

Chapman S. and Cowling T.G., 1970, *The Mathematical Theory of Non-Uniform Gases*, Cambridge University Press, Cambridge.

Dixon-Lewis G., 1990, in: *Twenty-Third Symposium (International) on Combustion*, pp305-24, The Combustion Institute, Pittsburgh.

Egolfopoulos F.N. and Law C.K., 1990, in: *Twenty-Third Symposium (International) on Combustion*, pp333-40, The Combustion Institute, Pittsburgh.

Hindmarsh A.C., 1977, *Solution of Block Tri-diagonal Systems of Linear Algebraic Equations*, LLL Report, UCID-30150.

Kee R.J., Rupley F.R. and Miller J.A., 1987, *the CHEMKIN Thermodynamic Database*, Sandia Report, SAND SAND 87-8215.

Kwon S., Tseng L.K. and Faeth G.M., 1992, *Combust. Flame*, 90:230-45.

Leung K.M. and Lindstedt R.P., 1995, *Combust. Flame*, 102:129-60.

Lindstedt R.P., Lockwood F.C. and Selim M.A., *Combust. Sci. Tech.*, 108:231-54.

Lindstedt R.P. and Maurice L.Q., 1995, *Combust. Sci. Tech.*, 107:317-53.

Lindstedt R.P. and Maurice L.Q., 1996, *Combust. Sci. Tech.*, 120:119-67.

Lindstedt R.P. and Skevis G., 1996, *Combust. Sci. Tech.*, 125:73-138.

Mason E.A. and Monchick L., 1962, *J. Chem. Phys.*, 36:1622.

Miller J.A., Smooke R.E. and Kee R.J., 1982, in: *Nineteenth Symposium (International) on Combustion*, pp181-96, The Combustion Institute, Pittsburgh.

Reid R.C. and Sherwood T.K., 1960, *Properties of Gases and Liquids*, McGraw Hill, New York.

Svehla R.A., 1962, *Estimated Viscosities and Thermal Conductivities of Gases at High Temperatures*, N.A.C.A. Report R-132

Warnatz J., 1992, in: *Twenty Fourth Symposium (International) on Combustion*, pp553-79, The Combustion Institute, Pittsburgh.

TIME DOMAIN ELECTROMAGNETIC SCATTERING SIMULATIONS ON UNSTRUCTURED GRIDS

P.J. Brookes, O. Hassan, K. Morgan, R. Said, and N.P. Weatherill

Department of Civil Engineering,
University of Wales, Swansea,
Singleton Park, Swansea SA2 8PP

ABSTRACT

An efficient solution procedure, simulating the interaction between plane electromagnetic waves and electrically large obstacles, is presented. The solution of Maxwell's curl equations is sought in the time domain by explicit timestepping. The spatial domain is discretised into a mesh of linear tetrahedral elements. To enhance the efficiency of the resulting computational procedure, an edge based representation of the mesh is employed. The Maxwell equations are spatially discretised using a Galerkin method, with stabilisation achieved by the adoption of a Lax-Wendroff numerical flux function. Approaches enabling the accurate modelling of electromagnetic scattering effects over a wide frequency range are investigated, highlighting the significance of the treatment of the mass matrix within the formulation presented. The development of a parallel environment, incorporating parallel mesh generation and parallel solution procedures is also discussed.

INTRODUCTION

Designers of aerospace vehicles are interested in the development of numerical methods for simulating the scattering of high frequency electromagnetic waves. The scattering problem, mathematically described by Maxwell's equations, has inspired the development of a broad base of solution techniques in both the frequency domain[1] and time domain.[2] However, the implementation of a solution procedure in a manner likely to produce results within acceptable timescales, for the wave frequencies of interest, is a significant problem. Solution methods based upon an unstructured tetrahedral discretisation of the solution domain appear suitable, since the geometries encountered are frequently complex in shape. Typical aircraft configurations for such wave propagation problems are defined as electrically large, often including internal material layers and electrically small features.

In this paper a solution of Maxwell's equations is sought in the time domain and employs explicit timestepping. However, the accuracy of such finite element, or finite difference, methods relies upon the use of a mesh which contains a significant number of nodes per wavelength. Clearly, attempts to solve the described problem place severe demands on current computational resources. Approaches enabling the accurate modelling of electromagnetic scattering effects over a wide frequency range are investigated. Initial efforts attempt to obtain a solution by applying the full power of parallel processing to the explicit edge based procedure.

THE NUMERICAL APPROACH

Maxwell's curl equations, defining electromagnetic wave propagation in free space, are considered. In this work, the external scattering problem of interest involves a single frequency time periodic incident wave and a perfectly conducting scatterer. The linearity of the governing equations allows a scattered field formulation of the problem. Such an approach avoids the need to simulate the propagation of the incident field. The computational domain is constructed by truncating the infinite solution region at a wave frequency dependent distance from the scatterer surface. Contaminated solutions produced by wave reflection at the far field boundary are prevented by applying a characteristic decomposition technique.[3] On the surface of the perfect electrical conductor, the scattered field components are subjected to conditions ensuring the removal of the tangential component of the total electric field and the normal component of the total magnetic field.

The chosen unstructured mesh based solution algorithm is constructed by an approximate variational formulation of the original problem. The spatial domain is discretised into an assembly of linear tetrahedral elements, using a Delaunay mesh generator.[4] The employment of a linear finite element interpolation function to define trial and weighting function sets allows a Galerkin approximation of the original problem. A discrete equation system can be constructed by adopting a standard finite element data structure and assuming the fluxes vary linearly over the elements between their nodal values. However, employing an edge based data structure, where each numbered edge of the tetrahedral mesh provides the numbers of the two associated nodes, both improves CPU efficiency and reduces data storage requirements. In this case the integrals are evaluated by summing individual edge contributions. The resulting central difference type discretisation is not well suited for practical computation. However, replacement of the actual fluxes by an appropriate consistent numerical flux function forms the basis for a practical solution algorithm. In this instance, a Lax-Wendroff flux function, $\mathcal{F}$, is employed to yield a discretisation of the form

$$\mathbf{M}\Delta\mathbf{U} = \sum_{edges} \mathcal{F} \tag{1}$$

The solution of equation (1) is advanced by explicit Euler timestepping. In the first instance the mass matrix $\mathbf{M}$ is diagonally lumped to directly obtain the solution increment $\Delta\mathbf{U}$, maintaining the explicit nature of the formulation.

To extract far field scattering data, necessary for the prediction of aircraft signatures, the solution is advanced over a prescribed number of cycles of the incident wave, such that periodic steady state conditions are achieved. The variation of the electric and magnetic scattered fields is then recorded over a further cycle, at the nodes on a closed surface S which encloses the scatterer surface. Typically, the surface S is chosen to coincide with the obstacle surface. A near to far field transformation yields the radar cross section of the obstacle.[3]

430

EXAMPLE SOLUTIONS
Perfectly Conducting Sphere

The solver is initially validated by considering the interaction between a time periodic plane wave and a perfectly conducting sphere of electrical diameter 2 wavelengths. On the scatterer surface a grid spacing of 25 nodes per wavelength is used to capture the scattering effects. The distribution of the radar cross section of the scatterer is computed from the phases and amplitudes output during the third cycle of the incident wave, and is in excellent agreement with the exact analytical distribution.[3]

Perfectly Conducting Ogive

A second test case is presented, involving an ogive of maximum electrical size twice the wavelength of the time periodic incident wave. The wave propagates in the direction of the grazing angle of the ogive. The mesh employed contains 300,747 elements and 57,278 points, with a grid spacing equivalent to 16 nodes per wavelength of the incident wave. The solution is again output after three cycles of the incident wave. The contours of a total E field component on a cut through the volume mesh and on the surface of the ogive are shown in Figure 1.

EXTENDING THE FREQUENCY RANGE

To accurately simulate the scattering effects at higher wave frequencies two approaches have been explored. The first investigation highlights a numerical solution technique to reduce the mesh size for a scattering problem at a fixed wave frequency, without degradation in the solution accuracy. The second methodology seeks to extend the frequency range by exploiting the current computational resources available, in particular massively parallel machines.

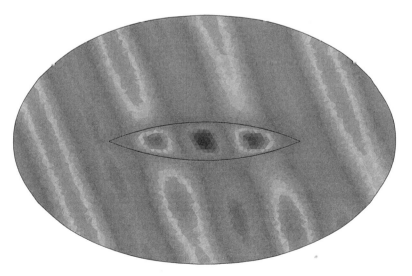

Figure 1. Contours of a total E field component on the ogive surface and on a cut through the volume mesh

Mass Iteration

In the numerical examples presented, the explicit nature of the scheme is maintained by diagonally lumping the mass matrix of equation (1). An alternative approach is to solve with the consistent mass matrix by employing an approximate factorisation procedure. In practice, an iterative formulation is applied at the end of each timestep. The effect of such an approach is demonstrated by considering the wave scattering properties of the perfectly conducting sphere described above. Using a mass lumping method, the computed RCS distributions of the obstacle for different mesh spacings (10, 15, 20 nodes per wavelength) are presented in Figure 2. It is noticeable that a mesh density of 15 nodes per wavelength is necessary to accurately capture the correct RCS distribution. However, application of the mass iteration technique (3 iterations) on a grid of spacing 10 nodes per wavelength predicts a cross section of accuracy comparable with the mass lumped approach on a grid of spacing 15 nodes per wavelength. See Figure 3.

The Parallel Environment

The described algorithm has been validated for scattering problems involving electrically small scatterers of simple geometrical shape.[3] Attempts to solve electrically large problems, regardless of geometry complexity, lead to rapidly increasing mesh sizes. One method of approach is to adapt the solution algorithm to enable parallel execution in a multi-processor parallel environment. However, an efficient parallel implementation of the basic procedure requires an effective procedure for generating partitioned sub-domains, to ensure a balance of the computational load between processors, and optimisation of the necessary inter-processor communication.

Partitioning Techniques Recursive Spectral Bisection (RSB)[5] and a Recursive Bandwidth Minimisation (RBM) techniques have been employed to achieve the de-

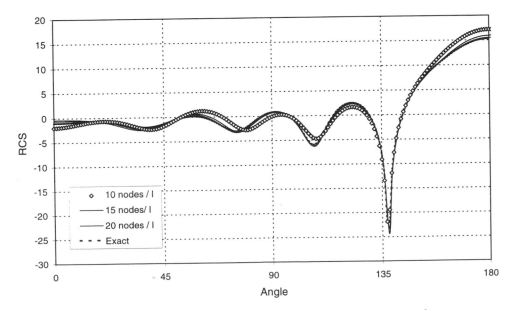

Figure 2. Computed RCS of a sphere using a lumped mass approach

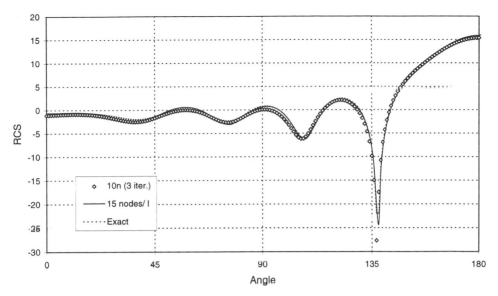

Figure 3. Comparison of the RCS of a sphere for mass lumping and mass iteration approaches

composition of a general unstructured mesh. The approaches are used to provide a colouring of the mesh nodes into an appropriate number of subdomains. On a Cray YMP-EL98, with 215MWords of usable memory, the sequential mesh generation, RSB and RBM procedures are limited to mesh sizes of the order of 13 million, 8 million and 16 million tetrahedral elements, respectively. Projected grid sizes currently necessary for typical aerospace high frequency problems show such procedures to be substantial bottlenecks. The extent of the limitations is clear when considering that the parallel implementation of the described solver is capable of solving on grids of up to 200 million elements on a Cray T3D (64Mbytes of memory associated with each of the 512 processors).

Three approaches to providing a capability for producing larger meshes have been investigated. The first two approaches involve the application of the standard h–refinement technique to a generated mesh. The h–refinement technique adds a new node to each edge in the mesh, sub–divides the existing tetrahedra, and forms new elements by appropriate reconnection of nodal points. In the first approach a grid is produced by performing h–refinement on a generated mesh. RBM is employed to decompose the resulting grid. Such an approach shifts the preprocessing bottleneck from the grid generation stage to the mesh partitioning algorithm. For the second procedure, it is necessary to initially work with a tetrahedral grid of a size which can be generated and decomposed within the limits imposed by the available computational resources. This mesh is decomposed into the same number of regions as there are processors available to perform the equation solution. A mesh of the desired size is then produced by h–refinement of each of the decomposed regions separately. It is clear that this technique removes the associated memory constraints. However, a disadvantage of such h–refinement based techniques is that the user forfeits the ability to determine the location of the available nodes on the finest mesh. An additional complication is that, when boundary edges are considered, the added nodes need to be located on the boundary surface.

The third and most general partitioning procedure is one in which a parallel strategy is applied to a sequential Delaunay mesh generator, allowing the simulation of problems on arbitrary large meshes.

Parallel Mesh Generation A parallel strategy is applied to the sequential mesh generation stage, employing a SPMD model within a master/slave program structure. A brief description of the procedure is presented. The starting point is the subdivision of the initial workload, in the form of the boundary data, into n closed subdomains by the master program. This is achieved by applying a greedy type algorithm on a generated initial Delaunay grid, followed by a triangulation of the internal subdomain interfaces. The n closed subdomains are then distributed to the slave programs for Delaunay grid generation. Since no communication exists between slave programs, the procedure is capable of generating n mesh partitions on a single processor sequential platform. To date, partitioned unstructured grids of up to 50 million elements have been generated on an 8 processor SGI Challenge machine, with 512 Mbytes of memory.[6]

Solver Parallelisation The parallel solution algorithm adopts a single program multiple data model and inter-processor communication is achieved using standard PVM routines. The adopted data structure is such that edges are classified as interior edges if and only if the two nodes associated with the edge are of the same colour. An edge with associated nodes of colour I and J, is stored as an interface edge in subdomain minimum(I,J). This leads to a duplication of nodes at the subdomain boundaries, and no duplication of edges. Within each subdomain there is a local numbering of the edges, nodes and boundary faces. At the start of each timestep, the interface nodes obtain contributions from the interface edges. These partially updated interface nodal contributions are then broadcast to the corresponding interior nodes in the neighbouring subdomains. A loop over the interior edges is followed by the receiving of the interface node contributions and the subsequent updating of all interior nodal values. The sending of the updated values back to the interface nodes completes a time step of the procedure.

HIGHER FREQUENCY SIMULATIONS

The interaction between a complete aircraft and a time periodic incident wave of frequency 300 Mhz is presented. This represents a maximum electrical dimension of

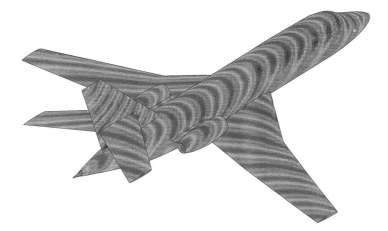

Figure 4. Solution contours of a scattered E field component on the surface of the aircraft.

the obstacle of 18 wavelength. The mesh consists of 15,182,752 elements and 2,553,495 nodes. The solution is output after 36 cycles of the incident wave, with a run time of 3 hours on the Cray T3D utilising 256 processors. Figure 4 shows the scattered E_x field component on the surface of the aircraft.

ACKNOWLEDGEMENTS

The authors wish to thank the UK Engineering and Physical Sciences Research Council for providing access to the CRAY T3D at the Edinburgh Parallel Computer Centre, under Research Grant GR/K42264, and for supporting the parallelisation activity under Research Grants GR/J12321 and GR/J91234. The authors also thank Dr. H. Simon, for providing access to his recursive spectral bisection software.

REFERENCES

1. I. Harari, K. Grosh, T.J.R. Hughes, M. Malhotra, P.M. Pinksky, J.R. Stewart and L.L. Thompson, Recent developments in finite element methods for structural acoustics, *Archiv. Comput. Methods Eng.* 3:131-309 (1996).
2. K. Morgan, O. Hassan and J. Peraire, A time domain unstructured approach to the simulation of electromagnetic scattering in piecewise homogeneous media, *Comput. Methods Appl. Mech. Eng.* 134:17-34 (1996).
3. K. Morgan, P.J. Brookes, O. Hassan, N.P. Weatherill, Parallel processing for the simulation of problems involving scattering of electromagnetic waves, *Comput. Methods Appl. Mech. Eng.* 152:157-74 (1998).
4. N.P. Weatherill and Ö. Hassan, Efficient 3D Delaunay triangulation with automatic point creation and imposed boundary constraints, *Int. Jo. Numer. Methods Engrg.* 37:2005-39 (1994).
5. H. Simon, Partitioning of unstructured problems for parallel processing, *Comput. Syst. Engrg.* 2:135-48 (1991).
6. N. Verhoeven, N.P. Weatherill and K. Morgan, Dynamic load balancing in a 2D parallel Delaunay mesh generator, in: *Parallel Computational Fluid Dynamics: Implementations and Results Using Parallel Computers*, Elsevier Science, Amsterdam (1996).

LARGE-EDDY SIMULATION OF THE VORTEX SHEDDING PROCESS IN THE NEAR-FIELD WAKE BEHIND A SQUARE CYLINDER

F. di Mare, W. P. Jones

Department of Chemical Engineering and Chemical Technology
Imperial College of Science, Technology and Medicine
London SW7 2BY
United Kingdom

INTRODUCTION

Large Eddy Simulation (LES) is an extremely promising technique for predicting the properties of a wide range of engineering flows. It involves a direct three-dimensional time dependent computation of the large-scale turbulent motions responsible for turbulent mixing whilst those with scales smaller than the computational grid (the sub-grid stresses) are modelled. The approach is particularly appropriate to turbulent flows in which large scale organised and non-stationary turbulence structures are present, aspects which are not accessible by conventional time averaged approaches. In the present paper LES is applied to the calculation of such a phenomena, namely the vortex shedding process which occurs from the flow around a square cylinder. The configuration consider is that studied experimentally by Lyn[1] and for the sub-grid stresses both the standard Smagorinsky model and the dynamic model proposed by Germano et al [2], have been applied. From the LES results the identification and analysis of the vortex shedding process in the near field wake behind the bluff body has been achieved by means of a phase-averaged, two-component velocity analysis in the flow region between the free stream and the cylinder sidewall and the result compared with measurements. Comparisons are also presented for the wall boundary layers which develop around the surface of the cylinder. Some numerical features of the computations presented in this paper are summarised in Table 1.

Table 1. Numerical aspects of the simulations

Computation	C_s	Grid	Filter	CFL
Smagorinsky model	0.15	$145 \times 137 \times 13$	Top hat	0.5
Dynamic Model	–	$145 \times 137 \times 13$	Top hat	0.3

GOVERNING EQUATIONS AND SOLUTION PROCEDURE
Mathematical Model

In large-eddy simulation, the large-scale quantities are defined by the convolution of the velocity and pressure fields with a filter function; the *grid* filter G can be defined as:

$$\bar{f}(\mathbf{x}) = \int f(\mathbf{x}')G(\mathbf{x}, \mathbf{x}')d\mathbf{x}' \tag{1}$$

where the filter function satisfies:

$$\int G(\mathbf{x}, \mathbf{x}')d\mathbf{x}' = 1 \tag{2}$$

Applying the filtering operator to the equations of motion for a constant-density flow yields:

$$\frac{\partial u_i}{\partial x_i} = 0 \tag{3}$$

$$\frac{\partial \bar{u}_j}{\partial t} + \frac{\partial \bar{u}_i \bar{u}_j}{\partial x_j} = -\frac{1}{\rho}\frac{\partial \bar{p}}{\partial x_i} - \frac{\partial \tau_{ij}}{\partial x_j} + \nu \frac{\partial^2 \bar{u}_i}{\partial x_j \partial x_j} \tag{4}$$

The effects of the small unresolved scales appears in the sub-grid-scale stress term

$$\tau_{ij} = \overline{u_i u_j} - \bar{u}_i \bar{u}_j \tag{5}$$

which must be modeled.

For the sub-grid stresses two models have been used. These are the standard Smagorinsky sub-grid scale model[3, 4] and a dynamic model, Germano *et al.*[2] with approximate localization[5, 6]. In the latter case the model parameter C_s was clipped to prevent the sub-grid viscosity from becoming negative. Negative viscosities represent 'explosive' type phenomena and if allowed would result in the solution of the filtered equations becoming unstable.

Flow Configuration and Solution Procedure

The partial differential transport equations have been discretised on a cartesian mesh stretched around the bluff body with 6165 nodes, (145×137 grid nodes), in the $x-y$ plane, and 13 nodes in the spanwise (z) direction. The grid nodes are algebraically distributed inside the domain by means of transfinite interpolation[7]. The dimensions of the computational domain represent a compromise between CPU time restrictions and storage constraints and the need to provide an adequate spatial resolution. The geometrical configuration chosen for the simulation is that used in the experimental study of Lyn[1, 8] and is shown in fig. 1. The coordinate origin is located at the centre of the cylinder. The upstream region of the solution domain extends from $-4.5D$ to $-0.5D$, while downstream of the cylinder it extends from $0.5D$ to $10D$. In the spanwise direction the domain extends from $-2.16D$ to $2.16D$. It was necessary to reduce the extent of the domain downstream of the cylinder in order to keep the mesh expansion ratios at an acceptable level. On the other hand, it was also necessary to ensure that the downstream boundary was located sufficiently far from the cylinder to prevent the influence of boundary conditions at the outlet boundary from propagating upstream. The grid expansion ratio was 1.05 in the streamwise direction, and the maximum cell length was $0.59D$ and $0.54D$ in the streamwise and normal directions respectively. The blockage caused by the presence of the cylinder is equal to 0.07, and it's aspect ratio is 4.32. The fluid properties are those of air at $25C$, and the Reynolds number, based on the characteristic dimension of the cylinder, D, and the inflow velocity, U_0, is 22000.

438

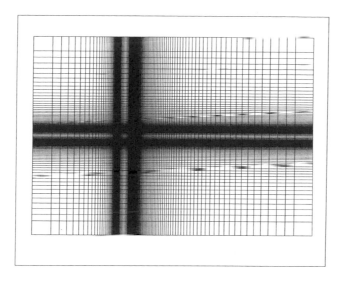

(a)

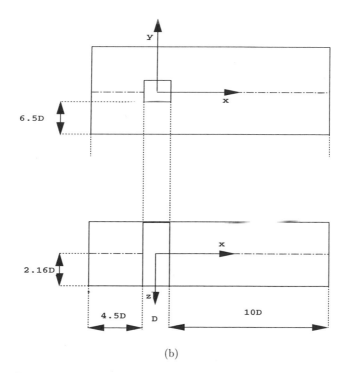

(b)

Figure 1. Computational mesh,(a), and computational domain, (b).

The filtered transport equations are discretised using finite volume methodology. The method utilises general curvilinear coordinates and a co-located mesh structure is employed whereby velocities and pressure are defined at mesh node points. The formulation is fully implicit and an approximate factorisation technique is adopted to determine the pressure. Spatial derivatives are evaluated using second order order accurate central differences and for convection terms these were also energy conserving. For time derivatives a second order accurate three point backward difference scheme is used. To avoid the odd/even node uncoupling of the pressure field that arises with this kind of grid arrangement, pressure smoothing, similar to that devised by Rhie and Chow[9] is adopted. The method is second order accurate in space and time and is described in detail in Jones and Wille[10, 11].

The time step was chosen so that the maximum Courant number was 0.5 in the computation using the standard Smagorinsky model whereas with the Dynamic model the maximum Courant number was 0.3.

Initial and Boundary Conditions

The initial conditions at time $t = 0$ within the domain are $U(x, y, z) = U_0$, $V(x, y, z) = 0$, $W(x, y, z) = 0$, with no disturbances superimposed to the mean field. At the inflow plane a constant streamwise velocity component, U_0, is imposed. Free slip conditions are implemented in the transverse direction, and periodic conditions are applied in the spanwise direction. Recent large-eddy simulations computations, performed by Arnal and Friedrich[12], have demonstrated that the channel width can influence the flow statistics when periodic boundary conditions are used in the spanwise direction. However they have also shown that incomplete grid resolution can lead to large discrepancies between experimental and numerical results. The number of grid-nodes in the spanwise direction has thus been chosen to ensure an optimum grid resolution compatible with storage and time constraints. At the same time the width of the domain in the spanwise direction has been chosen to be sufficiently large to provide meaningful flow statistics. At the outflow a zero-gradient/convective outflow condition is applied. In the co-located storage arrangement used the pressure values on the boundaries are not computed directly and are consequently obtained by extrapolation from the interior values. The treatment of the wall boundaries requires some attention, as resolution of the flow in the near wall region is prohibitively demanding in terms of the number of grid points needed to fully resolve the viscous sublayer. As a consequence the wall shear stress is determined by an approximate condition relating the instantaneous wall shear stress to the tangential velocity at the first grid point adjacent to the wall[13]:

$$\tau_w = \frac{< \tau_w >}{< \bar{u}_t(x, y_1, z) >} \bar{u}_t(x, y_1, z, t) \tag{6}$$

The wall shear stress has been obtained by requiring that the averaged velocity field at the wall adjacent grid node satisfies the logarithmic-law-of-the wall:

$$u^+ = 2.5 \ln y^+ + 5.50 \tag{7}$$

where

$$u^+ = \frac{< \bar{u}_t >}{\sqrt{< \tau_w > /\rho}} \tag{8}$$

In both 6 and 7, the overbar denotes spatial filtered quantities, while the brackets denote time-averaging; the subscripts t indicate that the velocity component tangential to the wall is being considered. It has been pointed out[14] that the extent to which the logarithmic law-of-the-wall provides an accurate representation of the velocity at the wall adjacent grid nodes depends on the size of the near wall grid cell; the grid size must be such that the first grid node lies well into the semi-logarithmic region. Consequently a van Driest-type damping function has been introduced in the computations using the standard Smagorinsky model in order to reduce the eddy viscosity in the viscous sublayer region:

$$\nu = \nu_0(1 - e^{y^+/A^+}) \tag{9}$$

where ν_0 denotes the eddy viscosity evaluated according to the Smagorinsky model, y^+ is the distance from the wall in wall units and A^+ is the van Driest constant, set equal to 26.

The wall shear stress and pressure are used to evaluate the lift and drag forces exerted on the cylinder surface. The total drag and lift coefficients, defined as:

$$C_d = \frac{F_d}{\frac{1}{2}\rho A U_0^2} \tag{10}$$

and

$$C_l = \frac{F_l}{\frac{1}{2}\rho A U_0^2} \tag{11}$$

where A is the area of the cross section of the cylinder normal to the streamwise direction and U_0 is the inlet velocity, are calculated by integrating the instantaneous frictional and 'pressure' forces over the surface of the cylinder[15].

RESULTS AND DISCUSSION

The results of the computations are presented and compared with measurements and previous calculations in figs. 2 to 9: where appropriate quantities and distances have been normalised with respect to the inlet velocity U_0 and the cylinder's characteristic dimension, D,

In fig. 2 the time averaged and instantaneous velocity fields in the plane of symmetry $z = 0$, obtained with dynamic model, are shown. These show the presence of regions of recirculating flow and the development of vortex structures that are then convected downstream from the front corners of the cylinder. Though not shown here almost identical velocity field contours were obtained with the standard Smagorinsky model.

In fig. 3, the time-averaged isolines of the model parameter C_s in the symmetry plane $z = 0$ resulting from the dynamic model are presented. The values of the parameter C_s are vary strongly with position and very low values, compared to the "optimal" Smagorinsky constant[2], are evident. These suggest the presence of large regions of quiescent flow[5].

The calculated and measured time-averaged velocity profiles and velocity gradients on the walls of the cylinder are shown in figs. 4 and 5. As is evident the agreement between the calculated profiles and the experimental results of Lyn and Rodi[1] is reasonable and, in accord with the measurements, boundary layer separation is predicted with both model. With the dynamic model the strength of the resulting recirculation is somewhat over predicted as is the boundary layer thickness at the downstream corner with both sub-grid models. In fig. 6 the mean velocity profile on the upper surface at the front corner is plotted in wall units. A clear semi-logarithmic region is evident in the computations though the dimensionless sublayer thickness (the additive constant)

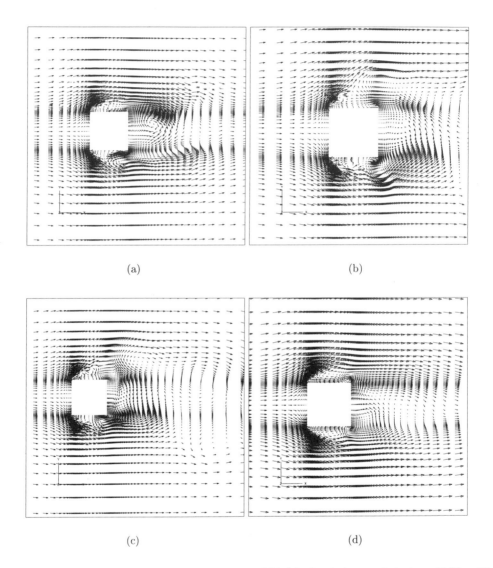

Figure 2. Instantaneous and time-averaged velocity field; (a) after 9 characteristic times $D/U_0 = 20$ sec.; (b) after 12 characteristic times $D/U_0 = 30$ sec.; (c) after 16 characteristic times $D/U_0 = 40$ sec.; (d) time-averaged velocity field.

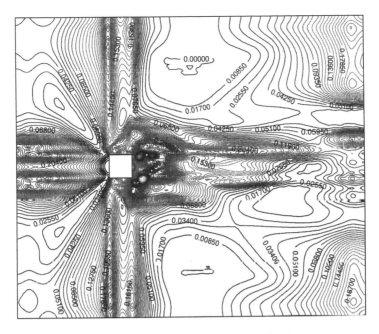

Figure 3. Model parameter time-averaged isolines for the dynamic model.

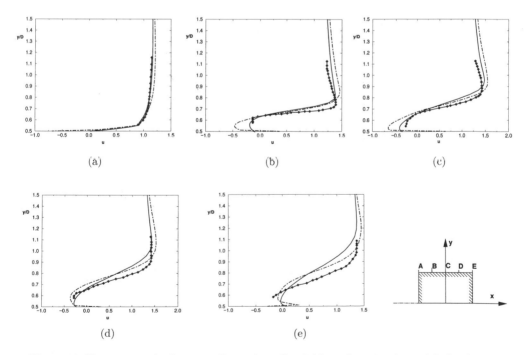

Figure 4. Time-averaged velocity profile on the wall; solid line: Smagorinsky model; dot-dashed line: Dynamic model; ◇: Lyn and Rodi[1].

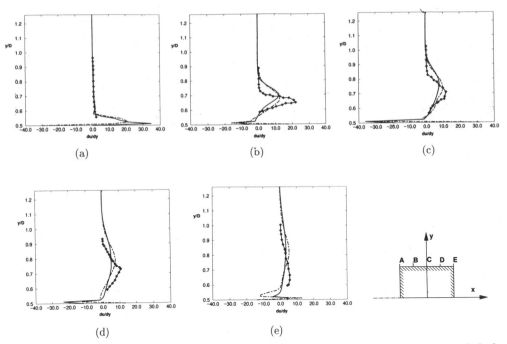

(a)   (b)   (c)

(d)   (e)

Figure 5. Time-averaged velocity gradients on the wall; solid line: Smagorinsky model; dot-dashed line: Dynamic model; ◇: Lyn and Rodi[1].

is lower than that implied by equation 7. (It must be noted that the local equilibrium conditions required for the law of the wall are not satisfied here). As is to be expected the mean velocity profiles at x−locations downstream of the front corner (not shown here) do not display a clear semi-logarithmic law of the wall in the region of flow recirculation.

In order to attempt a more detailed characterization of the wall boundary layer, the growth of the vorticity thickness, defined as:

$$\delta_\omega = \frac{U_{max} - U_{min}}{(dU/dy)_{max}} \tag{12}$$

along the cylinder's surface has been evaluated and the results are shown in fig. 7. . The previous calculated results of Djilali et al.[16, 17] are also included for comparison purposes. Consistent with the mean velocity profiles, (fig. 4) the calculations with both the standard Smagorinsky model and the Dynamic model over predict the growth of the vorticity thickness along the surface. However in the case of the Smagorinsky model the agreement with measured values is quite good.

The mean pressure coefficient along the cylinder surface is presented in fig.8 and compared with the experimental results obtained by Bearman and Obasaju[18], and Lee[19], and to the numerical results of Breuer et al.[20]. The results obtained with the Dynamic model seem to be in better agreement with the available experimental data, even though the presence of larger spatial oscillations must be remarked upon. This feature may be due to the low values assumed by the model parameter C_s, compared to the Smagorinsky constant, in some regions of the domain.

The frequency spectra of both lift and drag coefficients are presented in fig. 9. Because of the different scaling variously used for power spectrum estimation[21, 22, 23]

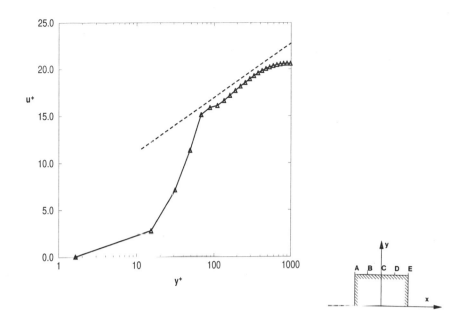

Figure 6. Mean velocity profile in wall units: $\triangle$: Dynamic model; dashed line: $u^+ = 2.5 \ln y^+ + 5.5$.

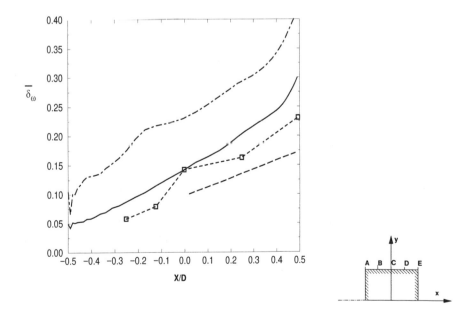

Figure 7. Time averaged vorticity thickness growth; solid line: Smagorinsky model; dot-dashed line: Dynamic model; $\square$: Lyn and Rodi[1]; dashed line: Djilali *et al.*[16, 17].

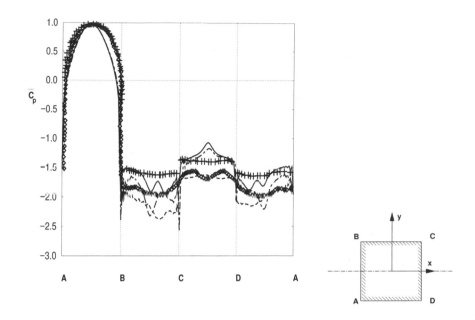

Figure 8. Mean pressure coefficient on the cylinder's surface; $\diamond$: Bearman and Obasaju[18]; +: Lee[19]; dashed line Breuer *et al.*[20]; solid line: Smagorinsky model; dot-dashed line: Dynamic Model.

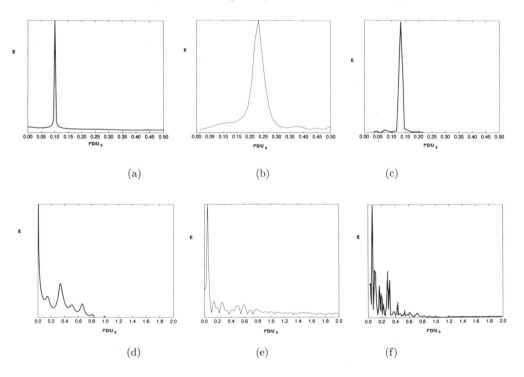

Figure 9. Lift and drag spectra vs. Strouhal number; (a) lift spectrum: Smagorinsky model; (b) lift spectrum: Dynamic model; (c) lift spectrum obtained by Breuer *et al.*[20]; (d) drag spectrum: Smagorinsky model; (e) drag spectrum: Dynamic model; (f) drag spectrum obtained by Breuer *et al.*[20].

Table 2. Bulk coefficients: St, $\overline{C}_l$, $C_{l,rms}$, $\overline{C}_d$, $C_{d,rms}$, l_r

Computation	$\overline{C}_l$	$C_{l,rms}$	$\overline{C}_d$	$C_{d,rms}$	St[a]	l_r[d]
Smagorinsky model	–	–	–	–	0.101	–
Dynamic model	–	–	–	–	0.199	2.54
Breuer et al.[20]	−0.04	1.15	2.30	0.14	0.13	1.46
Experimental results[1]	–	–	2.1	–	0.132	1.38

[a]Estimated from peak in lift spectrum
[d]Correlation length in the homogeneous direction

the spectra shown in fig.9 only provide an indication of the location of frequency peaks.

The results for the bulk coefficients obtained in the current work together with the experimental results[1] and previous calculations[20] are summarised in Table 2. As can be seen that the present estimated Strouhal number is in reasonable agreement with the experimental and numerical results chosen for comparison; the drag and lift statistics have not been calculated. The computed lateral correlation length confirms that the hypothesis of homogeneity of the flow in the spanwise direction is sufficiently accurate for the domain dimensions chosen for the present calculations.

CONCLUSIONS

The results of Large-Eddy simulation of turbulent isothermal flow around a square cylinder at Re= 22000 have been presented. For the sub grid stresses the standard Smagorinsky model and a Dynamic model have been used and both give results which are in reasonable accord with measurements. No appreciable differences between the results obtained with the two models are evident. However in the case of the Smagorinsky model it is necessary to introduce a van Driest type near-wall damping function to account for the presence of the viscous-sublayer. This is a somewhat *ad hoc* procedure which is not entirely consistent with the concept of modelling sub-grid stresses purely in terms of local flow properties. In addition the approach has been found to lead to a substantial under prediction of Reynolds averaged turbulence statistics in the present case. In the case of the Dynamic model no special near- wall treatments were found necessary.

The simulations with both models utilised approximate near wall boundary conditions which have previously been demonstrated to yield excellent results for plane channel flows. However in the present flow they gave rise to significant errors in the wall shear stress in the vicinity of stagnation points, a feature particularly evident in (the over predicted) pressure coefficient around the rear stagnation point. This is almost certainly associated with a basic limitation of the current approximate wall condition formulation; 6 implausibly implies that the instantaneous wall shear stress is zero wherever the Reynolds averaged value is zero.

The computations presented here also suggest that the Dynamic model is more sensitive to the coarseness of the computational mesh than the standard Smagorinsky model combined with damping functions. However, as observed by Franke, Rodi and Schönung[24], in both cases an inadequate resolution of the near-wall region can result in over prediction of shedding frequencies. It must also be remarked that the low values attained by the model parameter C_s when the Dynamic model was used have been found to affect the numerical stability of the calculation[20]. It can be concluded that if LES is to become a useful engineering tool then further effort will be necessary to improve the SGS models and to devise more accurate approximate wall boundary treatments for wall-bounded flows in complex geometries. In addition benchmarks against which numerical methods, SGS models and other factors can be tested need to be established.

ACKNOWLEDGMENTS

We gratefully acknowledge the support provided by EPSRC under grant numbers GR/K43903 and GR/K41601 and the CEC under Brite-Euram project LES/PDF-ECT, number BE 95-1927.

REFERENCES

1. D. A. Lyn and W. Rodi, The flapping shear layer formed by flow separation from forward corner of a square cylinder, *J. Fluid Mech.* 353:376 (1994).
2. M. Germano, U. Piomelli, P. Moin and W. H. Cabot, A dynamic sub-grid-scale eddy viscosity model,*Phys. of Fluids* 1760:1765 (1991).
3. J. Smagorinsky, General circulation experiments with the primitive equations.I - The basic experiment, *Month. Wheat. Rev.* 99:165 (1963).
4. R. Peyret. "Handbook of Computational Fluid Mechanics," Academic Press, London (1996).
5. U. Piomelli and J. Liu, Large-Eddy simulation of rotating channel flow using a localized dynamic model,*in:* "Large-eddy simulation of turbulent flows",TAM Report n.767, UILU-ENG-94-6023.
6. S. Ghosal, T. S. Lund, P. Moin and K. Akselvoll, A dynamic localization model for large-eddy simulation of turbulent flows,*J. Fluid Mech.* 229:255 (1995).
7. W. P. Jones. "BOFFIN: a Computer Program for Flow and Combustion in Complex Geometries," manual, London (1991).
8. AA. VV. "Workshop on Large Eddy Simulation of Flows past Bluff Bodies," Univ. of Karlsruhe, Tegernsee (1995).
9. C. M. Rhie and W. L. Chow, Numerical study of the turbulent flow past an airfoil with trailing edge separation, *AIAA J.* 1525:1533 (1983).
10. W. P. Jones and M. Wille, Large Eddy Simulation of a Plane Jet in a Cross Flow, *Int. J. Heat Fluid Flow*, 296:306 (1996).
11. W. P. Jones and M. Wille, Large Eddy Simulation of a Round Jet in a Cross-Flow, *in:* "Engineering Turbulence Modelling and Measurements," Eds. G. Bergeles and W. Rodi, Elsevier , (1996)
12. M. Arnal and R. Friedrich, Large-eddy simulation of a turbulent flow with separation, *in:* "Turbulent Shear Flows 8," F. Durst, R. Friedrich and B. E. Launder ed, Springer-Verlag, Berlin (1993).
13. U. Schumann "Subgrid scale model for finite difference simulations of turbulent flows in plane channels and annuli," Journal of Computational Physics, 376:404 (1975)
14. U. Piomelli, J. H. Ferziger and P. Moin, New approximate boundary conditions for large-eddy simulations of wall bounded flows, *Phys. of Fluids* 1061:1068 (1989).
15. K. Karamcheti. "Principles of Ideal-Fluid Aerodynamics," John Wiley and Sons, New York (1966).
16. N. Djilali and I. S. Gartshore, Turbulent flow around a bluff rectangular plate. PartI: experimental investigation, *J. Fluid Eng.* 51:59 (1991).
17. N. Djilali, I. S. Gartshore and M. Salcudean, Turbulent flow around a bluff rectangular plate. Part II: numerical predictions, *J. Fluid Eng.* 60:66 (1991).
18. P. W. Bearman and E. D. Obasaju, An experimental study of pressure fluctuation on a fixed and oscillating square-section cylinders, *J. Fluid Mech.* 297:321 (1982).
19. B. E. Lee, The effect of turbulence on the surface pressure field of a square cylinder, *J. Fluid Mech.* 263:282 (1975).
20. M. Breuer, M. Porquie and W. Rodi, LES of flows past bluff-bodies: case A, *in:* "Workshop on Large Eddy Simulation of Flows past Bluff Bodies," Univ. of Karlsruhe, Tegernsee (1995).
21. S. Lawrence Marple Jr. "Digital Spectral Analysis," Prentice-Hall, New York (1987).
22. T. W. Parks and C. S. Burrus. "Digital Filter Design," John Wiley and sons, New York (1987).
23. IEEE. "Programs for Digital Signal Processing," John Wiley and sons, New York (1979).
24. R. Franke, W. Rodi and B. Schönung, Numerical calculation of laminar vortex-shedding flow past cylinders, *J. Wind Eng. and Ind. Aero.* 237:257 (1990).

SELF-ADAPTIVE, PARALLEL SOLUTION METHODS FOR COMPLEX FEM PROBLEMS IN CFD AND RADIATION MODELLING

Xiao Xu[1], Christopher C. Pain[1], Cassiano R.E. de Oliveira[1], Adrian P. Umpleby[1], and Antony J.H. Goddard[1]

[1]Applied Modelling and Computation Group
Centre for Environmental Technology
Imperial College of Science, Technology and Medicine
Prince Consort Rd, London SW7 2BP, UK

INTRODUCTION

This paper describes ongoing research into self-adaptive, parallel solution methods for CFD and radiation problems of interest to the nuclear industry. The main thrust of the research is reviewed. These methods apply to unstructured conforming finite element meshes in 2D (currently) and ultimately 3D. Benchmark applications to CFD and radiation propagation problems are presented.

The design and operation of nuclear reactors require the solution of very complex physical problems describing fluid flow, heat transfer and radiation transport. When translated to numerical models, say through the use of finite element discretisations, these give rise to very large computational problems which need to be solved accurately and in a realistic time scale. Only the combination of adaptivity and parallel computing is potentially capable of achieving this.

Research is underway to develop self-adaptive solution methods which enable a massively parallel processing (MPP) environment (unbalanced or otherwise) to be exploited fully in the numerical solution of complex 3-dimensional steady-state or time-dependent field problems using finite elements - meeting the multi-disciplinary need for efficient parallel numerical solution methods. The parallel adaptive framework envisaged has the following features: dynamic load balancing; dynamic reduction of the communication costs per iteration; dynamic reduction of the domain decomposition method (DDM) iterations required for convergence. It will fulfil, in a parallel context, a role similar to that of current serial mesh adaptivity methods.

The remainder of this paper is structured as follows: in the next section the domain decomposition method is briefly described. Section 3 presents the communication arrangement for the parallel computation. Section 4 describes the mesh adaptive technique and the adaptivity solutions for the fluid flows and radiation problem. Section 5 will outline the parallel adaptive solution strategies and lists some parallel computation results, which show the high parallel efficiency achieved.

High Performance Computing
Edited by R. J. Allan *et al.*, Kluwer Academic / Plenum Publishers, New York, 1999

DOMAIN DECOMPOSITION USING NEURAL NETWORK
GRAPH PARTITIONING TECHNIQUES

Domain Decomposition Methods (DDM) are being actively exploited to solve parallel sets of simultaneous equations formed with the application of a Finite Element Method (FEM) or other grid-based discretisation. To use a DDM the mesh must first be partitioned into a number (usually equal to the number of processors) of balanced subdomains such that the interprocessor communication is minimised. A simple method of balancing subdomains is to ensure that the number of nodes of a FEM mesh, in each subdomain, is approximately equal. To minimise the interprocessor communication the number of edges of the graph associated with the FEM mesh between the subdomains is minimised, subject to the complementary constraint of balancing the load.

A graph-partitioning algorithm has been developed which decomposes unstructured finite element/volume meshes as a precursor to a parallel domain decomposition solution method [1]. It does this by first constructing a coarse graph approximation using an automatic graph coarsening method. The coarse graph is partitioned and the results are interpolated onto the original graph to initialise an optimisation of the graph partition problem. In practice, a hierarchy of (usually more than two) graphs are used to help obtain the final graph partition. A mean field theorem neural network is used to perform all partition optimisation. The partitioning method is applied to graphs derived from unstructured finite element meshes and in this context it can be viewed as a multi-grid partitioning method.

Fig. 1 shows the decomposed FE model for a radiation benchmark consisting of a cask designed for transportation of spent nuclear fuel assemblies from pressured water reactors. The nuclear fuel assemblies are of square cross-section and are housed within a cylindrical flask, which is surrounded by neutron and gamma ray shielding.

AUTOMATIC ARRANGEMENT OF PARALLEL COMMUNICATION

In this section we describe a method of enhancing the parallel efficiency associated with the message passing model when applying a domain decomposition method (DDM) of an unstructured mesh by the Delaunay triangulation algorithm [2]. It is assumed that the mesh has already been decomposed into a number of subdomains and each subdomain is assigned to a unique processor, as described in [1]. A new method is proposed for automatically organising the communication between subdomains (processors). The objective is for each subdomain to send information synchronously to all its neighbouring subdomains in a minimum number of message passing steps. This is achieved with the proviso that a subdomain can exchange information with only one neighbouring subdomain (processor) at a time.

The automatic organisation of the synchronous communication between subdomains is achieved as follows: (i) generate a graph, where a vertex represents a communication link between subdomains (processors), i.e. the message passing between a pair of processors and an edge links two vertices if the two associated communications cannot be initiated simultaneously; (ii) colour the graph with a minimum number of colours, n say, using a suitable graph colouring algorithm, such that a vertex of a given colour has edges linking it to other vertices of different colours (figure 2); (iii) set an order for the message passing according to the colour map so that messages are exchanged simultaneously when these communication links have the same colour. In this fashion, any communication exchange between neighbouring subdomains can be completed within n steps.

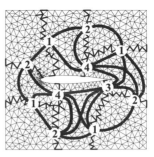

Figure 2. Graph with its vertices coloured.

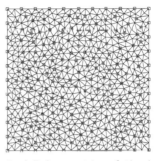

Figure 3. A Delaunay triangulation inside a square with colour 1 marked

Figure 1. A finite element mesh representation of a shipping cask radiation transport problem decomposed into 32 subdomains.

ADAPTIVITY IN INCOMPRESSIBLE FLOW AND RADIATION PROBLEM WITH AUTOMATIC REFINEMENT AND COARSENING

A common problem encountered by numerical practitioners is that the mesh used in a numerical simulation has to be generated a priori to the solution procedure. It is therefore difficult to resolve adequately the local physical features at a first attempt, and often the mesh needs modification for the solution procedure to satisfy the resolution requirements. In recent years considerable effort has been invested in developing error measures which serve as a criterion for mesh modification (remeshing)[3]. Two basic kinds of modifications may be needed: refinement and coarsening. The former will add new nodes to the current mesh in certain regions, enhancing the quality of the solution, while the latter will delete nodes to remedy excessive resolution in other regions.

A method has been developed for the coarsening and refinement of Delaunay triangulations. In a region to be coarsened, nodes are chosen to be retained that have a minimum (>0) number of non-retained nodes between them (see figure 3). It chooses these nodes using a node colouring (graph colouring) based technique. The method can be applied in a local (deleting a single node at a time) or a global sense. The mesh is coarsened without sacrificing the integrity of the boundary (geometry) by not allowing certain nodes on the boundary of the domain to be deleted. More details of the coarsening method can be found in [2].

The refinement method inserts a node in the middle of each edge for all edges in the region to be refined. All these nodes will be inserted one-at-a-time by using the node insertion process [4] of the Delaunay triangulation algorithm. In this way, a two-dimensional triangle element will be divided into four similar triangles.

The incompressible Navier-Stokes equation are discretised using mixed finite el-

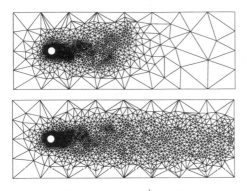
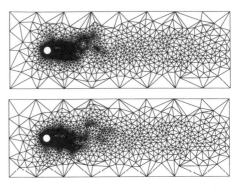

Figure 4. Dynamic adaptive meshing - following features of the flow past a cylinder at Re = 400, t = 20, 60 units respectively

Figure 5. Dynamic adaptive meshing following features of the flow past a cylinder at Re = 2500, t = 81, 82 units respectively

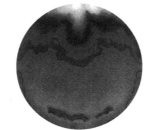
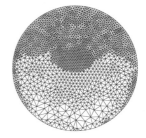

Figure 6. Radiation flux field using an initial mesh

Figure 7. Error distribution across the initial mesh, which is used to form an adapted mesh

Figure 8. Adapted mesh

ement formulations with different finite element expansions for pressure and velocity. The momentum equations are discretised by using a space-time Petrov-Galerkin method [5] in which across a space time element the weighting function is lent over in the direction suggested by the flow of information (characteristics). The associated dissipation is chosen to provide the 'correct' quantity at steady state.

The overall method, is applied to time dependent 2-D incompressible flow past a cylinder (figures 4, 5) at moderate Reynolds numbers.

Figure 6 shows the scalar flux of a one speed radiation calculation with no scattering but with absorption in which the radiation propagates through a 2-D cylinder. The radiation source emits isotropically from a number of surface elements at the top of the cylinder. Figure 7 presents the error distribution calculated from the flux shown in figure 6 for the mesh also shown in figure 6. With this error distribution a new mesh is obtained as seen in figure 8 and the radiation field is re-calculated using this grid.

PARALLEL ADAPTIVE STRATEGIES AND COMPUTATION RESULTS

Parallelisation will allow the solution method to utilise distributed computing resources more efficiently, such as a cluster of workstations, PCs, or even a mixture of different machines with widely varying speeds.

Dynamic load balancing is required to assign the amount of computations equally across the processors, i.e. such that each completes its job in the same time. The load may become imbalanced through mesh adaptivity and/or the processor capabilities.

Table 1. Shipping Cask benchmark parallel results

Number of processors	Wall clock time (sec)	speed up	max storage Mwords	nodes in subdomains	halo nodes
4	1115.1	1.00	5.64	7820-7892	3625
8	676.2	1.64	3.46	3883-3983	6674
16	360.4	3.09	2.18	1881-2046	9313
32	203.9	5.47	1.41	892-1086	12823
64	112.8	9.9	1.2	472-522	15281
128	82.6	13.5	1.2	230-270	18196

Parallel adaptive solution strategies

This strategy can be described as:

(1) generate an initial coarse global mesh using one processor;

(2) carry out a domain decomposition, which will assign each subdomain to a processor;

(3) perform parallel computation, where each processor will carry out computation and communication, if final solution is achieved then stop;

(4) implement self-adaptive algorithm on each processor (subdomain) according to error indicator, but do not change the inner boundary between processors - this will result in different loads on different processors;

(5) re-partition the mesh in parallel - this procedure assigns each node (or element) a mark which indicates migration (or non-migration) to a new processor;

(6) migrate nodes, edges, faces, elements and etc. among the processors - after this procedure each processor will only have the mesh data which belongs to it;

(7) go to stage (3).

Parallel computation results

Table 1 shows the parallel solution performance results for only the most demanding first energy group (containing localised neutron sources in the fuel) for the Shipping Cask benchmark (see fig.1) with 33565 elements and 31428 nodes, using a P_3 (6 unknowns per node) angular approximation. Performance is very encouraging, bearing in mind that solution in all subdomains will be strongly linked to those subdomains containing the sources within the fuel elements.

Table 2 shows a selection of relevant data for solution of a buoyancy driven flow problem in a cube (125000 elements, 132651 nodes). The relatively poorer parallel performance of the temperature equation solution, compared to that of the momentum and continuity equations, is due to the fact no attempt was made to optimise internal and external tolerances or the choice of solvers. The CPU time shown is for the first three time steps of this problem, as a function of number of processors. These time steps are the most expensive in CPU. There are $51 \times 51 \times 51$ temperature equations, three times this number of velocity equations and $50 \times 50 \times 50$ pressure equations. Notice, as expected, that the initialisation time increases with the number of processors.

Table 3 compares the performances of a number of solution methods. The multiwavefront methods 1-4 (Table 3) use a set of interface nodes (situated between the subdomains). In forming the parallel wavefront preconditioner the equations centred on small sets of interface nodes must be solved together with the equations internal to each subdomain. The convergence properties of the wavefront methods increase with number given in the table 3.

Table 2. The first three time steps of a buoyancy driven flow problem.

	8 processors	32 processors	128 processors
total run time including initialisation (sec)	256.9	97.2	55.7
CPU (sec) for assembly and soln of Navier Stokes	205.1	63.2	19.3
CPU (sec) for assembly and soln of temp eqn	36.4	14.2	5.0
No of reals used on processor one	2041638	478525	161200
No of integers used on processor one	1171590	268904	88472

Table 3. Pressure solution times (128 processors) for the first time step of a 3 D recirculating flow problem.

	No of DDM Iterations	Total internal iterations	CPU (sec) for pressure soln
multi-block-explicit	92	523	12.3
multi-wave front solver 1	82	512	14.5
multi-wave front solver 2	64	465	14.4
multi-wave front solver 3	76	467	13.9
multi-wave front solver 4	60	378	12.4

CONCLUSIONS

Domain partitioning methods, automatic arrangement of parallel communication, and self-adaptive solution methods have been developed with the overall aim of producing a more efficient parallel FEM solver. The quality of the mesh partitioning for parallel processing has been demonstrated and a new self-adaptive Delaunay triangulation method has been presented along with successful application to a range of 2-D transient fluid flow problems and a radiation transport problem. Parallel computations show the high parallel efficiency achieved for a range of radiation and CFD problems. Our current research focusses on parallel computation with dynamic meshing and partitioning.

ACKNOWLEDGEMENTS

This work is supported by the Engineering and Physical Sciences Research Council, UK, through Grant GR/K73466.

REFERENCES

1. C.C. PAIN, C.R.E. DE OLIVEIRA, AND A.J.H. GODDARD, *A neural network graph partitioning procedure for grid based domain decomposition*, Submitted for publication, Jan. 1997.
2. X. XU, C. PAIN, A. GODDARD AND C. DE OLIVEIRA, *An automatic adaptive meshing technique for Delaunay triangulations*, to be published in Comp. Meth. Appl. Mech. Eng.
3. O. ZIENKIEWICZ AND J. ZHU, *A simple error estimator and adaptive procedure for practical engineering analysis*, Int. J. Numer. Meth. Engng, 24 (1987), pp. 337–357.
4. P. GEORGE, *Automatic Mesh Generation*, Wiley, 1991.
5. J. JOHNSON, *Numerical solution of partial differential equations by the finite element method*, Cambridge Uni. Press, 1995.

HUMAN SYSTEMS AND INFORMATION

HPC AND HUMAN GEOGRAPHIC SOCIAL SCIENCE RESEARCH

Stan Openshaw

Centre for Computational Geography
School of Geography
University of Leeds
e-mail: *stan@geog.leeds.ac.uk*

INTRODUCTION

It is now widely accepted that faster computers are stimulating new ways of doing science and that computation is a scientific tool of equal importance to theory and experimentation. What is far less widely appreciated is that computation also provides a paradigm for geographers to do geography better and that a similar claim is generally applicable to most of the social sciences. The problem is that these disciplines have yet to discover that an HPC world exists and that they have many suitable applications for it. HPC is a most important tool that may well become even more useful as efforts are made to come to terms with the vast data riches that the computerisation of the modern world and all its management and administrative systems have created. The problem is that most geographers and social scientists seemingly still do not understand what HPC can deliver and as a result few make any use of it. This is a great pity. If that PC on your desk had access to a HPC that was 10,000 times faster and had 5,000 times more memory what would you do with it? Currently most geographers and social scientists would not be able to answer this question other than to say they do not need such powerful computers. It is not so much the idea of computation that is new but it is the notion of using HPC hardware to do it that is quite alien. Geographers and social scientists have learnt to cope without it and as a result see no need for it. The speed-up of PC's to offer levels of performance once only provided by mainframes has made it easy to continue with this fallacy. The new research opportunities engendered by the availability of almost unlimited amounts of computing power requires a whole new research agenda; Openshaw (1994).

Geographers have a key role to play here since by showing what can be done they may be able to aid the further development of a computational culture that is transferable to other social sciences. Geography has an advantage in that there is a long tradition of quantitative and mathematical modelling and analysis that could be ported into a HPC domain. There is also a strong applied component and several of the topics studied are of considerable practical significance; for example, the analysis of geographically referenced data or methods for finding the best spatial locations for facilities. There are also many potential end-users in government and commerce who may well be interested in the results. Finally, the Geographic Information Systems (GIS) revolution that started 10 years ago has created a data environment in which most computer data bases relating to land, property, and people now have geographical references attached to them. This huge explosion in the coverage and volumes of geographic data is a problem for all the social sciences, for industry, and commerce as these data constitutes a vast and relevant information resource that currently they are ill equipped to handle and exploit. Yet these data resources are the future raw materials for many social sciences in the new millennium. Its here where new discoveries, new concepts, and new results will have to be found.

The World of people about us is becoming increasingly data rich but poor in theory and understanding. There are large amounts of data but much of the applicable analysis and modelling methods still assume N (the number of observations) and M (the number of variables) are small. There is really very little idea of how to analyse databases of tens (soon 100's) of terabytes of data. New technologies are needed if good use is to be made of these immense data resources. However, it is also important to be confident that a computational route will deliver new science and that if we invest 100,000 or several million times more computational effort in some of these human systems modelling and society understanding projects that the benefits will be worthwhile and of comparable scientific importance or practical significance to computational research in other areas of science.

PROBLEMS

The consortium faced six major problems of working in a human geographical social science domain. They are:
1. lack of HPC awareness amongst the potential end-users,
2. lack of trained researchers who could use HPC,
3. lack of research initiatives from the Economic and Social Science Research Council in HPC,
4. a very small share (a few percent) of the total UK's HPC time allocation,
5. the rapid speed of the changes that are occurring in this area in terms of languages and hardware, and
6. underdevelopment and underfunding as the ESRC has historically regarded HPC as of interest to a very small minority, reflecting a fallacy that most social scientists only need a PC; although the ESRC has kept the HPC door open albeit by only a crack.

Other problems relate to the nature of HPC. In particular the need to develop explicitly parallel algorithms and the requirement to learn new parallel programming skills in addition to developing the application domain. No matter how you look at it, parallel processing is harder than serial programming. A task made difficult by the current reluctance of many social sciences to teach any kind of programming as part of post graduate research training.

OBJECTIVES

The consortium objectives were therefore:
1. to port a set of existing serial codes on to the Cray T3D in a form that is reasonably future proofed and preserves their portability to future parallel machines;
2. to raise awareness of HPC in geography and the social sciences via conferences, papers in journals, and book writing;

3. to focus on a small number of high profile demonstrator projects;
4. to develop some new scientific applications that are computationally intensive to illustrate practical benefits that can be gained from HPC; and
5. to take steps to preserve the investment in code and expertise.

None of this was easy due to massive under-funding and lack of resources. There were several other problems that were encountered. Unlike all the other consortium there was not one area of science but several different ones which had to developed in tandem. This is really quite different from focusing on a single code with an existing attached user community to service. Additionally, many of the areas covered were poorly developed and made extensive use of serial algorithms and legacy technologies. Porting often involved completely re-designing the serial algorithms by developing entirely new parallel ones. Maybe we were too ambitious but we did not merely want to convert existing code that could run poorly but create codes that would scale well and squeeze the highest level of performance out of the parallel hardware. As newcomers to HPC we felt that we may be we were being watched to see what we could manage to do and therefore we sought to produce codes that were highly efficient.

Our objective was to maximise the amount of science that could be done per hour of machine time and not merely to maximise the total gigaflops. They are not related. For example, the GAM (see later) in its original form required an estimated 9 days of single T3D node compute time. It parallelises well and you could gain a factor of 500 or so speed-up on the Cray T3D, reduce wall clock run times to 25 minutes and achieve high levels processor efficiency. However, to demonstrate new science we wanted to run this code 1,000 times to investigate error propagation and various statistical issues not previously considered. This would take 17 days on a 512 processor T3D; which means it cannot be run. So its back to the original algorithm. Develop a new one and after a while the 9 days reduces to 700 seconds and it would run on a workstation. The 1,000 replicates needed to improve the science now takes 8 hours on 512 processors and this is achievable but the cost in terms of research time, effort, and expertise is very high. The problem was not the lack of implicit parallelism (many geography problems that relate to a map are ideally placed for spatial decomposition) but mainly that the degree of parallelism was too fine and the associated algorithms needed to be completely re-designed in a more coarsely grained way; see for example, Openshaw and Turton (1996). Porting via fairly straightforward code translation was a rare luxury. Other problems were of our own making, in particular the desire to squeeze maximum levels of performance from the parallel codes resulted in too much effort being spent on codes that were probably not sufficiently important by themselves to justify the effort involved. However, this was part of the learning curve and our experiences have been used to write a book on parallel programming for geographers; Openshaw and Turton (1999).

A final characteristic was the lack of resources and also (it has to be said) of support from the ESRC. The ESRC has no long term view of how to deal with HPC and they appear to view it merely as a legacy small scale commitment to a low level of historical usage by an economist and a geographer. At any moment in time any future budgets for ESRC HPC appear uncertain and could be raided to support other areas of research. It needs to be put onto a more secure footing and related to a survey of likely future needs for HPC within the social sciences. So far this has not been done and as a result no strategy for the development of HPC within the social sciences has yet emerged. However it is an important topic because of its general applicability across the social sciences and because of the increasing use being made of HPC by many of the social science related end-user communities particularly commercial organisations. It is not just an opportunity for social scientists to study others making use of HPC but it also needs to be recognised as a powerful tool of general social science significance in its own right.

The Consortium's total resource base consisted of 1 full-time research associate, the author as the Principal Investigator, and variable inputs from about 10 other staff and researchers in Centre for Computational Geography and Computer Studies at Leeds. A pool of 8 postgraduates also contributed in various ways. HPC in a social science context is a small team science (in terms of research resources) rather than big science and this affects what can be done. However, the computational intensity is quite extraordinary. For example, some of the prototype applications developed by this consortium could easily have used 100% of the entire UK HPC resource.

WHAT THE CONSORTIUM PRODUCED

The short-term objective was to port serial codes from a range of illustrative application areas on to the Cray T3D (see Turton and Openshaw (1998). Table 1 provides a brief summary of the area of geographic science and their vital characteristics. The intention was to write research papers relating to each code and this process is still continuing. To preserve the investment a textbook on parallel programming for beginners in geography and the social sciences has been written. This contains many worked examples and some of the codes developed here. Finally, a social science code for benchmarking the performance of serial, vector, and parallel machines has been produced; see Openshaw and Schmidt (1996). This is important as a means of measuring HPC hardware performance on a typical memory hungry and low compute intensity geography code.

DEVELOPING A NEW PHILOSOPHY

The lack of a computational culture was dealt with by inventing a new word called "GeoComputation". This is defined as the application of HPC to develop computational methods and AI tools able to exploit the spatial data riches and hence discover new ways of doing geographic, social science, and applied research in this area of science. The first International Conference (organised by the HSM Consortium) was held in Leeds in 1996, the second in New Zealand in 1997, and next one is in Bristol 1998. GeoComputational theme sessions now occur at an increasing number of International Conferences in geography and regional science; for example, at the European Regional Science Congress in Vienna in August 1998. GeoComputation provides a focal HPC related rallying point for many computational scientists in geography and related disciplines and it is helping to foster a computational culture.

Table 1. Summary of codes ported to Cray T3D

Description of Application	Short-Name-	Original Algorithm	Parallel Codes written in
Spatial Classification	KOHONEN	re-written	Craft, HPF
Interaction Modelling	SIMS	re-written	F90, MPI, HPF
Spatial Optimisation	IRP2	re-written	Craft, HPF, MPI
Population Forecasting	POPMOD	ok	MPI
Zone Design	ZDES	re-written	MPI
Spatial Analysis	GAM	re-written	MPI
Spatial Modelling	GEM	re-written	MPI
New Modelling System	AMS	re-written	MPI
Concepts Discovery	CON	new	HPF
Social Science Benchmark Code	SSB/1	new	F90., MPI, HPF, BSP

WHAT IS THE ATTRACTION OF HPC?

There are a number of general opportunities. They include the following.

1. Improve legacy models and statistical methods by substituting more robust computational procedures for older technologies; for example, the use of Monte Carlo significance testing instead of more classical approaches than are invalidated by the nature of spatial data. Once there was no alternative, now there is if you can afford the computation costs.

2. Scale up existing methods to handle far larger and finer spatial resolution data and thus gain some benefits from the spatial data explosion.

3. Application of more robust and better parameter estimation and optimisation algorithms that are far less assumption dependent.

4. Investigate the error propagation properties of models that are widely used; for instance, to estimate confidence intervals for models that previous lacked any; for example, population forecasting.

5. Apply new approaches based on computational intelligent methods and develop new analysis and modelling tools that require large amounts of computation as a replacement for simplifying assumptions and lack of knowledge. There is an increasing tool-kit of practical and available Artificial Intelligence methods that can be used to model and analyse geodata. Particularly important here are: artificial neural networks, evolutionary programming, and fuzzy logic modelling.

6. Investigate novel computational technologies used in robotics and computer vision that have equivalent applications in geography. There is a wonderland of novel computational technologies that have yet to be applied to geodata and yet offer considerable future promise. Mention can be made of: computer vision as pattern detectors, artificial life, and intelligent agents; see Turton (1997) and also his chapter in this book.

SOME BRIEF CASE STUDIES

Parallel Spatial Interaction Modelling

The most obvious application of HPC is to take an existing code and scale it up so that it can use far larger data sets than is possible on a workstation. The use of finer resolution data may be expected to improve the quality of the results. The spatial interaction model is one example. This mathematical model is widely used to model the flows of people or goods or traffic from origin zones to destinations. Small problems run on a PC but compute times increase as NM where N is the number of origins and M is the number of destination zones. Larger datasets and, or, better geographic resolution result in larger N and M values. For example, the 1991 census provides journey to work data for 10,764 wards in GB (equivalent to a table with 10,764 rows and 10,764 columns) disaggregated in many different ways. Even larger interaction data sets exist; for example, credit card retail sales have a potential size of rank 1.6 million and telephone traffic about 32 million.

This model can be parallelised in various ways; see Turton and Openshaw (1997). Additionally, the availability of HPC codes allows the use of more flexible and robust optimisation technologies. Diplock and Openshaw (1996) report the failure of conventional non-linear optimisation methods to always work when faced with functions that contain exponential and other maths functions capable of causing arithmetic problems. Many statistical modelling methods probably suffer from similar risks it is just that few realise that

IEEE arithmetic creates false local optima before becoming a NaN black hole that will totally wreck a conventional non-linear optimiser. The solution is to use more robust optimisation methods based on genetic algorithms but the cost is a 100 to 1000 times more computational effort.

Spatial Location Optimisation

A related application is that of location optimisation modelling. Basically, a spatial interaction model is embedded in a non-linear optimisation procedure that seeks to identify the optimal distribution and size of some facility of interest; e.g. shops or car show rooms or banking machines or cellular transmitter locations to provide maximum coverage of a demand surface. Its a complex non-linear combinatorial optimisation problem for which there are unimaginably large numbers of possible solutions; e.g. find the best 60 locations from a set of 800 possible ones. The quality of the results depends on the performance of the optimisation heuristic and the number of alternative solutions that can be evaluated in a fixed time interval. A serial code was parallelised and a new coarsely grained location optimisation procedure developed. The entire program scales linearly so that a 256 processor T3D run ran about 256 times faster than on a single node. However, it was then discovered that the single node performance can be greatly improved by re-writing the model to minimise the amount of arithmetic being performed providing a speed-up of 1076 times! Running this code on a Cray T3D with 256 processors adds another factor of 256 producing a remarkable total speed-up of 2.8 million times compared with the performance of the original code on a workstation. The results are also twice as good, because many more candidate solutions can now be investigated (Turton and Openshaw, 1996). Sadly the end-user who provided the original code was uninterested because they lacked access to HPC and they believed that good solutions (half the performance of the best) were good enough!

A Computational Approach to Model Building

Other applications were concerned with trying to "create" better performing new generation models of spatial interaction. Once you have the capability of evaluating millions of models per hour then you can start to do the unthinkable and search the universe of alternative possible model formulations; see Openshaw (1988). A crude approach would be to randomly create and test millions of plausible model equations. Far smarter is the use of a Genetic Algorithm or Genetic Programming approach to find the best performing ones whilst only evaluating a small part of the universe of alternatives. If you can afford the computation costs and HPC is sufficiently fast then you may be able to dramatically improve model performances. Diplock (1996) and Turton et al (1996, 1997) report the results of using genetic programming to create new spatial interaction models with about 50% lower error rates. There are two outstanding problems here: first there is still a widespread prejudice against computationally derived models even if they do work well and second a single run of this procedure is equivalent to almost the entire ESRC annual allocation of Cray T3D time. Maybe this technology will be more important in the future when HPC is faster.

Engineering Geographical Zoning Systems

A zone is another word for geographical area or region or place or polygon that appears on a map. It is essentially a two-dimensional map based object with a boundary; e.g. Local Authorities, Towns, regions, census areas, sales territories, constituencies, zones used for reporting statistical data. Traditionally, users have had to use whatever zones existed and lacked the technology to change their design. The GIS revolution changes this provided there is sufficient compute power to solve the consequential zone design problem. Zone design can be viewed as a special type of combinatorial optimisation problem with integer variables, non-linear and possibly discontinuous objective functions, with constraints; see Openshaw and Rao (1995). The first algorithms were first produced 20 years ago to work on small problems because there was little suitable data. The technology was revived in the mid 1990s when it was realised that there was a real and urgent need for tools able to allow users to engineer their zoning systems; for example, to design data reporting zones for the 2001 census that will preserve the confidentiality of the data without doing too much aggregation damage to it. The best methods are based on simulated annealing but they are highly serial in nature and slow. It was necessary to create a parallel simulated annealing algorithm that would work well, offer excellent load balancing, scale over large numbers of processors and yield a real speed-up. The result was a hybrid Genetic Algorithm Simulated Annealer that used the GA to determine optimal temperatures in an adaptive and dynamic manner; see Openshaw and Schmidt (1996). The zone design code is attracting considerable interest because of its general applicability.

Geographical Data Mining or Exploratory Spatial Analysis

The explosion of geographically referenced data has greatly stimulated the demand for exploratory geographical analysis of a rapidly emerging geocyberspace of information. Increasingly once a data set exists there is an imperative for its analysis; e.g. crime data and health data. There is need for new highly automated geographical analysis tools that are safe and can be used and understood by end-users, which offer basic answers to questions such as "are there any patterns or relationships that stand out" when the user has not the foggiest idea of where on the map to look for them or what to look for.

The Mark 1 Geographical Analysis Machine (GAM) of Openshaw et al (1987) solves this problem by looking everywhere for evidence of localised patterning and then shows their locations as a three dimensional distribution. It is now attracting global interest. However, once you have found some possible patterns you then need to try and explain them. The Geographical Correlates Exploration Machine of Openshaw et al (1990) is a start in this direction. It looks at 2^{M-1} permutations of map coverages to define clusters that could be "explained" by local spatial associations. Both methods used to run on Cray X-MP hardware and these have been ported for use on the Cray T3D. They will also run on smaller problem sizes with the advanced options switched off on a workstation. The GAM is now available on the WWW; see http://www.ccg.leeds.ac.uk/smart/intro.html.

CONCLUSIONS

In many ways this HSM Consortium has skirted around two main outstanding issues of Human Systems Modelling. The funding from EPSRC was for porting existing codes and we supplemented this with additional resources provided by the School of Geography at Leeds in its support for a Centre for Computational Geography. The task now is to make these codes (or some of the more generally applicable ones) "available" to others to use. The problem is that most geographers (and social scientists) who might be interested are complete 100% HPC novices, they are probably terrified at the thought of the complexity, and only want codes that will work on workstations or PCs. There is a solution that is now being put in place and that involves hiding everything behind WWW interfaces. Even the most HPC illiterate social science user can now cope with the idea of sending data to a web site that later sends back the results. So far there is a GAM available in this fashions and others may follow. The machine that is used could be a workstation or HPC depending on problem sizes and whether the user has an HPC account.

The second issue is more long term and involves developing new generations of Human System Model. HPC developments, computational devices such as distributed intelligent agents, allied to data richness create the prospect of being able to build computer models and laboratories of entire artificial societies with levels of complexity starting to match those found in the observed world. Some of this work has started but there is need for a sustained and large scale Programme of Research if much progress is to be made, see Openshaw (1995). Additionally, like many of the potential applications of HPC in geography, it needs some assurance of access to the next two or three generations of leading edge HPC machines. Do not be surprised that if by 2010 (or so) some of the major users of the world's largest computers will be computational geographers and social scientists.

ACKNOWLEDGEMENTS

The research reported here was supported by EPSRC Grant GR/K43933 Research Grant. The author wishes to acknowledge the work reported done by all involved in this in the Human Systems Modelling Consortium at the University of Leeds.

REFERENCES

Diplock, G., 1996, *The Application of Evolutionary Computing Techniques to Spatial Interaction Modelling*. Unpublished Ph.D., University of Leeds.

Diplock, G. and Openshaw, S., 1996, Using simple genetic algorithms to calibrate spatial interaction models, *Geographical Analysis* 28:262-279.

Openshaw, S., 1988, Building an automated modelling system to explore a universe of spatial interaction models, *Geographical Analysis* 20:31-46.

Openshaw, S., 1994, Computational human geography: towards a research agenda, *Environment and Planning A* 26:499-505.

Openshaw, S., 1995, Human Systems Modelling as a new grand challenge area in science, *Environment and Planning* A 27:159-164

Openshaw, S., 1998, Supercomputing in geographical research, in Proceedings of *Symposium on Information Technology and the Scholarly Disciplines*, British Academy, London (forthcoming)

Openshaw, S. and Rao, L., 1995, Algorithms for re-engineering 1991 census geography, *Environment and Planning A* 27:425-446.

Openshaw, S. and Schmidt, J., 1996, Parallel simulated annealing and genetic algorithms for re-engineering zoning systems, *Geographical Systems* 3:201-220

Openshaw, S. and Schmidt, J., 1997, A social science benchmark (SSB/1) code for serial, vector, and parallel supercomputers, *Int. Journal of Geographical and Environmental Modelling* 1:65-82

Openshaw, S. and Turton, I., 1996, A parallel Kohonen algorithm for the classification of large spatial datasets, *Computers and Geosciences* 22:1019-26.

Openshaw, S. Charlton, M. Wymer, C. and Craft, A., 1987, A mark I geographical analysis machine for the automated analysis of point data sets, *Int. Journal of GIS* 1:335-58.

Openshaw, S. Cross, A. and Charlton, M., 1990, Building a prototype geographical correlates exploration machine, *Int. Journal of GIS* 3:297-312.

Openshaw, S., Turton, I., 1999, *High Performance Computing for Geographers*. Routledge (forthcoming).

Turton, I., 1997, *Application of Pattern Recognition to Concept Discovery in Geography,* M.Sc. Thesis, School of Geography, University of Leeds

Turton, I. and Openshaw, S., 1996, Modelling and optimising flows using parallel spatial interaction models, in: L. Bouge, P. Fraigniaud, A. Mignotte and Y. Roberts eds., *Euro-Par '96 Parallel Processing* Vol 2, Lecture Notes in Computer Science 1124, pp270-275 Springer-Verlag, Berlin.

Turton, I. Openshaw, S. and Diplock. G. J., 1996, Some geographical applications of genetic programming on the Cray T3D supercomputer, in: C. Jesshope and A. Shafarenko eds., *UK Parallel '96*, pp135-150 Springer-Verlag, Berlin

Turton, I. Openshaw, S. and Diplock, G. J., 1997, A genetic programming approach to building new spatial models relevant to GIS, in: Z. Kemp ed., *Innovations in GIS 4* pp89-102 Taylor and Francis, London

Turton, I and Openshaw, S. 1998, High performance computing and geography: developments, issues and case studies, *Environment and Planning A* (forthcoming)

Turton, I and Openshaw, S., 1997, Parallel spatial interaction models, *Geographical and Environmental Modelling* 1:179-197

APPLICATION OF PATTERN RECOGNITION TO CONCEPT DISCOVERY IN GEOGRAPHY

Ian Turton

Centre for Computational Geography,
School of Geography,
University of Leeds, Leeds, LS2 9JT
e-mail: *ian@geog.leeds.ac.uk*

ABSTRACT

This paper explores the ways that pattern matching techniques implemented on parallel machines can be applied to raster datasets to develop new concepts for geography. It is argued that it is becoming increasingly important for geographers to develop new ways of generalising datasets that will allow them to overcome the increasing data richness of the geocybersphere and that one of the ways they should consider going about this is to make use of parallel computing.

The data set used for this study is a population density surface derived from the 1991 census of population by Bracken and Martin (1989). Using this data set the aim is to take the data-poor geographical theories of urban social structure of the first half of the century and make use of the data-rich environments of the 1990s to test the theories in a general and robust manner. To achieve this parallel pattern matching techniques used in computer vision and other fields will be applied to raster data of population density and social and economic variables for Great Britain. Initially the raster data is segmented by the application of image analysis techniques to identify 129 urban areas in Britain. These urban areas are then compared to templates of the theoretical models of Burgess (1925) and Hoyt (1939) using methods developed in the fields of computer vision and medical imaging and here parallelised to allow the use of the larger geographical datasets. Several urban areas are found that are similar in social structure to the theoretical models developed earlier in the century. The urban areas are then compared to themselves to determine if there were any other groupings of modern British cities that can be made in terms of their social structure. Several such groups were discovered and will be briefly discussed.

INTRODUCTION

Openshaw (1994) argues that as the amount of data that is collected as a result of the GIS revolution increases geographers must start to apply new methods to these new data riches. It is no longer enough to merely catalogue the data and draw simple maps of it. It is also no longer acceptable to use crude statistical measures that average over a whole map or region and in so doing throw away the geographical content of the data. In other words geographers must generalise or drown in the flood of spatial data that has increased many fold during the 1980s and 1990s and which will continue to grow into the next century. As the amount of data grows, it becomes increasingly difficult for humans to find the time to study and interpret the data; the only solution is to pass more of the routine analysis to computers leaving the researcher with more time to study the truly interesting parts of the output.

This paper is a first attempt to apply these ideas to a geographical data set. One data set will be studied in detail though the ideas and methods developed will be applicable to many other data sets. The data set selected for this study is a population density surface derived from the 1991 census of population by Bracken and Martin (1989). Using this data set the aim is to take the data-poor geographical theories of urban social structure of the first half of the century and make use of the data-rich environments of the 1990s to test the theories in a general and robust manner. To achieve this parallel pattern matching techniques used in computer vision and other fields will be applied to raster data of population density and social and economic variables for Great Britain.

REVIEW OF URBAN SOCIAL STRUCTURE

Over time as a city grows different areas of cities become associated with different types of population and this leads to systematic relationships between geographic and social space. Bourne (1971) says "All cities display a degree of internal organisation. In terms of urban space, this order is frequently described by regularities in land use patterns." A seasoned traveller will soon notice that while each city is obviously different in its precise layout there are striking similarities between them, for instance they will all have an area of shops and offices, usually in the centre which is well served by transport links while residential areas tend to locate around this area with the areas of better housing being found farthest from the centre of the city. These observations lead to questions about how these patterns can be modelled to allow comparisons to be made between cities and to attempt to give insights into the growth and formation of these patterns.

Much of the defining work on these questions was undertaken by the Chicago ecologists who where concerned with the differences in environment and behaviour in different parts of the city. From this descriptive research into behaviour in various parts of the city grew an interest in the general structure of the city and its evolution over time. In this work they introduced terms from the classical ecology such as dominance, succession and invasion. The majority of this work was confined to the study of the large industrial city of Chicago. The work of the Chicago group can be traced back to the work of Hurd (1903) who developed several theories of urban expansion which stressed two main methods of growth: central and axial growth.

The Zonal Model of Burgess

The zonal model developed by Burgess (1925) is based solely on the central growth element and radial expansion. This is closely linked to the general assumptions of impersonal competition of ecological theory. The zonal model came about almost as an aside in the discussion of how urban areas expand.

The innermost and smallest zone is the central business district (CBD). This is the centre of the city's commercial, social and cultural life. This is the area of highest land values and so can only be used by activities that can generate the profits necessary to pay the high rents and taxes charged in this area. Thus this area contains the retail district with large department stores and other expensive outlets. Therefore in terms of residential type this is a sparsely settled area with few people being able to afford, or in fact choosing, to live in the central zone.

Zone II, the "zone in transition", is the remnant of the city's first suburban area where originally merchants and other successful citizens lived. However as the city has grown with businesses encroaching from the CBD the area has deteriorated and the once fine houses have been converted into industrial units in the inner area of the zone and subdivided into flats and bed-sits in the remainder of the zone. This is an area of first generation immigrants and social misfits. It is a very mixed population with high crime rates and a highly mobile population. As members of the population prosper they move outward to better areas of the city, leaving behind the elderly and "less fit" members of the population.

Zone III is the zone of independent working men's homes, providing housing for the families of shop and factory workers who have prospered sufficiently to move out from zone II. This area is still easily accessible to the central business district where the majority of the population work. Burgess characterised this area as being predominately working class. All age groups are represented in this area with traditional family forms being dominant.

Zone IV is an area of better residences, a zone of middle class population living in large private houses or good apartment blocks. Within this zone satellite shopping centres are found to compete with the more expensive central shops. Burgess suggests that women outnumber men in this zone.

Farthest out is zone V, the commuters' belt within a journey time of 30-60 minutes from the CBD. This is suburbia, characterised by single family dwellings. The majority of the men work in the CBD leaving the area as a dormitory town in the day. Burgess however points out that zone V is the least homogeneous zone of the city.

The Sectoral Model of Hoyt

The sectoral model developed by Hoyt (1939) has a much narrower focus than the zonal model of Burgess being primarily concerned with the distribution of rental classes. Based upon a block by block analysis of changes in a variety of housing characteristics in 142 US cities, Hoyt concludes that: "The high rent neighbourhoods do not skip about at random in the process of movement, they follow a definite path in one or more sectors of the city. ... the different types of residential areas tend to grow outward along rather distinct radii, and a new growth on the arc of a given sector tends to take on the character of the initial growth of that sector." Again the spatial expression of the model is an aside to the main thrust of the model which is the dynamics of rental patterns. Movement of the areas of high rent provide the main driving force of the model. As rich neighbourhoods move outward they are replaced by intermediate and lower class groups moving into the abandoned housing. Hoyt notes: "The high-grade residential neighbourhoods must almost necessarily move outward towards the periphery of the city. The wealthy seldom reverse their steps and move backwards into the obsolete houses which they are giving up. On each side of them is usually an intermediate rental area so they cannot move sideways. As they represent the highest–income group, there are no houses above them abandoned by another group. They must build new houses on vacant land. Usually this vacant ground lies available just ahead of the line of march of the area because, anticipating the trend of fashionable growth, land promoters have restricted it to high–grade use or speculators have placed a value on the land that is too high for the low–rent or intermediate rental group. Hence the natural trend of the high–rent area is outward, towards the periphery of the city in the very sector in which the high–rent area started."

The high status area is assumed to have started near the CBD which provides the employment and services for the residents of the high–rental area. It is on the far side of the city from the factories and warehouses. This can be affected by the prevailing wind conditions, for example many British cities have a rich area to the west of the city since the prevailing wind is from the west, so blowing clean fresh air into the best areas of the city and then blowing away any pollution from the industrial area in the east.

IMAGE ANALYSIS AND COMPUTER VISION METHODS

There are many possible definitions of image processing and computer vision, one of the most common that will be used in this paper is that image processing refers to processing an image (usually by a computer) to produce another image, whereas computer vision takes an image and processes it into some sort of generalised information about the image, such as labelled regions. However Ballard and Brown (1982) say that "Computer Vision is the enterprise of automating and integrating a wide range of processes and representations for visual perception." They go on to include the term image processing within this definition, there by implying that image processing is one step in computer vision. Niblack (1986) describes image processing as "the computer processing of pictures" whereas "Computer Vision includes many techniques from image processing but is broader in the sense it is concerned with a complete system, a 'seeing machine'."

The aim of using image processing and computer vision within the context of this work is twofold. First the computer must take a raster population density map of Great Britain, process it to provide a clean image for later processes to work with, then segment the image to find the urban areas. A second stage must extract these areas and use the locations of the towns and cities discovered in the first stage for classification against existing theoretical models of urban structure and also as inputs to a classifier that attempts to discover new structures within the British urban environment.

A technique that can be applied to the detection of urban areas with a population density surface is the determination of the gradient of the surface. This is analogous to the differentiation of a mathematical function. To determine where the maxima of a function are, the function can be differentiated. Where dy/dx is zero the function is either at a maximum, minimum or a point of inflection. To distinguish between these the function is differentiated again to give the second differential (d^2y/dx^2) which is negative at maxima. A similar process can be applied to images. The usual way to calculate a gradient of an image is to apply a pair of Sobel filters (see figure 1) to the image. Figure 2 shows an example of how this pair of filters are convolved with an image to give an X and Y gradient, which are then combined by using the sum of the absolutes to give the gradient of the image.

In the upper left hand section of the figure we can see the input image which contains a peak in the centre of the image. In the upper right hand corner we see the Y or vertical gradient of the input image. The lower left hand section contains the X or horizontal gradient of the image. In the lower right hand corner is the sum of the absolutes of the two gradients. Both the horizontal and vertical gradients show a zero crossing at the centre, however this information is lost in the sum of the absolutes of the gradients.

To discover if the zero crossings found in the gradients are maxima or minima it is necessary to calculate the second derivative of the image. This is done by applying the same Sobel templates to the X and Y gradients. The two second derivatives can be combined by summing

X Gradient filter				Y Gradient filter		
-1	0	1		-1	-2	-1
-2	0	2		0	0	0
-1	0	1		1	2	1

Figure 1. Sobel filters

Top-left grid:

1	1	1	1	1	1	1	1
1	2	3	4	4	3	2	1
1	2	3	4	4	3	2	1
1	3	4	5	5	4	3	1
1	3	4	6	6	4	3	1
1	3	4	6	6	4	3	1
1	3	4	5	5	4	3	1
1	2	3	4	4	3	2	1
1	2	3	4	4	3	2	1
1	1	1	1	1	1	1	1

Top-right grid:

-	-	-	-	-	-	-	-
-	4	8	11	11	8	4	-
-	3	4	4	4	4	3	-
-	3	5	7	7	5	3	-
-	0	1	3	3	1	0	-
-	0	-1	-3	-3	-1	0	-
-	-3	-5	-7	-7	-5	-3	-
-	-3	-4	-4	-4	-4	-3	-
-	-4	-8	-11	-11	-8	-4	-
-	-	-	-	-	-	-	-

Bottom-left grid:

-	-	-	-	-	-	-	-
-	6	6	3	-3	-6	-6	-
-	9	8	4	-4	-8	-9	-
-	11	9	5	-5	-9	-11	-
-	12	11	7	-7	-11	-12	-
-	12	11	7	-7	-11	-12	-
-	11	9	5	-5	-9	-11	-
-	9	8	4	-4	-8	-9	-
-	6	6	3	-3	-6	-6	-
-	-	-	-	-	-	-	-

Bottom-right grid:

-	-	-	-	-	-	-	-
-	10	14	14	14	14	10	-
-	12	12	8	8	12	12	-
-	14	14	12	12	14	14	-
-	12	12	10	10	12	12	-
-	12	12	10	10	12	12	-
-	14	14	12	12	14	14	-
-	12	12	8	8	12	12	-
-	10	14	14	14	14	10	-
-	-	-	-	-	-	-	-

Figure 2. First derivative of an image using a Sobel filter

the absolutes of the two gradients. If the sign of the output is required then the sign of the sum can be set to the sign of the largest absolute value of the two gradients.

By combining the information obtained from the two steps in the gradient process peaks can be detected. Areas that have a low gradient and a high negative second derivative are easily selected after processing; thresholding can be applied to limit the areas selected.

This makes use of the fact that, following aggregation, the population surface is relatively smooth and that urban areas are maxima of the surface. This means that no assumptions about the maximum of population density of the urban area are required and that both large cities and smaller towns can be detected by the same process. As can be seen in figure 3 a range of city sizes from London to Aberdeen can be seen, in the map of the second derivative. This figure was constructed by calculating the second derivative of the population density surface as described in above and then applying a threshold that removed all cells with a value greater than -45.

The next step was to convert the boxes extracted and to determine which town was within the box. A gazetteer sorted by population size was used in conjunction with a point in polygon program to determine the largest town in each box. The output was then modified by hand where two large urban areas had been conglomerated by the exaction program, for example Leeds and Bradford are a single output box, but only Leeds was inserted by the point in polygon method, so Bradford was inserted by hand. Most of the urban areas selected by this method are single distinct locations, however some are connurbations formed by the merging of two large towns or cities, such as Birmingham and Wolverhampton and Leeds and Bradford. Any area with an area of less than one kilometre square was eliminated from the study since it was felt that these areas would be too small to show any internal structure even at the 200 metre resolution. Bounding boxes for each area were then calculated so that census data could be extracted at a 200 metre resolution for the urban areas.

Once features have been extracted the process moves on to the detection of patterns within the features and the classification of the features discovered.

At this stage it is important that possible matches between an observed urban area and a theoretical template or another observed area are not overlooked due to changes in orientation, size or possible distortions caused by local topology. It is therefore necessary that the second stage of the process makes corrections for these differences without losing sight of the underlying structure of the area.

Figure 3. Second Derivative of GB Population, thresholded at -45

To achieve this aim it is possible to make use of many techniques that are based on template matching. This technique takes a pattern template that is being sought in the image and compares it to the image making use of a metric such as the sum of the absolute errors or the sum of the square of the errors between the template and the image. When an image closely resembles the template then these measures will be small whereas if the template is very different from the image then the error measures will be larger.

There are many problems that have to be overcome to make this technique work well in a general case. For instance if the pattern to be detected is non-symmetric then consideration must be taken as to how to compare the template in different orientations or to preproccess the image in such a way that the asymmetry is always in the same orientation (Schalkoff 1989). It is also necessary to consider the effects of scaling on the image under consideration since it is nearly always important to recognise the target pattern regardless of its size.

In this study the Fourier–Mellin transform will be used. This algorithm uses a two dimensional fast fourier transform (FFT) to give the spectral magnitude of the image which is translation invariant. This is then transformed to polar coordinates, and the radial element is logarithmically scaled. A second two dimensional FFT (a Mellin transform) produces a rotational, translational and scale invariant template (for further details see Schalkoff (1989), Turton (1997)).

The approach consists of calculating the Fourier–Mellin Invariant (FMI) for each image to be matched and then correlating the FMI descriptors. A high correlation indicates that the two images being compared are similar. Since the FMI is rotation, scale and translation invariant there is no need to carry out any extensive pre–processing before classifying images.

Since the method works on images, which are most conveniently stored as two dimensional arrays the parallel version was implemented using high performance Fortran (HPF). This has the advantages that it is easy to divide the arrays across processors and is not substantially different from the serial implementation of the code. The choice of HPF did not incur the performance penalties normally associated with the use of high level parallel programming methods, since the actual 2D FFTs were carried out using the optimised Cray library routine PCCFFT2D.

Before any patterns of social structure can be considered it is necessary to determine a variable or series of variables available in the 1991 census of population that can be considered to be a proxy for social status. There have been many attempts in recent years as to what constitutes deprivation both for an individual and for an area, and how best to measure this using variables that are collected in the censuses of population. It is generally recognised that it is important to avoid focusing on groups who are vunerable to deprivation but not of themselves deprived. The question also arises are the deprived always poor? or are the poor always deprived? (Townsend and Gordon 1991). There has also been considerable work carried out on the links between deprivation and ill health (e.g. Campbell, Radford, and Burton (1991), Morris and Cairstairs (1991), Jarman (1984)). This study however requires a less specific definition since a proxy for social class is required rather than a specific indicator of deprivation. So for the purposes of this study it was decided to use a combination of local authority rented households, unemployment, overcrowded households (more than one person per room) and households lacking a car and lacking central heating. The variables were normalised so that the highest cell value in each city is 100 and the remainder are scaled relative to the value of that cell.

To generate a pattern of the social structure of the selected urban areas the normalised variables are added together to give a single "deprivation" variable. This can be seen for Leeds in figure 4. The pattern can be clearly seen with the concentration of low social status seen to the north east of the city centre and a smaller centre of low status seen in the Headingly area to the north west of the centre. The "deprivation" variable surface was then created for each of the 129 urban areas extracted above. Once this operation was completed the Fourier–Mellin invariant transform was calculated for each urban area.

The first step for the pattern recognition exercise was to investigate the theoretical patterns of urban social structure discussed above. This was undertaken by creating synthetic patterns that matched the structures that were predicted by theory. Figure 5a shows a simple radial decay from high deprivation in the centre to low deprivation at the outskirts, figure 5b shows a stepped radial decay pattern as proposed by Burgess (1925) and figure 5c shows one of the sectoral models used. Sectoral patterns with 4, 6 and 8 sectors were used, as discussed by Hoyt (1939). For each of the synthetic patterns used the Fourier–Mellin invariant transform was calculated.

The next step was to discover which, if any, of the selected urban areas had any similarity with the theoretical models by calculating the root mean square error between the FMI of each urban area and each theoretical pattern.

The results of this process were then ranked by level of similarity. It was found that the stepped radial decay model was most similar to any of the urban areas tested, but that the simple linear decay model was also similar to a similar group of urban areas. These towns and cities are listed in table 1. The figure in brackets that follows each town name is the level of similarity between that town and the model. As can be seen the same towns can be seen to match the stepped and linear models of radial decay although the levels of similarity are better for the linear models compared to the stepped model. This is to be expected due to the similarity of the two theoretical models. However the sectoral model is similar to a very different group of towns and the levels of similarity are much worse than for the radial models.

The towns that are most similar to the radial models all tend to be small, inland towns, the exceptions being Doncaster, Scarbourgh and Corby. It seems likely that the majority of these towns grew slowly around markets before the industrial revolution and have retained their compact shape since.

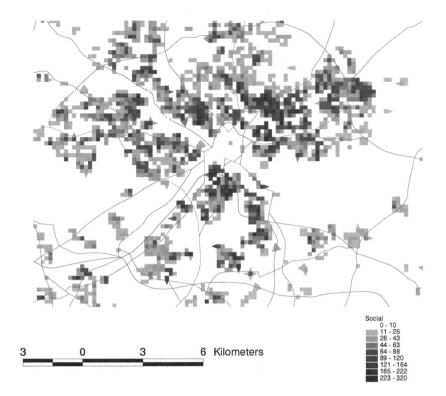

Figure 4. Social Structure of Leeds, see text for details

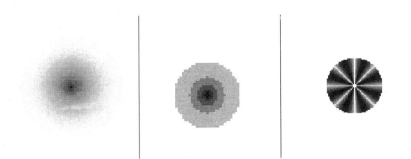

Figure 5. Three synthetic cities used for pattern detection: a) Simple radial decay; b) Stepped radial decay; c) Sectoral Model

Table 1. Towns and cities ranked in order of similarity to
theoretical models (see text for discussion)

Stepped Radial Decay Model	Linear Radial Decay Model	Sectoral Model
Skipton (0.43)	Corby (0.48)	Sheffield (3.80) 8
Corby (0.44)	Skipton (0.50)	Leicester (3.28) 8
Harrogate (0.46)	Harrogate (0.54)	Coventry (3.83) 8
Doncaster (0.49)	Doncaster (0.57)	Nottingham (3.97) 6
Blackburn (0.51)	Blackburn (0.59)	Bristol (4.13) 4
Colchester (0.52)	Colchester (0.61)	Weston–Super–Mare (4.23) 4
Scarbourgh (0.56)	Scarbourgh (0.64)	Derby (4.35) 8
Chesterfield (0.57)	Redditch (0.65)	Hull (4.36) 4
Gateshead (0.57)	Gateshead (0.66)	Bury (4.39) 8
Redditch (0.58)	Chesterfield (0.67)	Plymouth (4.40) 6
Wakefield (0.59)	Wakefield (0.67)	Aberdeen (4.42) 6
Macclesfield (0.60)	Macclesfield (0.68)	Cardiff (4.50) 8

Table 2. Groups of similar urban areas

Group 1	Group2	Group3	Group 4	Group 5	Group 6	Group 7
Airdrie	Motherwell	East Kilbride	Dundee	York	Aberdeen	Cardiff
Woking	Grimsby	Southport	Preston	Darlington	Petersfield	Portsmouth
Macclesfield	Scunthorpe	Bath	Reading	Hartlepool	Preston	Coventry
Aylesbury	Stevenage	Accrington	Northampton	Edinburgh	Warrington	Farnborough
Chester	Barrow–in–Furness	Crewe	Newport	Gateshead	Reading	Nottingham
High Wycombe	Runcorn	Lowestoft	Petersfield	Carlisle	Swansea	
Corby	Leamington Spa	Bracknell	Ipswich		Northampton	
Scarbourgh	Bedford	Hastings	Wigan		Dewsbury	
Harrogate	Barnsley	Basingstoke	Lowestoft		Leicester	
Skipton	Chesterfield	Ellesmere Port			Peterborough	
Redditch	St Albans	Mansfield				
Aldershot	Oxford	Halifax				

In contrast, the sectoral towns are larger industrial towns and also include a number of seaports (Bristol, Weston–Super–Mare, Hull, Plymouth, Aberdeen and Cardiff). There is a possibility that this was caused by the similarity of a half radial model to a sectoral model. This possibility was tested by calculating the cross correlation of a half radial model with the sectoral models and the radial model. The sectoral model had a correlation coeficiant of 9.20 with the half radial model while the radial model had a correlation coeficiant of 5.93. If the correlation was spurious it would have been expected that the sectoral model would have been more similar to the half radial model than the radial model was to the half radial model. Therefore the patterns are really sectoral and it is probable that their location on the sea has forced a more sectoral growth pattern since growth is constrained to be away from the sea and also that seaports are more likely to require distinct transport routes to their hinterland. The remaining industrial towns in the sectoral group are also likely to have developed strong transport links to allow both the import of raw materials and the export of finished goods to the surrounding areas.

To investigate the hypothesis that urban areas in Great Britain show similarity amongst themselves the similarity of each of 129 selected areas and all the other urban areas were calculated. Then for each town or city it is possible to select the other urban areas that the most similar to this town. From these lists it is possible to determine groups of towns and cities that are more similar to each other than to other groups.

The seven groups with the highest internal similarity are shown in table 2. In each group the largest difference in similarity between two urban areas within the group is 0.5 and in most cases is much smaller. Therefore it can be seen that the similarity levels within these groups is much smaller than the levels reported in the experiment above that compared theoretical models and urban areas where the lowest level of similarity found was 0.43.

Group 1 contains mostly small inland towns with the exception of Scarbourgh which is on the coast and Chester which is larger than the remainder. The group appears to be a mix of (ex–)industrial towns (Corby, Macclesfield), and towns where service industries predominate (High Wycombe and Skipton). Group 2 is a more homogeneous group being predominately towns that have grown up around a single industry: Motherwell and steel, Grimsby and fishing, Runcorn and chemicals, Barnsley and coal. The exceptions to this are St. Albans and Oxford, although Oxford does have a car manufacturing area on the outskirts of the historic town. Group 3 is a mixture of new towns (East Kilbride and Bracknell) which were either planned or underwent less structured but rapid growth in the 1980s, ports such as Lowestoft, Ellesmere Port and Hastings, and northern manufacturing towns (Accrington, Crewe, Mansfield). The common feature seems to be that all the towns in this group under went some sort of rapid growth at some time in their history. Group 4 seems to be a very mixed group of towns with no common feature readily apparent, whereas group 5 is a group of larger towns and cities though none of them is the largest in its immediate area. They are also all towns with major railways passing through them which may account for the similarity of social structure between them. Group 6 is a grouping of smaller towns which have not experienced the rapid growth of some of the other groups but that have grown steadily from markets and ports that date back to medieval times. The only exception to this is Warrington which is a major industrial centre. Group 7 is the smallest group and comprises cities that were extensively rebuilt after the second world war; they are also all transport nodes and may have had their development shaped by these two factors.

CONCLUSION

The aim of this work was to investigate the potential for the use of computerised generalisation as a solution to the problem of large geographic datasets. It sought to demonstrate that image processing methods that had been developed for the use of other disciplines could be successfully transfered to geography and to investigate the use of computer vision methods in the specific context of the social structure of urban areas.

The first step has been demonstrated how, after some suitable preprocessing of the raster dataset, the SURPOP population surface could be successfully segmented into urban and rural areas. This led to the extraction of 129 urban areas for Great Britain. The second stage showed how the urban areas of Great Britain could be compared to theoretical models of social structure that were developed predominantly in the United States. It should also be noted that this method of analysis allows the researcher to ignore many of the problems that have occurred in previous statistical studies of social structure, such as where is the centre of a city? where are the boundaries of the city? How wide should the rings in the model be? and such like.

Therefore, in conclusion, it appears that large geographical datasets can be processed by computer to extract both simplifications of the dataset and to carry on beyond this point to extract new concepts which can lead to new theory formulation with relatively little input from the human researcher. This process, if widely adopted, could prevent geographers drowning in the sea of new data that is being collected both by industry and new satellites every day.

ACKNOWLEDGEMENTS

This work was partly funded by EPSRC grant GR/K43933, who also provided the Cray T3D time used. I am grateful to David Henty of EPCC for his help in parallelising the FMI algorithm and to him and David Martin for their helpful comments on an earlier version of this work. I must also thank Stan Openshaw for his valuable input to early stages of the work. The census data used in this project is crown copyright and was purchased for academic use by the ESRC and JISC.

REFERENCES

Ballard, D. and Brown C., 1982, *Computer Vision*. Prentice Hall, NJ.

Bourne, L., 1971, *Internal Structure of the City*. Oxford University Press, New York.

Bracken, I. and Martin D., 1989, The generation of spatial population distributions from census centroid data, *Environment and Planning A* 21:537-43.

Burgess, E., 1925, The Growth of the City, in: R. Park, E. Burgess, and R. McKenzie eds., *The City*, pp37-44. Chicago.

Campbell, D., Radford J., and Burton P., 1991, Unemployment rates: an alternative to the Jarman index? *British Medical Journal* 303:750–755.

Hoyt, H., 1939, *The Structure of Growth of Residential Neighbourhoods in American Cities*, Washington.

Hurd, R., 1903, *Principles of City Land Values*, New York.

Jarman, D., 1984, Underprivileged areas: validation and distribution of scores. *British Medical Journal* 289:1587-1592.

Morris, R. and Cairstairs V., 1991, Which deprivation? A comparison of selected deprivation indexes. *J. of Public Health Medicine* 13:318-326.

Niblack, W., 1986, *An Introduction to Digital Image Processing* 2nd edn., Plenum Press, New York.

Openshaw, S., 1994, A Concepts–rich Approach to Spatial Analysis, Theory Generation, and Scientific Discovery in GIS using Massively Parallel Computing, in: M. Worboys ed., *Innovations In GIS*, pp123-138, Taylor and Francis, London.

Schalkoff, R., 1989, *Digital Image Processing and Computer Vision*, Wiley and Sons, New York.

Townsend, P. and Gordon D., 1991, What is enough? New evidence on poverty allowing the definition of minimum benefit, in: M. Alder, C. Bell, and A. Sinfield eds., *Sociology of Social Security*, Edinburgh University Press.

Turton, I., 1997, Application of Pattern Recognition to Concept Discovery in Geography. M.Sc., School of Geography, University of Leeds.

HIGH PERFORMANCE COMPUTING IN BANKING

J.A. Keane

Department of Computation, UMIST, Manchester, UK.

INTRODUCTION

There has long been a distinction between engineering and science applications with their compute-intensive requirements, and commercial applications with their data-intensive requirements. Engineering and science applications are characterised by the term *number crunching*, and the use of FORTRAN; commercial applications are characterised by the term *data processing*, and the use of COBOL and database systems.

Traditionally, high performance parallel computing (HPC) has been synonymous with compute-bound engineering and science applications. Whilst such applications remain important, there is an increasing realisation that this market for HPC has not grown since the mid-1970s, and if anything, has declined with less investment in military applications.

The commercial sector has traditionally used mainframes - over 80% of the world's mainframe sales are to commercial organisations - and is viewed as being conservative when considering new technology, whereas engineering/science has used whatever experimental technological improvements are offered. A consequence is that HPC, and particularly systems with very large numbers of processors, has largely been restricted to the engineering market. The potential HPC commercial market is estimated to be 2 to 5 times larger than the engineering market (Wray, 1995).

Commercial applications require increasingly complex *information* processing and larger data requirements, at the same time the rate of increase in mainframe performance is both declining and becoming more costly. In contrast, parallel HPC systems offer higher processing capabilities and larger upgrade potential: mainframe performance increases at the rate of 15% per year, whereas commodity processor power doubles every 18 months; the mainframe upgrade path is 20:1, the parallel systems upgrade path is 1000:1 in terms of processor power, memory size, and disk capacity (Teskey, 1996).

Earlier work has considered the use of parallel and distributed systems in financial organisations primarily to support operational data processing (Keane et al., 1992, 1995; Keane, 1993, 1996, 1998). In this paper the potential for HPC in banking applications is considered primarily from the emerging information requirements of decision support that impact performance, and in particular considers issues related to data mining for customer profiling.

The paper is structured as follows: the general area of banking applications is discussed; the utility of HPC in such areas is considered; following this background, the HYPERBANK project, investigating data mining for customer profiling in the banking sector, is discussed; finally conclusions are presented.

BANKING APPLICATIONS

The ever-increasing importance of IT to the banking and financial sector in terms of profitability, service provision, globalisation, and responsiveness to market change, increasing competition and customer needs is well illustrated in a series of recent papers (Teskey, 1996; Channon, 1997; Holland et al., 1997; Kulijis and Scoble, 1997; Sandbiller et al., 1997; Marshall and Nolan, 1997).

As with all application areas, there are two fundamental reasons for banks to be interested in HPC:*increased availability* and *higher performance*.

Commercial applications are business-critical and thus require very high availability. High availability has traditionally been achieved by duplication of resource, hence parallel systems directly address this fundamental requirement of commercial applications. The interested reader is directed to Watson and Catlow (1995) for details on how a commercial parallel system has addressed such high availability requirements.

The requirement for high performance, and the focus of this paper, is also related to the business-criticality of commercial information systems. Such systems have for decades provided support for the operational data activity of banks, such as on-line transaction processing. In recent years, with the impact of greater competition, business margins are becomingly increasingly tight. In such a climate, banks are addressing ways to more efficiently utilise their operational data to more effectively do business.

These challenges are summarised by the following: "banks are struggling to respond to the rapidly changing expectations of personal customers, the demands of regulators, the opportunities afforded by new technologies and the threat of new, potentially better-placed competitors", (*Angus Hislop, European Banking Partner, Cooper and Lybrand, The Banker 1995*).

CUSTOMER PROFILING

There is a perceived necessity that to compete in today's market, organisations need to more precisely and completely *profile* their customers. This is reflected in the following: "Marketers increasingly are recognising that past customer behaviour, as recorded in actual business transactions, is by far the best indicator of future buying patterns", (*Business Week, 1994*).

Within a banking context, customer profiling is becoming paramount, as is reflected in the following: "The first key to sustained success is a superior understanding of what tomorrow's customers want from their prime provider of financial service.", (*Angus Hislop, European Banking Partner, Cooper and Lybrand, The Banker 1995*); "Customer requirements are set to change significantly; if banks do not meet the challenge, non-banks will", (*Edmund Jensen, CEO Visa International, The Banker 1995*).

Customers represent the major opportunity for banks to respond effectively to competition and narrower profit margins. In such a market, the aim is to encourage the acquisition and retention of "good" customers, where "good" customers are defined as those who are most profitable for the bank. The first step then is to effectively segment the customer space such that the value of customers can be assessed quantitatively.

Acquiring and obtaining "good" customers requires a paradigm shift in the focus of banking information systems from an account-oriented focus to a customer-centric focus.

If this can be achieved, then a number of issues become possible, for example:

- more effective customer targeting - identify the attributes of customers that buy a certain product or products and only target that set of customers.

- prediction of customer behaviour - patterns of behaviour may be used to predict when a customer may require a product, or will cease to need a product etc;

- more accurate risk assessment, a consequence of this is that products can be priced differently for different customer segments as they represent different risks to the bank.

- better customer service - provision of information and advice that is closely associated with the individual circumstances of a customer based on their lifestyle, their life-stage, their intentions, family circumstances etc.

Ultimately certain customers may be regarded as "no risk" in which case, there is no need to charge any more for a product than the bank's profit plus internal risk. In turn, this represents a very attractive proposition to the customer. The whole process relies on having effective access to data relating to the customer - and similar customers - that can enable such.

Identifying "good" customers and relating them to products etc, may result in targeting such that a product is never offered to a customer unless the bank is prepared to accept that customer for that product if they apply, i.e. the offer is *guaranteed*, thus reducing mail costs and potential annoyance of customers who are targeted and then turned down.

In the past, given the cost of computing power and the paucity of data, only a rough analysis of statistically aggregate data was possible (Decker and Focardi, 1995). Recently, with improvements in data capture and network capacities, and the reduction in storage and processor costs, companies are able to build very large datastores. The problem is no longer the capture or storage of information, but its efficient exploitation.

Banking customer data is often in diverse sources spread across a number of databases, for example, a current account, a deposit account, a mortgage, a car loan etc. This is particularly an issue when different financial institutions merge and (attempt to) integrate their IT systems.

More fundamentally, it is difficult to provide a precise answer to the question *"what is a customer?"* , for example, does the owner of an account constitute a customer, what if a customer holds multiple accounts? One this has been addressed then other questions can be posed such as *"what are the attributes of a customer that influence them being regarded as good or bad?"* , *"what are the attributes that influence customer profitability?"* , *"what are the influences of one product on another?"* , and *"what are the attributes that influence the purchase of products by a customer?"* . A way to address these issues is to consider domain knowledge, for example, the knowledge that financial experts use to decide whether customers are good or bad risks, or whether a product is worth bringing to market etc.

What is ultimately required is a consistent user-based view of all the information concerning a customer, accessible in a timely and consistent manner. This accessibility will ultimately enable a market segmentation at the level of an individual customer.

THE HYPERBANK PROJECT

THE HYPERBANK project is funded by the EC ESPRIT Framework IV 6.8 initiative. The project involves banks from the UK, Greece and Sweden (CAPITAL BANK, the National Bank of Greece, and POSTGIROT BANK), IT suppliers from the UK, Greece and Belgium (Datel Advanced Systems, 01-Plioforiki, and Carleton Europe), and universities from the UK and Sweden (UMIST and KTH).

The aim of the project (Keane, 1997) is to provide the banking sector with the requisite toolset for increased understanding of existing and prospective customers, and better tailoring of products and services for those customers.

The planned approach to exploiting banking customer information is an holistic integration of business knowledge modelling, data warehousing and data mining, along with the enabling technology of high performance parallel computing. This has two major aspects:

- the business goals and rules that express the objectives of the banks in which context the information can be utilised;

- the pragmatic issues of accessing and manipulating this data in a timely fashion, this involves both bringing disparate data sources together - data warehousing (Mattison, 1996) - and analysing this data to determine implicit relations - data mining (Berry and Linoff, 1997).

Business models are usually derived from a number of sources such as interviews, observation, and existing procedures etc, operational data has often been neglected as a source of information for business models. This has partly been a result of the difficulty in accessing and abstracting the sheer volume of operational data.

HYPERBANK represents an interesting synergy of approaches in obtaining, storing, manipulating and utilising business knowledge. Organisational data is seen as representing the *actual* way that an organisation behaves and how it interprets its rules and processes. Information derived from operational data may be used to validate, question or improve the associated processes and rules represented in the business model; it may even derive previously unknown knowledge leading to new business processes and rules.

The focus of the rest of this section is the processing of the operational data and how high performance systems contribute; the interested reader is referred to Locopoulos and Kavakli (1995) for the approach to business modelling, and to Filipidou et al. (1998) for details of the integration of business modelling with operational data sources.

Data Warehousing

A *data warehouse* is defined as a single integrated data store which provides the infrastructural basis for information in an enterprise. Data warehousing is the application of replication and other technologies to bring data from a variety of sources into one or more collections that are designed to improve information access.

There is often potential for using high performance systems in the actual bringing together of data in this way. In addition, high performance is necessary both for accessing this data effectively, and for maintaining the currency of the warehouse with updates. The access issues are similar to those to any large dataset.

The updating and currency issues relate to how often data attributes need to be refreshed, and what are the propagation effects of these updates (sometimes termed *trigger processing*). It is estimated that above 5 gigabytes, parallel systems are necessary for effective results in data warehousing.

As mentioned earlier, to achieve effective customer profiling requires a customer-centric approach, hence rather than the present account-based operational databases, a customer-centric data warehouse is necessary. To establish such a warehouse requires the bringing together of many disparate operational databases with different scheme definitions where customer information is currently held, into a coherent whole.

As also mentioned earlier, the very concept of a customer is often unclear, and the business modelling activity aims to elicit a common understanding and representation of a customer and of customer profiling within the banking sector. As a result, the generation of the data warehouse will itself be related directly to the business model via an object repository.

Data Mining

Data Mining is the process of extracting implicit information from a data store, by the combination of various techniques, including data analysis, machine learning, knowledge based systems, and neural networks. Data mining is likely to be more effective when carried out on the data held in a data warehouse, as the data warehouse will provide clean and consistent data.

Current data mining tools do not embody the domain knowledge which could suggest appropriate derived attributes e.g. risk, profitability etc. Indeed it is unlikely that one approach to data mining would suffice for the diversity of banking concerns: Decision Support Systems may be projected in at least four distinct spaces: *Data, Aggregation, Influence* and *Variation*, and it is application-specific which is appropriate (Parsye, 1996).

Elsewhere we have discussed the types of area in which data mining of banking data could be applicable, such as customer retention, product marketing and pricing, and credit card fraud (Keane, 1996).

Previous experiences of using data mining suggest that extensive banking domain knowledge is required to steer the data mining to reduce the search space and thus the processing requirements. The integration of data mining with the business model (representing domain knowledge) is one of the aims of HYPERBANK and some progress has been made both using the business model to drive data mining and using data mining to validate and extend the business model.

The major influence of data mining on business models is via the derived rules. Most data mining results can be naturally represented as rules (e.g. classification, association), and the others appear to permit a rule-based expression at least (e.g. clustering). These rules can be validated by a domain expert as to their usefulness and, where appropriate, they can be integrated with the rules of the business model. In turn the business model rules are related to the processes and goals etc.

Data mining provides the greatest benefit when the data to be mined is large and complex. Relational database technology is optimised for on-line transaction processing and thus the data space mentioned above. The other three spaces (aggregation, influence and variation) are much more computationally complex and often require multiple scans of the dataset.

As a result, with large data sets, and in particular with large numbers of attributes, data mining becomes data and compute intensive. It is difficult to predict the performance of data mining on a large database by extrapolating from the performance on a small database, as for many data mining algorithms the required amount of computation and input-output increases more than linearly with data size (Berry and Linoff, 1997).

Parallel systems addresses both the data and compute intensive nature of data warehousing and data mining by offering increased main memory space, the capability of utilising larger numbers of disks thus providing higher input-output capacity, and higher processing capabilities.

High Performance Systems

As indicated, earlier, experience suggests that parallel processing is required for effective data warehousing and data mining. In particular, it is estimated that to convince business users to exploit data on a daily basis requires interactive response times (3-5 seconds).

Nonetheless, because of the cost and complexities of introducing parallel systems, allied to the lack of experience with HPC in the banks involved in the project, the initial data mining activity has used single (pentium) processors following the advice "don't go to parallel systems unless it is necessary".

On 10% of the customer marketing database using rule induction to classify loan accept/reject, a single processor provides adequate response times. However, on 2 million customer records representing the early settlement of loans we find rule induction taking over 24 hours on a PC.

We also wish to look at deriving association rules where relations between product sets are identified. This requires multiple scans of the entire dataset and association rules generally require exponentially more computational effort as the problem size grows (Berry and Linoff, 1997).

As a result of this necessity of using parallel systems, the project has begun to investigate the effectiveness of using IBM SP2 high performance systems for efficient data mining in three modes:

- in MPP mode, with single processor-memory nodes connected via a switch;

- in SMP mode, with 4 processors per memory nodes;

- in joint MPP-SMP mode, with multiple SMP nodes connected via a switch.

In each mode the data warehouse will be distributed across a number of disks. The experimentation will initially be done using parallel database software, with an intention to identify a parallel data mining API.

In this sense, HYPERBANK represents a proving ground for high performance computing in the potentially highly significant banking segment. In addition, as "customers" represent a pervasive aspect of almost all business segments, the project has an even wider potential impact.

CONCLUSIONS

Commercial organisations are engaging in an increasingly competitive market place, where more effective and efficient utilisation of information is becoming vital. The complexity and performance requirement of commercial applications are increasing to meet these requirements. As a consequence, HPC, with its potential for more processors and larger disk spaces, is gaining increasing significance for such applications. Indeed, such applications represent by far the largest potential market for HPC systems.

In banking and financial systems generally, there is a recognition that one area where information is available and potentially useful, is that of customer data. Customer profiling can enable better understanding of customer segments and ultimately individual customers. This increased understanding can result in better targeting of products to customers and better identification and quantification of risk, which should lead to increased market share and enhanced customer loyalty and service in a very competitive market. The HYPERBANK project is investigating the utility of HPC for banking, and specifically its use in the area of data mining for customer profiling.

ACKNOWLEDGEMENTS

The author wishes to thanks all former colleagues, and, in particular, current colleagues on HYPERBANK - this work is partly supported by ESPRIT HPCN Project 22693.

REFERENCES

Berry, M.J.A. and Linoff, G., 1997, "Data Mining Techniques", John Wiley & Sons, New York.

Channon, D.F., 1997, The strategic impact of IT on the retail financial services industry, *Proc. HICSS-30 Vol.III*, IEEE Press.

Duruma, F., 1996, On the parallel characteristics of engineering/scientific and commercial cpplications, *in:* "Parallel Information Processing", J.A. Keane, ed., Stanley Thornes Publishers, London.

Decker, K.M. and S. Focardi, S., 1995, Technology overview: a report on data mining, CSCS-TR-95-02, CSCS-ETH, Swiss Scientific Computing Center, Switzerland.

Fillipidou, D., Keane, J.A., Svinterikou S. and Murray, J., 1998, Data mining for Business process improvement: applying the HYPERBANK approach, *2nd International Conference on The Practical Application of Knowledge Discovery and Data Mining -PADD'98*; Practical Application Company, London, to appear

Holland, C.P., Lockett , A.G. and Blackman, I.D., 1997, The impact of globalisation and information technology on the Strategy and Profitability of the Banking Industry, *Proc. HICSS-30 Vol.III*, IEEE Press.

Holsheimer, M., Kerten, M. and Siebes, A.P.J.M., 1996, Data Surveyor: searching the nuggets in parallel, *in: Advances in Knowledge Discovery and Data Mining*, U.M. Fayyad *et al.* (Eds.), MIT Press, Boston.

Keane, J.A. 1993, Parallelising a financial system, J.A. Keane, *Future Generation Computer Systems*, 9(1), 41:51.

Keane J.A. et al., 1993, Commercial users' requirements for parallel systems, *Proc. of 2nd DAGS Parallel Computation Symposium*, F. Makedon et al. eds., 15-25.

Keane J.A. et al., 1995, Benchmarking financial database queries on a parallel machine, *Proc. of 28th Hawaii International Conference on System Sciences Vol II*, IEEE Press, 402:411.

Keane, J.A., 1996, Parallel systems in financial information processing, *Concurrency: Practice and Experience*, 8 (10), 757:768.

Keane, J.A. 1997, High performance banking, *Proc. of RIDE'97*, IEEE Press, 66:69.

Keane, J.A., 1998, A co-ordination framework for distributed financial applications, *Concurrency: Practice and Experience*, to appear.

Kulijs, J. and Scoble, C., 1997, Problems of management and decision making in multinational banking, *Proc. HICSS-30 Vol.III*, IEEE Press.

Loucopoulos, P. and Kavakli, E., 1995, Enterprise Modelling and the Teleological Approach to Requirements Engineering, *International Journal of Intelligent and Cooperative Information Systems*.

Marshall, C. and Nolan, R., 1997, IT-enabled transformation: lessons from the financial services, *Proc. HICSS-30 Vol.III*, IEEE Press.

Mattison, R., 1996, "Data Warehousing", McGraw-Hill, New York.

Parsaye, P., 1996, Surveying decision support, *Database Programming and Design*, pp. 27-33.

Sandbiller, K., Will, A., Buhl, H.U., and Nault, B.R., 1997, IT-enabled incentive schemes in telephone banking, *Proc. HICSS-30 Vol.III*, IEEE Press.

Teskey, F.N., 1996, Parallel processing in a commercial open systems market, *in:* "Parallel Information Processing", J.A. Keane, ed., Stanley Thornes Publishers, London.

Watson, P. and Catlow, G., 1995, The Architecture of the ICL GOLDRUSH MegaSERVER, *in:* "Advances in Databases ", C.A. Goble and J.A. Keane, eds., Lecture Notes in Computer Science Vol. 940, Springer Verlag, Berlin.

Wray, F., 1995, "High Performance Computing: Status and Markets", UNICOM Applied Information Technology Report, UNICOM Publishers, London.

LEGACY SYSTEMS - THE FUTURE OF HPC

J.A. Keane[1], M.F.P. O'Boyle[2], R. Sakellariou[3]

[1]Department of Computation, UMIST, Manchester, UK.
[2]Department of Computer Science, University of Edinburgh, UK.
[3]Department of Computer Science, University of Manchester, UK.

INTRODUCTION

A *Legacy Information System* (IS) is defined as any IS that significantly resists modification and evolution to meet new and constantly changing business environments (Brodie and Stonebraker, 1995). Legacy systems are exemplified by 10s of millions lines of COBOL. There are around 200 Billion lines of COBOL code operating in the world, representing 30 million person years of development (McFarland, 1996). Single applications of over 6 million lines are not considered unusually large (Sayles, 1996).

Legacy IS maintenance takes 80-90% of the IT budget (Brodie and Stonebraker, 1995). 40% of maintenance is perfective, of this approximately half, i.e. 20% of all maintenance, is general optimisation (performance improvement).

High performance parallel computing (HPC) is usually seen as the way that performance improvement can be achieved. There is an increasing appreciation that if HPC is ever to become common technology it needs to address commercial applications (Keane et al., 1993; Fox, 1994; Darema, 1996; Lilja, 1997). When commercial applications are discussed in the context of HPC, database performance is focused upon. Despite the money, time and effort involved in their optimisation, there has been little discussion of the potential for executing legacy systems, and hence COBOL programs, in parallel.

This paper considers the requirements of COBOL applications by addressing typical code examples and considering how these requirements can be addressed by making use of existing parallel compilation technology. The structure of the paper is as follows, the present almost non-existent relation between COBOL and HPC is discussed, characteristics of COBOL applications are considered, a technical outline is presented considering different styles of COBOL loops and how existing compiler technology could be exploited for each, and finally the conclusions are presented.

COBOL AND HPC - THE PRESENT

Various tools support migration of COBOL applications to HPC systems. There is potential benefit in overall throughput of programs by running the parallel system as n different processors. However, single applications still run on one processor. In practice, applications moved from IBM's MVS to UNIX usually require more execution time. A program that makes database calls may exhibit speedup as the calls may execute in parallel; if a program does not - the vast majority - it may exhibit parallel speed-down as the files accessed may be partitioned onto a number of disks causing increasing latency*, and there is no parallel execution to make use of the increased availability of data.

Commercial users who ask for a migration path whereby their existing legacy COBOL systems - without database calls - can exploit a parallel system, are told that they can be re-written to call the database or the underlying operating system to execute user created threads. For example, MICROFOCUS offer user callable APIs and syntax for direct support of user level threading. MICROFOCUS also have a COBOL/SQL transparency system that enables applications to access relational databases using standard COBOL I/O syntax. However, this "rewriting solution" ignores at least two fundamental issues:

- commercial application programmers have expertise in COBOL and (usually) in the application domain in which they work; in general they do not have system-level or parallelisation skills;

- the nature of legacy systems means that many systems that are performance-bound are not well understood; they work, but defy attempts to modify them, and remain crucial to the organisation.

It is unrealistic to parallelise legacy systems in this way: The sheer volume of code prohibits such rewriting, and where attempts have been made there are more reports of failure than success (MacFarland, 1996). In particular, rewriting functioning business-critical applications is extremely high-risk.

The sheer number of COBOL programmers - 3 million - mitigates against retraining, allied to the realisation that these programmers' application knowledge represent significant investment and company knowledge (MacFarland, 1996; Sayles, 1996). Staff and installed systems represent massive investment and major assets for organisations. For most commercial IT departments the major investment is in staff skills, followed by investment in legacy applications (Teskey, 1996). In addition, a significant amount of new code is also being written in COBOL to address increasing demands for management information and decision support. As a result it is likely that COBOL will retain a primary position in commercial applications for many years (Bradley, 1996).

One solution is to apply compiler-based auto-parallelising techniques to COBOL as are currently available for FORTRAN, SQL and object-oriented languages. The data manipulation aspects of, and correspondence between, SORT and SEARCH in SQL and COBOL is obvious, as is the potential for using concurrent object techniques for the emerging COBOL-97. In the same way techniques to auto-parallelise FORTRAN DO loops can be applied to COBOL PERFORM loop.

Applying auto-parallelising techniques in this way is very attractive for commercial users as their software investment in COBOL is much larger than the corresponding investment in FORTRAN, and even modest benefits will have great impact.

CHARACTERISTICS OF COBOL APPLICATIONS

Analysis suggests that the two areas of standard COBOL that take up most run-time are input-output (access to disk) and loops (formed by PERFORM and GO TO

---

*The cost of waiting to get access to an item held in a remote memory or on disk is termed latency.

statements). Input-output is most crucial, however, once that is addressed there must be parallelism within the code to exploit the I/O parallelism.

Most COBOL applications fit into one of two categories:

1. On-line transaction processing (OLTP): where inter-transaction parallelism is exploited to allow several instances of the program to run concurrently (subject to database locks etc.) on independent transactions.

   OLTP programs are typically executed in parallel by the Transaction Processing monitor (e.g. CICS, Tuxedo, TPMS) within which they are developed and executed. In these cases intra-transaction parallelism often comes as a by-product of the underlying RDBMS. As a result, there is little performance improvement likely for such applications. Nonetheless, there is potential for improvement.

2. Batch: where the larger programs tend to handle serially a large number of records, subjecting them to similar processing. This is the read-modify-write loop used in many COBOL programs. Analysis suggests that most execution time is spent within these loops, as a result. they represent a major target for performance-enhancement.

   For many organisations, the overall runtime of their batch programs is a major operational issue. It means they need to size their systems on the basis of single stream performance rather than aggregate performance and to avoid this they often split work (manually) into separate runs (e.g. by initial letter of customer surname) which can be executed concurrently. In many case this "batch processing window" takes up all the time available between OLTP work. A relatively small improvement here would be significant.

The similarity between database processing and COBOL applications is indicated by their access to large datasets held on disk. Techniques used by parallel database implementations to minimise disk access by caching, and using distributed file-stores to increase concurrency will be of use for parallel COBOL implementation.

COBOL Program Characteristics

An analysis of a COBOL benchmark (Iliffe, 1978) indicated that seven verbs accounted for 95% of executed statements, while the same seven accounted for almost 90% of stored statements. These verbs are MOVE, IF, GO TO, ADD, PERFORM, READ and WRITE, which can be grouped as follows:

- memory access: MOVE, IF and ADD, with 70% of both static and dynamic usage;

- loops: PERFORM and GO TO, with 18% of dynamic and 27% of static usage;

- I/O: READ and WRITE, with 7% of dynamic and 5% of static usage.

Loop and I/O statements together cover 25% of execution time. Obviously within loops a large percentage of time is spent on memory accesses.

TECHNICAL OUTLINE

There has been much work parallelising both query languages and FORTRAN. The database work provides exactly a performance-enhancing migration route to HPC.

While there remains much to be done to achieve the performance of hand-tuned implementations, auto-parallelising FORTRAN compilers can achieve significant speed-up without user intervention. FORTRAN auto-parallelisation has been the subject of research for over twenty years, following the trend in computer architecture from vector supercomputers to massively parallel distributed memory architectures to today's symmetric multiprocessors (Banerjee et al., 1993; Wolfe, 1996).

There has been comparatively little work investigating the auto-parallelisation of COBOL. The possibility of using SPMD parallelism has been discussed (Darema, 1996), and an approach using COBOL-97 to develop independent objects by wrapping up existing code and using a client-server architecture for increased performance (Flint, 1996). Other work includes (Richter 1993), and the precursor of this investigation (Sakellariou and O'Boyle, 1996).

Many COBOL legacy systems typically have unstructured control-flow. Much effort has gone into the re-structuring of such applications to enable easier maintenance (Miller and Strauss, 1987; Sneed, 1991; Haimut et al., 1995). This re-structuring effort essentially provides rules to turn unstructured code into structured code. Such re-structuring rules can also help to make the code more amenable to parallelism.

Many performance developments in computer architecture, e.g. pipelining and vector processors, have been driven by the engineering sector's need for higher computation speeds. As a result, in recent years, the focus for FORTRAN compilers has been ensuring that data is made available to exploit these features. Optimising compilers for scientific applications are predominantly concerned with utilising main memory and cache. A disk is just an extra element in the memory hierarchy: cache, on-processor, off-processor, disk, with increasing latency. The read to disk is no different conceptually, though slower, to the read from a remote memory. Indeed parallel FORTRAN compilation is beginning to address disk locality as an important issue.

Most current research in automatic parallelism is targeted around FORTRAN loop structures: parallelism resulting from repeated execution of the same instructions on different data: data parallelism. A different form is when different parts of a program can be executed concurrently: task parallelism.

Data Parallelism

In the context of COBOL applications, data parallelism can be distinguished into *file parallelism*,where a program runs in parallel against a number of files, and *record parallelism*, where different records of the same file can be processed in parallel. In general by viewing COBOL files as arrays of records - the first element of an array equates to the first element of a file - many of the techniques for FORTRAN array parallelism can be adapted (Sakellariou and O'Boyle, 1996).

Two statements can be executed in parallel only if there is no relationship between them which constrains their execution order. Such constraints are identified by means of a data dependency analysis (Banerjee et al., 1993). It is important to identify dependencies between different loop iterations. If iteration I1 precedes iteration I2 then three types of data dependence may exist:

- a data flow dependence, where a variable written in I1 is read in I2;

- a data anti-dependence , where a variable read in I1 is written in I2;

- a data output dependence, where a variable is written in both I1 and I2.

A frequent occurrence in COBOL applications is that of updating a master file with the contents of a transaction file; this of course involves reading from two files and writing to one. Consider the following code: record parallelism can be exploited via repeated execution of the UPDATE paragraph.

```
    SET EOF TO FALSE
    PERFORM UPDATE UNTIL EOF
    ...
UPDATE.
    READ TRANS-FILE AT END SET EOF TO TRUE
    END-READ
    IF NOT EOF
        (... read master according to trans-rec; update master fields)
        WRITE MASTER-REC
    ...
```

The fields of the master record define a data output dependency as they are computed in any two iterations of the loop. By applying the privatisation technique (Banerjee et al., 1993), where each processor has its own copy of the variables, the output dependency can be removed. Essentially the problem relates to whether a master record written in an earlier iteration may be read in a subsequent iteration; this is eliminated if no two transaction records update the same master record. If this does not hold, then updates to the same master record should be executed on the same processor or synchronisation is necessary. The transaction file is often sorted before update, in this case it can be split into sub-files, no two sub-files can refer to the same master record. In general, detecting parallelism based on the content of data is difficult.

Not all COBOL loops have independent iterations, for example many have shared variables (checks, totals, sequence numbers, controls etc) for which sequential semantics need to be preserved. FORTRAN dependency analysis can be used to determine what ordering has to be enforced for sequential consistency, and the query decomposition approaches in SQL, where sub-query results are correlated by a master query, can be used.

As "what-if" analysis and forecasting models become more important to decision support and information management, the amount of computation within each iteration will be substantially more complex than above: in this case many computation optimisations may be useful.

Re-Structuring, Task Parallelism and COBOL-97

It may not always be possible to detect directly the parallelism present in programs. Consider again the example of a transaction file and a master file, and an alternative coding:

```
    READ TRANS-FILE AT END SET EOF TO TRUE
    PERFORM UPDATE UNTIL EOF
    ...
UPDATE.
    (... read master according to trans-rec; update master fields)
    WRITE MASTER-REC
    READ TRANS-FILE AT END SET EOF TO TRUE
    ...
```

Now the fields of the transaction record show a data flow dependence as they are being read at the iteration preceding the one in which they are used. To uncover the parallelism the code must be re-structured to the earlier form; similar transformations have been studied in the FORTRAN context (Banerjee et al., 1993).

Re-structuring may involve a series of transformations to increase the clarity of the source code, such as the removal of GO TO statements. In legacy COBOL programs, loops are often effected by GO TO backward branches.

Memory Hierarchies

In reality, COBOL files and tables are too large for a single memory, hence data required by one processor may be on disk or held in a remote memory in a parallel system. One way of handling of memory hierarchies is to use latency tolerance: hide latency by overlapping communication with computation. Pre-fetching is the basic technique. For COBOL if required data is held in a remote memory pre-fetching can be used. The more interesting and applicable case is to generalise this approach to handle disk access.

The following generic COBOL loop is ubiquitous in many applications: it reads every record in a salary file, updates each record and writes it back to the file. This style of loop is performance-bound on the read-from and write-to disk:

```
PERFORM UNTIL EOF
    READ SALARY-FILE
    AT END SET EOF TO TRUE
    NOT AT END
        PERFORM PROCESS-SALARY-REC ...
        WRITE SALARY-REC
    END-READ
END-PERFORM
```

Adding pre-reading, this becomes:

```
READ SALARY-FILE INTO SALARY-REC(1) AT END ...
PERFORM VARYING I FROM 2 BY 1 UNTIL EOF
    READ SALARY-FILE INTO SALARY-REC(I) AT END ...
    NOT AT END
        PERFORM PROCESS-SALARY-REC (I-1)
    WRITE SALARY-REC (I-1)
    END-READ
END-PERFORM
```

This loop peeling is similar in effect to software pipelining used in processors to reduce access latency to cache. In particular, when disk I/O is considered, there are techniques in the parallel database community and the shared-disk and shared-nothing models (DeWitt and Gray, 1992; Miller et al. 1995). Other techniques that can be used involve latency avoidance, such as exploiting temporal and spatial locality.

Searching and Sorting

Many commercial applications involve sorting and searching. SEARCH is part of the underlying database functionality although in many cases, particularly in the MVS community, COBOL applications still use raw or indexed-sequential files rather than a proper database. SORT is a valuable problem to parallelise because sizes are such that it invariably involves backing store, usually disk.

The correspondence between SORT and SEARCH on COBOL tables and on SQL relations is obvious, and there is much previous work in this area from the RDBMS community, hence where possible, calls will be planted to the underlying RDBMS (Kumar et al., 1995).

CONCLUSIONS

The argument presented here is not that users require their COBOL applications to run across thousands of processors, or that such applications would generally benefit from such systems. The argument is that now and for the foreseeable future the investment in legacy systems and the lack of a performance-enhancing migration route effectively limits the spread of HPC into the commercial sector.

It appears the only cost-effective way to address these issues is to provide automatic parallelisation of legacy COBOL applications to exploit HPC. Ultimately these improvements are likely to alleviate much of the need for optimisation in software maintenance thus allowing programmers to focus on improved decision support and information management.

It is also important to recognise that new commercial applications are constantly being developed and new COBOL standards are emerging. It is important to further note that many fourth-generation tools generate COBOL code.

The COBOL-97 standard, incorporates objects. Partitioning an application into encapsulated objects should give a natural basis for decomposition of an application for concurrent execution. Initial analysis suggests that MICROFOCUS, with OBJECT COBOL, have adopted something similar to the SMALLTALK Object model. This means that classes have an active and singular role at run-time which makes partitioning more difficult. From the perspective of parallel implementation, initial analysis suggests that OBJECT COBOL needs some re-design to make it distributable and so enable the natural partitioning for concurrency.

The investigation discussed here hopes to proceed with the development of the outlined techniques and show automatic performance improvement for COBOL legacy systems using HPC. In addition, the demonstration of technologies and approaches that show the performance-enhancing capability of parallel HPC systems should increase the commercial sector's acceptance of HPC.

ACKNOWLEDGEMENTS

The authors wish to thank all colleagues who have commented on the work, in particular, the partners in the REAPER and PARCOB projects, and Nic Holt of ICL.

REFERENCES

Banerjee U. et al., 1993, Automatic Program Parallelisation, *Proc. of the IEEE*, 81(2), pp. 211-243.
Bradley, J., 1996, COBOL in 2010: mainstream or memory, *American Programmer*, September.
Brodie M.L. and Stonebraker, M., 1995, "Migrating Legacy Systems", Morgan Kaufmann Publ., San Francisco

Darema, F. 1996, On the Parallel Characteristics of Engineering/Scientific and Commercial Applications, *in:* "Parallel Information Processing", J.A. Keane, ed., Stanley Thornes Publishers.

DeWitt D. and Gray, J., 1992, Parallel Database Systems, *CACM* 35(6), pp. 85-98.

Flint, E.S., 1996, COBOL Legacy Programs Serve the Future, *American Programmer*, September.

Fox, G.C., Williams., R.D., Messian, P.C., eds., 1994, "Parallel Computing Works", Morgan Kaufmann Publ., San Francisco

Haimut, J-L. et al., 1995, Requirements for Information System Reverse Engineering, *Proc. IEEE Working Conference on Reverse Engineering*, IEEE Press.

Iliffe, J.K., 1978, Interpretive Machines, *in:* "The Microprocessor and its Applications", D. Aspinall, ed., Cambridge University Press, Cambridge.

Keane J.A. et al., 1993, Commercial Users' Requirements for Parallel Systems, *2nd Annual DAGS Parallel Computation Symposium*, pp. 15 25.

Kumar, A., Lee, T., and Tsotras, V., 1995, A Load-Balanced Parallel Sorting Algorithm for Shared Nothing Architectures. *Distributed and Parallel Databases*, 3 (1), pp. 37–68.

Lilja, D., 1997, Position Statement, Architectural Trends for Shared-Memory Multiprocessors, *Proc. 30th HICCS, Vol. I*, IEEE Press.

McFarland, D.E., What's Past is Prologue: The Future of COBOL, *American Programmer*, September.

Miller, J.C. and Strauss, B.M., 1987, Implications of Automated Restructuring of COBOL, *ACM SIGPLAN Notices*, 22 (6), pp. 76-82.

Miller, L.L. et al., eds, 1995, "Parallel Architectures for Data/Knowledge-Based Systems", IEEE Press.

Richter, L., 1993, Restructuring COBOL Applications for Coarse Grained Parallelism, *Proc. Commercial Applications of Parallel Processing Systems - CAPPS*, MCC Technical Report.

Sakellariou R. and O'Boyle, M.F.P., 1996, Towards a Parallelising COBOL Compiler, *Proc. 14th IASTED Conference on Applied Informatics*, IASTED-ACTA Press, pp. 190-192.

Sayles, J., 1996, COBOL and the Real World, *American Programmer*, September.

Sneed, H., 1991, Bank Application Reengineering and Conversion at the Union Bank of Switzerland, Proc. Conference on Software Maintenance, pp. 60-72, IEEE Press.

Teskey, F.N., 1996, Parallel processing in a commercial open systems market, *in:* "Parallel Information Processing", J.A. Keane, ed., Stanley Thornes Publishers, London

Wray, F., 1995, "High Performance Computing", UNICOM Publishers, London.

Wolfe, M., 1996, "High Performance Compilers for Parallel Computing", Addison-Wesley, Reading.

ASTROPHYSICS AND COSMOLOGY

SIMULATIONS OF LATTICE QUANTUM CHROMODYNAMICS ON THE CRAY T3D AND T3E

UKQCD Collaboration, presented by David Richards[1]

[1]Dept. of Physics and Astronomy
University of Edinburgh
Mayfield Road
Edinburgh EH9 3JZ, UK

INTRODUCTION

Quantum Chromodynamics (QCD) is the theory of the strong interaction. The strong force is felt by hadrons, such as the proton, neutron, kaon, but is not felt by leptons, such as the electron, neutrino, muon. It is a local gauge theory, similar to Quantum Electrodynamics (QED) but with a non-Abelian interaction.

QCD is formulated in terms not of the observed hadrons, but rather of more fundamental particles, the quarks and gluons. Quarks come in six *flavours*, from the lightest "up" (u) and "down" (d) quarks of which most everyday matter is constructed, to the heaviest, the "top" (t) quark, nearly 200 times the mass of the proton, and recently discovered at Fermilab in the USA. In addition, each quark comes in three "colours", analogous to charge in electromagnetism. The gluons, which come in eight colours, mediate the force, and play the rôle of the photon in QED. Gluons couple to any particle carrying colour charge. Thus they also interact with each other, in contrast to QED where the photon is neutral. This non-Abelian interaction leads to an anti-screening of colour charge, and accounts for the very different behaviours exhibited by QCD and QED.

QCD exhibits asymptotic freedom: with increasing energy, the strength of the interaction decreases and a quantitative study can be performed by a perturbative expansion in the strong-interaction coupling α_s. At low energies, characterised by the proton mass, the coupling strength is too large to permit a perturbative expansion. Here the non-linear behaviour of the theory is fully manifest, and the only *ab initio* means we have of a quantitative investigation is a lattice gauge theory (LGT) simulation. In particular, such a simulation can:

- explain the confinement of coloured quarks into colourless hadrons,

- predict the ratios of hadron masses,

- provide values for the hadronic form factors and matrix elements crucial to the experimental study of the Standard Model of particle physics.

In summary, a lattice gauge theory simulation provides the connection between the fundamental quarks and gluons and the observed hadrons.

LATTICE GAUGE THEORY

Continuum space-time is replaced by a four-dimensional lattice, or grid, of points. The gluons are represented by SU(3) matrices, U, and are associated with the links joining the sites of the lattice. The quarks are represented by anti-commuting Grassmann variables, ψ, associated with the sites of the lattice. The expectation value of an observable $\mathcal{O}$ is given by the Feynman path integral

$$\langle \mathcal{O} \rangle = Z^{-1} \int \mathcal{D}\psi \mathcal{D}\bar{\psi} \mathcal{D}U \mathcal{O}(U, \psi, \bar{\psi}) e^{-S_U(U) - S_F(U, \psi, \bar{\psi})} \tag{1}$$

where

$$Z = \int \mathcal{D}\psi \mathcal{D}\bar{\psi} \mathcal{D}U e^{-S(U, \psi, \bar{\psi})} \tag{2}$$

and the action is the sum of two parts, the first, S_U, depending only on the gauge fields, and the second, S_F, depending on both the gauge and matter fields. Because the ψ are anti-commuting variables, we can perform the ψ and $\bar{\psi}$ integration exactly, to obtain

$$Z = \int \mathcal{D}U \det M(U) e^{-S_U(U)} \tag{3}$$

where

$$S_F(\psi, \bar{\psi}, U) = \bar{\psi} M(U) \psi. \tag{4}$$

A LGT simulation comprises the generation of gauge configurations $\{U\}$ distributed according to eqn. (3).

Both $S_U(U)$ and $M(U)$ comprise terms local or nearest-neighbour in space-time. However, the evaluation of det M is non-local, reflecting the need for the quarks to satisfy Fermi-Dirac statistics, and requires the use of a sparse linear solver. Furthermore, det M must be evaluated every time a gauge field on any link is updated. Thus, theorists have employed the Quenched Approximation: the configuration space is sampled according to eqn. (3), with det $M \equiv 1$. We can interpret this as the quarks propagating in a background gluon field, and the suppression of quark loops.

There are both theoretical and phenomenological reasons for believing that, for many quantities, the quenched approximation is good at perhaps the 10% level. The benefit is a saving of a factor of many thousands in computer time. Thus, most high-precision calculations, including the bulk of the material in this talk, have been in the quenched approximation. However, simulations are now sufficiently precise that the need to correctly include fermions is apparent. At the end of this talk, I shall present the most recent results from the UKQCD Collaboration in "full" QCD.

Even a LGT calculation in the quenched approximation is a very onerous task. The lattice has to be sufficiently big that finite-size effects are under control, whilst sufficiently fine that results are insensitive to graininess of the lattice. The state-of-the-art quenched simulation of UKQCD has:

- $24^3 \times 48$ lattice,

- A lattice spacing a of around 0.7 fm,

- A spatial box length L of around 1.7 fm

498

Table 1. Lattice parameters and ensemble sizes used in the simulation.

β	Lattice size	# Cfgs.	$C = $ TAD	$C = $ NP
			Ensemble size	
5.7	$12^3 \times 24$	499	499	
	$16^3 \times 32$	145	145	
6.0	$16^3 \times 48$	499	499	496
	$32^3 \times 64$	72		72
6.2	$24^3 \times 48$	218	218	216

In both the quenched approximation, and for "full" QCD, the calculations are dominated by the inversion of the large, sparse matrix $M(U)$. The solver is implemented very efficiently; on 96 processor elements of the Cray T3E, each of 900 MFlops peak speed, our code sustains a speed of 25-30 GFlops.[1]

RESULTS IN THE QUENCHED APPROXIMATION

The UKQCD Collaboration has generated gauge configurations at three different values of the lattice spacing, corresponding to inverse couplings of $\beta = 5.7, 6.0$ and 6.2; the parameter values used in the simulations are listed in Table 1.

Static Quark Potential

One of the most striking results of lattice gauge simulations is the confirmation of a linear force between static, heavy quarks, with the consequence that quarks and gluons are confined into colour-singlet hadrons. In Figure 1, we show the force between static colour sources vs. the separation for each of our three values of β. The force is a constant at large distances. Furthermore, the results from the three simulations lie on a single curve, though the simulations span a factor of two in the lattice spacing; the data scale.

Light Hadron Spectrum

QCD determines the ratio of hadron masses precisely, up to relatively small electroweak effects, and the masses of hadrons containing light quarks are well known. Thus the measurement of the light-hadron spectrum is the benchmark calculation of lattice QCD.

We have seen that the discretisation errors in the static-quark force are small; formally, they are $O(a^2)$. Unfortunately, the naïve Wilson fermion action introduces $O(a)$ discretisation errors. In order to reduce these errors, we employ the $O(a)$-improved Sheikholeslami-Wohlert (SW) action:[3]

$$S_F = S_F^W - i\frac{c_{\text{SW}}\kappa}{2} \sum_{x,\mu,\nu} \bar{q}(x) F_{\mu\nu}(x) \sigma_{\mu\nu} q(x), \tag{5}$$

where S_F^W is the standard Wilson action, κ is the hopping parameter, and $F_{\mu\nu}$ is a lattice definition of the field-strength tensor. Using tree-level perturbation theory, the "clover" coefficient c_{SW} is unity, and the discretisation errors in hadron masses and, with the appropriate choice of operators, on-shell matrix elements are formally $O(ag^2)^4$. UKQCD has performed an extensive investigation of the hadron spectrum and hadronic

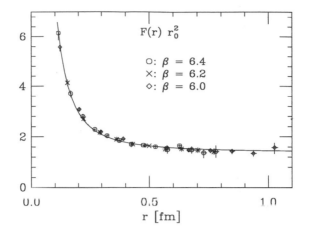

Figure 1. The force between two static colour sources vs. the separation at each value of β. Here r_0 is a characteristic scale derived from potential models for quarkonia, introduced in reference 2.

matrix elements using this value of c_{SW}, and the discretisation errors, particularly for systems containing heavy quarks, can be substantial[5].

Recently, two prescriptions for determining c_{SW} have been proposed with the aim of reducing discretisation errors still further. In the first, the clover coefficient is constrained to its mean-field improved, or tadpole, value[6]

$$c_{SW} = \text{TAD} = \frac{1}{u_0^3} \qquad (6)$$

where

$$u_0 = \langle \frac{1}{3}\text{Tr}U_\square \rangle \qquad (7)$$

is an estimate of the mean-value of the link variable U_μ. Though formally the discretisation errors remain $O(ag^2)$, this prescription recognises the poor behaviour of naïve lattice perturbation theory arising from the "tadpole" contributions, and attempts to resum much of the higher order contributions.

The second prescription[7] determines c_{SW} non-perturbatively in such a way as to remove *all* $O(a)$ discretisation errors from hadron masses, and, with an appropriate choice of operators, from all on-shell matrix elements:[8]

$$c_{SW} = \text{NP} = \frac{1 - 0.656g_0^2 - 0.152g_0^4 - 0.054g_0^6}{1 - 0.922g_0^2}, \quad g_0^2 \leq 1, \qquad (8)$$

where $g_0^2 = 6/\beta$.

We have investigated the hadron spectrum and matrix elements for both these prescriptions, and by performing the simulations at a variety of lattice spacings, investigated scaling and the approach to the continuum. The ensembles used for $c_{SW} = \text{TAD}$

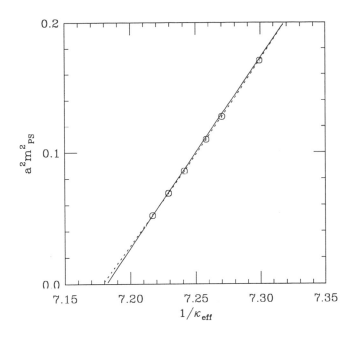

Figure 2. The chiral extrapolation of the pseudoscalar meson mass on the $16^3 \times 48$ lattices at $\beta = 6.0$, with $c_{SW} = $ TAD. The solid and dashed lines represent linear and quadratic fits to the data, as discussed in the text.

and $c_{SW} = $ NP are listed in Table 1. Note that for both the $16^3 \times 48$ lattices at $\beta = 6.0$ and the $24^3 \times 48$ lattices at $\beta = 6.2$ the sample size for $c_{SW} = $ NP is a little smaller than for $c_{SW} = $ TAD. This is due to the observation of "exceptional configurations" for $c_{SW} = $ NP; none were encountered in the case of $c_{SW} = $ TAD. We chose not to include these configurations in the $c_{SW} = $ NP analysis.

Hadron masses and amplitudes were obtained from single- and multi-exponential fits to the available matrices of hadron correlators; results from the different methods were generally consistent. In the following, quoted results are from the smaller of the lattices, of spatial extent 1.7 fm; results from the larger lattices are generally consistent, albeit with larger statistical errors. We obtain values at the physical light-(up/down) and strange-quark masses by interpolation and extrapolation, respectively, in the unrenormalised quarks mass

$$am_q = \frac{1}{2\kappa} - \frac{1}{2\kappa_c} \qquad (9)$$

where κ_c is the critical value of the hopping parameter, corresponding to vanishing pseudoscalar mass. The data for m_{PS}^2 favour a quadratic, rather than a linear, fit to m_q; an example is shown in Figure 2. The data for the other channels are consistent with a linear chiral extrapolation. Note that the u and d quark masses (assumed equal) are the value of m_q at with $m_{PS}/m_V = M_\pi/M_\rho$, the physical mass ratio.

We now discuss the extrapolation of our results to the continuum, using m_ρ as an example. The chirally extrapolated ρ masses on our various ensembles are shown in Figure 3, with the mass scale set from the string tension $\sqrt{K}$. The results with $c_{SW} = $ TAD have, in principle, $O(a)$ discretisation errors, though much reduced compared to

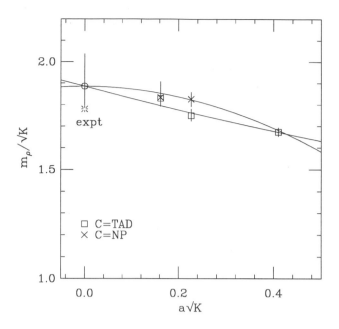

Figure 3. m_ρ vs. the lattice spacing, with the scale set by the string tension. The curves are from a simultaneous fit to the $c_{\mathrm{SW}} = \mathrm{TAD}$ and $c_{\mathrm{SW}} = \mathrm{NP}$ data, as described in the text.

the Wilson and tree-level-SW actions. Those with $c_{\mathrm{SW}} = \mathrm{NP}$ are completely free of $O(a)$ errors. Thus we assume the following form for the continuum extrapolations

$$
\frac{m_\rho}{\sqrt{K}} = \begin{cases} C_0 + C_1^{\mathrm{TAD}} a + C_2^{\mathrm{TAD}} a^2 & c_{\mathrm{SW}} = \mathrm{TAD} \\ C_0 + C_2^{\mathrm{NP}} a^2 & c_{\mathrm{SW}} = \mathrm{NP} \end{cases} \tag{10}
$$

A simultaneous four-parameter fit of eqn. (10) to the $c_{\mathrm{SW}} = \mathrm{TAD}$ and $c_{\mathrm{SW}} = \mathrm{NP}$ data is shown as the solid line in Figure 3.

The strange quark mass m_s can be determined by requiring that any one of m_K, m_{K^*}, and m_ϕ attain its physical value, using, say, m_ρ to set the scale. There is some evidence of inconsistency between these prescriptions in quenched QCD[9]. In order to investigate this, we show in Figure 4 the ratio m_{K^*}/m_ρ versus the lattice spacing, in units of m_ρ, using both m_K and m_ρ to determine m_s. Whilst there is an inconsistency between the prescriptions for both actions at a fixed β, this inconsistency decreases as we move nearer the continuum.

The picture of consistent continuum extrapolations continues into the baryon sector. In Figure 5 we show the continuum extrapolation of the nucleon mass with the scale taken from the ρ mass.

Heavy Quark Matrix Elements

Measurements of the hadron spectrum enable us to benchmark our lattice calculations. We will now examine the predictive power of LGT simulations to provide phenomenological input crucial to our interpretation of experiment, in particular in the study of systems containing heavy quarks.

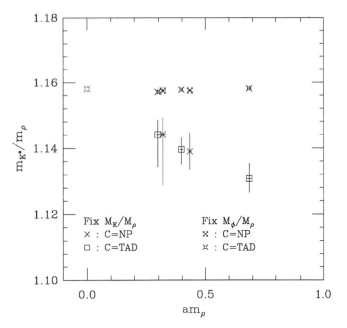

Figure 4. m_{K^*} versus the lattice spacing a, with the scale set by m_ρ.

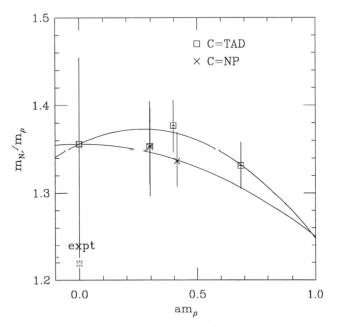

Figure 5. m_N versus the lattice spacing, with the scale set by m_ρ. The curves are from a simultaneous fit to the $c_{\mathrm{SW}} = \mathrm{TAD}$ and $c_{\mathrm{SW}} = \mathrm{NP}$ data, as described in the text.

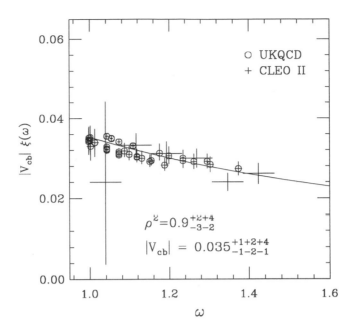

Figure 6. Least-χ^2 fits to CLEO II experimental data for $|V_{cb}|\xi(\omega)$. The first error on $|V_{cb}|$ is experimental, the second from lattice statistical uncertainties, and the third from lattice systematic errors on ρ^2.

The Standard Model of the fundamental interactions has many free parameters, in particular the quark masses and the elements of the CKM matrix determining the mixing under the weak interaction between the various quark flavours. The CKM matrix elements are constrained by the requirement that the matrix be unitary. It is important to verify the extent to which these constraints are satisfied, both as a check on the model and as a theatre in which to seek evidence of physics beyond the Standard Model. The determination of the poorly known CKM matrix elements involving the b quark is particularly important, since this will provide the first evidence of CP-violation in processes not involving the s quark.

The "B-factories" currently being constructed, such as BaBar at SLAC in which the UK particle physics community is an important participant, are charged with determining these matrix elements. However, a vital prerequisite is a quantitative description of the low-energy QCD effects that mask the weak interactions of the quarks. Lattice QCD simulations are able to describe these in a model-independent way.

As an example of the power of LGT simulations to aid experiment, we will consider the determination of the CKM matrix element V_{cb} through the semi-leptonic decay $\bar{B} \to D^*\ell\bar{\nu}$. The decay rate as a function of $\omega = v \cdot v'$, where v and v' are the four-velocities of the $\bar{B}$ and D^* mesons respectively, may be written, in the heavy-quark limit, as

$$\frac{d\Gamma(\bar{B} \to D^*\ell\bar{\nu})}{d\omega} = |V_{cb}|^2\xi(\omega)^2K. \tag{11}$$

Here V_{cb} is the CKM matrix element, $\xi(\omega)$ is the Isgur-Wise function, and K is a product of calculable factors. The UKQCD Collaboration has computed $\xi(\omega)$[10], and

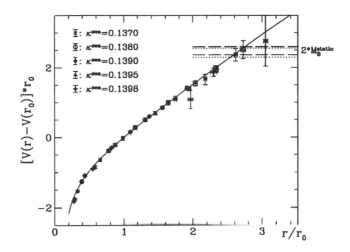

Figure 7. The scaled and normalised potential as a function of r/r_0 on the $12^3 \times 24$ lattices at $\beta = 5.2$, for various values of the sea-quark mass. The solid curve is a prediction using a simple linear and Coulomb term for the potential.

fitted it to $s\zeta_{\text{NR}}(\omega)$, where

$$\xi_{\text{NR}}(\omega) \equiv \frac{2}{\omega + 1} \exp\left(-(2\rho^2 - 1)\frac{\omega - 1}{\omega + 1}\right), \tag{12}$$

ρ^2 is the slope of $\xi(\omega)$ at $\omega = 1$, and s accounts for the uncertainty in the normalisation of the lattice data. By comparing the measured decay rate with the form of eqn. (11), we are able to extract V_{cb}. This is illustrated for the case of data from the CLEO II experiment[11] in Figure 6.

FULL QCD

Improvements in algorithms, together with increasing computational power, are enabling calculations in "full", unquenched QCD. Indeed, the statistical and systematic uncertainties on quenched calculations are such as to enable estimates of the systematic effects due to quenching on, for example, f_B[12]. UKQCD's use of the Cray T3E at Edinburgh is dedicated to full QCD simulations, and the reliable calculation of the hadron spectrum and matrix elements as outlined above. As a preliminary, we have investigated the force between static colour sources discussed earlier.

In the presence of dynamical quarks, we expect the simple linear behaviour of the potential at large distances, responsible for the constant force of Figure 1, to be modified. In particular, when the potential is equal to the mass, $2M_B^{\text{static}}$, of a pair of "mesons" comprising a static quark, and light antiquark, the colour string should break and the potential flatten. In Figure 7, we show the static potential at $\beta = 5.2$ as a function of the separation between the colour sources for various values of the sea-quark mass. Our results are neither in disagreement with the simple linear behaviour, nor with string breaking at the scale $2M_B^{\text{static}}$.

CONCLUSIONS

Lattice Gauge Theory simulations provide our only *ab initio* method for studying the low-energy properties of QCD. There are now reliable, and precise, measurements in the quenched approximation of the light-hadron spectrum and hadronic matrix elements. Indeed, the computations are now sufficiently precise that the need to correctly include the effects of sea quarks is apparent. These calculations, and the provision of the appropriate hadronic matrix elements, are crucial to the effective use of the "*B*-factories" and phenomenology.

Acknowledgements

This work was supported by EPSRC through grant GR/K41663, and PPARC through grants GR/L22744 and GR/L56336. The author acknowledges PPARC through the award of an Advanced Fellowship.

REFERENCES

1. Z. Sroczynski, S.M. Pickles and S.P. Booth, Nucl. Phys. B (Proc. Suppl.) 63 (1998), 949.
2. R. Sommer, Nucl. Phys. **B411** (1994) 839.
3. B. Sheikholeslami and R. Wohlert, Nucl. Phys. **B259** (1985) 572.
4. G. Heatlie *et al.*, Nucl. Phys. **B352** (1991) 266.
5. C.R. Allton *et al.* (UKQCD Collaboration), Phys. Lett. **B292** (1992) 408.
6. G.P. Lepage and P.B. Mackenzie, Phys. Rev. **D48** (1993) 2250.
7. K. Jansen *et al.*, Phys. Lett. **B372** (1996) 275; M. Lüscher *et al.*, Nucl. Phys. **B478** (1996) 365.
8. M. Lüscher *et al.*, Nucl. Phys. **B491** (1997) 323.
9. R. Gupta and T. Bhattacharya, Phys. Rev. D55 (1997) 7203.
10. S.P. Booth *et al.* (UKQCD Collaboration), Phys. Rev. Lett. **72** (1994) 462; Phys. Rev. **D52** (1995) 3536.
11. "Semi-leptonic *B* Decays", S. Stone, in "*B* Decays", ed. S. Stone, 2nd Edition (World Scientific, 1994).
12. C. Bernard *et al.*, Nucl. Phys. B (Proc. Suppl.) 63A-C (1998) 362.

TOWARDS AN UNDERSTANDING OF GALAXY FORMATION

F.R. Pearce,[1*] C.S. Frenk,[1] A.R. Jenkins,[1] J.M. Colberg,[2]
P.A. Thomas,[3] H.M.P. Couchman,[4] S.D.M. White,[2] G.P. Efstathiou[5]
and J.A. Peacock[6]

[1]Physics Department, South Road, Durham DH1 3LE, UK
[2]Max-Plank-Institut fur Astrophysik, 85740 Garching, Germany
[3]Astronomy Centre, University of Sussex, Falmer,
 Brighton BN1 9QH, UK
[4]Department of Physics and Astronomy, Univ. of Western
 Ontario, London, Ontario N6A 3K7, Canada
[5]Institute of Astronomy, Madingley Road, Cambridge CB3 0HA, UK
[6]Royal Observatory Edinburgh, Blackford Hill,
 Edinburgh EH9 3HJ, UK

INTRODUCTION

Computer simulations have played a central role in defining the modern approach to physical cosmology. Simulations provide the means to calculate the formation history of cosmic structure, given a set of assumptions regarding the physics of the early universe and the values of the cosmological parameters. Several key concepts have been elaborated with the aid of N-body simulations. For example, the cold dark matter cosmogony and the attendant idea of biased galaxy formation – the focus of cosmological discussion during the 1990s – were first explored in detail in a series of N-body simulations[1]. Twenty years after the first cosmological N-body simulations were performed, this technique is now firmly established as the main theoretical tool for calculating the non-linear phases in the evolution of structure and for testing theories of the early universe against observational data.

The improvement in the scope and quality of cosmological simulations over the past two decades has been impressive. For example, the N-body simulations that first rejected light massive neutrinos as viable candidates for the dark matter followed a total of 1000 dark matter particles, each of mass 10^{14} $M_\odot$[2]. Today, the best simulations have over 20000 times as many particles and can attain subgalactic mass resolution in cosmological volumes[3]. For the most part, this progress has been due to improvements in computer technology, but there have been conceptual advances as well. These include improved N-body algorithms and the development of efficient techniques to calculate hydrodynamic processes alongside the evolution of collisionless dark matter. The latter

---

*e-mail: *F.R. Pearce@durham.ac.uk*

High Performance Computing
Edited by R. J. Allan *et al.*, Kluwer Academic / Plenum Publishers, New York, 1999

is, of course, essential for modelling the formation of the visible constituents of our universe.

The quest for a detailed understanding of galaxy formation has become one of the central goals of modern astrophysics. On the observational side data from the Keck and Hubble Space telescopes have revolutionised our view of the high redshift Universe and there are claims that the epoch of galaxy formation has already been observed[4]. From the theoretical point of view the problem is fundamental because its resolution involves the synthesis of work from a wide range of specialities. A full treatment requires consideration of the early Universe processes that create primordial density fluctuations, the microphysics and chemistry that precipitate star formation within giant molecular clouds[5], the energy exchange present in supernova feedback[6, 7], the dissipational processes of cooling gas, the dynamics of galaxies moving within a dense environment[8, 9] and the large scale tidal torques which determine the angular momentum profiles of the resultant objects[10].

Analytic approaches to the problem founder partly because of the lack of symmetry. As high redshift observations show, real galaxies do not form in a smooth, spherically symmetric fashion but rather form as complicated collections of bright knots which merge and evolve into normal galaxies[11]. Because of its complexity this problem is ideally suited to numerical simulation but at first sight a fully self-consistent model appears impossible, at least with current computers and algorithms. However, the advent of parallel supercomputers coupled to the fastest numerical techniques does allow realistic simulations of certain aspects of the problem to be attempted[12, 13, 14, 15, 16]. These simulations are targeted towards the most intractable areas where a more detailed understanding will allow generalisations to be made which are applicable to the wider problem.

Hydrodynamical simulations, although intrinsically more difficult and computationally expensive, are more useful than collisionless models because it is much easier to compare the results to observations. Bias, the way in which the visible matter relates to the underlying density field, severely limits the ability of collisionless simulations to distinguish between competing cosmologies [17, 3, 18]. To model gas within cosmological simulations we use smoothed particle hydrodynamics (SPH), a technique ideally suited to this problem because of its large dynamic range.

The Virgo Consortium[19] is a large international collaboration interested in supercomputer simulations, a subset of whose members are also involved in the Hydra consortium[20], a smaller group specialising in galaxy formation and simulation algorithms. Both teams have played a key role in the development and implementation of astrophysical supercomputer codes. In particular, we made a successful port of the main, originally serial, production code to the Cray T3D[15] and have participated in the design and development of a new message passing version of the same code tailored to the T3E[21]. During the past few years we have gained experience on a variety of supercomputers and have access to world class computing resources.

In the following we will map out a long-term programme that leads to a proper treatment of galaxy formation. With our background of using simulation as a tool targeted at well defined science goals, code development, computer access and extensive collaboration, we are uniquely placed to make significant advances in this competitive area.

THE CLUSTERING OF GALAXIES

Observationally there are several galaxy surveys[22, 23, 24] underway at the present time. These studies will produce unprecedented three-dimensional maps of the structure in our Universe and provide a powerful diagnostic for cosmological models. On very large scales, we have begun the "Hubble Volume" project (see below), the largest N-body simulation ever attempted, with over 1 billion particles in a $2000h^{-1}$ Mpc box. On smaller scales, simulations that include a gaseous component are best placed for making observational comparisons because they do not suffer from the overmerging effects present in the dark matter only models. A suitable simulation requires the ability to resolve bright galaxies within a big enough volume that large scale effects are correctly modelled. In addition, each galaxy must contain sufficient particles that overmerging and tidal disruption are not major problems in the densest environments [9].

One of the main goals of the Virgo consortium is to carry out simulations of large cosmological volumes including a gaseous component which is subject to radiative cooling. As galactic dark matter halos form, gas is expected to cool inside them to make galaxies. With these simulations we aim to address one of the key outstanding issues in studies of large-scale structure: the relation between the distributions of galaxies and mass. Our first suite of simulations, carried out during 1996/97, contained 2 million dark matter and 2 million gas particles, with a gas particle mass of 1.9×10^9 $M_\odot$, a dark particle mass ten times larger, and a spatial resolution of $\sim 25h^{-1}$ kpc. These simulations were the first to resolve individual gaseous "galaxies" in volumes large enough that their spatial distribution can be calculated reliably. Typically, 2000 galaxies formed in each of the simulated volumes. In most models, the clustering pattern of these "galaxies" turned out to be biased relative to the mass, in a complicated, non-linear fashion. In the $\Omega = 1$ models the bias is always positive, whereas in the low-Ω models there is an antibias on small scales which gradually turns into a positive bias on large scales. The sign of these effects is just that which is necessary to reconcile these models with observations[3], although the amplitude fell short of the required strength. These results are encouraging but, on closer inspection, there are some important discrepancies between the simulations and the real world that merit further consideration. Most notably, the mass function of galaxies does not show the exponential cut-off at the high mass end characteristic of the galaxy luminosity function. Instead, anomalously large galaxies that have no counterpart in the real universe tend to form in the centres of groups and clusters. This problem had been hinted at in earlier simulations[9], but it has only become clearly manifest in the Virgo simulations, which are the first to fully explore this physical regime.

We are currently investigating in detail the cooling properties of the gas at the centre of our simulated clusters from various perspectives. We first need to establish whether the behaviour seen in the simulations is a genuine hydrodynamical effect or whether it is an algorithmic artifact produced by the smoothing inherent in the SPH technique which precludes it from modelling a multiphase medium correctly.

THE HUBBLE VOLUME PROJECT

The main programme of the Virgo consortium in the area of large-scale structure is the "Hubble volume" project, a set of simulations larger than any attempted to date. Each "Hubble volume" will be a region of linear, comoving dimension 2000-3000 h^{-1} Mpc, one third the Hubble length of an $\Omega = 1$ cosmology. In this cosmology, the side of the simulation cube corresponds to a redshift of $z \simeq 1.25$, and the length of the diagonal extends to $z = 4.6$, nearly the size of the "observed" universe. The simulations will follow a billion particles, over one order of magnitude more than the largest existing simulations (which have also been carried out by the Virgo consortium). These simulations will be performed on 512 processors of the Munich T3E and will be the first production runs to be undertaken with the new "message-passing" Virgo code.

These simulations are designed with a number of specific problems in mind. We will be able to determine accurately, for the first time, the abundance and statistical properties of the cluster population and to compare these estimates with analytic predictions based on the Press-Schechter theory. Of particular interest is the behaviour of the high mass end of the distribution since the frequency of these rare events provides a practical way to estimate cosmological parameters[25, 26]. This regime has never been tested numerically before because even the largest volumes simulated to date produce only a few hundred clusters[27]. By contrast the Hubble volume simulations will produce tens of thousands of clusters. Another novel aspect of these simulations will be output data along a past light cone. This will allow us to study the time evolution of clustering, including the evolution in the abundance of clusters above a fixed mass scale to $z \simeq 1.25$. We will also use the light cone data to model gravitational lensing effects due to the large-scale structure, using techniques similar to those developed for other gravitational lensing applications[28]. The advantage of using the Hubble volume simulations for lensing studies is that projection effects along the line-of-sight are correctly taken into account. Another important application of these simulations will be the construction of mock galaxy catalogues. With the Hubble volume simulations we will be able to extract many large independent volumes. Proposed surveys include faint extensions (~ 20000 galaxies to $B_j \simeq 22$ in the 2dF) for which clustering evolution, which is modelled in our simulations, will be important. Associated with these galaxy surveys are large quasar surveys. Our simulations will also be useful in interpreting these provided we can find a plausible way of relating the formation sites of quasars to the properties of the dark matter density field. Finally, our Hubble volume simulations will be ideal for making mock deep "pencil beam" surveys.

The Hubble volume project will test the limits of state-of-the art computer technology. The runs are the largest calculation that can fit into the 512-processor T3E. The analysis of the data, however, will be even more challenging. For example, positions and velocities for the particles in each timeslice will generate roughly 12 Gbytes of data in compressed form (2 byte integers) while the light cone output will generate 48 Gbytes. After our initial analyses, we intend to make a substantial fraction of the dataset publicly available to the cosmology community. Apart from the scientific applications just discussed, a huge simulation like this will have considerable value as an educational tool.

STAR FORMATION AND FEEDBACK

Before studying the effects of star formation and feedback in more detail we plan to examine the so called "cooling catastrophe" by generating a sequence of simulations of the same section of the Universe at increasing resolution. The cooling catastrophe is a major problem for hierarchical clustering models which predict that present day objects form via the merger of a large number of smaller pieces[29]. At earlier times the Universe was denser, making cooling more efficient, the natural consequence of which is that in hierarchical models all the available gas is expected to cool into small subgalactic lumps at high redshift[30]. The existence of extended disks in spiral galaxies and hot gas in clusters[31] demonstrates that the cooling catastrophe did not occur in the Universe we observe.

The same effect manifests itself in numerical simulations. As the resolution is increased, objects can be discerned at earlier times and a larger fraction of the gas cools into small, cold knots at high redshift[32]. Even at moderate resolution the fraction of gas in collapsed objects exceeds that observed today. Once the gas has cooled into dense clumps it is very difficult to reheat it or spin it up sufficiently to produce extended spiral disks. One possible solution to this problem is feedback: in the observed Universe the collapse of gas into cold, dense knots promotes star formation and some of these stars are massive enough to undergo supernovae explosions that reheat the surrounding gas.

The physics of star formation is a complicated field which is not yet well understood. At present one needs to employ physically reasonable assumptions to model this process within cosmological simulations [6, 7, 33]. A stable method for placing stars within a simulation is required before the more complicated process of the feedback of energy due to supernova explosions can be considered. We have been examining the effect on the subsequent cooling rate of the gas produced by replacing gaseous particles with collisionless star particles[34] along with examining the physical motivation behind the Jeans criterion for star formation and the effect of different star formation algorithms on the star formation rate.

Currently one of the most intractable problems on the road to galaxy formation is the implementation of energy feedback due to supernovae[6, 7, 33]. This is observationally very uncertain because of our as yet poor understanding of the stellar initial mass function (IMF)[35] and of the effects of a large injection of energy and momentum on the surrounding gas. The region of interest for feedback mechanisms, the high mass end of the IMF, is not very well constrained even before possible environmental factors are taken into account. In addition the timescales involved are very short: the heating and snowplough effects caused by a supernova explosion are over within 10000 years or so, whereas simulation timesteps may be 10 million years, longer than the exploding star's entire main sequence lifetime. Resolution is also a problem as galactic star forming regions are small, with giant molecular clouds ranging in mass up to $10^6 M_\odot$, comparable to single particle masses in even the highest resolution simulations.

Exactly what to do with the returned supernovae energy is a further complication. At the densities required to form stars, cooling is very efficient and so any energy fed back as heat quickly radiates away and has no effect upon subsequent evolution[6]. Imparting a kinetic energy kick to neighbouring gas on the other hand has an appreciable effect because the energy takes several steps to dissipate and if the object is small the gas can escape and disperse[7]. This has the effect of destroying small objects within the simulation volume and retarding the global star formation rate. Unfortunately this process is resolution dependent, although with an appropriate choice of parameters it is possible to reproduce similar star formation rates to those observed[36].

One of the main goals of the early stages of this programme is to design a resolution-independent feedback mechanism, a vital building block for any successful study of galaxy formation.

HIGH-REDSHIFT OBJECTS

Heating, perhaps due to the radiation fields of distant quasars, ionises most of the gas at high redshift, preventing it from cooling and retarding the build-up of structure. Some of the gas, however, remains neutral and this is detected in absorption as the

Lyman-α forest in high resolution spectroscopic observations of quasars[37, 38]. Each feature in the spectra is formed by the line-of-sight to the quasar intersecting a clump of gas. These clumps are the building blocks out of which galaxies form and their clustering properties provide a powerful discriminant amongst rival cosmologies[39, 40, 41].

Hydrodynamical simulations are the ideal tool to examine the evolution of the physical state of these clouds. We have been employing high resolution simulations to study the distribution of column densities as a function of redshift, the detailed shape of the line profiles, the internal structure of the clouds and their merger history. As part of this programme we have completed the highest resolution hydrodynamical simulation to date, containing 8 million particles in a $22h^{-1}$Mpc box.

GALAXY FORMATION

A full treatment of galaxy formation would address such questions as the origin of the angular momentum that supports the disks of spiral galaxies and the physical basis for the Hubble sequence of galaxy morphologies, along with such fundamentals as the Tully-Fisher relation and the colour-magnitude relation of elliptical galaxies.

For this purpose the cosmological simulations detailed above are of little use because no internal information is recovered. For more detailed morphological information several orders of magnitude more mass resolution is required. This can be attained by using nested box techniques and variable particle masses but remains very difficult. Galactic disks are elusive and to this date, despite numerous valiant attempts, no one has managed to successfully simulate their formation[42, 43, 40]. The origin of the angular momentum within the disk remains an unsolved problem.

On these scales all of the problems encountered in modelling star formation and feedback are exacerbated because the resolution is very high. The precise hydrodynamical algorithm employed is also very important because the effects of angular momentum transport are vital for disk stability.

We are proposing a new supercomputer programme of gas dynamic simulations of individual clusters which extends the exploratory work[9]. The general goal is to investigate the phase space structure of the galaxy population within clusters and to provide estimates of both spatial and velocity biasing. Earlier work[9] simulated the formation of a single cluster in which galaxies were identified as knots of cold gas. The final structure of this "galaxy" distribution was highly biased with respect to the dark matter, but certain aspects of the solution seemed unrealistic, particularly the fact that over half the final galactic mass in the cluster ended up in a central, dominant galaxy. This is reminiscent of the problems found in the clusters that formed in the Virgo gasdynamic cosmological simulations discussed above. Real galaxies, of course, are made of stars, not of cold gas. To increase the realism of the simulation and perhaps to solve this "overcooling" or "overmerging" problem we intend to follow star formation *in situ* within the cold, collapsed knots. The necessary modifications to the simulation code are already in place and we plan to undertake a systematic study, to build up an ensemble of cluster realisations. This will explore simple parameterizations for star formation, and determine the sensitivity of the resulting galaxy structure to the star formation scheme and parameters. Eventually, we will single out a specific parameterization from this study and examine the resultant galactic structure in a set of cluster realisations spanning a moderate richness range.

With the ability to resolve bright galaxies in clusters, our ensemble of cluster simulations will provide a unique sample to address several important unresolved issues. These include determining the orbital characteristics of galaxies in clusters, examining the mechanism of "galaxy harassment" to explain some of the morphological properties of cluster galaxies, calculating the importance of ram-pressure stripping of disk gas by the hot intracluster medium, quantifying the amplitude and establishing the origin of spatial and kinematical biases in the galaxy populations, and testing the accuracy of cluster mass determinations from the dynamics of cluster galaxies. The high resolution of our simulations will allow us to investigate the properties of the intracluster medium with higher accuracy than has been possible so far. Thus, we will be able to test previous claims regarding the convergence of the X-ray luminosity in the cores of simulated clusters, to determine the relation between X-ray luminosity and X-ray temperature, and to measure the temporal evolution of the X-ray cluster luminosity and temperature functions. Finally, we are optimistic that our simulations will be accurate enough to tackle, for the first time by direct simulation, the difficult problem of cooling flows at the centres of clusters.

INDIVIDUAL GALAXIES

One of the stated long-term aims of the Virgo programme is to carry out simulations of individual galaxies incorporating the physical effects thought to be relevant: dark matter clustering, gas cooling, star formation and feedback and photoionisation by an external UV field. Over the past two years we have incorporated all these effects into our simulation codes. A stellar population synthesis model will shortly be implemented: as stars are formed, the synthesis model calculates the time evolution of their spectrophotometric properties so that, at any epoch, we can calculate synthetic spectra, and thus the colours and luminosities of the stellar populations. In this way we are able to make "visible galaxies" in the computer and measure their properties as a function of time.

Even though the physical processes thought to be responsible for galaxy formation are relatively straightforward, it is still the case that no one has succeeded yet in making a realistic disk galaxy in the computer. The obstacle is a process of angular momentum transfer inimical to hierarchical clustering. In gravitational instability theories angular momentum is acquired as a result of tidal torques acting on the protogalaxy, a process that has been calculated in detail using N-body simulations[44, 45, 46]. Recent N-body/SPH simulations show that, in the absence of star formation, protogalactic gas loses much of its angular momentum as it cools into pregalactic fragments that subsequently merge, transferring their (orbital) angular momentum to the outer dark matter halo[47,40]. Thus, most of the gas in the simulations ends up in a large clump at the centre of the final galaxy and the disk that forms is much too massive and slowly rotating compared to observed disks. The solution to this problem remains unclear, but feedback is likely to play a crucial role, preventing some of the gas from cooling into dense fragments and allowing it to settle, instead, in a less inhomogeneous fashion and on a longer timescale than the dark matter fragments.

These simulations offer the possibility of making direct contact with recent data on high redshift galaxies, such as the images and spectra from HST and Keck. Some of the issues that we plan to address are:

(i) The properties of protogalaxies. The Keck and Hubble Space Telescopes are now providing quantitative information on properties of galaxies at intermediate and high redshift such as morphologies, sizes, colours, and star formation rates[11]. These data leave little doubt that substantial evolution in the galaxy population has occurred since redshift $z \simeq 1$ and suggest that the epoch of galaxy formation may be as recent as $z \simeq 3$, (as expected in CDM models[48, 30, 49]. Our simulations will help to establish

the conditions and cosmological parameters required for this evolution to occur and the physical processes that determine the observable properties of high redshift galaxies. In particular, they will quantify the clustering of galaxies at high redshift, the merger rates of galaxies, the signatures of recent mergers, and their effect on galaxy morphology.

(ii) The visible effects of galaxy mergers. The observational community is beginning to accept the view, long held by many theorists, that mergers are the key process that drives galaxy formation. This growing acceptance is due largely to direct evidence from HST images of clumpy and clearly interacting galaxies at intermediate redshifts, but also to the increased sophistication of numerical simulations. The simulations that we plan to perform will quantify the merger rates of galaxies in different cosmologies, the signatures of recent mergers, and the effect of mergers on galaxy morphology. By selecting galaxies from a range of environments, our procedure for laying down initial conditions automatically accounts for environmental effects. In this way, we hope to gain some understanding of the origin of the Hubble sequence of morphological types

(iii) Galactic chemical evolution. In principle, it is straightforward to model in the simulations the injection of metals processed by stellar evolution into the intergalactic medium. The effects of enrichment on the cooling rates are well understood, but the extent to which the metals mix with the surrounding gas is not. A further factor that has hampered progress in this area is the lack of population synthesis models for stars with a mixture of metallicities. Such models, however, have recently become available (Charlott, Worthey, private communication).

Although such a simulation of galaxy formation, including the effects of photoionisation, star formation and feedback with a resolution high enough to allow realistic, stable disks to form is not yet possible on today's generation of computers the road is clear and with the speed of advance of computer architectures significant inroads into this fundamental astrophysical problem can now be made.

ACKNOWLEDGEMENTS

The work presented in this paper was carried out using codes developed and made available by the Virgo Supercomputing Consortium (*http://star-www.dur.ac.uk/ fraz-erp/virgo/virgo.html*) using computers based at the Computing Centre of the Max-Planck Society in Garching and at the Edinburgh Parallel Computing Centre.

REFERENCES

1.   Frenk, C.S., Physica Scripta, T36:70 (1991).
2.   Frenk, C.S., White, S.D.M. and Davis, M., *Astrophy. Journ.*, 271:417 (1983).
3.   Jenkins, A.R., Frenk, C.S., Pearce, F.R., Thomas, P.A., Colberg, J.M., White, S.D.M., Couchman, H.M.P., Peacock, J.A., Efstathiou, G. and Nelson, A.H., *Astrophy. Journ.*(1997) submitted.
4.   Baugh, C.M., Cole, S., Frenk, C.S. and Lacey, C.G., *Astrophy. Journ.*, (1997) submitted.
5.   Lada, E.A., Evans, N.J. and Falgarone, E., *Astrophy. Journ.*, 488:286 (1997).
6.   Katz, N., *Astrophy. Journ.*, 391, 502 (1992).
7.   Navarro, J.F. and White, S.D.M., *Mon. Not. Roy. Ast. Soc.*, 265:271 (1993).
8.   Moore, B., Katz, N., Lake, G., Dressler, A. and Oemler, A., *Nature*, 379:613 (1996).
9.   Frenk, C.S., Evrard, A.E., White, S.D.M. and Summers, F.J., *Astrophy. Journ.*, 472:460 (1996).
10. Katz, N. and Gunn, J.E., *Astrophy. Journ.*, 377:365 (1991).
11. Steidel, C.C., Pettini, M. and Hamilton, D., *Astron. Journ.*, 110:2519 (1995).
12. Cen, R. and Ostriker, J., *Astrophy. Journ.*, 464:270 (1996).
13. Dubinski, J., *New Astronomy*, 1:133 (1996).
14. Tsai, J.C., Katz, N. and Bertschinger, E., *Astrophy. Journ.*, 423:553 (1994).
15. Pearce, F.R., Couchman, H.M.P., *New Astronomy*, 2:411 (1997).

16. Governato, F., Baugh C.M., Frenk C.S., Cole, S., Lacey, C.G., Quinn, T. and Stadel, J., *Nature*, (1997) submitted.

17. Carlberg, R.G., *Astrophy. Journ.*, 433:486 (1994).

18. Thomas, P.A., Colberg, J.M., Couchman, H.M.P., Efstathiou, G.P., Frenk, C.S., Jenkins, A.R., Nelson, A.H., Hutchings, R.M., Peacock, J.A., Pearce, F.R. and White, S.D.M., *Mon. Not. Roy. Ast. Soc.*, (1997) in press.

19. Virgo Consortium *http://star-www.durham.ac.uk/~frazerp/virgo/virgo.html*

20. Hydra Consortium *http://phobos.astro.uwo.ca/hydra_consort*

21. MacFarland, T., Pichlmeier, J., Pearce, F.R., Couchman, H.M.P., 1997 European Cray/SGI MPP Workshop (1997).

22. Colless, M., Boyle, B., astro-ph/9710268 (1997).

23. PSCz survey *http://www-astro.physics.ox.ac.uk/~wjs/pscz.html*

24. Loveday, J., astro-ph/9605028 (1997).

25. White, S.D.M., Efstathiou, G. and Frenk, C.S., *Mon. Not. Roy. Ast. Soc.*, 262:1023 (1993).

26. Eke, V.R., Cole, S. and Frenk, C.S., *Mon. Not. Roy. Ast. Soc.*, 282:263 (1996).

27. Cole, S. and Lacey, C.G., *Mon. Not. Roy. Ast. Soc.*, (1996) submitted.

28. Wilson, G., Cole, S. and Frenk, C.S., *Mon. Not. Roy. Ast. Soc.*, 282.501 (1996)

29. Press, W.H. and Schecter, P., *Astrophy. Journ.*, 187:425 (1974).

30. White, S.D.M. and Frenk, C.S., *Astrophy. Journ.*, 379:52 (1991).

31. Jones, C. and Forman, W., *Astrophy. Journ.*, 276:38 (1984).

32. Katz, N., Hernquist, L. and Weinberg, D.H., *Astrophy. Journ.*, 399:L109 (1992).

33. Mihos, C.J. and Hernquist, L., *Astrophy. Journ.*, 464:641 (1996).

34. Pearce, F.R., *Mon. Not. Roy. Ast. Soc.*(1997) submitted.

35. Larson, R.B., *Mon. Not. Roy. Ast. Soc.*, 218:409 (1986).

36. Madau, P., astro-ph/9612157 (1997).

37. Prochaska, J.X. and Wolfe, A.M., *Astrophy. Journ.*, 487:73 (1997).

38. Fernandez-Soto, A., Lanzetta, K., Barcons, X., Carswell, R., Webb, J. and Yahil, A., *Astrophy. Journ.*, 460.L85 (1996).

39. Weinberg, D.H., Hernquist, L. and Katz, N., *Astrophy. Journ.*, 477:8 (1997).

40. Navarro, J.F. and Steinmetz, M., *Astrophy. Journ.*, 478:13 (1997).

41. Gardner, J.P., Katz, N., Hernquist, L. and Weinberg, D.H., *Astrophy. Journ.*, 484:31 (1997).

42. Steinmetz, M., *Mon. Not. Roy. Ast. Soc.*, 278:1005 (1996).

43. Summers, F.J., Davis, M. and Evrard, A.E., *Astrophy. Journ.*, 454:1 (1995).

44. Barnes, J. and Efstathiou, G. *Astrophy. Journ.*, 319:575 (1987).

45. Frenk, C.S., White, S.D.M., Davis, M. and Efstathiou, G., *Astrophy. Journ.*, (1988). 327:507

46. Warren, M.S., Quinn, P.J., Salmon, J.K. and Zurek, W.H., *Astrophy. Journ.*, 399:405 (1992).

47. Navarro, J.F., Frenk, C.S. and White, S.D.M., *Mon. Not. Roy. Ast. Soc.*, 275:56 (1995)

48. Frenk, C.S., White, S.D.M., Efstathiou, G. and Davis, M., *Nature*, 317:595 (1985)

49. Cole, S., Aragon-Salamanca, A., Frenk, C.S., Navarro, J.F.and Zepf, S., *Mon. Not. Roy. Ast. Soc.*, 271:781 (1994)

MASSIVELY PARALLEL SIMULATIONS OF COSMIC STRINGS

Mark Hindmarsh[1] and Graham Vincent[1]

[1]Centre for Theoretical Physics
University of Sussex,
Brighton BN1 9QJ U.K.

INTRODUCTION

A cosmic string is an ultra-thin tube of exotic matter which may have been left over from a very early hot, dense phase of the Universe's history[1, 2]. Certain Grand Unified Theories, which are theories purporting to describe all the elementary particles and their interactions on an even footing, predict the existence of strings. If the temperature was ever high enough, the strings would have inevitably formed in a tangled network, and would still be present today, albeit greatly diluted, but still with possible observational effects. Thus the study of their properties and dynamics helps us get information about the earliest phases of the Universe's history.

By early we mean any time between 10^{-36} s and 10^{-11} s, when the temperature of the Universe was decreasing from 10^{29} K (10^{16} GeV in Natural Units) to 10^{15} K (10^2 GeV), at which point the average thermal energy of the particles was about that acheivable in the most powerful accelerators today. As the Universe cooled, some spinless particles (Higgs particles) may have condensed into a zero-momentum state, in a similar manner to the condensation of atoms in superfluid ^{4}He. It is thought that a condensate of Higgs particles is responsible for the generation of the masses of all the known elementary particles. A Higgs condensate is also involved in making cosmic strings, although it should be stressed that it is not the same one as is implicated in the mass generation of known particles. There are reasons for believing that such extra condensates exist: the theory describing the currently known particles, the Standard Model, is almost certainly not the end of the story, and it is widely thought to be part of a larger "Grand Unified" Theory, or GUT, which would have many more types of Higgs condensates.

Consider a wave function $\phi(t, \boldsymbol{x})$ describing the zero momentum condensate. In most places, $|\phi|^2$ is constant, signalling a uniform condensate. However, there may also be lines along which the wave function vanishes, and changes phase by 2π around an infinitesimal circle enclosing the line of zeroes. Such zeroes cannot be removed by small perturbations of the field, and so are structurally stable. The modulus of the wave function regains its equilibrium value at a distance $\hbar/m_h c$ from the centre, where m_h is the mass of Higgs field itself. Associated with this departure from the equilibrium

High Performance Computing
Edited by R. J. Allan *et al.*, Kluwer Academic / Plenum Publishers, New York, 1999

value is an increase in the local energy density. This thin tube of energy – in a typical GUT the width is of order 10^{-29} m – is the cosmic string.

Condensed matter physicists will immediately notice the close similarity with vortices in superfluid Helium, and in the case where the Higgs field is charged, with flux tubes in a Type II superconductor (although it should be noted that the charge is not ordniary electric charge, but another type of charge allowed by many GUTs). We consider strings which are more like superconductor flux tubes, in that their energy density is highly localised, and their interactions very short range.

Cosmic strings have several types of observable effects today. Firstly, they may be as massive as 10^{21} kg m^{-1}, which immediately suggests gravitational effects. As a string with mass per unit length μ and typically causes fluctuations in the gravitational potential of order $(G\mu)/c^2$, which is at a level of about 10^{-6} for strings as massive as the figure quoted above. A lot of effort has recently gone into working out the effect of these fluctuations, particularly in the Cosmic Microwave Background. Another possible effect is that the strings can spit out Higgs particles with masses of up to 10^{15} GeV, which would quickly decay into known particles and appear as highly energetic cosmic rays. It is also possible that the strings could act as superconducting wires, which would give rise to spectacular effects if the string passed through the magnetic field of a galaxy. However, the length density of string today would be very low, and the nearest string would be of order 1 Gpc away. Their effects are therefore hard to detect.

STRING DYNAMICS

The serial code which accomplishes the initialisation, evolution and analysis of a string network was originally written by Sakellariadou[3, 4, 5] and developed by the authors[6, 7]. The parallel version was developed by N. Floros and I. Wolton of the Southampton HPCI centre. Related codes, using different algorithms which can be applied to strings in an expanding Universe, have been written by Albrecht and Turok[9], Bennett and Bouchet[10, 11], and Allen and Shellard[12].

Evolution

If we neglect the small effect of the expansion of the Universe, the equation of motion for a cosmic string is the relativistic wave equation

$$\mathbf{X}'' = \ddot{\mathbf{X}}, \tag{1}$$

where $\mathbf{X}$ is a position three-vector, $\dot{\mathbf{X}} = \frac{\partial \mathbf{X}}{\partial \eta}$, $\mathbf{X}' = \frac{\partial \mathbf{X}}{\partial \sigma}$, η is time and σ is a spacelike parameter along the string. This simple form is achieved by choosing a convenient coordinate parametrisation which is implicit in the following constraints:

$$\mathbf{X}' \cdot \dot{\mathbf{X}} = 0, \tag{2}$$

$$\mathbf{X}'^2 + \dot{\mathbf{X}}^2 = 1. \tag{3}$$

The physical content of the first constraint is that there are no physical tangential velocities (a string is invariant under boosts along its tangent) and equation 3 ensures that the energy of a segment of string is proportional to its invariant length.

The wave equation can be evolved with the Smith-Vilenkin algorithm[3]. This algorithm uses the exact finite difference solution to equation 1,

$$\mathbf{X}(s, \eta + \delta) = \mathbf{X}(s + \delta, \eta) + \mathbf{X}(s - \delta, \eta) - \mathbf{X}(s, \eta - \delta). \tag{4}$$

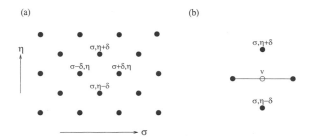

Figure 1. (a) The worldsheet as a diamond lattice. Note that at half the string points are defined at one time step and the other half at the next. (b) The velocity defined as a time centred finite difference can be though of as the velocity of a link between defined positions.

The string worldsheet must be discretised in some way. From equation 4 one can see that it is appropriate to represent the worldsheet of the string by a diamond lattice, as in Figure 1a. Note that a consequence of this discretisation scheme is that at each time step, only half the string points have their positions defined. Thus if $X(\sigma')$ is defined at $\eta' - \delta$, it is undefined at η' but *is* defined at $\eta' + \delta$. At a given time step, these gaps can be considered to be filled in by the velocities of those string points with undefined positions at that time step. These velocities are time-centred finite differences

$$\mathbf{v}(\sigma',\eta') = \frac{1}{2\delta}[\mathbf{x}(\sigma',\eta'+\delta) - \mathbf{x}(\sigma',\eta'-\delta)], \tag{5}$$

and represented by the open circle (o) in Figure 1b. An evolution scheme in terms of positions and velocities can be derived from equations 4 and 5 requiring only data from the previous time step. The evolution equations are[3]

$$\mathbf{x}(\sigma,\eta+\delta) = \frac{1}{2}[\mathbf{x}(\sigma+\delta,\eta) + \mathbf{x}(\sigma-\delta,\eta)] + \mathbf{v}(\sigma,\eta)\delta, \tag{6}$$

$$\mathbf{v}(\sigma,\eta+\delta) = \frac{1}{2}[\mathbf{v}(\sigma+\delta,\eta) + \mathbf{v}(\sigma-\delta,\eta)] + \frac{1}{4\delta}[\mathbf{x}(\sigma+2\delta,\eta) - 2\mathbf{x}(\sigma,\eta) + \mathbf{x}(\sigma-2\delta,\eta). \tag{7}$$

The algorithm will then consist of choosing appropriate initial conditions for the positions and velocities of *all* string points at time steps of η_0 and η_1 and evolving them using a "leap-frog" method: updating the velocity and positions at alternate time steps. The added complication is that at a given time step, half the string points update their positions and the other half update their velocities.

It is useful to consider a velocity $\mathbf{v}(\sigma',\eta')$ as residing on a string *link* between $\mathbf{x}(\sigma'+\delta,\eta')$ and $\mathbf{x}(\sigma'-\delta,\eta')$. Links are of invariant length 2δ and have a tangent vector

$$\mathbf{u}(\sigma',\eta') = \frac{1}{2\delta}[\mathbf{x}(\sigma'+\delta,\eta') - \mathbf{x}(\sigma'+\delta,\eta')]. \tag{8}$$

The link velocities and tangents are subject to the discrete versions of the gauge conditions equations 2 and 3.

$$\mathbf{u} \cdot \mathbf{v} = 0, \tag{9}$$

$$\mathbf{u}^2 + \mathbf{v}^2 = 1. \tag{10}$$

The discrete gauge conditions are satisfied by the evolution equations 6 and 7 (or alternatively equation 4), thus if the discrete gauge conditions are satisfied by the initial conditions, they will be satisfied at all subsequent time steps. This is achieved in the Smith-Vilenkin algorithm by discretising spacetime as well as the worldsheet

Link type	velocity v	tangent u	picture of example on 2-d lattice
	example	example	
Stationary	0	1	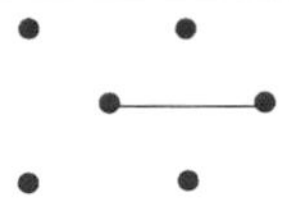
		(1,0,0)	
Diagonal	1/sqrt(2)	1/sqrt(2)	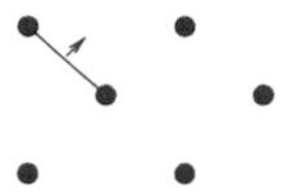
	(1,1,0)	(1,-1,0)	
Cusp	1	0	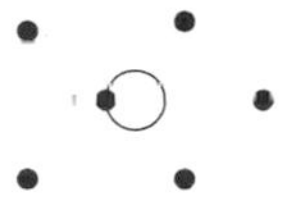
	(-1,0,0)		

Figure 2. Three types of links allowed in the Smith-Vilenkin algorithm and examples represented on a *spatial* lattice.

which restricts the possible values of $\mathbf{u}$ and $\mathbf{v}$ from which initial configurations can be constructed. Note that if the string points are initially defined on a cubic lattice then, by equation 4, they will stay on a cubic lattice for all successive time steps δ which implies that the *components* of $\mathbf{u}$ and $\mathbf{v}$ are restricted to integral or half integral values from their definitions as centred differences in equations 5 and 8, and then to 0, $\frac{1}{2}$ and 1 from the worldsheet constraint equation 10. The resulting set of possible vectors $\mathbf{u}$ and $\mathbf{v}$ (there are not many), are further constrained by equation 9 so that there are three basic types of link, given in Figure , together with a representation of the link on a spatial lattice (note that the cusp is actually of zero spatial extent). If a string network is constructed from links of these three types, it will satisfy the gauge conditions. In three dimensions the links restrict the string points to a face-centred cubic lattice of cell size 2δ.

In the computer code, the positions of the string points and their velocities are determined from the initial conditions (see below), and evolved using equations 6 and 7. The only extra feature to be introduced is reconnection. As the string points lie on the sites of a lattice, identifying crossing events is simple. When two strings cross, they reconnect the other way with a probability which is set to P_I. For most of our work $P_I = 1$, subject to the condition that reconnection does not create a loop smaller than a threshold value for the energy (or equivalently invariant length) E_c. Loops with energy greater than or equal to E_c are allowed to leave the network, although loops with energy of E_c are forbidden to reconnect. This allows energy to leave the network fairly efficiently; otherwise it takes much longer for the effect of the initial conditions to wear off. This feature may also model more realistic networks as, in an expanding Universe, small loops will decouple from the expansion and are highly unlikely to reconnect. If $E_c = 2\delta$, then the smallest loops produced are cusps and the cutoff becomes a lattice cutoff.

Periodic boundary conditions are imposed throughout, and the evolution time is restricted to half the box size, as after this time causal influences have propagated around the box.

520

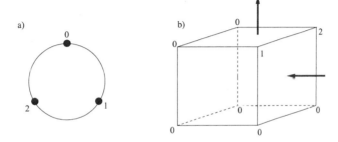

Figure 3. (a) The phase of the zero-momentum wave function is discretised into 3 values, $2\pi n/3$, with n 0, 1 and 2. (b) Phases are randomly placed on the sites of a lattice and the winding around each face decided using the geodesic rule which states that the phases will interpolate along the shortest path between the discrete group elements. In the example shown, there is a string (arbitrarily defined as a clockwise winding) coming in from the right and leaving through the top of the cell.

Formation

The Smith-Vilenkin algorithm, as it stands, provides an exact evolution for a string network defined on a face-centred cubic lattice using a restricted set of links. Initial string configurations are generated using the Vachaspati-Vilenkin algorithm[8], which mimics the random values that the phase of the wave function ϕ takes when the condensate forms. In this approach, the continuous phases are approximated by the values $2\pi n/3$, which are placed randomly on the sites of a cubic lattice of cell size ξ_0 as in Figure . Strings (or anti-strings) are identified as passing through a face with a net winding about the manifold. The strings are then considered to join up within the lattice cells to from a network of Brownian strings with step length ξ_0 on a cubic lattice. There is a possible ambiguity in the link up. A cubic lattice cell can have at most two strings entering and two emerging[8]. The strings must then be linked up randomly.

It is easy to define this network on a face-centred lattice by treating each section ξ_0 as made up of $\xi_0/2\delta$ stationary links defined on a cubic lattice with cell size 2δ. Then the strings are free to move on a face-centred cubic lattice also with cell size 2δ.

DATA STRUCTURES

Dynamical evolution

A data structure for a string must reflect the fundamental dynamical object. The obvious structure is a linked list, with forward and backward pointers. The list carries the dynamical information of each element of string, its position and velocity in 3 dimensions. We call such an element a "node" (see Figure 4).

There are also data structures associated with the search for crossings, when two different nodes have the same position. This requires some kind of sorting based on

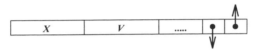

Figure 4. The basic data structure is a linked list, which carries the position X and velocity bfV of a string node. The ellipsis indicates extra "flags of convenience" used when scanning the list.

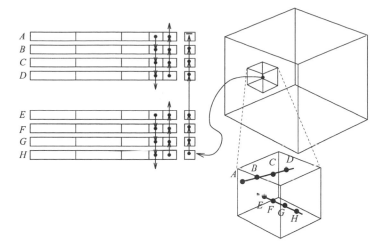

Figure 5. Data structures used in searching for string crossings. The simulation volume is divided into cubic subdomains, each containing a small amount of string. Each node has another backward pointer which chains together nodes in the same subdomain. A 3D array of pointers, one for each subdomain, points to the ends of the chains.

the positions of the nodes. Our search algorithm divides the spatial lattice into cubic subdomains, each of which is searched separately for crossings. Another set of backward pointers is required to chain together nodes in the same subdomain, plus a 3D array of pointers to the end of the chain in each subdomain (see Figure 5).

On parallelisation, we are immediately presented with the choice of whether to decompose the data in the linked list structure or in the spatial lattice. Each choice has its disadvantages.

(a) *Linked list decomposition.* The parallelisation of the dynamical evolution is straightforward, but the search algorithm is hard, as nodes which are local in real space need not be on the same processor.

(b) *Spatial decomposition.* The parallelisation of the search algorithm is almost trivial, as each processor can search independently. However, the dynamical evolution now becomes the difficult part to implement, as nodes can cross processor boundaries, and both the dynamical data and connectivity information has to be propagated constantly between processors.

We chose route (b), as the complexities associated with implementing the search algorithm across processors in message-passing seemed greater than those associated with transferring data between processors.

Our decomposition is 1-dimensional: each processor looks after one slice of the simulation volume V. This again is easier to deal with than a proper 3D decomposition, as each processor only has two neighbours. However, it reduces scalability, as the volume per processor inevitably increases as $V^{2/3}$, and eventually one always runs out of memory.

Given this decomposition, one must develop extra data structures to cope with the fact that strings are distributed across processors. Consider Figure 6. In order to evolve node B by one time step, processor π needs the dynamical information from local node A and also from nodes C and D on processor $\pi + 1$. We call C and D *halo nodes* for processor π, and keep copies of their dynamical data x^i and v^i. Conversely, processor $\pi + 1$ needs the dynamical data from node B on π as well as that from local nodes D and E. Node B is therefore a halo node for processor $\pi + 1$, and copies of its dynamical data are required locally on $\pi + 1$.

We introduce a data structure associated with the halo nodes, which contains pointers to where the copy of the halo node data can be found locally, as well as to where the original is stored remotely. A diagram of how this works is contained in Figure 7.

For a node like B whose forward neighbour is on a remote procesor, the forward pointer points to a halo data entry instead of a to another node. This contains a pointer to the copy C' of the dynamical data of the neighbouring node C. Again, as C's forward neighbour, D, is also a halo node, its forward pointer also points to a halo data element. The chain terminates at D', as halo nodes are no more than two deep. A similar chain can be traced back from node C on processor $\pi + 1$, although there need only be one halo node in the reverse direction, because of an asymmetry in the evolution algorithm.

These then are the main data structures associated with the network and its evolution. There are also flags associated with each node, which are used when searching through the list, to check whether the node has already been visited, and also to mark nodes which are on small closed loops (we call these *flags of convenience*). "Small" here means less than or equal to a user-defined threshold size of one, two, or three links.

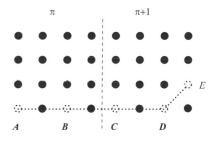

Figure 6. A string, with nodes labelled A to E, crossing the boundary between processors π and $\pi + 1$. The data structures associated with this configuration are shown in Figure 7.

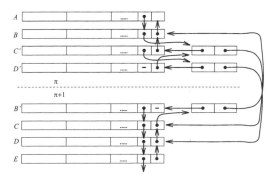

Figure 7. Data structures associated with a string crossing a processor boundary. On the right are the halo data structures, which contain pointers to the local copies (indicated by primes) of nodes belonging to a neighbouring processor. The copies are necessary for the dynamical update of the nodes' positions and velocities.

Initial conditions

Generating the initial conditions requires a 3D lattice of phases, which again follows a 1D decomposition. Recall that the formation algorithm generates a network of stationary strings. A stationary link is two lattice units in length, so it would seem that the formation lattice must be half the linear dimension of the evolution lattice. We could also map one link on the formation lattice to any integer multiple of string links on the evolution lattice, so in fact the ratio of linear dimensions of the two lattices could be any even number. This allows one to control the initial step length of the network, and hence the initial density.

ALGORITHMS

Formation

The algorithm for generating the initial conditions has not been parallelised. Only one processor at a time is active, and so only one string is traced and hooked up at a time. When the string is traced to a processor boundary, the work is handed on to the adjacent processor. When all sites on the lattice of phases have been visited, the evolution phase can begin.

This is not a serious problem for code performance, as the generation of initial conditions takes a small fraction of the total time for a typical simulation. Nonetheless, successfully parallelising this part of the code would be an interesting problem, and would eliminate an inelegance. A useful feature of the code is that it can also start by reading in the state of a string network generated and evolved by another run.

Evolution

With the correct data in the data structures, the first step of the update is trivial, and merely involves each processor scanning through its nodes and updating according to the ordinary serial algorithm. The difficulty arises when nodes have to be moved. This happens when the evolution takes a node across a processor boundary. It is not just a simple matter of transferring the node from the local data linked list to the halo data linked list on the processor which it leaves, and performing the reverse operation on the processor at which it arrives, for there are many cases depending on whether the node's neighbours are halo nodes or not. The logic is too complex to be itemised here, and will be detailed elsewhere[13]. Moving nodes after reconnection is clearly one of the major hurdles to be overcome in this scheme. However, if the algorithm is wrong it is likely to be catastrophically wrong, which eases the pain of debugging slightly.

Analysis

The analysis part of the code is not yet fully developed. One can output the number of loops of string with 1, 2, 3, or greater than 3 links, and one can also output a user-defined selection of any or all of the components of the energy-momentum tensor. In the evolution, one finds that very many degenerate loops consisting of one link, whose ends are at the same spatial point, are split off from the network.

One can define a scale length, $\xi(t)$, for the rest of the network, not including small loops, by considering the total length $L(t)$ and taking

$$\xi(t) = \sqrt{V/L(t)}, \qquad (11)$$

Table 1. Timings from various size runs of the parallel string evolution code run on the EPCC Cray T3D. ‖ cost is the percentage of the CPU time spent in the extra routines introduced by parallelisation, which is sensitive both to the percentage of halo nodes and to the total number of nodes.

Procs.	Total nodes	Nodes/Proc.	Halo Node %	CPU time per step	‖ cost
4	466k	116k	3.2	6.3s	43%
8	466k	58k	6.2	3.9s	53%
16	466k	29k	12.5	2.7s	65%
32	466k	15k	24.3	2.2s	73%
32	3.7M	116k	12.5	26.3s	83%

where $V = L^3$ is the simulation volume.

RESULTS AND TIMINGS

It turns out that very little time is taken up with the inter-processor communication. The biggest penalty comes from the extra processing associated with the new data structures, particularly scanning through the linked lists when string segments are transferred. It is therefore best to minimise the proportion of string which is liable to be transferred, by making the domain that each processor controls as thick as possible (it must in any case be thicker than 8 lattice points in order for it to be impossible for any node to be simultaneously a halo node on the processors either side). One can see from Table 1 how the cost of the parallelisation increases as the fraction of halo nodes increases. By cost we mean the time the code spends in the extra routines imported by the parallelisation.

Figures 8 and 9 come from a test run with evolution lattice size $L = 1024$, $\xi_0 = 8$, and 64 processors. There are a total of 10^6 nodes in the simulation. These days a run with 10^6 nodes is easily feasible on a workstation. However, this is nearly the largest feasible on the T3D with its limited 64 Mb per processor, and illustrates the drawback of the 1D decomposition. Much larger runs can be done on machines with more memory per processor.

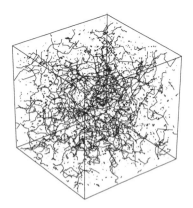

Figure 8. A box of string taken from the end of a run with the parallel cosmic string on the EPCC T3D. String in the form of loops with only one link are not plotted.

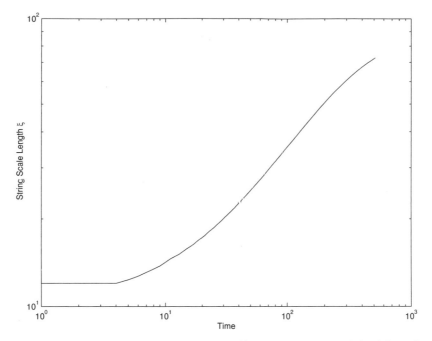

Figure 9. Log–log plot of the density length scale $\xi(t)$ against time t. ξ is defined from the total length of string *not* in one-link loops $L(t)$, according to $\xi(t) = \sqrt{V/L(t)}$.

Figure 8 shows the network at time 512, which is when we stop the simulation (we have not plotted the loops of length 2δ for clarity). Figure 9 is a log–log plot of the length scale $\xi(t)$ defined from the total length of string in loops of length greater than 2δ. The algorithm preserves the invariant length of string, so what we are seeing is how the string network transfers the energy in the form of long string, stretching across the entire simulation volume, to energy in the form of tiny degenerate loops. Conventionally, dimensional analysis would indicate that at late times, $\xi(t)$ should be proportional to t, as there is no other scale in the dynamics other than the time. However, it is clear that the behaviour is not very linear, nor is it some other simple power law. It is possible that there is some kind of subtle lattice effect here: ξ is much bigger than the lattice spacing at the end of the simulation, but the energy loss from the network is to loops whose size is of order the lattice size. However, there is no sign of the lattice scale on the long string. For more discussion on this point see Vincent, Hindmarsh and Sakellariadou[7].

SUMMARY AND FUTURE WORK

Cosmic strings may have been formed in the very early Universe. If so, they are predicted still to be around, although their density would be very low. They are thought to behave more or less like free relativitic strings, which reconnect every time they pass through each other. Simulating these objects directly is a challenging parallel programming problem which we can claim to have only partly solved. The 1D decomposition is unsatisfactory and should be improved to a properly 3D one, and the formation phase of the code should also execute in parallel. Nonetheless, our code can still perform much larger runs than are feasible on a serial machine.

An interesting recent development is that with the increased availability of parallel resources we are also able to simulate the underlying theory describing the dynamics of the wave function itself[14]. This enables us for the first time to check the assumptions that went into the "ideal string" approximation.

For the immediate future, the main requirement is to port more of the analysis routines of the serial code, in particular those associated with calculating the Fourier transforms of the components of the energy-momentum tensor. These are required for calculations of the gravitational perturbations induced by the strings. A secondary goal is to be able to output the distribution of the sizes of loops. It is useful to know this to be able to check how the string is distributed between small loops and the few long strings which occupy most of the simulation volume. Serial simulations indicate that the scale $\xi(t)$ defined from the total length of *long* string only is much closer to linear in t. It may be that the small loops at or near the lattice size do not have anything to do with the physics of real strings. This claim is reinforced by the results of Reference 14, in which linear behaviour of $\xi(t)$ is observed with little or no loop production, the energy being transferred directly from the network into radiation, which is not included in the ideal string approximation. This somewhat surprising and controversial result indicates that simulations of the ideal string type, although possessing a far greater dynamic range, should be interpreted carefully. In particular, loop production is to be treated as an approximation to the real energy loss mechanism.

ACKNOWLEDGEMENTS

This work was done as part of the Particle Cosmology HPCI Consortium, using resources made available by the EPSRC High Performance Computing Initiative, and was also funded by PPARC under grants B/93/AF/1642 and GR/K55967, and by a PPARC studentship (GV).

REFERENCES

1. A. Vilenkin and E. Shellard, *Cosmic Strings and other Topological Defects*. Cambridge: Cambridge University Press (1994).
2. M. Hindmarsh and T.W.B. Kibble, Cosmic strings, *Rep. Prog. Phys.* 58:477 (1995).
3. A.G. Smith and A. Vilenkin, Numerical simulation of cosmic string evolution in flat space-time, *Phys. Rev.* D36:990 (1987).
4. M. Sakellariadou and A. Vilenkin, Numerical experiments with cosmic strings in flat space-time, *Phys. Rev.* D37:885 (1988).
5. M. Sakellariadou and A. Vilenkin, Cosmic-string evolution in flat space-time, *Phys. Rev.* D42:349 (1990).
6. G.R. Vincent, M. Hindmarsh, and M. Sakellariadou, Correlations in cosmic string networks, *Phys. Rev.* D55:573 (1997).
7. G.R. Vincent, M. Hindmarsh, and M. Sakellariadou, Scaling and small scale structure in cosmic string networks, *Phys. Rev.* D56:637 (1997).
8. T. Vachaspati and A. Vilenkin, Formation and evolution of cosmic strings, *Phys. Rev.* D30:2036 (1984).
9. A. Albrecht and N. Turok, Evolution of cosmic string networks, *Phys. Rev.* D40:973 (1989).
10. D.P. Bennett, High resolution simulations of cosmic string evolution: Numerics and long string evolution, in: *Formation and Evolution of Cosmic Strings* G. Gibbons, S. Hawking and T. Vachaspati, eds., Cambridge University Press, Cambridge (1990).
11. F.R. Bouchet, High resolution simulations of cosmic string evolution: Small scale structure and loops, in: *Formation and Evolution of Cosmic Strings* G. Gibbons, S. Hawking and T. Vachaspati, eds., Cambridge University Press, Cambridge (1990).

12. E.P.S. Shellard and B. Allen, On the evolution of cosmic strings, in: *Formation and Evolution of Cosmic Strings* G. Gibbons, S. Hawking and T. Vachaspati, eds., Cambridge University Press, Cambridge (1990).

13. N. Floros, M. Hindmarsh, M. Sakellariadou, G. Vincent and I. Wolton, (1998) in preparation.

14. G. Vincent, N.D. Antunes, and M. Hindmarsh, Numerical simulations of string networks in the abelian higgs model, *Phys. Rev. Lett.* 80:2277 (1998).

THE UK MHD CONSORTIUM: GOALS AND RECENT ACHIEVEMENTS

Alan Hood,[1] Tony Arber,[1] Klaus Galsgaard,[1] Axel Brandenberg,[2]
Steve Brooks,[2] Chris Jones,[3] Graeme Sarson,[3] and Gavin Pringle[4]

[1]University of St Andrews
[2]University of Newcastle
[3]University of Exeter
[4]Edinburgh Parallel Computing Centre

INTRODUCTION

The UK MHD Consortium was formed in October 1996 and brings together the prominent MHD groups in the UK. The research topics are in Astrophysics with particular emphasis on the Solar Corona (St Andrews University), the Earth's Geodynamo (Exeter, Glasgow and Newcastle Universities) and dynamo action in turbulent plasmas (Newcastle University).

Before describing some of the projects being run by the MHD Consortium, it is important to discuss the MagnetoHydroDynamic equations that unify the research of these different research groups.

$$\rho \left(\frac{D\mathbf{v}}{Dt} \right) = -\nabla p + \mathbf{j} \times \mathbf{B} + \rho \mathbf{g} + \rho \nu \nabla^2 \mathbf{v}, \tag{1}$$

$$\frac{\partial \rho}{\partial t} + \nabla \cdot (\rho \mathbf{v}) = 0, \tag{2}$$

$$\frac{\partial \mathbf{B}}{\partial t} = \nabla \times (\mathbf{v} \times \mathbf{B}) + \eta \nabla^2 \mathbf{B}, \tag{3}$$

$$p = \rho RT / \tilde{\mu}, \tag{4}$$

$$\frac{\rho^\gamma}{\gamma - 1} \left(\frac{D}{Dt} \right) \left(\frac{p}{\rho^\gamma} \right) = \nabla \cdot (\kappa \nabla T) - \text{radiation} + \text{heating}, \tag{5}$$

$$\mathbf{j} = \frac{\nabla \times \mathbf{B}}{\mu}, \qquad \nabla \cdot \mathbf{B} = 0, \tag{6}$$

where ρ is the plasma density, $\mathbf{v}$ is the plasma velocity, p the gas pressure, $\mathbf{j}$ the electric current density, $\mathbf{B}$ the magnetic induction, $\mathbf{g}$ the gravitational acceleration, T the temperature, R the gas constant, $\tilde{\mu}$ the mean molecular weight, γ the ratio of

High Performance Computing
Edited by R. J. Allan *et al.*, Kluwer Academic / Plenum Publishers, New York, 1999

specific heats, κ the anisotropic thermal conductivity, η the magnetic resistivity and μ the magnetic permeability. The total time derivative is defined as

$$\frac{D}{Dt} = \frac{\partial}{\partial t} + (\mathbf{v} \cdot \nabla). \tag{7}$$

In some cases the energy equation (5) is replaced by a simpler form, either adiabatic or specified temperature profile. The addition of the magnetic terms to the usual fluid equations introduces many new effects of which the most important is *magnetic reconnection.*

The three main classes of problems being run on the T3D are (i) the MHD modelling of the Solar Corona, (ii) the Geodynamo and (iii) dynamo action in turbulent plasmas

THE EARTH'S GEODYNAMO

The geodynamo project at Exeter University is interested in numerically modelling the MHD processes responsible for the production of the geomagnetic field. Previous work by the group has focussed on the 'mean field' approximation,[1] where only the axisymmetric field and a single non-axisymmetric mode are considered. This simplification of the true 3D problem has allowed the importance of a number of different physical processes, and the dependence on the relevant non-dimensional parameters, to be investigated. There is still much that can be done with this simplified model. In particular it is important to investigate the effect that the use of hyper-diffusivities – adopted by many workers to avoid the numerical difficulties associated with the low viscosities pertaining to the Earth's core – has had on these solutions. Although the bulk of this work has been carried out on workstations to date, an HPF version of the code has been developed for the T3D, allowing runs to be carried out at the higher resolutions needed for these low viscosity runs.[2] Some work has also been put into an MPI version of this code, which should ultimately prove more efficient.

To truly assess the limitations of the mean-field work, it is necessary to relax the two-mode approximation, however, and move towards a fully 3D treatment. The development of a serial code for this is largely complete; it has many similarities to the mean-field code, so that, with our current experience, porting this code to the Cray should proceed without too many difficulties.

The main constraint for running these codes is CPU time, rather than memory. For an expansion with N modes in each dimension, the time taken per time-step varies as N^3 for the mean-field code. A Courant-type constraint on the step-size means that the total time-dependence for realistic runs is nearer to N^4, however. For the 3D code, the CPU requirements will effectively scale as N^5.

In order to cover a reasonable time-span, a low resolution 3D run will require approximately 100 single-processor hours; the higher truncation runs envisaged will each require several thousand hours. To conduct a proper survey of 3D dynamo behaviour, a substantial number of runs will be required. An initial estimate for a reasonable project, of 60 low resolution runs and 15 high resolution runs, would require of order 60000 single-processor hours.

Projects that could be carried out with the mean-field code might be computationally somewhat less expensive, but the higher resolution required for the low viscosity calculations of current interest in this approximation largely offsets this.

DYMAMO ACTION IN IRROTATIONAL TURBULENCE

Fluid motions that are generated by density fluctuations (for example in the early Universe) are irrotational. The motivation behind our work is to investigate the relationship between such irrotational turbulence and the dynamo. The governing equations are the continuity equation, a Bernoulli-type equation and the induction equation. The equations are solved using sixth order centred finite differences with a third order time step in a periodic box. The code is written in HPF for the Cray T3D. The initial condition for the velocity potential is a random field. We find that the velocity decays with time to the power -2/3 and the correlation length of the flow increases with time to the power +1/3. In order to understand these results we need to carry out calculations modifying some of the physics in the code to isolate the relevant effects. As a long term goal we want to investigate the nature of turbulence under general relativistic conditions, because the bulk motions in the early Universe are comparable with the speed of light. The code has been tested successfully in HPF v2.1 with a mesh of 128^3.

THE SOLAR CORONA

There are several areas of coronal research that are using or have used the T3D. One of the major outstanding problems in Solar Physics is to explain the high temperature of the solar corona. At the photosphere the temperature is the order of 6,000K but it rises through the chromosphere and transition region to reach about 2×10^6K in the corona. This necessitates the need for a heating mechanism to keep the corona at such a high temperature. The heating mechanism has to be magnetic in nature and both current sheet (narrow, intense, current concentrations) and wave heating mechanisms depend on the resistive terms in the induction equation, Equation (3). Many stars are now known to have a hot corona and there is a pressing need to understand how the solar corona is heated. A numerical experiment run on the T3D involved the braiding of the coronal magnetic field by random photospheric motions.[3] The plasma at the photospheric boundaries was sheared in one direction (say x direction) and then sheared in the other, y direction. After a few repetitions, current sheets began to form and the magnetic field lines started to reconnect, releasing magnetic energy in the form of Joule heating. The current sheets are clearly seen in the isosurfaces of current shown in Figure 1. The time evolution of the Joule dissipation and the Poynting flux (Figure 2) show that the Joule dissipation grows exponentially as the initially uniform field is stressed. Then it saturates at the value of the Poynting flux. Thus, the magnetic energy introduced through the motions of the boundary is converted into heating the plasma. A typical high resolution run required 10,000 processor hours to simulate 100 Alfvén times.

Another interesting coronal problem is the formation, support, stability and eruption of quiescent solar prominences. An example of a prominence in shown in Figure 3. Stellar prominences have now been observed and, while there are many similarities between the stellar and solar cases, there are also many differences. An important problem is to understand how they form. Observations suggest that the old proposed mechanisms of either thermal condensation or ballistic injection are not valid. Instead, prominences are observed to form when there is

1. a magnetic polarity inversion line at the photospheric surface, separating pre-existing magnetic flux systems and newly emerging flux regions,

2. flux convergence, as two separate flux systems approach one another and

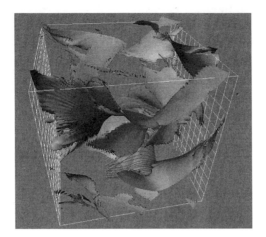

Figure 1. Iso-surfaces of the current density. The initial field contained no current concentrations and this simulation confirms that photospheric motions can generate current sheets in the overlying corona.

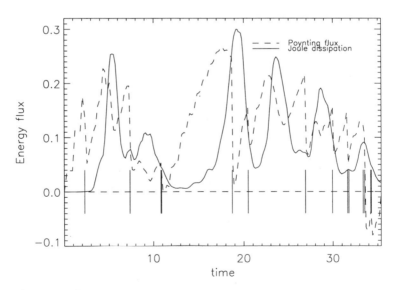

Figure 2. The time evolution, in units of the Alfvén crossing time, of the Joule dissipation and the Poynting flux.

Figure 3. A quiescent prominence containing cool, dense plasma surrounded by the hot tenuous corona.

3. flux cancellation, where opposite polarities interact and cancel.

This last condition suggests that magnetic reconnection could occur at the chromospheric level and dense chromospheric plasma would be trapped on the newly reconnected field lines and then lifted up into the corona by magnetic forces. A numerical simulation was performed in which an old flux system is represented by two lines of opposite magnetic flux.[4] A new bipolar region is shown as a pair of small circular flux regions. The bipolar region is moved towards the old flux region. Figure 4 shows that the convergence of the flux system results in an increase in plasma density. Subsequent magnetic reconnection occurs in this high density region and lifts the plasma up into the corona.

The third solar project follows the non-linear evolution of an unstable twisted coronal loop. An important stabilising feature for coronal magnetic fields is the inertial line-tying, provided by the dense photosphere, that effectively anchors the magnetic footpoints. The code for this project is a constrained transport model[5] using Van Leer limiters[6] on the gradient terms in the numerical flux functions. This is based on a TVD MUSCL approach and is designed to maintain $\nabla.\mathbf{B} = 0$ to machine precision as well preserve the monoticity properties of the advection equation. Results from this code (Figure 5) show the nonlinear formation of narrow current sheets around the unstable current column.[7] Such current sheets are expected from linear theory. Non-linear simulations are needed to determine saturation levels, late time development and heating resulting from this explosive release of magnetic energy. This code is written in HPF and performance of the code is shown in Figure 6. Here the tests were for 81^3 and 161^3 grids on both the T3D and T3E. Efficient parallelism is clearly obtained for this code with typical speed ups of 1.9 on doubling the number of processors.

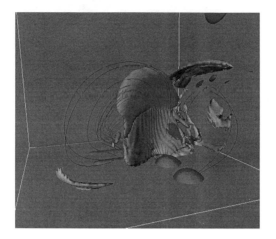

Figure 4. The two bipolar flux systems are moved towards each other. Magnetic reconnection occurs above the photosphere and lifts the dense plasma up into the corona.

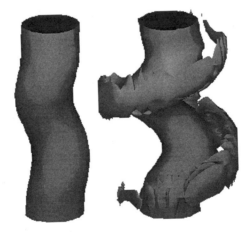

Figure 5. Displacement of central current column has generated a current sheet at the outer mode rational surface. Snapshots were at time $t = 5$ and $t = 10$.

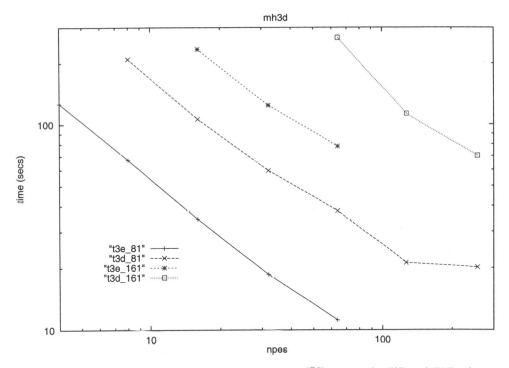

Figure 6. Runtime vs. number of processors (npes) for HPF code on the T3D and T3E using different size grids.

ACKNOWLEDGEMENTS

The T3D computing time is made available to the consortium by PPARC. We would also like to acknowledge the invaluable help of the support staff at EPCC.

REFERENCES

1. R. Hollerbach and C. A. Jones, A geodynamo model incorporating a finitely conducting inner core, *Phys. Earth Planet Inter.* 75:317 (1993).
2. G. R. Sarson, C. A. Jones and A. W. Longbottom, Convection driven geodynamo models of varying Ekman numbers, *Geophys. Astrophys. Fluid Dyn.* (submitted 1998).
3. K. Galsgaard, and Å. Nordlund, Heating and Activity of the Solar Corona: 1 Boundary Shearing of an Initially Homogeneous Magnetic Field, *J. Geophys. Res.* 101:13445 (1996).
4. K. Galsgaard and A. W. Longbottom, Formation of Prominences by Flux Convergence, *Astrophys. J* (submitted 1998).
5. C. R. Evans and J. F. Hawley, Simulation of magnetohydrodynamic flows: a constrained transport method, *Astrophys. J.* 322:659 (1988).
6. B. van Leer, Towards the ultimate conservative difference scheme V. A second-order sequel to Godunov's method, *J. Comp. Phys.* 32:101 (1979).
7. T. D. Arber, A. W. Longbottom and R. A. M. Van der Linden, Current sheet formation and the effect of anomalous resistivity in unstable coronal loops. In preparation.

N–BODY SIMULATIONS OF GALAXY FORMATION ON A CRAY T3E

P. R. Williams and A. H. Nelson

Department of Physics and Astronomy
University of Wales College of Cardiff
PO Box 913, Cardiff CF2 3 YB

In this article we present results from a galaxy formation simulation carried out on the Cray T3E at the Edinburgh Parallel Computing Centre, as part of the Virgo Consortium's time allocation. The simulation used in excess of 10^5 particles, larger than any simulation of this type in the literature to date.

In the following sections we give a brief introduction to the problem of galaxy formation, discuss the numerical methods used in particle simulations of galaxy formation, briefly review previous approaches to parallelisation, and describe the new algorithm used here.

INTRODUCTION

Galaxies are bound agglomerations of gas, stars and non–luminous (dark) matter. Understanding the physical mechanisms which drive their formation is one of the central problems faced by astrophysicists today.

The currently most popular and successful paradigm for how galaxies form focuses on the gravitational collapse of a region in the early Universe. The collapse of such a region is a highly non–linear and violent event, incurring large density contrasts and complex geometries in the proto–galactic objects which form.

To simulate such physical systems requires numerical techniques that can model the coupled evolution of dark matter, stars and self–gravitating gas. This involves calculating long range gravitational interactions, short range hydrodynamical interactions in the gas, and the conversion of gas mass into stars.

Furthermore, the numerical techniques used must be able to resolve the large density contrasts inherent to a collapse scenario; and must be free from imposed geometric constraints.

High Performance Computing
Edited by R. J. Allan *et al.*, Kluwer Academic / Plenum Publishers, New York, 1999

NUMERICAL TECHNIQUES

The combination of two numerical techniques, Smoothed Particle Hydrodynamics [1] and Treecode Gravity [2, 3] has been used extensively to study a wide range of astrophysical scenarios. The particulate, Lagrangian nature of these methods make them ideally suited to the problem in question.

Treecode is an efficient method of calculating the gravitational acceleration of a particle. This is achieved by devolving the calculation into short and long range interactions. Near interactions are calculated by direct summation of particle–particle Newtonian accelerations, while distant interactions are coarsely resolved. Using Treecode, the number of operations in calculating the gravitational accelerations at N particle positions is $\propto N \log N$. Hence, for large N, Treecode is vastly more efficient than direct summation, which scales $\propto N^2$.

Smoothed Particle Hydrodynamics (SPH) is a numerical interpolation method which can also approximate the spatial derivatives of a variable at any position, given a set of neighbouring points at which the variable is known. SPH is used to interpolate gas densities, and calculate spatial derivatives (e.g. velocity divergence $\nabla.\vec{v}$) required in the calculation of hydrodynamic time derivatives. To do this, a list of neighbouring particles is required: this is obtained for each particle by examining a linked–list, which has already been constructed for use in the gravity calculation.

The large range in densities (typically $>10^{11}$) present simultaneously in a proto–galaxy implies that a large range in dynamical timescales are also present simultaneously. Therefore, to improve efficiency, an individual particle timestep time integration scheme is commonly used [4, 5]. This adds significant complexity to the algorithm, but can reduce the number of expensive acceleration calculations required by a factor >15.

Methods of effectively implementing star formation in particle simulations are in their infancy. The method used here models a star formation rate $\dot{\rho}_* \propto \rho_g^n$, where ρ_g is the gas density and ρ_* is the stellar density [6]. This is commonly known as a Schmidt star formation law [7]. To model the conversion of gas mass into stars on large scales, star mass particles are then regularly created to statistically represent the spatial distribution of recent star formation.

METHODS OF PARALLELISATION

There have been a number of successful implementations of Treecode on parallel machines, mostly intended for the massively parallel CM machine [8, 9, 10]. The most efficient version to date is the Orthogonal Recursive Bisection (ORB) algorithm [11, 12] which has recently been modified to include SPH [13].

The method is to allocate a spatial domain to each processor such that the work is shared equally between the processors. Each processor must then construct a local linked list. In order to calculate the gravity accelerations and find particle neighbour lists, relevant linked list information must then be communicated between neighbouring processors. When SPH is included, an overlap region between spatial domains must be included, in order to find neighbours for particles near the domain boundary. As the system evolves in time, the domain boundaries move to maintain load balancing.

The strengths of this algorithm are that the data is distributed amongst the processors, little duplication of data is required, and there are no serial operations. Its disadvantages are that the efficiency in SPH tests was found to be $<50\%$ when using >16 processors [13], the algorithm is very complicated, and there is significant book keeping required to deal with the storage and communication of the individual linked lists on each processor.

The most computationally expensive part of the calculation in Treecode–SPH algorithms is the acceleration evaluation. This involves calculating the gravitational acceleration, finding SPH particle neighbours and evaluating the relevant quantities using SPH. In tests of serial Treecode–SPH, these operations take typically ~95% of the total CPU time. The new algorithm used here concentrates on parallelisation of only this, the most costly part of the algorithm. Constructing the linked list and integrating particle variables in time remain serial operations.

Each processor has a copy of all particle data, is assigned an equal fraction of the total particle set for the duration of the simulation, and calculates their accelerations and SPH quantities when required. Some quantities are not known globally on all processors, and are required on all other processors before further calculation can continue. At each timestep, processors must globally communicate 8 words per particle for all particles within it's own subset for which an acceleration calculation has been carried out. To avoid message bottlenecks when large numbers of processors are used, the processors are configured in a ring structure. Messages are passed around the ring, ensuring that each processor only ever has one message at any one time.

Such an approach has the advantages of minimizing complexity in the algorithm; while still making possible simulations containing large particle numbers and attaining a significant reduction in run time. Communication overheads are minimal, and the star particle formation method discussed above fits easily into the algorithm. Also, an individual particle timestep integration scheme can be used, giving a further significant decrease in run time.

Its disadvantages are that the algorithm does not scale well for large numbers of processors and small numbers of particles. This is because the fraction of total time spent in the serial parts of the algorithm increases dramatically as the number of particles $N \to 0$ and the number of processors $p \to \infty$; and communication speed is increased when optimally large messages are sent. Also, the size of simulations is limited by the memory available to each processor, since all processors must have a copy of all particle variables.

RESULTS AND DISCUSSION

The simulation discussed below was carried out with this new code, and took ~15 days on 32 processors of the Cray T3E at The Edinburgh Parallel Computing Centre. In total 128,000 particles were used. This is larger than any simulation of the formation of a single galaxy in the literature to date. The high spatial resolution has for the first time resolved unambiguously the details of how a disk galaxy might form.

Such a simulation could only have been calculated on a high performance parallel computer such as a Cray T3E: it is estimated that to carry out the simulation on a workstation would have taken ~8 years. In total, $\sim 2 \times 10^6$ timesteps were carried out. The speed–up was found to be ~16. Load–balance was found to be excellent: processors spent typically <1% of their time idle. The percentage time spent in communication was ~2%.

Initially, mass was distributed in a uniform density sphere, in solid body (clockwise) rotation about the z–axis, and expanding radially about its centre (in Hubble flow). At the start of the simulation there were no star particles, only gas and dark matter.

The simulation evolved as follows: the initial expansion was halted by gravity, and the gas and dark matter began to collapse. At this time only a small amount of gas had been converted into stars, due to the low gas density. The collisionless dark

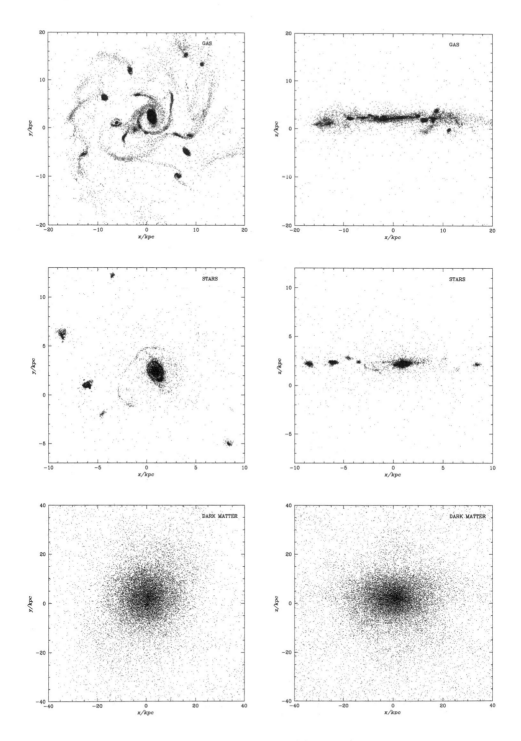

Figure 1. Particle plots taken during the most dynamically active phase of the simulation. Hydrodynamic shocks, thin linear features and dense knots of matter are common in the distribution of gas and star particles. Already, the collisionless dark matter particles have reached a smooth equilibrium configuration in the central regions. From top to bottom: gas, star and dark matter particles are shown.

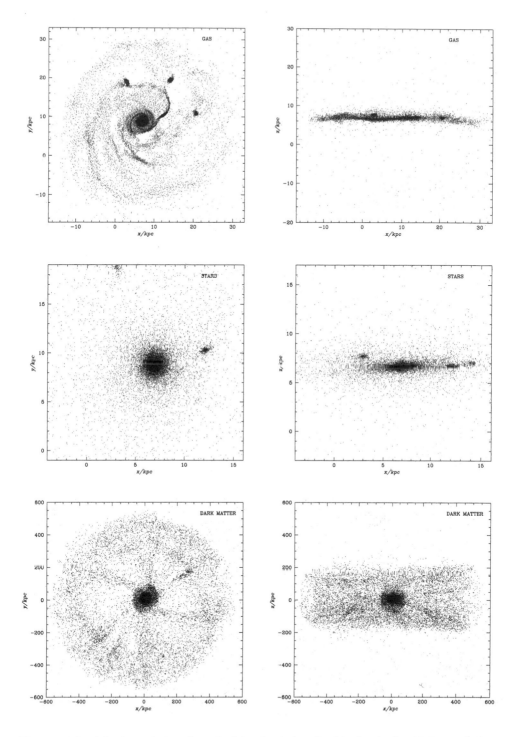

Figure 2. Particle plots taken at the end of the simulation. By this time in the simulation (~2 Gyr after figure 1), gravitational interactions have smoothed out most of the highly complex and irregular geometries, tending to circularize the particle orbits. In this figure, the full extent of the dark matter distribution is shown. From top to bottom: gas, star and dark matter particles are shown.

matter passed through the origin of collapse and relaxed to a flattened spheroidal distribution. However, gas is able to viscously dissipate energy. Therefore, on collapse, the gas shocked in the x–y plane and formed a thin, rotating disk. This phase in the formation of the galaxy was characterized by highly complex geometries, thin linear features, shocks, and fragmentation in the gas distribution (see figure 1). The high densities within the gas fragments induced high star formation rates. Subsequent evolution involved the disruption of these gas rich fragments through tidal interaction, and progressive smoothing and circularization of the disk matter (see figure 2).

CONCLUSIONS

The only way to improve our understanding of the physical processes important in forming a galaxy is to carry out numerical simulations of greater resolution and to model more sophisticated physics. Parallel computing is the only tool available at present which can hope to achieve this aim.

The parallel algorithm used here is simple to implement, and was found to give good efficiency (>80%) for small numbers of processors (<16) and large numbers of particles (>5×10^4). The algorithm does not scale well for very large numbers of processors, and the number of particles that can be used is limited by the memory available to each processor (the memory requirements are ~2000 particles per Mb). However, the simulation discussed here demonstrates that the algorithm is still a very useful tool in carrying out simulations of this type.

The additional time and effort required to implement a fully parallel algorithm such as ORB would not result in a comparative increase in performance on the present algorithm: for a numerical astrophysicists, the detailed computer science of scaling, portability and efficiency must take second place to practicality.

ACKNOWLEDGEMENTS

This calculation was carried out under the Virgo Consortium's time allocation on the Cray T3E at the Edinburgh Parallel Computing Centre.

REFERENCES

1. Monaghan, J.J., *Ann. Rev. Astron. Astro.* 30:543 (1992)
2. Barnes, J.E. and Hut, P., *Ap. J. Supp.* 64:715 (1989)
3. Hernquist, L., *Ap. J. Supp.* 64:715 (1987)
4. Hernquist, L. and Katz, N., *Ap. J. Supp.* 75:419 (1989)
5. Navarro, J.F. and White, S.D.M. *Mon. Not. Roy. Astro. Soc.* 265:271 (1993)
6. Williams, P.R., *Ph.D. Thesis* Univ. Wales College Cardiff (1998) in preparation
7. Schmidt, M., *Ap. J.* 129:736 (1959)
8. Hillis, W.D. and Barnes, J., *Nature* 326:27 (1987)
9. Salmon, J.K., *Ph.D. Thesis* CalTech (1990)
10. Makino, J. and Hut, P., *Comp. Phys. Reports* 9:196 (1989)
11. Warren, M.J. and Salmon, J.K., *Supercomputing '92* 570 (1992)
12. Warren, M.J. and Salmon, J.K., *Supercomputing '93* 12 (1993)
13. Antonioletti, M., *Ph.D. Thesis* Univ. Wales College Cardiff (1997)

NOVEL METHODS AND APPLICATIONS

EARLY EXPERIENCE WITH THE TERA MTA SYSTEM

Richard M. Russell

Vice President, Marketing,
Tera Computer Company,
Eastlake Ave E.,
Seattle, WA 98102-3027
E-mail: *russell@tera.com,* WWW: *http://www.tera.com*

OVERVIEW

This paper is organized in four parts:
- Company status
- Multithreaded architecture
- Early results from SDSC
- Tera's price/performance environment

COMPANY STATUS

Tera celebrated its tenth anniversary in 1997. The company was founded by Jim Rottsolk and Burton Smith who had previously worked together at Denelcor, the maker of HEP data flow architecture systems. Initial work focused on compiler development and software validation of Tera's multithreaded architecture design.

After four years, hardware development began in 1991 with a small team of people, roughly the same number that worked on Manchester's Mark 1 in the late 1940s [1]. A prototype multithreaded architecture (MTA) system was first demonstrated in November, 1996.

The hardware design of the Tera MTA system is a consequence of the company's early investment in software design and research into automatic parallelization techniques, scalable computing, and scalable networks. Hardware design of Tera MTA did not begin until a comprehensive software environment for parallel programming had been articulated and tested against problems of interest via simulation.

A total of about $60M was spent in the company's first ten years. $20M of this was funding from the U.S. Government, principally DARPA. The balance was obtained from private investors and from a successful stock offering to the public in 1996. The company had approximately $14M on hand at the end of 1997 with revenue expected in 1998. During 1998, the company expects to transition from development into full-scale production and is cautiously ramping up MTA system manufacturing.

The company installed its first Tera MTA system at the San Diego Supercomputer Center (SDSC) at the University of California, San Diego (UCSD) at the end of 1997. Initial funding for this system was provided by the National Science Foundation with further funding expected from the Department of Energy and the Department of Defense. DARPA funding to UCSD is underwriting a further evaluation focusing on defense-related applications. Partners in the DARPA-funded evaluation are The Boeing Company, CalTech, JPL, Sanders, Tera, and UCSD.

MULTITHREADED ARCHITECTURE

Tera MTA systems are true shared memory systems that are architecturally scalable. They are designed to address three issues that have been inhibiting the growth of high performance computing:

- Inability of vector supercomputers to cost effectively scale to support an order of magnitude increase in high performance computing without a change of programming model — Tera's current MTA implementation scales to 256 processors and its uniform shared memory architecture scales essentially without limit.
- Difficulty of programming highly scalable distributed memory machines — Tera provides a high productivity programming environment that lets programmers concentrate on finding and exploiting parallelism without having to worry about data placement issues.
- Difficulty of obtaining scalable performance on problems with large irregular, dynamically changing grids or poor data locality due to algorithmic constraints — Tera has shown that such problems can scale to the limits of problem parallelism on a multithreaded architecture system.

Each Tera processor has up to 128 hardware streams. Each hardware stream can be thought of as a virtual processor with its own "state". That is to say, each has a program counter, 31 general registers, a stream status word, and target and trap registers.

Tera processors are configured as part of Resource Modules. Each Resource Module may have an MTA processor, an I/O processor, and memory boards.

The Table 1 shows Tera MTA system characteristics.

Communication paths between MTA processors, I/O processors and memory units is via a full bandwidth communications network, even in the case of a single processor system. Memory addresses are randomized at system startup time by the hardware. Contiguous addresses in memory usually reside on different memory boards.

Threads of computation are bound to individual hardware streams by user-level instructions that reserve/create/quit streams. Tera's optimizing compilers can generally find

Table 1. System Characteristics

System size, processors	16	32	64	128
Peak performance, gigaflop/s	14.4	28.8	57.6	115.2
Memory capacity, gigabyte	16-32	32-64	64-128	128-256
Bisection bandwidth, gigabyte/s	153.6	153.6	307.2	614.4
I/O bandwidth, gigabyte/s	6.4	12.8	25.6	51.2

enough loop-based parallelism in only moderately size programs (for example, structural analysis problems with only 30,000 finite elements) to ensure that an MTA-16 system can be kept busy.

A different hardware stream is activated every clock period. A stream is selected from a pool of streams that are ready to execute. Each stream may issue up to 8 memory references using an instruction lookahead feature. Figure 1 shows how a succession of active streams can keep an MTA processor busy while hiding memory latency.

A stream can stall while waiting for a memory transfer to complete if there is a dependency on the outcome of the transfer. If there is no dependency the stream can continue in the ready pool issuing instructions. As long as there are at least about 25 active streams per processor at all times, all of the remaining streams can be stalled waiting for memory operations to complete without causing the processor itself to stall. When a processor stalls it issues phantom instructions which can be counted. In total, there are eight hardware counters which can be used to collect performance measurement information. These counters are accessible to the user.

Hardware mechanisms manage the stream pool. A runtime operating system manages the interface between hardware and user programs, and Tera's Operating System.

A particular type of memory operation is used to synchronize threads. There are four additional bits associated with every 64-bit word. One of these bits is the full/empty bit which can be toggled by variants of the normal load and store instructions to show whether or not a

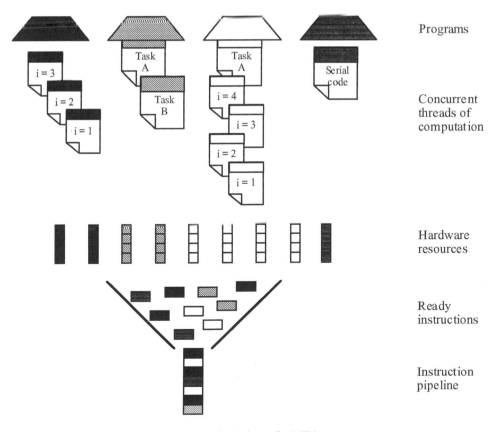

Figure 1 — A logical view of a MTA processor

547

word has been updated. Like all other memory operations, this very lightweight synchronization occurs asynchronously with other stream operation.

A benefit of Tera's lightweight synchronization is the ability of Tera's compilers to automatically optimize update operations of the form

$$Y(INDX(K))=Y(INDX(K))+X*A(K)$$

and run them in parallel. Such updates are a common feature of finite element analysis codes and generally cannot be run in parallel on other parallel architectures.

Tera supports TeraOS, a UNIX-based operating system, Fortran 77, Fortran 90 (planned for 1998), C, and C++. All of these languages share a common "back end" compiler that provides mixed language computability. Tera's compilers support Cray extensions in addition to a few Tera specific extensions that express parallelism. Support for OpenMP and MPI is planned.

References 2-4 give further information on Tera's hardware and software architecture.

EARLY RESULTS FROM SDSC

A summary of results measured on the Tera MTA system at SDSC can be found at *http://www.cs.ucsd.edu/users/carter/Papers/tera.ps* .

At the time of this writing, Tera has reported results on several applications including AMBER, SWEEP3D, and LSDYNA3D, all using one processor of the MTA system installed at SDSC.

AMBER is a production code used in rational drug design. The Tera MTA version optimized by Tera is 45% shorter than the AMBER code optimized for distributed memory systems.

Initial AMBER results are shown in Table 2. In cases without water, the Tera MTA system is slower than a T90. However, when water is present, the Tera system outperforms the T90. Tera believes that the MTA's performance of about 180 megaflops per processor is problem independent and will scale linearly with the number of processors.

SWEEP3D is a 3D neutral particle transport code that was used by the Department of Energy's ASCI program to benchmark competing systems. The sequential version of the code that Tera's compiler optimizes automatically is one fifth the size of the MPI version. Additionally, Tera's version will perform at about the same rate if it was modified to deal with irregular grids and contact problems. This is not true of distributed memory systems.

A single MTA processor computes a SWEEP3D ijk cell at 107ns compared to about 70 ns for a T90. Simulations show that an MTA-16 would outperform a T3E-512 on this

Table 2. Results from AMBER.

Problem	Tera MTA time (seconds)	SGI T90 time (seconds)
Pastocyanin 99 residues 1457 atoms	5.01	3.92
DNA in vacuo 274 residues 4282 atoms	17.82	8.09
DNA in water 2592 residues 7682 atoms	26.77	37.84

Table 3. LSDYNA3D Performance Scaling.

System size	Elements	Gflop/s	MFE/s
MTA-16	60,000	2	2
MTA-32	125,000	4	4
MTA-64	250,000	8	8
MTA-128	500,000	16	16

problem by a factor of about 1.67 and that an MTA-128 would outperform the so-called ASCI Red Intel-based system which has 9000 processors.

LSDYNA3D is a structural analysis application that is widely used to simulate car crashes. On a 27,000 finite element benchmark, Tera has shown that the MTA system can compute 4.2 timesteps a second at a computing rate of 135 megaflops per second. Tera believes this performance is scalable as shown in the Table 3. It is noteworthy that Tera obtains this performance on a large production code with minimal code changes, mainly the insertion of compiler directives. Given the freedom to tune the code, Tera believes that a factor of two increase in performance could be obtained.

TERA'S PRICE/PERFORMANCE ENVIRONMENT

Tera believes the MTA is significant in that it offers both high performance and ease-of-use. The current implementation of MTA uses a number of proprietary GaAs ASICs and thus Tera's systems are priced to compete with high end vector supercomputers like the SGI/Cray T90.

Tera expects that its performance advantage will be most marked on applications that do not scale on today's MPP, SMP, and PVP architectures. Where users have applications that scale well on today's architectures, Tera's advantage will be less clear, although in the case of SWEEP3D it is clear that scalability should be measured in terms of absolute performance and not speed-up.

In the future, Tera will reimplement the MTA system using CMOS technology and expects to replace its current MTA processor with a one- or two-chip microprocessor package. We believe that the technology curve for multithreaded architecture will be very exciting and that it will be possible to extend the benefits of uniform shared memory to systems with hundreds, and thousands of processors.

ACKNOWLEDGEMENTS

AMBER, and SWEEP3D results were submitted by John Feo, Tera's on-site representative at SDSC; Simon Kahan and Tera's application group did the work on LSDYNA3D.

REFERENCES

1. S.H. Lavington, *History of Manchester Computers*, NCC Publications (1975).
2. R. Alverson, D. Callahan, D. Cummings, B. Koblenz, A. Porterfield and B. Smith, The Tera Computer System, in: *Proceedings AlCa:Tera, International Conference on Supercomputing,* pp1-6 (June 1990).
3. G. Alverson, R. Alverson, D. Callahan, B. Koblenz, A. Porterfield, and B. Smith, Exploiting Heterogeneous Parallelism on a Multithreaded Multiprocessor, in: *Proceedings of the 6th ACM International Conference on Supercomputing* (July 1992)
4. G. Alverson, P. Briggs, S. Coatney, S. Kahan, R. Korry, Tera Hardware-Software Cooperation, in: *Proceedings of Supercomputing 97* (November 1997)

HPC ON DEC ALPHAS AND WINDOWS NT

Denis Nicole, Kenji Takeda and Ivan Wolton

Southampton HPCI Centre
Department of Electronics and Computer Science
University of Southampton
Southampton SO17 1BJ

ABSTRACT

We have obtained a dedicated computational cluster of eight DEC Alpha systems interconnected by 100 Hz switched Ethernet and running Digital Visual FORTRAN on Windows NT. This is an 8 Gflop/s (peak) system with 2 Gbytes of memory. The total cost was under £50,000. We have just finished porting MPI onto this environment and are now able to run mainstream UK HPC codes such as ANGUS. We believe that our system is a highly cost-effective environment for the development and medium-scale execution of science and engineering codes; it currently represents the biggest single computational resource at Southampton.

We present our early experiences with this leading edge medium-scale resource and some early performance results for sequential, MPI and PVM FORTRAN codes.

INTRODUCTION

The consolidation of the microprocessor market has resulted in there now being little distinction between the PC and workstation markets. By taking advantage of the inherent economies of scale in processors, memory, switching technology and software it is possible to utilise commodity components to build supercomputer-level machines at low cost. The Beowolf project [1] has concentrated on using Intel-based machines running Linux to provide very cost-effective production machines for a number of applications.

We have recently purchased a dedicated computational cluster of DEC Alpha workstations. These compete on a node for node basis with systems from IBM and SGI/Cray for many scientific and engineering applications, but by using commodity components the cost is lower by a factor of at least three. By also utilising the Windows NT operating system further savings can be made. Our long term goal is to provide a full, remote access service on this HPC system.

High Performance Computing
Edited by R. J. Allan *et al.*, Kluwer Academic / Plenum Publishers, New York, 1999

SYSTEM CONFIGURATION

The exact configuration of the installed DEC Alpha cluster is given in Table 1. As this is a compute cluster only two 21" and two 15" monitors were purchased. The other four main compute nodes share a single monitor through a manual switchbox. The total system cost was £50,000.

The server node uses Linux to allow good cross-compatibility with a variety of UNIX systems via Samba. We have used Debian Linux as it was the most up-to-date common distribution. The DLT drive represents a long-term investment for data backup.

ADVANTAGES

There are several advantages in opting to buy a DEC Alpha Windows NT system over other similar commodity supercomputing solutions. The 500 MHz DEC Alpha AXP21164 processor offers at least twice the performance of a Pentium system at only 50% higher cost (at Q3 1997 prices). By utilising fewer more powerful nodes, communication overhead can be significantly reduced which is the critical bottleneck in commodity supercomputer systems. These processors compete against Intel in the NT server market and are therefore priced aggressively. Additionally, the standard motherboard architecture means that true commodity components can be used, such as ordinary SIMMs, PCI bus cards and EIDE disks. In step with Intel's processor roadmap, Digital are continuing to keep pace and the onset of Samsung manufactured Alpha chips with enhanced clock speeds means more competition in the marketplace.

While the hardware is a considerable cost, software costs can also be a major issue. In order to run the best FORTRAN compilers the choice of operating system is limited to Digital UNIX and Windows NT.

Digital UNIX is expensive and this is exacerbated by the need to use SCSI rather than the cheaper EIDE disks. Windows NT Workstation is considerably cheaper than Digital

Table 1. Configuration of DEC Alpha cluster

Eight nodes, each with
500 MHz Alpha 21164 processor
256 Mbyte RAM
2.5 Gbyte EIDE drive
Two additional 5Gbyte drives to support Windows NT 5.0 and Linux
Windows NT version 4 (service pack 3)
Server node
200 MHz Pentium
32 Mbyte RAM
4 x 5 Gbyte EIDE drive
30 Gbyte DLT backup
Debian Linux
Network connectivity
100 Mbit Ethernet
100 Mbit twelve port Ethernet switch
Compilers
Digital Visual FORTRAN
Visual C++ v4.1 (RISC, shortly to be upgraded to v5.0)

UNIX, and the excellent Digital FORTRAN compiler is available complete with IMSL libraries at significantly lower cost than the UNIX version. Digital recently licensed Microsoft Developer Studio for use with its FORTRAN compiler and so programmers can benefit from this powerful, user-friendly environment. Additionally most Windows x86 packages run out of the box under the FX32! emulation package [2].

A fundamental problem of any 32-bit operating system, such as Windows NT 4.0, is that the maximum addressable memory space is limited to 2 Gbytes. Linux has a memory limit of 3-4 Gbytes. However, Digital's alliance with Microsoft in developing 64-bit Windows NT 5.0 means that Alpha systems will be able to overcome this limit as soon as the software becomes available.

We are pursuing the long-term goal of delivering an effective remote and local parallel computing service directly under Windows NT. Windows NT is the wave of the future, whether we like it or not.... and it runs Microsoft Office.

Uniprocessor Performance

For most small benchmarks the 500 MHz DEC 21164A Alpha is a 100 Mflop/s system. Linpack gives figures of 110 Mflop/s (201) and 97 Mflop/s (200). The Whetstone benchmark delivers 528926 kwhetstones per second. A performance breakdown running Livermore loops is given in Table 2. Enabling debugging which entails disabling compiler optimisations hurts performance badly, as shown in Table 3.

This level of performance is encouraging considering the price and is very respectable compared with what are traditionally regarded as "high-end" workstation systems such as those based on RS/6000, MIPS and UltraSPARC processors.

These benchmark figures translate into good real application performance. We have used the Alpha cluster to perform partitioning of a fifteen million element unstructured tetrahedral grid, which requires a two Gbyte memory region. Initially we performed this on one node of the Southampton IBM SP2 with 256 Mbytes of RAM and reconfigured to page off five SCSI disks simultaneously. This took nine hours to complete and required us to run in the overnight queue, and only on the one specifically configured node.

The same job took six hours to complete on one AlphaNT node. This used 256 Mbytes RAM and paged off a single EIDE drive. Setting up the necessary swapfile took just six mouse clicks and a reboot. We were therefore able to do eight partitioning jobs in parallel, overnight and without having to fight through any queues.

Table 2 Performance breakdown of 500 MHz DEC Alpha 21164 system running Livermore loops with full compiler optimisations (level 5).

Measurement	Mflop/s
Maximum rate	796.34
Quartile Q3	153.27
Average rate	141.33
Geometric mean	102.52
Median Q2	86.09
Harmonic mean	81.88
Quartile Q1	62.24
Minimum rate	29.76

Table 3. Performance breakdown of 500 MHz DEC Alpha 21164 system running Livermore loops with debugging enabled and no compiler optimisation.

Measurement	MFlop/s
Maximum rate	31.59
Quartile Q3	20.42
Average rate	15.11
Geometric mean	13.11
Median Q2	12.95
Harmonic mean	11.40
Quartile Q1	7.87
Minimum rate	4.63

DISADVANTAGES

There are many disadvantages in using Windows NT as it certainly is not as mature as UNIX. We can expect some improvements with version 5 which we are currently running under beta on a couple of machines. Other problems are being addressed at Southampton.

Filenames under Windows NT are not case-sensitive[1], ie: `prog.f` = `prog.F`, which can cause problems for some preprocessing makefiles. Resource leaks can occur, particularly in DLLs when processes are killed. NT is really only a one-at-a-time multi-user system as all users share, and can modify, the same drive map. It is only designed to support one interactive user at a time.

A significant area for improvement is in remote access. Remote logins using the Microsoft `telnet` daemon (in beta) are not very stable. Currently the only facility for remote windows graphical applications is WinVNC [3]. While initial tests demonstrate that this allows full graphical remote access from Win32 and X clients and performs well under Windows 95, its performance on the Alpha is unacceptably slow at present. This is being addressed; such remote access is also a feature touted in Windows NT version 5. We have also successfully tested a freely available `rlogin` daemon. `Edlin` is, however, the only editor we can run remotely through `telnet` presently. We are tackling all of these problems as they are crucial in being able to offer a remote, multi-user, high performance computing service.

In terms of networking, NT does not seamlessly integrate into a UNIX environment. We use a Samba server to bridge this gap but individual file security is not propagated through Samba, NISGina provides a hack to support NIS logins but this has not been tested on Alphas at present, and domain logins currently require NT Server software.

MPI Performance

At present the only implementation of MPI available for Windows NT running on DEC Alphas is that developed by Mississippi State University based on MPICH and known as WinMPICH [4]. We have ported this fully to Digital Visual FORTRAN and we believe we are the only group running FORTRAN MPI on Alpha NT. WinMPICH is still in Beta (as of January 1998). While the performance between MPI processes on a single machine is reasonable for a layered OS-based implementation, its performance is disappointing between machines as can be seen in Table 4. We are working to enhance the performance of MPI between computers as a matter of urgency both by layering over Windows Named Pipes and by implementing our own protocols at the NDIS level.

---

[1] Except in the POSIX box which is otherwise almost useless.

Table 4. COMMS1 MPI benchmark performance.

Measurement	Between two processes	Between two machines
rinf (kbyte/s)	10800	59.2
nhalf (byte)	9070	130
Startup (µs)	842	2200

MPI under Windows NT is seriously immature, not only in terms of its inter-processor performance. As it was designed primarily for shared memory, when operating across machines it runs under Administrator accounts (equivalent to UNIX superuser) with full system privileges. It may also leave dead processes hanging when not exiting properly, this also occurs with some UNIX-based MPI implementations.

PVM Performance

We have also acquired a public domain implementation of PVM [5] and rlogin/rlogind for the DEC Alphas. We have not yet ported PVM for Digital Visual FORTRAN. As can be seen from Figures 1 and 2, the performance of PVM is much better than WinMPICH between machines.

This port of PVM for Windows NT will be included in the general distribution of PVM version 3.4. However it still is far from perfect. It fails to redirect i/o and requires manual starting of pvmd3 daemons on remote processors.

HPC Performance

Initial tests of the real application performance of this system have been carried out. The benchmark version of the combustion code ANGUS was compiled to run under Windows NT with only a single modification to the main source code; replacing the UNIX "/dev/null" with "NUL" for dummy file output. ANGUS is a finite-difference code which uses a regular grid and straightforward domain decomposition. The most intensive part of the program is solving the Poisson equation for the pressure. A number of solution algorithms have been implemented and the multigrid solver was used in this test. Performance for a 40x40x40 grid with 2x2x1 processor decomposition is given in Table 5 for running on a single DEC Alpha and on two machines. For comparison T3D performance using MPI is quoted. Note that in the production code Cray-specific SHMEM libraries replace the MPI calls to optimise performance on Cray T3D and T3E machines.

Table 5. Performance of ANGUS code on Alpha NT cluster, with T3D performance for comparison.

Configuration	time per iteration (s/iteration)	comms time (s)
One CPU, one process	20	2.4
One CPU, four processes	38	21.3
Two CPUs, four processes	285	263.0
T3D using MPI (four nodes)	61.6 (nodesecs/iteration)	14.0 (nodesecs)

PVM bwtest (between processes)

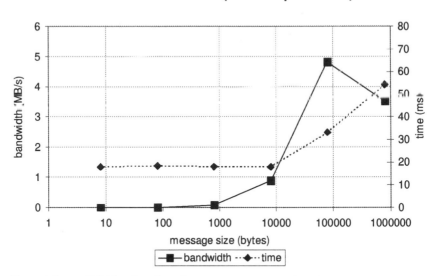

Figure 1. bwtest PVM benchmark performance on a single machine between two processes.

PVM bwtest (between machines)

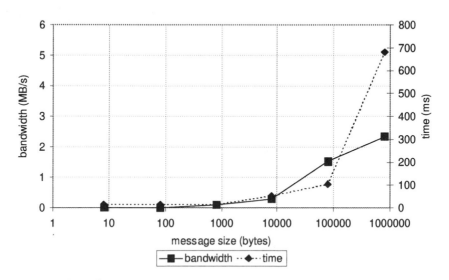

Figure 2. bwtest PVM benchmark performance between two machines (note timescale differs from Figure 1).

This demonstrates the current discrepancy between shared and distributed memory performance of WinMPICH in its current beta stage. A promising result is that it shows that uniprocessor MPI can be used for application development now.

Performance results for other parallel CFD codes are published in Reference 6.

CONCLUSIONS

In this paper we present our early experiences of using DEC Alpha workstations running Windows NT for high performance scientific computing. Using Digital Visual FORTRAN, this is a good development environment and offers reasonable shared memory MPI and PVM performance for testing purposes. However, parallel programming between machines is at the bleeding edge; we still have a lot of work to do before this becomes a suitable platform for real users. For production work a combination of Digital UNIX compilation nodes and Linux compute nodes can provide cost effective, medium-scale commodity supercomputing today.

ACKNOWLEDGMENTS

The authors would like to thank Stewart Cant (Cambridge) and David Emerson (Daresbury) for the ANGUS code, and Ken Morgan (Swansea) for the unstructured grid and related software.

REFERENCES

1. The Beowolf Project, http://cesdis.gsfc.nasa.gov/beowolf
2. R.J. Hooway and M.A. Herdeg, DIGITAL FX!32: Combining emulation and binary translation, *Digital Technical Journal*, 9(1) (1997).
3. *WinVNC* from Olivetti & Oracle Research Lab, http://www.orl.co.uk/vnc
4. A. Skjellum, B. Protopopov, S. Hebert, P.J. Brennan and W. Seefeld, *MPI on Windows NT. 0.92 Beta release* (1997). Currently available at: http://www.erc.msstate.edu/mpi/mpiNT.html
5. *PVM for Windows NT*, http://www.epm.ornl.gov/pvm/NTport.html
6. D. Emerson, D.A. Nicole and K. Takeda, An Evaluation of Cost Effective Parallel Computers for CFD, to be presented at the *10th International Conference on Parallel CFD*, Taiwan (May 1998).

A PROGRAMMING ENVIRONMENT FOR HIGH-PERFORMANCE COMPUTING IN JAVA

Vladimir Getov,[1] Susan Flynn-Hummel,[2] and Sava Mintchev[1]

[1]School of Computer Science, University of Westminster
Harrow Campus, Northwick Park, Harrow HA1 3TP, U.K.
[2]T.J. Watson Research Center, IBM Research Division
P.O. Box 218, Yorktown Heights, NY 10598, U.S.A.

INTRODUCTION

It is generally accepted that computers based on the emerging hybrid shared/distributed memory parallel architectures will become the fastest and most cost-effective supercomputers over the next decade. This, however, makes the search for the most appropriate programming model and corresponding programming environments more important than ever before. Arguably the most serious obstacle to the acceptance of parallel supercomputers is the so-called *software crisis*. Software, in general, is considered the most complex artifact in high-performance computing; since the lifespan of parallel machines has been so brief, their software environments rarely reach maturity and the parallel software crisis is especially acute. Hence, portability, in particular, is a critical issue in enabling high-performance parallel computing. Application programmers need flexible yet comprehensive interfaces which cover both the shared memory and the distributed memory programming paradigms.

The Java language has several built-in mechanisms which allow the parallelism inherent in scientific programs to be exploited. Threads and concurrency constructs are well-suited to shared memory computers, but not large-scale distributed memory machines. Although sockets and the Remote Method Invocation (RMI) interface allow network programming, they are rather low-level to be suitable for SPMD-style scientific programming, and thus, codes based on them would potentially underperform platform-specific implementations of standard communication libraries like MPI. Nevertheless, as a programming language, Java has the basic qualities needed for writing high-performance applications. With the maturing of compilation technology, such applications written in Java will doubtlessly appear. Fortunately, rapid progress is being made in this area by developing optimizing Java compilers, such as the IBM High-Performance Compiler for Java (HPCJ), which generates native codes for the RS6000 architecture.[10] Since the Java language is fairly new, however, it lacks the extensive scientific libraries of other languages like Fortran-77 and C. This is one of the major obstacles towards efficient and user-friendly computationally intensive programming in Java.

In order to overcome the above problems, we have applied a Java-to-C Interface (JCI) generating tool[12] to create Java bindings for various legacy libraries such as MPI, BLAS, BLACS, PBLAS, LAPACK, ScaLAPACK, etc., which are compatible with the library specifications.[7] With performance-tuned implementations of those libraries already available on different platforms, the potential exists for high performance parallel programming in Java. In this article we show that this is indeed possible using HPCJ, native communication and scientific libraries, and bindings automatically created by the JCI tool, which have been the basic components of our programming environment. Finally, we present performance results with two kernels from the NPB and PARK-BENCH suites, which demonstrate the efficiency of our approach.

BINDING NATIVE LEGACY LIBRARIES TO JAVA

The JCI tool

At first sight it appears that the binding of a native library to Java should not be a problem, as Java implementations support a *native interface* via which C functions or Fortran subroutines can be called. There are some hidden problems, however. First of all, native interfaces are reasonably convenient when writing new C code to be called from Java, but rather inadequate for linking legacy codes. The difficulty stems from the fact that Java generally has different data formats to C, and therefore existing C codes cannot be called from Java without prior modifications, or without providing an additional interface layer.

Binding a native library to Java is also accompanied by portability problems. The native interface is not part of the Java language specification,[8] and different vendors offer incompatible interfaces. Furthermore, native interfaces are not yet stable and are likely to undergo change with each new major release of a Java implementation. Currently, the Java Native Interface (JNI) in Sun's JDK 1.1 is regarded as the definitive native interface, but it is not yet supported in all Java implementations on different platforms. Thus to maintain the portability of the binding one may have to cater for a variety of native interfaces.

In addition to the above problems, a large legacy library like MPI can have over a hundred exported functions, therefore it is preferable to automate the creation of the additional interface layer. In order to call a C function from Java, the JCI tool has to supply for each formal argument of the C function a corresponding actual argument in Java. Unfortunately, the disparity between data layout in the two languages is large enough to rule out a direct mapping in general. For instance:

- primitive types in C may be of varying sizes, different from the standard Java sizes;

- there is no direct analog to C pointers in Java;

- multidimensional arrays in C have no direct counterpart in Java;

- C structures can be emulated by Java objects, but the layout of fields of an object may be different from the layout of a C structure;

- C functions passed as arguments have no direct counterpart in Java.

Table 1. Mapping of compound C types into Java types

C type	Java type	
char *	ObjectOfChar	,if at top level and not the type of a function
	String	,otherwise
struct *name* *	class *name*	
void *	Object	
c_type *	ObjectOf*j_type*	
char []	String	
c_type []	*j_type* []	
struct *name*	class *name*	

Therefore, we have defined a specific mapping which is implemented in the JCI tool Table 1 shows the scheme currently used to map C types onto Java types. Primitive types are not listed in this table because they are to be found in the documentation of each JVM's native interface. C pointers are represented in a type-safe way by a family of Java classes generated by JCI. Each such class is named ObjectOf*j_type*, and contains a field val of type *j_type*. Pointer objects can be created and initialized by Java constructors, or by the overloaded function JCI.ptr; they can be dereferenced by accessing the val field. In general, the defined mapping is not unique; on the contrary – there is a number of different mappings to choose from and the selection represents an important trade-off between the extent of the performance overhead introduced by the binding on the one hand, and the ease of use of the programming interface from Java on the other.

The JCI tool helps to make our bindings flexible and adaptable to different libraries, native interfaces, and platforms. It takes as input a header file containing the C function prototypes of the native library and outputs a number of files comprising the additional interface. In particular, the tool generates a C stub-function and a Java native method declaration for each exported function of the native library. Every C stub-function takes arguments whose types correspond directly to those of the Java native method, and converts the arguments into the form expected by the C library function. As different Java native interfaces exist, different code may be required for binding a native library to each Java implementation. We have tried to limit the implementation dependence of JCI output to a set of macro definitions describing the particular native interface. Thus it may be possible to re-bind a library to a new Java machine simply by providing the appropriate macros.

Binding C libraries (MPI, BLACS, PBLAS)

The largest native library we have bound to Java so far is MPI: it has in excess of 120 functions.[12] The JCI tool allowed us to bind all those functions to Java without extra effort. Since MPI libraries are standardized, the binding generated by JCI should be applicable without modification to *any* MPI implementation. As the Java binding for MPI has been generated automatically from the C prototypes of MPI functions, it is very close to the C binding. This similarity means that the Java binding is almost completely documented by the MPI-1 standard, with the addition of a table of the JCI mapping of C types into Java types.

Other libraries written in C for which we have created Java bindings are the Parallel Basic Linear Algebra Subprograms (PBLAS) and the Communication Subprograms (BLACS). The library function prototypes have been taken from the PARKBENCH

Table 2. Native Legacy Libraries Bound to Java

library	written in	Size of Java binding		
		functions	C lines	Java lines
MPI	C	125	4434	439
BLACS	C	76	5702	489
BLAS	F77	21	2095	169
PBLAS	C	22	2567	127
PB-BLAS	F77	30	4973	241
LAPACK	F77	14	765	65
ScaLAPACK	F77	38	5373	293

2.1.1 distribution at www.netlib.org/parkbench. Table 2 gives some idea of the sizes of JCI-generated bindings for individual libraries. In addition, there are some 2280 lines of Java class declarations produced by JCI which are common to all libraries.

Binding Fortran-77 libraries (BLAS, PB-BLAS, ScaLAPACK)

The JCI tool can be used to generate Java bindings for libraries written in languages other than C, provided that the library can be linked to C programs, and prototypes for the library functions are given in C. We have created Java bindings for a number of legacy libraries written in Fortran-77 including the Scalable Linear Algebra PACKage (ScaLAPACK)[3] and its constituent libraries (BLAS Level 1–3, PB-BLAS, LAPACK, etc.)

The bindings generated by JCI are fairly large in size (see Table 2) because they are meant to be portable, and to support different data formats. On a particular hardware platform and Java native interface, much of the binding code may be eliminated during the preprocessing phase of its compilation. As our experiments on IBM SP2 machines so far have shown, a negligible amount of time is spent in the binding itself during the execution of Java programs.

THE HIGH-PERFORMANCE PROGRAMMING ENVIRONMENT

The initial structure of our programming environment including all basic components is illustrated in Figure 1. The JCI tool takes as input the header file containing the C function prototype declarations of the native legacy library and generates automatically all files comprising the required binding:

- a file of C stub-functions;

- files of Java class and native method declarations;

- shell scripts for doing the compilation and linking.

Then, the bound libraries can be dynamically linked to the Java Virtual Machine (JVM) upon demand and used during the execution. So far we have done experiments with two varieties of the JVM – the Java Development Kit (JDK) for Solaris on a cluster of Sun workstations; and IBM's port of JDK for AIX 4.1 on the SP2.

Most JVMs contain a Just-in-Time (JIT) compiler to improve the execution performance. A JIT compiler turns Java bytecode into native code on-the-fly, as it is loaded into the JVM. The JVM then executes the generated code directly, rather then interpreting bytecode, which leads to a significant performance improvement. In this way, the best performance results using the programming environment in Figure 1 can

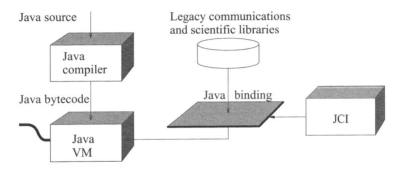

Figure 1. Programming environment using a conventional Java virtual machine

be achieved, but the execution time is still much longer in comparison with similar computations using conventional languages such as Fortran-77 or C and the corresponding compilers. The reason for this noticeable difference is twofold – firstly, the JIT translation adds an extra overhead to the execution time; and secondly, the compilation speed requirements constrain the quality of optimisation that a JIT compiler can perform. Therefore, the performance of this environment is relatively low as there is usually a large imbalance between the efficiency of the performance tuned implementations of legacy libraries and the rest of the code at execution time.

An optimising native code compiler for Java can be used instead of the JVM in order to overcome the above problem. Such a compiler translates bytecode directly into native executable code as shown in Figure 2. It works in the same manner as compilers for C, C++, Fortran, etc. and unlike JIT compilers, the static compilation occurs only once, before execution time. Thus, traditional resource-intensive optimisations can be applied in order to improve the performance of the generated native executable code. In our experiments, we have used a version of HPCJ, which generates native code for the RS/6000 architecture. The input of HPCJ is usually a bytecode file, but the compiler will also accept Java source as input. In the latter case it invokes the JDK source-to-bytecode compiler to produce the bytecode file first. This file is then processed by a translator which passes an intermediate language representation to the common back-end from the family of compilers for the RS/6000 architecture. The back-end outputs standard object code which is then linked with other object modules and the previously bound legacy libraries to produce native executable code. In this way, our programming environment conforms to the basic requirements for high-performance computing as the experimental results in the next section show.

EXPERIMENTAL RESULTS

In order to evaluate the performance of the Java binding to native libraries, we have translated into Java the Matrix Multiplication (MATMUL) benchmark from the PARKBENCH suite.[13] The original benchmark is in Fortran-77 and performs dense matrix multiplication in parallel. It accesses the BLAS, BLACS and LAPACK libraries included in the PARKBENCH 2.1.1 distribution. MPI is used indirectly through the BLACS native library. We have run MATMUL on a Sparc workstation cluster, and on the IBM SP2 machine at Southampton University (66MHz Power2 "thin1" nodes with

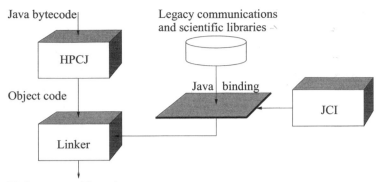

Java bytecode

Legacy communications
and scientific libraries

HPCJ

Object code

Java binding

JCI

Linker

Native executable code

Figure 2. Programming environment using the high-performance compiler for Java which generate native executable code

128Mbyte RAM, 64bit memory bus, and 64Kbyte data cache). The results are shown in Table 3 and Figure 3.

Further experiments have been carried out with a Java translation of a C + MPI benchmark – the Integer Sort (IS) kernel from the NAS Parallel Benchmark suite,[1] version NPB2.2. The program sorts an array of N integers in parallel; where the problem size (class A) is specified as $N = 8M$. The original C and the new Java versions of IS are quite similar, which allows a meaningful comparison of performance results.

We have run the IS benchmark on two platforms: a cluster of Sun Sparc workstations, and the IBM SP2 system at the Cornell Theory Center. Each SP node used has a 120 MHz POWER2 Super Chip processor, 256 MB of memory, 128 KB data cache, and 256 bit memory bus. The results obtained on the SP2 machine are shown in Table 4 and Figure 4. The Java implementation we have used is IBM's port of JDK 1.0.2D (with the JIT compiler enabled). The communications library we have used is the LAM implementation (version 6.1) of MPI from the Ohio Supercomputer Center.[4] We opted for LAM rather than the proprietary IBM MPI library because the version of the latter available to us (PSSP 2.1) does not support the re-entrant C library required for Java.[11] The results for the C version of IS under both LAM and IBM MPI are also given for comparison.

It is evident from Figure 3 that Java MATMUL execution times are only 5–10% longer than Fortran-77 times. These results may seem surprisingly good, given that Java IS is two times slower than C IS (Figure 4). The explanation is that in MATMUL

Table 3. Execution statistics for the MATMUL benchmark on the IBM SP2 machine at Southampton University

Problem size (N)	Lang	MPI imple-mentation	No of processors				
			1	2	4	8	16
Execution time (sec):							
	Java	LAM	—	17.09	9.12	5.26	3.53
1000	F77	LAM	—	16.45	8.61	5.12	3.13
	F77	IBM MPI	33.25	15.16	7.89	3.91	2.20
Mflop/s total:							
	Java	LAM	—	117.0	219.4	380.2	566.9
1000	F77	LAM	—	121.6	232.3	390.4	638.3
	F77	IBM MPI	60.16	132.0	253.6	511.2	910.0

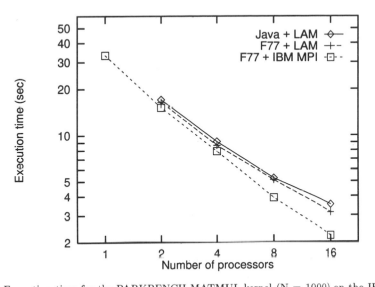

Figure 3. Execution time for the PARKBENCH MATMUL kernel (N = 1000) on the IBM SP2 at Southampton University

Table 4. Execution statistics for the NPB IS kernel (class A) on the IBM SP2 at Cornell Theory Center

Class	Language	MPI imple-mentation	No of processors				
			1	2	4	8	16
Execution time (sec):							
A	JDK	LAM	—	48.04	24.72	12.78	6.94
	hpj	LAM	—	23.27	13.47	6.65	3.49
	C	LAM	42.16	24.52	12.66	6.13	3.28
	C	IBM MPI	40.94	21.62	10.27	4.92	2.76
Mop/s total:							
A	JDK	LAM	—	1.75	3.39	6.56	12.08
	hpj	LAM	—	3.60	6.23	12.62	24.01
	C	LAM	1.99	3.42	6.63	13.69	25.54
	C	IBM MPI	2.05	3.88	8.16	14.21	30.35

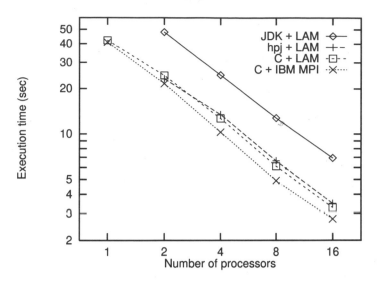

Figure 4. Execution time for the NPB IS kernel (class A) on the IBM SP2 at Cornell Theory Center

most of the performance-sensitive calculations are carried out by the native library routines (which are the same for both Java and Fortran-77 versions of the benchmark). In contrast, IS uses a native library (MPI) only for communication, and all calculations are done by the benchmark program.

It is important to identify the sources of the slowdown of the Java version of IS with respect to the C version. To that end we have instrumented the JavaMPI binding, and gathered additional measurements. It turns out that the cumulative time spent in the C functions of the JavaMPI binding is approximately 20 milliseconds in all cases, and thus has a negligible share in the breakdown of the total execution time for the Java version of IS. Clearly, the JavaMPI binding does not introduce a noticeable overhead in the results from Table 4.

RELATED WORK

Many research groups and vendors are pursuing research to improve Java's performance which would enable more scientific and engineering applications to be solved in Java. The need for access to legacy libraries is one of the burning problems in this area. Several approaches can be taken in order to make the libraries available from Java:

- Hand-writing existing libraries in Java. Considering the size of the available libraries and the number of years that were invested in their development, rewriting the libraries would require an enormous amount of manual work.[2]

- Automatically translating Fortran-77/C library code into Java. We are aware of two research groups that have been working in this area – University of Tennessee[5] and Syracuse University.[6] This approach offers a very important long-term perspective as it preserves Java portability, while achieving high performance in this case would obviously be more difficult.

- Manually or automatically creating a Java wrapper for an existing native Fortran-77/C library. Obviously, by binding legacy libraries, Java programs can gain in performance on all those hardware platforms where the libraries are efficiently implemented.

The automatic binding, which we are primarily interested in, has the obvious advantage of involving the least amount of work, thus reducing dramatically the time for development. Moreover, it guarantees the best performance results, at least in the short term, because the well-established scientific libraries usually have multiple implementations carefully tuned for maximum performance on different hardware platforms. Last but not least, by applying the software re-use tenet, each native legacy library can be linked to Java without any need for re-coding or translating its implementation.

The binding of native libraries to Java has certain limitations. In particular, for security reasons applets downloaded over the network may not load libraries or define native methods. The solution of this problem does not seem very difficult though. By using a virtual environment like IceT[9] both processes and data would be allowed to migrate and to be transferred throughout owned and unowned resources, under flexible security measures imposed by the users.

After the initial period when the first Java versions were built for portability, the Java compiler technology has now entered a second phase where the new versions are also targeting higher performance. For example, JIT compilers have dramatically improved their efficiency, and are now challenging mature C++ compilers. The developers of HPCJ have adopted the 'native compiler' approach in order to gain faster execution times. A different strategy has been chosen by the authors of Toba.[14] Toba translates Java bytecode into C source code, which is then compiled with the appropriate compiler optimisation flags for high performance. Another advantage of this approach is that it is as portable as any other C software.

CONCLUSIONS

The JCI tool for automatic creation of interfaces to native legacy libraries (whether for scientific computation or message-passing) improves the portability of such interfaces. In addition to the JCI-generated bindings, the basic components of out high-performance Java programming environment include performance-tuned implementations of scientific and communications libraries available on different machines, and a native Java compiler like IBM's HPCJ.

Very good evaluation results with kernels from both the NAS Parallel Benchmarks and the PARKBENCH suite using this environment are shown on SP2 machines. This must be a persuasive argument for automatic binding of native legacy libraries to Java, and for the advantages of the native Java compilers.

ACKNOWLEDGEMENTS

This work has been carried out as part of our collaboration with colleagues from the University of Southampton (U.K.) and the Cornell Theory Center (U.S.A.). In particular, we are grateful to Tony Hey for his continuous support and encouragement and to Ian Hardy for helping with the IBM SP2 experiments.

REFERENCES

1.   D. Bailey, E. Barszcz, J. Barton, D. Browning, R. Carter, L. Dagum, R. Fatoohi, S. Fineberg, P. Frederickson, T. Lasinski, R. Schreiber, H. Simon, V. Venkatakrishnan, and S. Weeratunga,

The NAS parallel benchmarks, Technical Report RNR-94-007, NASA Ames Research Center (1994).
http://science.nas.nasa.gov/Software/NPB .

2.   A. Bik and D. Gannon, A note on native level 1 BLAS in Java, *Concurrency: Pract. Exper*, 9:11 (1997).

3.   L. Blackford, J. Choi, A. Cleary, E. D'Azevedo, J. Demmel, I. Dhillon, J. Dongarra, S. Hammarling, G. Henry, A. Petitet, K. Stanley, D. Walker, and R. Whaley, ScaLAPACK: A linear algebra library for message-passing computers, in: *Proceedings of SIAM Conference on Parallel Processing* SIAM (1997).

4.   G. Burns, R. Daoud, and J. Vaigl, LAM: An open cluster environment for MPI, in: *Proceedings of Supercomputing Symposium '94*, Toronto (1994).
http://www.osc.edu/lam.html .

5   H. Casanova, J. Dongarra, and D. Doolin, Java access to numerical libraries, *Concurrency: Pract. Exper*, 9:11 (1997).

6.   G. Fox, X. Li, Z. Qiang, and W. Zhigang, A prototype of Fortran to Java converter, *Concurrency: Pract. Exper*, 9:11 (1997).

7.   V. Getov, S. Flynn-Hummel, and S. Mintchev, High-performance parallel programming in Java: Exploiting native libraries, *Concurrency: Pract. Exper*, (1998) in press.

8.   J. Gosling, W. Joy, and G. Steele. *The Java Language Specification, Version 1.0*, Addison-Wesley, Reading (1996).

9.   P. Gray and V. Sunderam, The IceT environment for parallel and distributed computing, in: *Scientific Computing in Object-Oriented Parallel Environments*, Y. Ishikawa, R. Oldehoeft, J. Reynders, and M. Tholburn, eds., LNCS 1343 (1997).

10.   IBM Corp., High-performance compiler for Java: An optimizing native code compiler for Java applications, (1997).
http://www.alphaWorks.ibm.com/formula .

11.   IBM Corp., Programming environment for AIX: MPI programming and subroutine reference (1997).
http://www.rs6000.ibm.com/resource/aix_resource/sp_books/pe .

12.   S. Mintchev and V. Getov, Towards portable message passing in Java: Binding MPI, in: *Recent Advances in PVM and MPI*, M. Bubak, J. Dongarra, J. Waśniewski, eds., LNCS 1332 (1997).

13.   PARKBENCH Committee (assembled by R. Hockney and M. Berry), PARKBENCH report-1: Public international benchmarks for parallel computers, *Scientific Programming*, 3:2 (1994).

14.   T. Proebsting, G. Townsend, P. Bridges, J. Hartman, T. Newsham, and S. Watterson, Toba: Java for applications – a way ahead of time (WAT) compiler, in: *Proceedings 3rd Conference on Object-Oriented Technologies and Systems (COOTS'97)* (1997).

HIGH PERFORMANCE DISTRIBUTED FDTD ELECTROMAGNETIC FIELD COMPUTATION FOR ELECTRONIC CIRCUIT DESIGN

C.J. Gillan and V.F. Fusco

The High Frequency Electronics Laboratory
Dept. Electrical and Electronic Engineering
Ashby Building, Stranmillis Road
Queen's University of Belfast
N. Ireland BT7 1NN, UK

INTRODUCTION

The typical operating frequency of modern high frequency electronic circuits constructed on gallium arsenide (GaAs) and silicon for radar and broadband wireless applications is several GigaHertz, that is in the microwave region, making the wavelength of the electromagnetic signals present comparable with the dimensions of the circuit itself. Under these conditions, the geometry of the circuit, and the packaging surrounding it, will significantly influence the operational characteristics of the device. Thus, a detailed knowledge of the electromagnetic fields present is essential to perform accurate circuit design.

These circuits may be modelled by solving Maxwell's equations however to complicate matters, the circuits have application specific geometries, active and passive loads and are composed of several different materials. For example, a GaAs monolithic integrated circuit (MMIC) can be created on a 200 micron thick substrate but may be connected to other microstrip circuits fabricated on 650 micron thick teflon material. Encapsulating the device often involves using a ceramic package providing a hermetically sealed enclosure with controlled impedance feed through lines providing the pathway for signals to enter and leave the device. Effects not important at lower frequencies, can become significant in the microwave region, for example, a small series inductance of the order of picoHenries in the path will produce a significant impedance for microwave signals.

Direct numerical solution appears to be the only feasible way to determine the EM fields in this kind of device. Hafner[1] noted, for example, in the era of mainframe computing, that:

> "Numerical computations of electromagnetic fields have always been one of
> the main applications for the biggest computers."

The finite difference time domain method (FDTD) is one numerical and storage intensive technique with wide applicability which has been extensively developed and

High Performance Computing
Edited by R. J. Allan *et al.*, Kluwer Academic / Plenum Publishers, New York, 1999

widely used since it was proposed by Yee[2]. It has the advantage that very little analytical preprocessing is required. Today, the FDTD technique is routinely interfaced, on a one off basis, to other electric circuit solver programs, for example the well known timedomain statespace electronic circuit simulation package SPICE[3] which is used to model particular current characteristics in particular sub-components of a device

In common with finite difference and finite volume solution techniques applied in other areas of engineering, some of which are discussed elsewhere in this volume, the FDTD method is ideally suited to parallel or distributed computation in which the geometric grid is decomposed over several processors. An analysis of the elapsed and CPU times for a sequential FDTD calculation yields a bottleneck index[4] approaching unity meaning that the computation is CPU bound. In recent years, the performance of the FDTD method has been investigated on transputers[5], on networks of distributed computers[6], using uniform memory architectures (UMA) on symmetric multiprocessors (SMP)[7] and also on non uniform memory architectures (NUMA) on massively parallel processors (MMP)[7]. Our group has concentrated on the creation of computational electromagnetics software which makes optimal use of heterogeneous networks of computers, a readily available and economic resource; as a first stage we have implemented, using the client-server paradigm[8], an FDTD solution engine which encompasses a load balancing tool.

CLIENT-SERVER IMPLEMENTATION OF THE FDTD METHOD

Today, even small laboratories and small businesses have a local network infrastructure (LAN) linking desktop computers, at least, together and on all these processors the Internet Protocol suite is readily available, regardless of the operating system in use. Moreover, these machines tend to be idle at evenings and weekends representing an enormous waste of CPU cycles. Our aim has been, therefore, to harness this resource. The client-server paradigm[8] is widely used in commercial distributed computing, for example by electronic mail systems and by worldwide web browsers already running on most networks, and we have applied this to design a suite of electromagnetics software. Actually, we have used the server-client[9] variation of this paradigm as employed by the X-windows system common on Unix desktop workstations. At the heart of our design are two features:

1. the specification of a detailed set of messages, composed of one of more data structures, which are exchanged between client and server processes in the system.

2. the requirement that all processes, clients and server alike, must be programmed to react, asynchronously, to any one of the predefined messages.

The program level implementation of the message passing is a secondary issue. All message passing takes place through levels three and four, the transport and network layers, of the seven layer OSI protocol stack. We have employed two options. On the one hand, we have direct calls to TCP/IP routines which correspond to levels four and three above and, on the other, we have used the PVM[10] message passing harness. Specifically, the key routine in the TCP/IP implementation is the *select()* call for I/O multiplexing with the *pvm_trecv()* call being the analogous routine in the PVM option. By employing an asynchronous mechanism alone, we can easily implement a simple timeout [8] mechanism which is a vital first step towards load balancing, full fault tolerance and failover redundancy between the client and server processes.

Our FDTD implementation consists of a server program and one client; in any given run only one copy of the server executes (on some designated processor) but multiple copies of the client execute concurrently, one on each available processor. At the outset of the computation, each client is oblivious to the existence of other clients; the role of the server is to decompose the total volume among clients, and then to direct and control the subsequent iteration over time steps where each client updates field values at points within its sub-volume. At the end of each timestep, clients must exchange electric field data along intersecting boundaries; this inter-client communication is direct, but under the control of the server process, thereby permitting dynamic reconfiguration of the number of clients and of the volume decomposition at the end of any timestep. The data that must be interchanged depends, of course, on the way in which the finite difference equations are written. Physically, this interchange of data reflects the passage of electromagnetics waves through the total volume.

The FDTD algorithm in our codes has been developed from the analysis presented by Kunz and Leubbers[11] in which the scattered field formulation is used. While most of our program is in the C language, the numerically intensive field update routines have been coded in Fortran in order to exploit the excellent optimization available with Fortran compilers.

DYNAMIC LOAD BALANCING

The critical hardware limitation on any distributed computation is the network, specifically its latency, bandwidth and topology. To a large degree, this can be overcome in FDTD computations as long as the surface to volume ratio of the sub-domains is small since the distribution of boundary field values may be overlapped with field updates[6]. In a calculation which is decomposed only along the Z axis, corresponding to processors connected by a linear network topology such as 10 Base 2 thin wire ethernet for example, the E_x and E_y field updates along the boundaries may be performed first and then dispatched across the network while the processor is busy with updating these fields at all internal points in the volume and the E_z field at all points in the volume.

In general the processors used in a distributed computing environment are not dedicated and may be servicing other users concurrent with the FDTD client. In previous work[12] we have illustrated the imbalance that this can create. Within the server-client paradigm, enhanced with the single writer-multiple reader shared address space paradigm, we have implemented a more powerful dynamic load balancing tool than in previous work. A load reporting client runs on each processor on which an FDTD client exists. At predefined intervals, the load report client interrogates the operating system and retrieves copious data on the processor status with this information being passed across the network to the load balance server running on the same processor as the FDTD server. The interrogation process invokes predefined system interfaces such as the *pstat()* function on UNIX and the *sys$getjpi()* function VMS. The *pstat()* function is what lies beneath well known UNIX commands such as *ps* and *top*. It must be stressed that direct use of *pstat()* yields a wealth of data not available from the command line.

The load balance server maintains, therefore, a table of processor states for every processor used in the FDTD computation. From this table, decisions may be made on how to distribute, or redistribute the FDTD computation. For example, the number of user processes, the percentage CPU usage and the amount of virtual page swapping are all listed in the processor states table. All multi-user operating systems provide mechanisms, even on single processors, for independent processes to share parts of

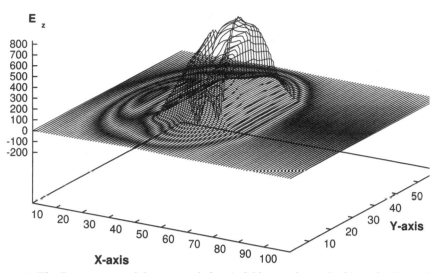

Figure 1. The Z component of the scattered electric field around a perfectly conducting strip ($25 \leq x \leq 75$, $30 \leq y \leq 40$) after 154 timesteps calculated using the distributed FDTD program as discussed in the text. The scale on each axis has been adjusted for plotting purposes.

their virtual address spaces (i.e. shared virtual memory), a feature which should not be confused with UMA design for SMP as mentioned above. By locating the processor states table in a region of shared memory, it is automatically available to be read by the FDTD server process, and other compute engines, without the degradation in performance that would potentially be associated with passing the entire table as a message.

RESULTS

In figure 1 we illustrate the results of a test calculation with our 3 dimensional FDTD codes in which a plane wave Gaussian pulse is incident over the entire XY plane on a very thin, perfectly conducting strip. The strip is located in the XY plane, parallel to the X-axis with an aspect ratio of 5:1, lying on top of a 0.8 mm thick dielectric substrate with $\epsilon_r = 4$; this arrangement is resonant at 7.4 GHz approximately. The plane wave Gaussian pulse travels in the positive X direction and has a frequency response of $0 < f < 10$ Ghz. Figure 1 shows the Z component of the electric field, in an XY plane close to the conducting strip, after 154 timesteps have elapsed. From figure 1, surface wave leakage in the positive and negative y directions, about the line y= 35, can be clearly seen. These surface waves are a significant loss component in planar microwave radiating structures giving rise to deleterious losses and parasitic coupling within the circuit environment. Backscatter due to the wave impedance mismatch between the incoming Gaussian pulse and the resonator, along the line x = 15, can also be seen.

The computation discussed here was performed, for 512 timesteps, on a Hewlett Packard 715/100 workstation having a 100 Mhz CPU and 128 Mbytes of main memory. In sequential mode, the *elapsed* time for the run was approximately 150 minutes. Using four identical HP 715/100 workstations linked by a 10-Base-2 thin wire ethernet, the distributed calculation took just under 60 minutes. The electromagnetic effects discussed above, can only be seen by using a full electromagnetic simulation coupled with a visualization tool. We are presently constructing a movie, of 29 seconds duration, from the computed results thereby visualizing the E_z field for the entire scattering process. Similar movies can be produced for other, non-zero, field components, of course.

CONCLUSIONS

We have described the implementation of high performance computational elec- tromagetics software in a heterogeneous distributed computing environment using the FDTD method as vehicle to illustrate our design concepts. The fundamental design features in our software are the use of the server-client paradigm coupled with asyn- chronous I/O and the use of shared virtual address spaces within the single writer- multiple reader model. We have made direct use of IP calls meaning that our suite may be used on operating systems, such as DEC VMS, where there is no version of PVM generally available. The extra programming effort is also repaid by the fact that we can easily incorporate fault tolerance and task migration into the software.

We have already used our suite[13] to evaluate the performance of the PVM[10] mes- sage passing harness as compared to direct use of the TCP/IP calls. Additionally, we investigated the influence of using heterogeneous hardware, that is different floating point representations, for different sub-domains of the problem.

In future we plan to enhance the suite to encompass more solution methodologies beyond FDTD, such an interface to SPICE, before tackling the fundamental circuit design problems discussed at the start of the chapter. Additionally, we shall replace PVM with use of MPI[14] routines.

ACKNOWLEDGEMENTS

This work was supported by the UK Engineering and Physical Sciences Research Council under contract GR/L23215.

REFERENCES

1. C. Hafner, Parallel Computation of Electromagnetic Fields on Transputers, *IEEE Ant. Prog. Soc. Neu ter*, 6-12:October (1989).
2. K. S. Yee, Numerical Solution of Initial Boundary Value Problems involving Maxwell's Equations in Isotropic Media, *IEEE Trans. Ant. Prop.* 14:302-7 (1900).
3. M.J. Picket-May, A. Taflove, V.A. Thomas, M.E. Jones and E. Harrigan, The use of SPICE lumped circuits as sub-grid models for fd-td analysis, *IEEE Microwave and Guided Wave Letters* 4:141 (1994).
4. M. Sakaki, H. Samukawa and N. Honjou, Effective utilization of IBM 3090 large virtual storage in the numerically intensive computations of ab initio molecular orbitals, *IBM Systems Journal* 27:528-540 (1988).
5. K.C. Chew and V.F. Fusco, A Parallel Implementation of the Finite Difference Time Domain Algorithm, *Intl. J. of Numerical Modelling: El. Networks, Devices and Fields* 8:293 (1995).
6. D.P. Rodohan, S.R. Saunders and R.J. Glover, A Distributed Implementation of the Finite Difference Time-Domain (FDTD) Metho,d *Intl. J. of Numerical Modelling: El. Networks, Devices and Fields* 8:283 (1995).
7. A. Taflove, *Computational Electrodynamics: The Finite Difference Time Domain Method*, pp545-84, Artech House, Boston, ISBN 0-89006-792-9 (1995).
8. K.P. Birman, *Building Secure and Reliable Network Applications*, chapter 4, Manning Publica- tions Co. (1997).
9. A. Nye, *The Xlib Programming Manual for Version 11*, O'Reilly and Associates, Inc., Sebastopol (1992).

10. G.A. Geist, A. Beguelin, J. Dongarra, W. Jiang, R. Manchek and V. Sunderam *PVM 3 User's Guide and Reference Manual*, Oak Ridge National Laboratory Technical Report ORNL/TM-11616 (1994).

11. K.S. Kunz and R.J. Leubbers, *The Finite Difference Time Domain Method for Electromagnetics*, CRC Press, Boca Raton, ISBN 0-8493-8657-8 (1993).

12. V. F. Fusco, J. Mullan and C. J. Gillan, Optimizing the Parallel Implementation of a Finite Difference Time Domain Code on A Multi-User Network of Workstations, *App. Comp. Electromagnetics Soc. J.*, Special Issue on Computational Electromagnetics and High Performance Computing (1998)

13. C. J. Gillan and V. F. Fusco, Optimizing FDTD Electromagnetic Field Calculation on Distributed Networks, *Intl. J. Num. Mod.: Dev., Net. and Fields* (1998) submitted.

14. J. Malard, *MPI: A Message Passing Interface standard*, Technical Report, Edinburgh Parallel Computer Centre. 1995

LIST OF PARTICIPANTS

Mr C J Adam, University of Edinburgh, Physics   *colin.adam@ed.ac.uk*
Mr M Alam, University of London, Engineering   *m.alam@qmw.ac.uk*
Dr R J Allan, CLRC Daresbury Laboratory, DCI   *r.j.allan@dl.ac.uk*
Dr R E Ansorge, University of Cambridge, Physics   *real@phy.cam.ac.uk*
Mr K Atkinson, IBM UK Ltd   *ken_atkinson@uk.ibm.com*
Dr M Baker, University of Portsmouth, CSM   *mab@sis.port.ac.uk*
Mr A Baldock, Hitachi Europe Ltd   *a.baldock@hpcc.hitachi-eu.co.uk*
Mr D Beagle, Digital Equipment Company
Mr M Beare, University of East Anglia, Mathematics   *m.beare@uea.ac.uk*
Mr T Beckers, NEC   *tbeckers@ess.nec.de*
Mr L Black, NEC
Dr R J Blake, CLRC Daresbury Laboratory, DCI   *r.j.blake@dl.ac.uk*
Dr S Booth, University of Edinburgh, EPCC   *s.booth@epcc.ed.ac.uk*
Dr S Breuer, University of Edinburgh, EPCC   *s.breuer@epcc.ed.ac.uk*
Dr C Bridgeman, University of Cambridge, Chemistry   *cate@atm.ch.cam.ac.uk*
Dr J M Brooke, University of Manchester, MCC   *j.m.brooke@mcc.ac.uk*
Mr P J Brookes, University of Wales Swansea, Civil Engineering   *cgbrook@swansea.ac.uk*
Dr M Bull, University of Manchester, Computer Science   *markb@cs.man.ac.uk*
Dr I J Bush, CLRC Daresbury Laboratory, DCI   *i.j.bush@dl.ac.uk*
Mr B J Byron, University of Exeter, Physics   *byron@excc.ex.ac.uk.*
Prof J Clarke, UMIST, Chemistry   *hjrc@umist.ac.uk*
Dr N Clarkson, EPSRC Research Facilities Team   *nicola.clarkson@epsrc.ac.uk*
Mr T Cooke, Progress Computing Ltd .   *t.cooke@progress.co.uk*
Dr J Coomer, University of Exeter, Physics   *coomer@excc.ex.ac.uk*
Mr K Corless   *100416.540@compuserve.com*
Prof D Crighton FRS, University of Cambridge, DAMTP
Prof M Cross University of Greenwich, Centre for Numerical Modelling   *m.cross@gre.ac.uk*
Dr A Dave, Cray Research/Silicon Graphics   *ameet.dave@cray.com*

Mrs B A de Cuevas, Southampton Oceanography Centre, James Rendell Division	beverley.decuevas@soc.soton.ac.uk
Prof L M Delves, NA Software Ltd.	delves@nasoftware.co.uk
Dr D A Dixon, Pacific Northwest National Laboratory, Environmental and Molecular Sciences	david.dixon@pnl.gov
Dr A J Dobbyn, University of Manchester, Chemistry	a.j.dobbyn@man.ac.uk
Dr B R Dobson, CLRC Daresbury Laboratory, SRD	b.r.dobson@dl.ac.uk
Mr J Docherty, Digital Equipment Company Ltd.	john.docherty@digital.com
Dr D Drikakis, UMIST, Mechanical Engineering	drikakis@umist.ac.uk
Dr I Drummond, University of Cambridge, DAMTP	itd@damtp.cam.ac.uk
Dr P J Durham, CLRC Daresbury Laboratory, DCI	p.j.durham@dl.ac.uk
Dr D Emerson, CLRC Daresbury Laboratory, DCI	d.r.emerson@dl.ac.uk
Dr R Evans, CLRC Rutherford Appleton Laboratory, DCI	r.g.evans@rl.ac.uk
Mr R W Ford, University of Manchester, Centre for Novel Computing	rupert@man.ac.uk
Dr M Foulkes, Imperial College, Blackett Laboratory	m.foulkes@ic.ac.uk
Dr L Freeman, University of Manchester, Computer Science	lfreeman@cs.man.ac.uk
Dr A Gadian, UMIST, Physics	gadian@mailhost.mcc.ac.uk
Dr R Gatten, NERC	bob.gatten@nerc.ac.uk
Dr A Geiger, University of Stuttgart, HLRS/RUS	geiger@hlrs.de
Dr V Getov, University of Westminster, Computer Science	getovv@westminster.ac.uk
Dr C J Gillan, Queen's University Belfast, Electrical and Electronic Engineering	cj.gillan@ee.qub.ac.uk
Prof M Gillan, University of Keele, Physics	pha71@keele.ac.uk
Mr D Goodman, Digital Equipment Company	
Dr M Grayson, University of Sheffield, Chemistry	m.grayson@sheffield.ac.uk
Dr M F Guest, CLRC,Daresbury Laboratory, DCI	m.f.guest@dl.ac.uk
Mr G Hamer, NERC	gbh@nerc.ac.uk
Dr P Hassid, Tera Computer Company	pierre@tera.com
Mr P Hatton, University of Birmingham, Information Services	p.s.hatton@bham.ac.uk
Prof I Hillier, University of Manchester, Chemistry	ian.hillier@man.ac.uk
Dr A Hinchliffe, UMIST, Chemistry	alan.hinchliffe@umist.ac.uk
Dr M Hindmarsh ,University of Sussex, Centre for Theoretical Physics	m.hindmarsh@sussex.ac.uk
Dr A Hood, University of St.Andrews, Mathematics and Computational Science	alan@mcs.st-and.ac.uk
Mr B Hourahine, University of Exeter, Physics	bg@excc.ex.ac.uk
Dr K Jenkins, University of Cambridge, Engineering	kwj20@eng.cam.ac.uk
Dr R Jones, University of Exeter, Physics	r.jones@exeter.ac.uk
Mr B Jones, NEC	bruce@patrol.i-way.co.uk
Dr W P Jones, Imperial College, Chemical Engineering	w.p.jones@ic.ac.uk
Mr J S Junday, University of Warwick, Engineering	
Dr S Jury, University of Edinburgh, Physics	sij@ph.ed.ac.uk
Dr C Keable, Cray Research Ltd	crispin.keable@cray.com
Dr J Keane, UMIST, Computation	jak@co.umist.ac.uk
Dr D J Kerbyson, University of Warwick, Computer	djke@dcs.warwick.ac.uk

Science
Dr K Kleese, CLRC Daresbury Laboratory, DCI *k.kleese@dl.ac.uk*
Prof P Knowles, University of Birmingham, *p.j.knowles@bham.ac.uk*
Chemistry
Mr M Kyle, University of Cambridge, Wolfson *mjak2@cam.ac.uk*
College
Dr D Laff, CLRC Daresbury Laboratory, DCI *d.laff@dl.ac.uk*
Dr D Lancaster, University of Southampton, *djl@ecs.soton.ac.uk*
Electronics &Computer Science
Dr C Lantwin, NEC *clantwin@ess.nec.de*
Dr C Lazou, HiPerCom Consultants Ltd. *ip61@cityscape.co.uk*
Mr J Lenihan, Digital Equipment Company, Galway
Mr P Lockey, NERC Bidston Observatory, Proudman *plo@pol.ac.uk*
Oceanographic Laboratory
Mr N MacLaren, University of Cambridge, Computer *nmm1@cam.ac.uk*
Laboratory
Mr K Maguire, CLRC Daresbury Laboratory, DCI *k.magurie@dl.ac.uk*
Mr S Mitchell, NEC *smitchell@ess.nec.de*
Dr I Morrison,University of Salford,Physics *i.morrison@physics.salford.ac.uk*
Dr K Murphy,Queen's University Belfast,Computer *k.murphy@qub.ac.uk*
Science
Dr M O'Neill, FECIT *meon@fujitsu..com*
Dr A H Nelson,University of Cardiff,Physics and *nelson@cf.ac.uk*
Astronomy
Dr D Nicole,University of Southampton,Electronics *dan@ecs.soton.ac.uk*
& Computer Science
Dr A Nisbet,University of Manchester,Computer *nisbeta@cs.man.ac.uk*
Science
Dr R Nobes,FECIT *nobes@fecit.co.uk*
Dr C J Noble, CLRC Daresbury Laboratory, DCI *c.j.noble@dl.ac.uk*
Prof S Openshaw,University of Leeds,Centre for *stan@geog.leeds.ac.uk*
Computational Geography
Dr E Papaefstathiou,University of *stathis@dcs.warwick.ac.uk*
Warwick,Computer Science
Mr D Paver, Digital Equipment Company
Prof S Pawley FRS, University of Edinburgh, Physics *g.s.pawley@ed.ac.uk*
Dr F R Pearce, University of Durham, Physics *f.r.pearce@durham.ac.uk*
Dr M Peters, Springer-Verlag Mathematics Editor *peters@springer.de*
Dr P Plechac, CLRC Rutherford Appleton *p.plechac@rl.ac.uk*
Laboratory, DCI
Dr A Price, Fujitsu Systems Europe *a.price@fujitsu.co.uk*
Dr R Proctor, NERC Bidston Observatory, Proudman *rp@pol.ac.uk*
Oceanographic Laboratory
Mr W H Purvis, CLRC Daresbury Laboratory, DCI *w.purvis@dl.ac.uk*
Dr G Rajagopal, University of Cambridge, Cavendish *gr115@phy.cam.ac.uk*
Laboratory
Dr B Ralston, IBM UK Ltd. *ben_ralston@uk.ibm.com*
Dr M Razaz, University of East Anglia, School of *mr@sys.uea.ac.uk*
Information Systems
Miss N Reszka, Univ. of Gdansk, Poland
Prof B Richards, University of Glasgow, Aerospace *b.richards@aero.gla.ac.uk*

Engineering

Dr D G Richards, University of Edinburgh, Physics and Astronomy	*dgr@ph.ed.ac.uk*
Dr G Riley, University of Manchester, Computer Science	*g.riley@cs.man.ac.uk*
Mr R M Russell, Tera Computer Company	*russell@tera.com*
Dr M Rutter, University of Cambridge, DAMTP	*mjr19@cam.ac.uk*
Mr V Sakthitharan, Imperial College, Mechanical Engineering	*v.sakthitharan@ic.ac.uk*
Dr J Schmidt, University of Leeds, University Computing Service	*j.g.schmidt@leeds.ac.uk*
Dr S G Seo, University of Newcastle upon Tyne, Computing Science	*s.g.seo@ncl.ac.uk*
Dr P Shellard, University of Cambridge, DAMTP	*p.shellard@damtp.cam.ac.uk*
Dr P Sherwood, CLRC Daresbury Laboratory, DCI	*p.sherwood@dl.ac.uk*
Dr A Simpson, University of Edinburgh, EPCC	*ads@epcc.ed.ac.uk*
Dr L Steenman-Clark, University of Reading, Meteorology	*lois@met.rdg.ac.uk*
Mr D Stephenson, Hitachi Europe Ltd.	*dstep@hitachi-eu.co.uk*
Mr A Sunderland, CLRC,Daresbury Laboratory, DCI	*a.sunderland@dl.ac.uk*
Dr Z Szotek, CLRC Daresbury Laboratory, DCI	*z.szotek@dl.ac.uk*
Dr K Takeda, University of Southampton, Electronics & Computer Science	*ktakeda@soton.ac.uk*
Prof K T Taylor, Queen's University Belfast, DAMTP	*k.taylor@qub.ac.uk*
Dr J Taylor, Quadrics Supercomputers World Ltd.	*johnt@quadrics.com*
Dr W M Temmerman, CLRC Daresbury Laboratory, DCI	*w.m.temmerman@dl.ac.uk*
Dr L A Thompson, EPSRC	*thompson@epsrc.ac.uk*
Mr K Turner, University of Manchester	
Dr I Turton, University of Leeds, Centre for Computational Geography	*ian@geog.leeds.ac.uk*
Dr N R Walet, UMIST, Physics	*niels.walet@umist.ac.uk*
Prof D W Walker, University of Wales Cardiff, Computer Science	*d.w.walker@cs.cf.ac.uk*
Dr A Wall, EPSRC Research Facilities Team	*a.wall@rl.ac.uk*
Dr S Wang, Taipei Representative Office in the UK, Science Technology Division	
Mr S Ward, EPSRC	*s.ward@epsrc.ac.uk*
Dr C Whelan, University of Cambridge, DAMTP	*cw18@damtp.cam.ac.uk*
Dr P Williams, University of Wales Cardiff, Department of Astrophysics	*peter.williams@astro.cf.ac.uk*
Prof B R Williams, DERA, Aerodynamics	*100723.2425@compuserve.com*
Dr M R Wilson, University of Durham, Chemistry	*mark.wilson@durham.ac.uk*
Dr M Wilson, EPSRC Research Facilities Team	*m.wilson@epsrc.ac.uk*
Dr I Wolton, University of Southampton, Electronics & Computer Centre	*i.c.wolton@ecs.soton.ac.uk*
Dr P Young, University of Manchester, Computer Science	*pyoung@cs.man.ac.uk*

INDEX

Action
  Sheikholeslami–Wohlert (SW), 499
  Wilson, 499
Advection scheme
  Piecewise parabolic, 360
  Prather Eulerian, 371
  Semi-Lagrangian transport, 371
Aircraft, radar cross section, 429
Airflow modelling, 348
Algorithm
  alternating direction implicit (ADI), 410
  genetic, 462
  greedy, 434
  Householder reflection, 70
  Krylov subspace, 277
  Lanczos/Arnoldi, 278–279, 282–283
  Metropolis, 168
  orthogonal recursive bisection (ORB), 538
  recursive bandwidth minimisation (RBM), 432
  recursive spectral bisection (RSB), 432
  simulated annealing, 462
  spectral, 409; see also Fast Fourier Transform
  Smith–Vilenkin, 518–521
  Vachaspati–Vilenkin, 512
Antarctic circumpolar current, 325, 331
Anti-ferroelectric phase, 198
Arakawa–B grid, 359
Assessment of Global Ocean Circulation (AGORA) project, 366
Atmospheric modelling, 34, 317
Atomic force microscopy (AFM), 140
Atomic physics code
  DVR3D, 24
  FARM, 294
  MOLSCAT, 23
  R-matrix propagator, 24
Azores current, 326

BCS theory, 168
Band gap, 203, 205
Benzene, 252
Bering Strait, 327
Bi-disperse liquid crystal mixtures, 199
Biochemistry, 231–235

Biphenyl and 4-cyanobiphenyl 256
Bogoliubov-de Gennes (BdG) equation, 147, 150
Boltzman simulation, 186
Boundary conditions, 168
  Lees & Edwards, 188
Brillouin zone, 153
British Atmospheric Data Centre (BADC), 322
Bryan–Cox–Semtner code, 326, 337
Business modelling, 183

CCSD(T), 218
CFD code
  CFX4, 91
  Eulerian formulation, 421
  FELISA, 23
  FLUENT, 91
  Lagrangian formulation, 421
  PHOENICS, 91
CKM matrix, 502
Car–Parrinello UK (UKCP) consortium, 26, 135–136, 167
Carbon dioxide, 317
Central Laboratory of the Research Councils (CLRC), 4, 21
Ceperley–Alderley exchange correlation, 204
Chapman–Enskog theory, 419
Chemical dynamics, 234
Chiral molecule, 196
Climate variability, 317
Coarsening, 186
Code analysis tool
  FORGE, 121
  VAMPIR, 88, 104
Code maintenance tool, CVS, 121
Code tracing tool
  PACE toolset, 59
  PASCO, 57
  PICL, 57
  Paragraph, 57, 66
  SDDF, 57
Code
  Aerolog, 105, 110
  CANT-3D, 105, 110
  DBETSY3D, 105, 109

Code (*cont.*)
  FORESYS, 104
  SEMC3D, 104, 108
Collaborative computational projects (CCPs), 4–6
Collision strength, 294, 297
Colloid hydrodynamics consortium, 14
Collosal magneto-resistance, 208
Commodity supercomputing, 38
Complex fluids consortium, 193
Computational chemistry, 215, 217, 229, 237, 259
Computational chemistry code
  CADPACK, 267
  COLUMBUS, 238, 242
  GAMESS-UK, 23, 30, 230, 267
  GAUSSIAN, 230, 267
  HONDO, 267
  MOLPRO, 24, 267
  MRCI, 234, 240
  NWChem, 221, 262
  Turbomole, 267
Computational combustion, 395, 417
Computational combustion code
  ANGUS, 23, 398, 555
  SENGA, 399
Computational geography, 457, 467
Computational structural mechanics, 383, 385
Computer: *see* Cray, IBM, SGI, DEC, Fujitsu, NEC,
      Hitachi, Transtech, Kendall-Square, Tera,
      Intel, Hewlett Packard, Meiko
Computing in banking, 479
Configuration interaction, 237
Configuration space, 169
Cooling catastrophe, 510
Cooper pair, 147
Cooperation in groups, 38
Coordinate
  Jacobi, 307–308
  Radau, 307–308
Cosmic string, 517
Cosmological simulations, 16
Courant number, 421, 440
Covalent bond, 176, 178, 180
Cray J90, 22, 220, 350
Cray T3D and T3E, 3, 11, 21, 230, 245, 260, 267, 45,
      76, 96, 113, 124, 320, 334, 343, 359, 375,
      380, 385, 407, 433, 462
Cray YMP, 433

DEC alpha cluster, 22, 37, 38, 96, 551, 340
Daresbury Laboratory, 5, 21; *see also* CLRC
Dark matter, 507, 537
Data handling, 317
Data mining, 484
Data warehousing, 482
Davidson procedure, 239
Defects, 166
Density functional theory (DFT) 136, 147–149, 156,
      167, 175, 181, 195, 207, 218, 249, 262, 265
Diffusion QMC, 168–170
Dirac equation, 302
Direct correlation function, 195

Discrete variable representation (DVR), 307–313
Dislocations, 166, 196
Dissipative particle dynamics (DPD), 185–189, 191
Donor level, 203, 205
Dynamic scaling, 185
Dynamical matrix, 178

ESPRIT, 26, 30, 267, 79, 103, 482
ESPRIT project,
  AIOLOS, 381
  HP-PIPES, 26, 30
  IMMP, 267
Ecological systems consortium, 15
Ecosystem modelling, 359
Edinburgh Parallel Computing Centre (EPCC), 11,
      45, 113, 170, 191, 230, 318, 369, 375, 395,
      407, 473
Edinburgh EPCC seminars, 18
Eigensolver
  BFG, 25
  Jacobi, 25
  PARPACK, 309–310
  PeIGS, 25, 76, 264, 275, 296, 309–310
  sequential truncation, 308
  symmetric, 69
Eight-band model, 148–149
El Niño, 321, 365
Electric circuit design, 567
Electromagnetic scattering, 429
Environmental Molecular Science, 215, 223
Environmental modelling code
  Clark model, 348
  European Regional Seas Ecosystem Model
      (ERSEM), 359
  Integrated Forecasting System (IFS), 34, 321, 371
  Ocean Circulation Advanced Modelling
      (OCCAM), 326, 344, 366
  SLIMCAT, 371
  Shelf sea model (POL-SSM), 359, 365
  Southampton-East Anglia (SEA), 338
Equatorial Pacific, 329
European Centre for Mid-range Weather Forecasts
      (ECMWF), 317, 327
Exchange correlation hole, 172
Exchange correlation interaction, 178, 207
Exchange mean field, 173
Explicit forward time centred space scheme, 359

Fast Fourier transform, 25, 35, 128, 472
Fermi resonance, 161
Ferroelectric phase, 198
Ferromagnetism, 210
Fine-resolution antartic model (FRAM), 325
Finite difference discretisation, 359
Finite element, 91, 449
Finite volume, 93
Finite-size errors, 168, 171
Flame, spherically expanding, 426
Flow in sqare duct, turbulence, 409
Flow past a cube, 412
Flow past a square cylinder, 437

Flow, velocity profile, 410
Fluid boundary layer, 408
Fluid dynamics, computational, (CFD), 91, 379, 382, 385; *see also* CFD, computational combustion, Fluid tube necking, 191
Fluid, direct numerical simulation (DNS), 35, 395, 407, 417
Fluid, direct numerical simulation consortium, 407
Fluid
    compressible, 414
    multi-phase, 186
    viscous incompressible, 186
Fluid simulation and statistical mechanics, 35
Fluid-structure interaction, 385
Fourier filtering, 326
Fourier–Mellin invariant, 472
Frank elastic constant, 195
Fujitsu VPP300, VPP700, 22, 320
Function
    Bessel, 301–304
    Hankel, 303
    Irregular Coulomb, 301
    Legendre, 295, 303–304
    Neumann, 301
    spherical Coulomb, 301–304

GW method, 168, 173
Gaussian basis, 204
General Circulation Model (GCM), 318, 333, 348
    resolution, 334
Generalised gradient approximation (GGA), 175, 178
Geodynamo project, 530
Geographic information systems (GIS), 457, 467
Geographical analysis machine (GAM), 463
Geosciences, 34
German research and industry, 379
Global array tools, 25, 238, 241, 263
Global Atmospherical Modelling Project (UGAMP) UK consortium, 34, 317, 371
Global information infrastructure, 384
Global ocean circulation, 34
Global summation, 116
Gluon, 497
Grain boundaries, 142, 166, 196–197
Grand Unified Theory (GUT), 166, 517
Graph colouring, 450
Grid partitioning tool, JOSTLE, 95
Adaptive grid, 420, 449
Grid
    Delaunay generation, 430, 450
    domain decomposition, 113, 188, 449
    Galerkin discretisation, 430
    partitioning, 25, 95
    refinement, 420, 451
    tetrahedral, 430
    unstructured, 450
Gulf stream, 325, 369

HPC policy, Scientific Working Group, 3
HPC steering committee, 7

HPCI consortia, 6–8, 12, 21, 33, 45
HYDRA consortium, 508
HYPERBANK project, 480
Hadron, 497
Hall effect, 203
Halo node, 522, 339
Hamiltonian Matrix, 217, 276, 281, 294–295, 308–313
Harmonic approximation, 175, 178
Harmonic oscillator, 157
Hartree–Fock equation, 218, 262
Heavy Fermion compounds, 208
Heavy quark matrix, 502
Hellman–Feynman theorem, 178
Hewlett Packard, 572
Hexagonal packing, 199
Higgs particle/condensate, 517
High Performance Computing Initiative (HPCI), 3, 6, 11, 12, 21, 318
High-Tc superconductivity: *see* Superconductivity
Hitachi SR2201, 23, 320, 375, 380
Hubble volume project, 509
Human geographical social science, 457
Human systems modelling consortium, 15, 457, 460
Hund's rules, 209
Hydrodynamic forces, 186
Hydrodynamic regime, 191
Hydrodynamical simulation, 508
Hydrogen bond, 176, 178, 181

I/O bottleneck, 363
IBM SP2, 22, 36, 37, 220, 245, 260, 267, 553, 564, 87, 96, 110, 123, 153, 380, 484
Ice sheet modelling, 39
Impurities, 155, 159–161
Incompressible flow, 451
Indonesian throughflow, 325
Infra-red spectrum, 155–157, 161–162
Integration
    contour, 303
    Feynman path, 498
    Gauss–Legendre, 304
    oscillatory, 285
    partial wave, 301
    6-dimensional, 28
Intel Paragon, 267, 380
Intel Pentium II cluster, 37
Interaction potential
    Gay–Berne, 194–199
    Lennard–Jones, 194–199
    finite-range, 302
    static quark, 499
Interdiffusion, 185
Interfaces, 185, 190
Internal combustion engine, 381
Interstitial, 155–159
Ionisation, double event, 286
Isotropic–Nematic transition, 195–196, 199

JANAF polynomials, 419
Jahn–Teller distortion, 203, 205

Janak's theorem, 203, 205
Jeans criterion, 511
Jenson number, 412
Joule dissipation, 531

Kato cusp condition, 287
Kendall Square KSR-2, 267
Kohn–Sham equation, 147, 178, 250, 266

LSDA+U scheme, 211
Laminar separation bubbles, 408
Landau-de Gennes theory, 196
Language directives, OpenMP, 55
Language
    CHHP3S, 60
    COBOL, 479, 487
    COBOL-97, 491
    Craft, 460
    High Performance Fortran (HPF), 103, 263, 460,
        473
    HPF Plus, 104
    JAVA 40, 559
    Occam, 40
    PERL, 297
    SQL, 488
    Smalltalk, 493
Large grain data flow graph (LGDF), 128
Lattice guage theory, 497
Lax–Wendroff flux function, 430
Legacy systems, 487
Lepton, 497
Library
    Cray SciLib, 473
    HARWELL, 25, 69
    LAPACK, 153, 265, 282
    MUMPS routine, 85
    NAG parallel, 69
    PARASOL, 79
    ScaLAPACK, 25, 69, 76, 79, 204, 264, 275–276,
        560
Libration, 161, 178
Linear muffin-tin orbital method, 147
Linear scaling methods, 144
Liquid crystal display, 193
Liquid crystal, 39, 135–136, 143, 193
Liquid metals, 142–143
Local density approximation 167, 173, 207, 210
Localisation, 207–209
Lottery, maximum entropy analysis, 38

MIMD programming, 263
Mach number, 414
Magazine, HPC
    HPCNews, 3
    HPCProfile, 3, 25
Magnetic moment, 208–209
Magneto hydrodynamics consortium, 17, 529
Materials Chemistry, 34
Materials modelling code
    AIMPRO, 156, 204–205

Materials modelling code (cont.)
    CASTEP, 176
    CETEP, 24, 29, 30, 136, 251
    CONQUEST, 24, 144
    CRYSTAL, 23, 29, 30
    GHMC, 113
    OF-AIMD, 144
Materials modelling, 14
Maxwell equations, 429
Maxwell institute, 20
Mean field theory, 167, 172–173
Mediterranean outflow, 332
Meiko CS2, 108
Meridional heat transport, 330
Mesh: see Grid
Meso-scale model, 348
Mesogens, 199–200
Message passing library
    Global array tools, 25
    MPI-2, 363
    MPICH, 554
    PVM, 123, 340, 408, 434, 555, 570
    SHMEM, 35, 48, 115
    TCGMSG, 238, 242
Metacomputing, 383
Microsoft Windows NT, 551, 554
Millenium dome, 165
Modular Ocean Model (MOM), 326, 337
Molecular dipole, 198
Molecular dynamics, 186, 188, 194, 229
Molecular dynamics code
    AMBER, 230, 231, 548
    CHARMM, 233
    DL_POLY, 23, 27, 30, 231
    GBMESO/GBMEGA, 35, 194, 199
    GBMOL, 199
    GULP, 35
Molecular mechanics, 229
Molecular orbital theory, 217, 239
Molecular orientational disorder, 176
Molecular quadrupole, 198
Monte-Carlo simulation, 186, 194, 229, 230
Multiconfiguration SCF, 238, 262

N-body simulations, 507, 537
NACA test problems, 388
NEC SX4, 55, 320, 380
NUMA programming/architecture, 241, 264, 570,
        53
Nanoindentation, 135–136, 140
Navier–Stokes equations, 91, 187, 359, 383, 409, 451
Nematic phase, 193, 199
Neutron scattering, 186
Non-equilibrium behaviour, 185
Normal modes, 175
North Sea, 360
Nucleation, 185

Object-oriented programming, 263, 493
Ocean Circulation Advanced Modelling (OCCAM)
        Consortium, 35, 326, 338

Ocean volume flux, 327
Odd-even effects, 199
Oilfield well analysis, 17
Orbital moment, 209
Orbital-dependent functionals, 208
Order parameter, 186, 189
Ordering transition, 196

Pair correlation function, 195
Parallel Load balancing, 25, 243, 425, 452, 571
Parallel benchmark
  Linpack, 54
  NAS, 567
  PARKBENCH, 561, 563
  STREAM, 54
  SpecFP, 54
Parallel task scheduling 25, 130
Parallelisation tool
  automatic, 97, 130
  CAPTools, 100
Particle cosmology code, STRING/ASTRING, 36
Particle cosmology consortium, 36
Pattern recognition, 467
PeIGS: see eigensolver
Performance estimation, 40
Performance optimisation, 45, 53
Performance prediction, 58
Perturbative methods, 172
Peterov–Galerkin method, 452
Photoemission, 150
Potential energy surface, 237
Power plant engineering, 381, 26, 30
Poynting flux, 531
Predictive tracing, 57, 62
Press–Schechter theory, 510
Procurement HPC'97, 8, 36
Propagator; see also Atomic physics code
  Burke–Baluja–Morgan, 276, 295
  Light–Walker, 276
  Taylor series, 278
Pseudopotential, 175–176, 204, 251

Quadrature: see Integration
Quantum chromodynamics (QCD), 497–506
Quantum chromodynamics UK consortium
  (UKQCD), 13, 113
Quantum electrodynamics (QED), 497
Quantum Hall effect, 168
Quantum Monte Carlo, 168, 171–172
Quantum Monte-Carlo (QMC) consortium, 14
Quantum field theory, 166
Quark, 497–506
Quenched approximation, 498

R-matrix theory, 293–300
Radar cross section, 431
Radiation transport, 450
Raman spectrum, 155, 157, 162
Random number generator, 114
Rayleigh instability, 186
Reimann solver, 387

Relativistic electron gas, 171
Remote sensing, 40
Research Councils, UK, 3–5, 7, 21
Reynolds number, 438, 407, 452
Rotation, 158
Runge–Kutta method, 388

SGI Challenge, 434
SGI Origin 2000, 23, 37, 260, 317
SUN workstation, 64
Samoa Passage, 329
Satellite altimeter data, 365
Scaling hypothesis, 189–190
Scanning tunnelling microscopy (STM), 140
Scattering cross section
  radar: see Radar cross section
  triple differential, 286
Schmidt start formation law, 538
Schrödinger equation, 157, 166–167, 207, 217, 239,
    307
  time-dependent, 276
Sea level variability, 325, 327
Sea surface temperature, 329
Self-interaction corrected LDA (SIC-LDA), 208–211
Semiconductor, 135–136, 141–142
Siggia mechanism, 191
Signal processing applications (DSP), 127
Slater's transition argument, 203, 205
Slater–Jastrow functions, 168
Smectic phase, 193, 198–199
Smooth particle hydrodynamics (SPH), 508, 538
Software development, 220
Software maintenance, 119
Software portability, 119
Solar corona, 530–535
Sommerfeld parameter, 286
Soret effect, 419
Southampton Oceanographic Centre, 317, 326
Spectrum, incoherent inelastic neutron, 175–176
Spectrum, light Hadron, 499
Spectrum, vibrational, 178, 181
Spin moment, 209
Spin-polarised DFT, 203, 205
Spinodal decomposition, 185
Steric effect, 198
Stochastic differential equations, 187–188
Stratospheric chemical transport, 371
Strouhal number, 447
Structural engineering code
  ABAQUS, 92
  ANSYS, 92
  LSDYNA3D, 549
  NASTRAN, 92, 386
  PHYSICA, 93
  SPECTRUM, 93
  SWEEP3D, 648
Sub-grid stress (SGS) flow
  dynamic Germano model, 438
  Smagorinsky model, 438
Superconductivity, 39, 147, 208
  high-temperature (HTSC), 89, 147, 152, 208, 212

Surface, 135–140
Symmetric multi-processing (SMP) programming,
  55, 484, 570

TOPEX/Poseidon altimeter data, 365
Tank waste processing, 216, 223
Technology foresight, UK initiative, 165
Tera MTA, 545
Terrain-following coordinate, 348
Thermal diffusion, 419
Thermotropic liquid crystal phases, 193
Titanium dioxide, 25
Tomography, electrical impedence, 39
Transition pressure, 209–210
Translational mode, 178
Transtech Paramid, 96
Treecode gravity, 538
Tully–Fischer relation, 512
Turbulence database, 411
Turbulence model
  Johnson & Coakly, 409
  Reynolds stress, 412
  large eddy simulation, 412, 437
Turbulence, 395, 407, 412
  irrotational, 531

UK Meteorological Office (UKMO), 317
US ASCI initiative, 379

US Department of Energy, 215
University of Edinburgh, 3, 9, 11
University of Manchester, 9
University of Southampton, 6, 33, 119
Urban social structure, 468

VIRGO consortium, 16, 508, 537
VLSI, 128
Validation of software and hardware, 40
Van Hove singularity, 150
Variational QMC, 168
Vector programming, 55
Vibration, 157–158
Visualisation system AVS, 189
Void, 155, 161–162
Volume collapse, 208
Vortex shedding, 412, 437

Wave function, Coulomb, 286
Wave function, Kinoshita, 287
Wilkes formula, 419
Workstation cluster, 452; *see also* DEC cluster, Intel
  Pentium cluster
World Ocean Circulation Experiment (WOCE), 327

X-ray luminosity, 513

Zeolite, 135–136, 140–141